Botany | An Introductory Survey Of The Plant Kingdom

Botany
An Introductory Survey Of The Plant Kingdom

JAMES D. HAYNES
PROFESSOR OF BIOLOGY
STATE UNIVERSITY
COLLEGE AT BUFFALO
BUFFALO, NEW YORK

JOHN WILEY & SONS, INC.
NEW YORK
LONDON
SYDNEY
TORONTO

"If a man does not keep pace with his companions, perhaps it is because he hears a different drummer. Let him step to the music which he hears, however measured or far away."

Thoreau: *Walden* XVIII

This book is dedicated to those people who have encouraged me to follow my own drummer.

JDH

This book was set in Memphis Light by Progressive Typographers, Inc. It was printed by Vail Ballou. The designer was Jerry Wilke.

Library of Congress Cataloging in Publication Data:

Haynes, James D 1934-
Botany: an Introductory survey of the plant kingdom.

1. Botany. I. Title.
QK47.H33 581 74-7084
ISBN 0-471-36550-5

Printed in the United States of America

10-9 8 7 6 5 4 3 2 1

preface

An introductory course to the plant sciences traditionally has been called "General Botany" and usually has contained physiology, anatomy, phylogeny, genetics, ecology, and a survey of the plant groups.

In any generalized introductory course, it is impossible to include a discussion of all aspects of the particular discipline. Therefore, students and teachers of the plant sciences often have ignored the survey of plant groups and spend their time on the underlying principles and generalities necessary to an understanding of plants as organisms functioning in a complicated environment.

This has resulted in principles courses such as general biology or general botany where cells, tissues, organs, metabolism, reproduction, and heredity have been discussed, leaving no time for placing these processes into an organized, integrated picture of plants functioning as organisms within the framework of the plant kingdom.

After the generalized principles approach, the student can then find courses in morphology and anatomy where he can specialize to prepare for a career in botany. These courses usually are structured to include an intensive investigation into the morphological and anatomical details of many members of each group of the plant kingdom. Also, the uninitiated student often faces problems of taxonomy and classification, which are extremely confusing, but with which he must deal in such a course.

It seems, then, that there is a place for a logical transition from the generalized principles approach to the more specialized approach of the morphologist and anatomist. Not only is this transition necessary for the future botanist, but many students in liberal arts colleges become biology majors who teach biology without an awareness of the major plant groups with which they must live and work.

This textbook is designed for a one-semester survey of the

plant kingdom, and its purpose is twofold. The student who has had a principles course in general biology will find the book useful as an introduction to botany and the more advanced areas of plant morphology and anatomy if he desires a career in any of the plant sciences. However, if his interests are in some other area of biology, he will find information to make him aware of the vast numbers, the complexity, and the intricate beauty of the members of the plant kingdom. He also will acquire an awareness of the ways in which plants are important to man.

This book does not discuss or settle any of the varied and interesting interpretations of taxonomy and classification. Grouping organisms into various taxonomic categories is done strictly for the convenience of discussion. However, the taxonomy of various authorities is followed, and the student will be able to take the information from this text and proceed, with very little effort, to a more critical taxonomic treatment.

I have attempted to include examples of all of the major groups of plant. They were chosen not only because they represent the group, but because of the ease with which they might be obtained from local field collections or through biological supply houses. In all cases, fresh, living specimens are desirable for laboratory investigations, and perhaps more time should be spent on these than on dried or preserved specimens.

The documentation is not exhaustive, but it provides ample opportunity for further investigation. Emphasis in the suggested references is on works that contain generalized introductions to more specialized fields.

James D. Haynes

acknowledgments

For any person to write a textbook that he hopes will be adopted by his colleagues and read by students is a presumption of the highest order. He soon finds that, although he considers himself to have a fairly good grasp of his field, questions and doubts about specific problems arise to plague his scientific honesty. Because of these doubts, I have found it a great source of comfort to be able to turn to the following friends and colleagues for their constructive criticism about the areas indicated.

Valentine J. Nadolinski, Professor of Biology, State University College at Buffalo, Buffalo, N.Y., for his criticism of the bacteria chapter.

Lois H. Tiffany, Professor of Botany and Plant Pathology, Iowa State University, Ames, Iowa, for her criticism of the other fungal chapters.

Robert A. Sweeney, Professor of Biology and Director of the Great Lakes Laboratory, State University College at Buffalo, Buffalo, N.Y., for his criticism of the algal chapters.

Wendel B. Wickland, Assistant Professor of Biology, State University College at Buffalo, Buffalo, N.Y., for his criticism of the chapters on lichens and bryophytes.

Carol R. Sweeney, Adjunct Professor of Biology, State University College at Buffalo, Buffalo, N.Y., for her criticisms of the chapters on vascular plants.

I am also thankful to the reviewers, unknown to me, to whom the text was sent by the editor. Their comments have certainly aided in the production of this work.

Although I am deeply indebted and personally grateful to all of the above botanists who took so much time on my behalf, I assume complete responsibility for any errors that appear in the work.

All of the line drawings were done by Donald A. Licht. Each of his drawings is signed with the Chinese symbol signifying his signature. Any success enjoyed by this text is certainly dependent to a large degree on the many hours he spent following my complicated and sometimes confusing directions. The other illustrative material was prepared by me except where indicated in the legend of the illustration. I am grateful to each of those people for allowing me to use their photographs.

I also thank Virginia Moreland, who translated my penmanship into readable manuscript. She also corrected some of the spelling and grammatical errors. Finally, I must thank those undergraduate and graduate students who have been in my classes over the years for providing me with the stimulation to write, and William Schefler, my former department chairman, who encouraged me to begin this task.

J.D.H.

contents

1 nonvascular plant groups

fungal groups | introduction

1

Although they cannot be completely delimited by any one characteristic, the fungal groups have much in common. Most of them produce a definite cell wall composed of **cellulose** and/or **chitin.** Their somatic body is usually filamentous or tubular with a growing point at the apex of each filament (**hypha**). There is little specialization of cells such as is found in higher plants, but the filaments may be grouped into large and intricate "fruiting" structures (**sporophores**).

Since all of the fungi lack chlorophyll, they require a preformed energy source. This requirement places them in ecological niches as **saprobes, parasites,** or **commensals.** Their occupancy of these niches often brings them into conflict with man, and it is this conflict that introduces them to many people for the first time. The molding of food, clothing, and wood, the killing of important agricultural and ornamental plants, and the disease relationships they form with man and other animals have often brought them to man's pessimistic attention.

It should be immediately mentioned, however, that the fungi are also quite beneficial. Their contributions to medicine, industry, agriculture, and food production are legion. They are extremely important in the **ecosystem** as decomposers of organic matter where they aid in the recycling of carbon, nitrogen, and other mineral ions. Not the least important of their beneficial qualities are their direct gastronomic appeal and their interest as fascinating organisms.

Although they must be provided with preformed carbohydrates, the food products they store are fats or oils, not starches. Many fungi can synthesize the vitamins they need, but some are deficient in vitamin production, especially in the ability to produce thiamine and/or biotin, and must get either these vitamins or their precursors from their environment. Fungi differ widely in their food requirements. Some have enzyme systems that allow them to subsist on a wide range of substrata while others require specific substrates or are obligate parasites with a very limited host range. Mycorrhizal relationships with the root systems of higher plants have been well established and commensal relationships with algae, forming lichens, are well known.

Fungi reproduce by many varied methods. Asexual reproduction is extremely important in most groups and, indeed, is the only method of reproduction known for some (at least at the present time). **Conidia** (exogenous asexual spores) may or may not be produced on specialized structures called **conidiophores,** which may be free or contained in various types of structures. Fragmentation of somatic structures is another common means of asexual reproduction and often a single cell from a hyphal filament is capable of reproducing the entire organism. Slightly modified bits of mycelium or single vegetative cells are often given specialized names such as arthrospores, oidia, or chlamydospores, but these names simply designate specialized fragments. Fission and budding of somatic cells also produce new individuals in an asexual manner in some fungi.

Sexual reproduction involves the union of two compatible haploid nuclei. The methods by which this task is accomplished in the fungi are extremely varied, but three distinct phases are present: **plasmogamy, karyogamy,** and **meiosis.** Plasmogamy is the fusion of two compatible protoplasts. This union brings the two haploid nuclei into proximity within the same cell. Karyogamy may take place immediately as in the lower fungi, but is often delayed. The delay may be prolonged in time and space in the higher fungi resulting in a dikaryotic condition. As the fungus increases in size, the two nuclei are perpetuated by mitotic duplications and are represented throughout the body of the fungus. Eventually the associated haploid nuclei fuse and produce a diploid nucleus (**zygote**). The diploid nucleus sooner or later divides by meiosis and reestablishes the haploid condition.

Probably nowhere else in the plant kingdom is it more important to understand that meiosis and mitosis are strictly nuclear phenomena. Within the fungi, **mitosis** (**karyokinesis**) is quite often not followed by cell wall formation and division of the cytoplasm (**cytokinesis**). The resulting multinucleate condition in many phases of the life history of many fungi provides interesting problems in the understanding of the genetics and sexuality of the group. Nuclear migration from cell to cell in the fungal body further complicates the problems. To confuse mitosis with cell division or cell duplication

5

fungal groups—introduction

would certainly lead one to misunderstand the wide variety of possibilities.

The sex organs (**gametangia**) are quite varied among the groups. They may produce differentiated sex cells (**gametes**) that are indistinguishable from one another (**isogametes**) in the process of **isogamy,** or that are morphologically different. Two types of gametes are produced among the latter group. In the first type (**heterogametes**) the gametes are both flagellated but one is larger than the other. The larger one is often called the female and the smaller one the male. The process is called **heterogamy.** Finally, in the second type, the larger gamete (**egg**) is produced in a specialized gametangium (the **oögonium**), and the smaller (**sperm**) is produced in an **antheridium.** This process is called **oögamy.** Many fungi produce no differentiated sex organs but the sexual function is assumed by the somatic hyphae, and "genetic" control of the reaction rests with "compatibility" loci.

Many ways have evolved by which the union of compatible nuclei (**fertilization**) takes place. However, these are best discussed with the groups in which they appear.

Sexual compatibility in the fungi is an extremely complicated topic that must be overly simplified in a treatment such as this. Two major types of compatibility exist. **Homothallic fungi** are those in which the thallus or strain is self-compatible: plasmogamy and karyogamy occurring between structures produced on the same thallus or within the same strain. **Heterothallic fungi** are those in which the thallus or strain is self-sterile and requires the presence of a separate thallus or strain bearing a compatible mating structure before fertilization can take place.

The use of the terms **monoecious** and **dioecious** to describe the distribution of sex organs in fungus thalli (as indeed on the **gametophytes** of other plants) is unfortunate. The terms were first coined to describe the distribution of spore-bearing organs on the asexual generation (**sporophyte**), and they will be so used in this text.

To say that fungal classification and taxonomy is in a state of flux is to understate an obvious condition. Fortunately, the beginning student need only concern himself with placing the fungi into large, convenient, and often arbitrary groups for ease of discussion. The taxonomic treatment of the fungi (as well as other plant groups) in this book is intended to fulfill this concern and nothing more. A generalized classification system is followed, but the student should be surprised only if, in later studies, he does not find many deviations from the one used here. For example, the bacteria, cellular slime molds, and myxomycetes (slime molds) have been included among the fungal groups although the differences between them and the so-called true fungi are many and varied.

Nevertheless, the aim of this text is to give the reader a good understanding of a few organisms that are representative of the major groups. If the ones selected lead to a more detailed study in greater depth at a later time, so much the better.

fission fungi: schizomycetes

2

Anton van Leeuwenhoek described bacteria that he found in rainwater, vinegar, saliva, and many other substances and sent records of his observations to the Royal Society of London in a series of letters from 1675 to at least 1683. His descriptions were accurate for his time, and present-day bacteriologists can make reasonable identifications using his work. His were the first observations to be so reported although we now know from the fossil record that bacteria have been on the earth in excess of at least three billion years. Indeed, most botanists would agree that the bacteria, possibly along with blue-green algae, were the first organisms present on earth. However, the significance of Leeuwenhoek's observations were not realized by the Royal Society and the observations made for many years after his were simply to satisfy individual curiosities about occurrence and morphology. It remained for Louis Pasteur, with his experiments in 1864 finally disproving **spontaneous generation** and his later work relating "germs" to disease, to make the study of bacteriology popular and advance it to the level of a science.

Because of the way in which bacteria derive their nutrition, they have often been considered fungi. However, some botanists are now classifying them with the blue-green algae in the kingdom Monera. This classification recognizes the similarity of structure of the nuclei of the blue-green algae and the bacteria to the detriment of their difference in nutrition. If pressed for a generalization, many botanists would still place the bacteria in the plant kingdom because of historical perspective, early training, convenience, and the similarity some bacteria have to "higher" plants. They are often classified with the fungi because they are heterotrophic and have many morphological, reproductive, and physiological similarities to certain groups of the fungi. Distinctions of taxonomic and phylogenetic placement of the group are best left to specialists in the field, and will not be argued in this treatment. In any case, the bacteria have now been recognized as an extremely important and diverse group of organisms and anyone who is a serious student of the group must specialize. This treatment is presented to give the student some basic ideas concerning the bacteria, and is barely introductory. Therefore, for the above reasons, the bacteria are classified with the fungi in the plant kingdom.

Bacterial taxonomy depends on culture and serologic characteristics and biochemical activity in addition to cellular morphology. As a result of classification using this information there have been described in excess of 190 genera and 1700 species. Because of their importance to mankind, they receive study out of proportion to their numbers when compared to most other plant groups.

MORPHOLOGY

The bacteria usually possess one of three shapes, although intermediate forms may be found. Also, there are some that produce fruiting bodies (e.g., the myxobacters) that may be visible to the unaided eye, and others that are filamentous (e.g., the chalmydobacteriales). However, this discussion will be limited to the spherical **(coccus)**, rodlike **(bacillus)**, and spiral **(spirillus)** forms (Figure 2.1).

The coccus bacteria may be found as single cells, but usually they are grouped together in a fairly distinctive pattern (Figure 2.1). The pattern may be explained by understanding their method of asexual reproduction which occurs by cell duplication **(binary fission)**. If a single cell duplicates in one plane and the two daughter cells remain attached, the bacterium is a **diplococcus** form (e.g., *Diplococcus pneumoniae*). If such duplication continues and the chain of cells that is produced is not broken, a **streptococcus** form is produced (e.g., *Streptococcus pyogenes*). However, if cellular duplication occurs in one plane, and then the next duplication occurs in the plane at right angles to the first one, a group of four cocci attached in a square is produced. These are called the **tetracocci** (e.g., *Gaffkya tetragena*). When duplication occurs in three planes at right angles to each other, cubes of cells are consistently produced. These are called the **sarcinae** (e.g., *Sarcina lutea*). Finally, duplication in three planes may occur in a rather random pattern producing irregularly shaped groups of cells called **staphylococci** (e.g., *Staphylococcus aureus*). One cannot always be sure of the type of arrangement he is seeing if he observes only one group of cells

9

fission fungi: bacteria—schizomycetes

Figure 2.1
The most common forms of bacteria.

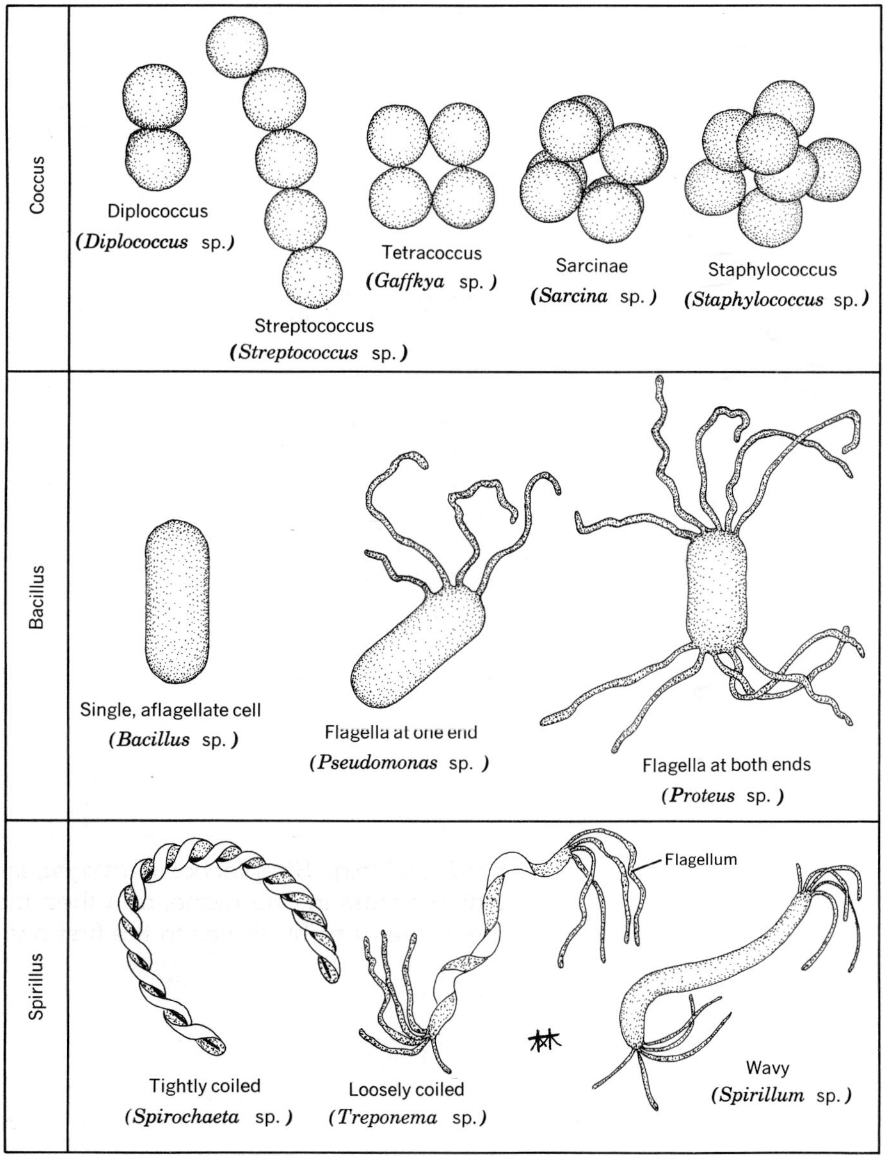

because the cells may be broken apart from each other, and there are several intermediate forms. However, by looking at a slide containing several of the bacteria, he can usually ascertain the group to which the bacterium belongs. Most bacteria are extremely small, and the cocci are the smallest. Their average size may range between 0.75 to 1.25 microns in diameter.

The bacillus bacteria (e.g., *Bacillus cereus*) also reproduce by binary fission but do not, as a rule, form characteristic groupings although they may produce chains or pairs of cells. In some cases, groupings of cells may occur under certain culture conditions but these groups are not usually considered to be true morphological characteristics. Mostly, then, the bacilli occur as single rodshaped cells. The bacilli have a much larger average size range (0.5 to 1 by 2 to 3 microns) than do the cocci but some species (e.g., *Clostridium tetani*) may be very long (0.3 to 0.8 by 2 to 5 microns). Many of the bacilli are flagellated. The **flagellum** is a very thin hairlike appendage that extends through the cell wall from the cytoplasm. The length of the flagellum is often many times the length of the cell, but is much thinner. There are many variations in the pattern of attachment of the flagella to the cell but the flagella are responsible for the motility of those bacilli that can move.

The spirilli also occur mostly as single cells, all of which are flagellated. They vary greatly in size (1 to 2 by 6 to 8 microns), and number of spirals. Some are tightly coiled organisms (e.g., *Spirochaeta* spp.) while others are loosely coiled (e.g., *Treponema* spp.), wavy (e.g., *Spirillum* spp.) or simply curved (e.g., *Vibrio* spp.).

A particular bacterial species, when grown under consistently uniform environmental conditions, exhibits a great consistency of morphological characteristics. However, it must be emphasized that if environmental changes such as nutritional substrate or temperature fluctuations are introduced into the environment, the morphology of the organism may be greatly varied. For example, *Serratia marcescens* is a bacterium that presents a striking response to changes in temperature. When grown at room temperature, the organism produces a group of pigments that collectively appear to be bright brick red. However, if the temperature is increased to about 37° C, the bacterium will produce no pigments at all and will thus appear colorless. Variations in size, shape, cell groupings, and even biochemical characteristics may be induced by regulating changes in various environmental parameters. Therefore, in any study of bacte-

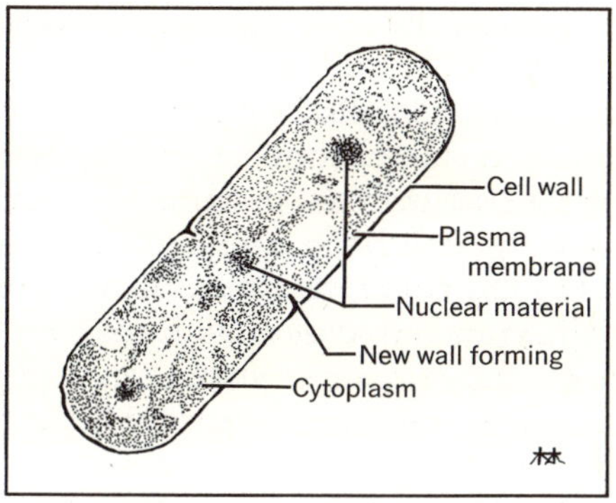

Figure 2.2
Drawing of bacterial cell from electronmicrograph. Not all structures are named.

ria, it is extremely important to work under carefully defined environmental conditions.

The anatomy of the bacterial cell (Figure 2.2) is simple although this should not be taken to mean that bacteria are uncomplicated organisms. The outermost boundary of the bacterial cell is the **cell wall.** It most bacteria, the wall is covered with a **capsule.** Just inside the wall is a **plasma membrane** that serves as a boundary for the **protoplasm** which may be separated into **cytoplasm** and **nuclear material.**

The cell wall is composed of a complicated arrangement of amino acids, carbohydrates, and lipids. Because penicillin inhibits an enzyme that stimulates the cross linking between two peptides on adjacent polysaccharide chains, the presence of penicillin in the environment of bacteria prevents cell wall formation. It is because of this action that penicillin is an effective antibiotic. Although differences in wall chemistry have been described, they all are rigid enough to enable the cell to maintain its shape and retain the cellular components in a particular relationship.

The capsule, when present, is a loose, slimy layer composed mostly of polysaccharides and polypeptides. However, the composition of the capsule is usually species specific. For example, the capsule that surrounds various species of *Pneumococcus* can be separated into more than 72 types on the basis of slight variations in chemistry. The materials that compose the capsule are secreted by the cell and, because of their viscosity, are not easily dissolved. Many bacteriologists believe that the capsule may be the repository of waste materials from the cell. It has been well established that the capsule aids in preventing dessication. It has also been observed that the capsule may be used as a stored food supply by some bacteria. The capsule certainly brings many bacteria into direct conflict with man, but this will be discussed in a later section.

The plasma membrane in bacteria is, like all other living plasma membranes, selectively permeable. The membrane is extremely important to the cell because it controls the passage of materials into and out of the cell. It has also been demonstrated that the majority of the enzymes associated with respiration are found in this structure. Any chemical or physical activity that interferes with the proper activity of the plasma membrane quickly causes the death of the cell.

The cytoplasm contains many types of **granules.** These include concentrated deposits of carbohydrates, fats, phosphates, and sulfur. The number and kinds of granules vary with the age and environment of the bacterium. **Ribosomes,** important in protein synthesis, are also found in the cytoplasm.

A great deal of scientific debate revolves around the bacterial nucleus. It was long thought that bacteria had no nuclei and, indeed, there is no evidence for its presence when using conventional staining and microscopic techniques. However, the fact that bacterial cells produce new cells exactly like themselves is circumstantial evidence that bacteria must have some method whereby inherited characteristics are passed from one generation to the next. The advent of the electron microscope and the development of new fixatives and stains have conclusively demonstrated that bacteria do possess the nuclear material **desoxyribonucleic acid (DNA).** However, this material does not seem to be contained in a well-organized nucleus bounded by a **nuclear membrane** as is found in the more complex plants. Thus, bacteria share their type of nuclear arrangement with the blue-green algae. This nucleus type is known as the **prokaryotic nucleus** and is the basis for putting the bacteria and the blue-green algae together into the kingdom Monera. There is speculation among biologists that perhaps the prokaryotic nuclei are better organized than has been previously thought; a great deal of research into the question is being generated.

LIFE HISTORY

By far most bacterial populations are produced by the asexual reproduction method of binary fission. When this occurs, there seems to be a reorganization of the cellular contents in such a manner that a constriction of the plasma membrane divides the cell equally into

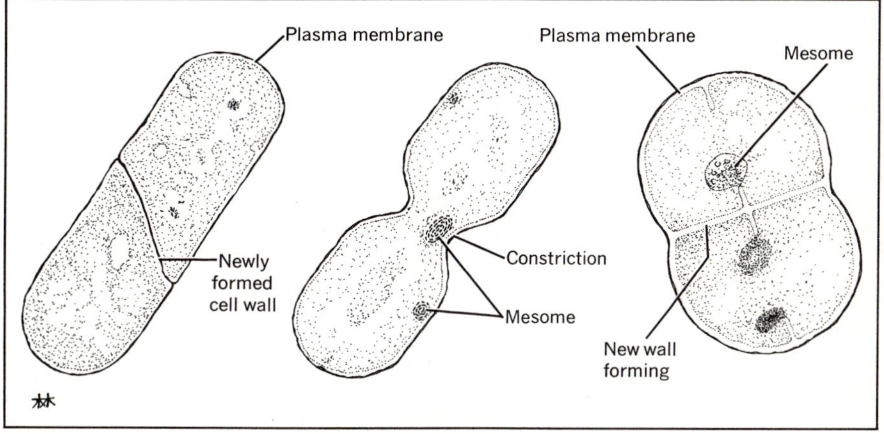

Figure 2.3
Binary fission.

two parts (Figure 2.3). After this occurs, a cell wall is secreted and the cells separate. This type of reproduction produces two daughter cells that are genetically alike and usually equal in size. As far as is now known, bacteria can continue in this method of reproduction without any sexual exchange.

Only recently has evidence become available for the belief that sexual events occur among the bacteria. This evidence was obtained from the careful examination of a common bacterium, *Escherichia coli*. Several strains of the bacterium have been described. When two carefully defined strains were grown together in the same culture and their offspring tested, it was found that many of the offspring contained a significant number of the combination of characteristics. Of course, many of the progeny were identical to one or the other of the parents, but enough combinations were found to believe that some sort of nuclear exchange had taken place. Further experimentation has established that nuclear recombination does occur, although the sexual events do not appear to be obligatory for the continuation of the species.

It should also be pointed out that sexuality among bacteria is not like sexuality among higher plants (or animals). There is no complete fusion of nuclear material among bacteria, therefore no zygote is produced. Instead, there may be a connection between two bacterial cells with some DNA being transferred from one to the other. The donor DNA may replace some of the recipient DNA or it may simply add DNA to the recipient. Thus genetic changes are introduced into the progeny of the recipient cells. Also, free DNA of a

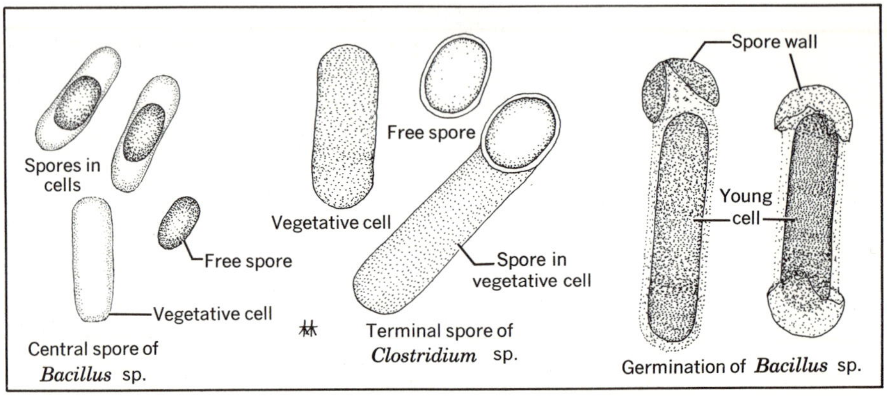

Figure 2.4
Endospore formation and germination.

particular species may be taken up by a proper recipient (**transformations**), or DNA may be transferred from one member of a species to another member of the same species by a "carrier" virus (**transduction**). Regardless of how the genetic changes are introduced, the new cells react like partial diploid organisms. Because sexuality among bacteria is so recently known, much investigation must proceed before it is fully understood.

For a long time it has been noted that many of the bacilli (e.g., *Bacillus* spp., *Clostridium* spp.) sometimes transform into oval bodies called **endospores** (Figure 2.4). The endospore may be produced centrally, terminally, or subterminally inside of the old bacterial cell wall. The protoplasm of the cell condenses into one of these positions and secretes a thick, highly resistent spore wall around itself. In proper environmental conditions, the wall of the spore will break open and a bacterial cell will be produced from it. Although this is not a means of population increase because only one bacterial cell is produced from each spore, it was long thought that endospores were produced only as a result of increasingly unfavorable environmental conditions. This led to the specualtion that the endospores served as "survival mechanisms." It is now known that endospores may be produced even when the bacterium is under favorable conditions. The phenomenon of the bacterial endospore, then, still needs a great deal of investigation.

NUTRITION

The bacteria, as with all other organisms, may be divided into two groups on the basis of the way in which they derive their nutrition.

That is, those organisms that can use carbon dioxide (CO_2) as their sole source of carbon are called **autotrophs.** Very few of the bacteria fall into this group, but those that do may be separated into **phototrophs** and **chemotrophs.** The phototrophs (e.g., *Chlorobium limicola*) are those that can use the energy of sunlight and the chemotrophs (e.g., *Thiobacillus thiooxidans*) are those that use the chemical energy of inorganic compounds during the synthesis of their own protoplasm. Phototrophic bacteria do not carry on the photosynthetic process as it is found in the green plants. The bacterial chlorophyll is different from the chlorophyll of green plants, oxygen is not liberated during the photosynthetic process, and the hydrogen source is not water but usually a sulfur compound (e.g., hydrogen sulfide). These two groups of bacteria represent only a very small percentage of the total number. By far most of the bacteria derive their energy from the organic compounds contained in or produced by other organisms. These are the **heterotrophs,** and it is their activity that makes the bacteria (as well as the fungi) significant to man.

The bacteria may also be roughly divided into two groups on the basis of their activity in the presence of free oxygen. The **aerobic bacteria** are those that grow in the presence of free oxygen, and the **anaerobic ones** are those that grow in the absence of free oxygen. There are some intermediate groups that grow either in the presence or absence of free oxygen; these are the **facultative anaerobic bacteria.** Another intermediate group includes those bacteria that require oxygen in concentration less than that found in air. These are known as the **microaerophilic bacteria.** Most bacteria are aerobic, but some of those that are most dangerous to man are anaerobic.

HUMAN RELEVANCE

The bacteria have a profound and far-reaching influence on man and his activities. Because of their nutritional requirements (particularly of the heterotrophs) the bacteria function in such a way as to decompose organic materials. This process involves the constant transformation of elements from one compound to another. For example, it has been estimated that 95% of the atmospheric carbon dioxide results from microbial respiration. This would lead one to believe that the total mass of microbes would be far greater than all other organisms. Because it has been demonstrated that the microbes are much more metabolically active than higher organisms, the ratio would be less than 95:1. However, the bacteria constitute a very significant portion of the total microbial mass.

Nitrogen is one element that is transformed almost exclusively by bacteria (some blue-green algae are also important) into a form usable by green plants in their synthesis. Because nitrogen is such an important constituent (about 16%) of proteins, the nitrogen "cycle" is a very important biological phenomenon. When nitrogen is in a protein, it is no longer available as a nutrient to plants. Therefore, when the organism containing the protein dies or rids itself of the protein in its waste products, it is ecologically important that the nitrogen be recycled for use in the construction of new proteins. This recycling process is a long and intricate one that requires many enzymatic reactions.

The first step in the process is the degradation of the protein. This involves the breakdown of proteins to their constituent amino acids. Although relatively few bacteria participate in the first part of this process (mostly done by fungi and actinomycetes), some members of the genera *Clostridium*, *Proteus*, *Pseudomonas*, and *Bacillus* are important. The final breakdown to the amino acids often involves many bacteria.

To go from the point where amino acids are released to the point where nitrates are once again available to the green plants requires the formation of ammonia, nitrites, and finally nitrates. Each of these steps requires specific bacterial action. Although the details of this process are beyond the scope of this discussion, it is important to stress that without the process, very little protein would ever be produced. Sulfur is another important element that is recycled by bacteria; carbon has already been mentioned. Therefore, as far as man is concerned, bacteria are extremely important as recycling agents for some of his more important elements.

It should be pointed out that this recycling process reduces the huge piles of waste materials and dead bodies of organisms that would otherwise present an impossible disposal problem. Bacterial decomposition is important, in this context, in sewage treatment plants.

The decomposition reactions of bacterial enzymes have produced many materials that man finds useful. Several dairy products are the result of the decomposition of milk by specific bacteria. The commercial buttermilk is a result of the inoculation of skim milk from a starter culture of *Streptococcus cremoris* or *S. lactis*, with

17

fission fungi: bacteria — schizomycetes

Leuconostoc citrovorum or *L. dextronicum*. Buttermilk is probably the most widely used fermented milk in the United States, but others are used here as well as in the rest of the world. For example, sour cream is simply cream that has been inoculated with the same bacteria used to produce buttermilk. Yogurt is produced by the action of *Streptococcus thermophilus* and *Lactobacillus bulgaricus* on milk from which some of the water has been evaporated. More exotic fermented milks would include kefir which is the action of two kinds of bacteria and one lactose-fermenting yeast on cow, goat, or sheep milk contained in goat leather bags. This is a lactic-acid and alcoholic fermentation. Another would be kumiss which is the action of the same microorganisms on mare's milk. Kefir is used throughout the Balkan countries and kumiss is native to parts of Russia. Butter is given its flavor by the same bacteria used in buttermilk but also with *Streptococcus diacetilactis*.

Cheese has been described as "rotten" milk. Several hundreds of cheese varieties are available and they get their distinctive flavors not so much from the milk from which they are made but from the microorganisms used during the curing process. Many cheese varieties are produced as a result of fungal action (see following chapters), but many others are produced from bacterial action. Many of the soft, unripened cheeses are produced directly from the curds of milk that have resulted from the inoculation of fresh milk with various species of *Streptococcus* and *Lactobacillus* along with rennet extract that has been prepared from the enzyme rennin which is obtained from the stomach of calves. This fresh curd may have cream added to produce cottage, cream, neufchatel, and ricotta cheeses.

The ripened cheeses are those to which bacteria (or "molds") have been added after the curd has been pressed into a form. The block of cheese is then placed under environmental conditions that favor the growth of the desired bacterium (or combination of bacteria) over other microorganisms. Members of the genera *Lactobacillus*, *Streptococcus*, and *Propionibacterium* are used in the production of Swiss cheese. *Brevibacterium linens* is an extremely important proteolytic bacterium and is believed to be the one that gives the flavor to Limburger. It is used also in combination with a number of other bacteria to flavor other cheeses. A few of the cheeses that depend upon bacteria for their flavors are Parmesan, Romano, American, Cheddar, Edam, Provolone, Swiss, Gouda, Munster, Hand, and Limburger. The use of fermented milk products is a very old method of preserving milk. The microbiological fermentations produce various acids that inhibit or prevent the growth of other microorganisms that would spoil the milk for human consumption.

Many food materials other than milk were, and still are, preserved by microbial activity. For example, lactic-acid bacteria are mostly responsible for the fermentations that transform cucumbers to pickles, cabbage to sauerkraut, and make olives edible. The products go through a number of stages where successions of bacteria promote the fermentation. For example, members of the genera *Aerobacter*, *Escherichia*, and *Bacillus* are mostly responsible for the early stages and *Leuconostoc* and *Lactobacillus* are mostly responsible for the later ones.

Many industrial products are also produced by bacterial fermentations, sometimes in cooperation with yeasts. Some of the products that require bacteria somewhere in the processing are lactic acid used in food products, textile and chemical manufacturing; vinegar; dextran used as a stabilizer in food products and as a blood-plasma substitute; acetone-butanol used as a solvent; and bacterial amylase used in sizing papers and desizing textiles. Many other products could also be included but the list would get much too long.

The nutritional requirements of bacteria have brought them into direct conflict with man. Many of the diseases of man and his plants and animals are really the relationship that they have with bacteria. A **pathogen** is any agent capable of producing a disease; **pathogenicity** is the ability of a pathogen to enter a host and cause the disease. A measure of the ability of a pathogen to cause disease (i.e., its degree of pathogenicity) is called its **virulence.** The virulence of a pathogen can vary with a number of factors inherent in itself, its host, and its environment. Many bacteria cause disease relationships with man. Man has discovered ways to prevent or cure many bacterial diseases. Proper methods of sanitation along with immunization and the production of antibiotics have greatly reduced the incidence of such diseases caused by bacteria as diphtheria (*Corynebacterium diphtheriae*), scarlet fever (*Streptococcus* spp.), rheumatic fever (*Streptococcus* spp.), tuberculosis (*Mycobacterium tuberculosis*), whooping cough (*Bordetella pertussis*), typhoid fever (*Salmonella typhosa*), syphilis (*Treponema pallidum*), and gonorrhea (*Neisseria gonorrhoeae*), at least in the western world.

Although these diseases, and many others, may not now be very important, it is interesting to speculate how they have affected the history of the world. Armies have been depleted and entire populations wiped out by some of them. However, when they have struck some of civilizations' leaders, chaos was often the result. For example, Mary Renault, in her book *The Persian Boy*, a story of the last 10 years of the life of Alexander the Great, expresses the belief that Alexander's lover, Hephaistion, died of typhoid fever and that Alexander himself died of pneumonia about a year later in Babylon after he had conquered most of the world that was known at that time. If he had lived, perhaps some of his ideas of government would have had a more lasting effect, even to the present time.

Typhoid fever is caused by a bacillus known as *Salmonella typhosa* which can live in the bloodstream of man. This disease produces high fevers and inflammation of the intestine. The spleen becomes enlarged and ulcers are produced in the intestine. The patient often begins to feel better and desires solid food before the intestinal ulcers have healed. Perforation of the ulcers soon leads to death. Hephaistion exhibited the high fevers and died soon after eating. Because typhoid bacilli can live for about a week in water, they are often spread in this fashion. Hephaistion could easily have gotten the bacterium along with his water. Incidentally, one of the reasons why the Greeks and other people added wine to their water was to make the water fit to drink. Of course, they had no knowledge of bacteria, but the alcohol in the wine probably served to sterilize the water. They did know that it was better to drink wine in water than to drink straight water although they did not know why.

Diplococcus pneumoniae is the most common cause of lobar pneumonia although the same organism may cause other diseases such as arthritis and meningitis. There are about 75 strains of the bacterium that are closely related to the chemistry of the capsule. If the capsule is removed, the host organism will not be infected. It seems, therefore, that the capsule of the bacterium protects it from the **phagocytes** of the host organism. The sulfa drugs were the first effective means of combating the disease, but many antibiotics are now used. The coccus is spread through the air by droplets expelled from the nose or throat by persons who have pneumococcal infections. Also, some persons seem to harbor the coccus without being themselves infected. These persons are called carriers. Alexander was much depressed after the death of Hephaistion and would not follow the advice of his doctors (indeed, he had Hephaistion's doctor hanged) when he became ill. Given the conditions under which he was living, it was easy for him to contract pneumonia.

The kinds of disease relationships that the above organisms form with man should not be confused with the toxins that are produced by bacteria and ingested by man. These are "food poisonings" that were sometimes called ptomaine poisoning. The latter is a misnomer because the ptomaines are a class of chemical amines and are not toxic. Many of the bacteria can cause gastrointestinal distress if they have grown in food that is ingested. However, the most feared food poisoning, and the one that produces one of the most potent toxins known to man, is botulism.

Botulism is caused by a spore-forming bacillus, *Clostridium botulinum*. This bacterium is widely distributed in the soil and the gastrointestinal tract of herbivorous animals. The bacterium cannot invade any type of tissue, whether healthy or dead, and will not grow except under anaerobic conditions. Because it produces spores that are highly resistant to heat, it may remain alive in cans of improperly processed, nonacid foods such as beans, peas, corn, carrots, or canned meats. If the can of food has not been placed under temperatures and pressures that are high enough to kill the spore, it can later germinate under the anaerobic conditions inside of the can. At that time it begins to produce the toxin without producing any odor or gas. If the food is later eaten without heating, **botulism** results. The toxin is very heat labile and is broken down after boiling for 10 to 15 minutes. Before strict government control of standards for the processing of canned food, botulism outbreaks were rather common. Because of the strictness now imposed upon the commercial food industry, botulism is rare. The latest outbreak was in 1971 when a man died because of eating a can of contaminated potato soup. Most botulism that occurs is because of improperly home-canned foods.

The toxin may circulate in the bloodstream for as long as three weeks before symptoms are noted, although this would indicate a small dosage and recovery is quite likely. Usually, however, symptoms are noticed within 18 hours after ingestion because of the amount of toxin consumed. Dosages at this level are commonly fatal. Eventually the toxin reaches the nervous system where it blocks nerve impulses, thus causing paralysis. As a measure of its toxicity, it has been calculated that a few milligrams of properly placed poison could wipe out the population of New York City. Although the number of molecules of the toxin may vary, it is known that 0.03 ml of juice containing the toxin can be a fatal dose for a healthy being. In one reported case, the mother of a family opened a jar of home-canned green beans. While pouring them into a saucepan, she tasted one to check for salt. Her daughter came into the kitchen after they had been cooking for a few minutes and tasted the beans. After the cooking process had been completed, the beans were served to the father and son with their evening meal. The mother later died of botulism, the daughter was extremely ill, but recovered; the father and son were unaffected. This story also illustrates the effect of heat upon the toxin.

Before methods of treatment with antitoxins were developed, death was usually the result. However, fatality has now been reduced to as little as 50%, but the effects may be present for several weeks or months until the body completely rids itself of the toxin.

Most bacterial food poisonings are caused from *Staphylococcus* spp. in foods that have been allowed to remain unrefrigerated after being contaminated with the bacterium. Such things as chicken in sauces, potato salad, gravies, and cream fillings are good media for the bacterial growth. While the toxins produced by these bacteria are not usually fatal, they may cause such severe gastrointestinal upset with vomiting and diarrhea that the victim may begin to consider death as a welcome alternative. Serious disability or death usually results only when the person is old or also ill from some other cause.

As one might expect, many bacteria cause disease relationships with the plants man uses. The symptoms expressed by the host plant may be many and varied. These are likely to include death from bacterial toxins of certain tissues, wilts, retarded growth, changes of color, increase in number of cell duplications (**hyperplasia**), and increase in cell size (**hypertrophy**). Hyperplasia and hypertrophy may be directed in such a way that leaves and stems may be twisted or otherwise distorted or new tissues (galls, cankers) be developed. Certain bacteria may also produce excessive numbers of short branches (witches' brooms) or roots (hairy root). Certain fungi also produce many of these same symptoms.

Bacterial spot diseases are those whose toxins cause tissue death in a small area of leaf, stem, or fruit. Examples of these are bacterial spot caused by *Xanthomonas vesicatoria* on tomatoes and halo blight caused by *Pseudomonas coronafaciens* of oats, barley, wheat, and rye. Wilt diseases are caused by slime-producing bacteria that plug the xylem of vascular plants. Examples of these abound, but some common commercial plants that are affected are tomato wilt (caused by *Bacterium solanacearum*) and tobacco wilt (caused by *Pseudomonas tabaci*). Galls are masses of disorganized tissues caused by excessive cell duplication and enlargement. These structures often interrupt the proper flow of water and nutrients through the host vascular plant. The production of excessive numbers of roots and stems may also result. The gall is often located near the place where the host comes out of the ground and is thus called a **crown gall.** These plant cancers or tumors are most often caused by *Agrobacterium tumefaciens* and affect many vascular plants as widely diverse as fruit trees, roses, blackberries, grapes, sugar beets, alfalfa, cotton, and walnuts.

A very serious disease of pears and apples is fire blight caused by *Erwinia amylovora*. This organism is a tiny bacillus with an average size of 0.6 × 1.5 microns although it may get a bit larger. It may be spread by any number of vectors such as squirrels, birds, plant lice, flies, bark borers, and honeybees. The tiny bacteria may also be spread by splashing rain. They gain entrance into the new host through stomata, lenticels, or any scratches or breaks in the bark.

Fire blight symptoms may be expressed in many ways although not all symptoms may be present at the same time. The blossoms may turn brown and wilt soon after they open. If young fruits have begun to be produced, they may acquire the same symptoms. The symptoms may then spread to adjacent leaves and stems. The leaves themselves may be directly invaded and large lesions may appear anywhere on the leaf. Blight of the twigs may result from passage of the bacterium into the stem from the leaf. This can cause twisting and blackening of the twig tips and the leaves which give the appearance of being exposed to a fire, hence the common name of the disease. Mature fruits may be affected if they are exposed to the bacterium. The bacterium may continue to spread throughout the plant by means of the vascular system until finally the large branches and even the trunk may be affected. Large lesions of a definite size (a **canker**) are usually produced on the branches. These cankers ooze sap that is filled with bacteria. Animals, particularly flies and honeybees are attracted to the oozing sap, pick up the bacterial cells on their legs and transfer them to new hosts. The cankers are areas of overwintering for bacterial cells, and the next spring, when they begin to ooze, they serve as a source of inoculum for the new season. Some research has indicated that honeybees may maintain the bacteria in their hives over winter and inoculate the young shoots during the next season even before the old cankers begin to ooze. Although pears seem to be the most susceptible host, apples are also easily affected. However, apricots, plums, prunes, cherries, quince, loquats, wild crabs, hawthorns, American mountain ash, strawberry, and spiraea among many others may also contract the disease.

Although fire blight was first observed in 1780 along the Hudson River, it was not described until 1794. It is a disease that is probably indigenous to North America and infected native hosts but was not noticed to any great extent until pears and apples began to be widely cultivated. The disease spread westward with the spread of the fruit trees and reached California in about 1900. Many theories were advanced as to the cause of fire blight, but in 1881, Professor T. J. Burrill proved that the disease was caused by a bacterium. This was probably the first time that it had been proved that bacteria can cause plant diseases. Preventative measures include removing blighted plant parts by surgery, removing blighted trees, and spraying with various copper mixtures.

Because bacteria are so very important to man's well-being, both constructively and destructively, they deserve intensive specialized study. Indeed, there are many biologists who do nothing but study the bacteria. This chapter has been but a bare introduction to this very important group of organisms.

3 the cellular slime molds | acrasiales

The Acrasiales is a small group of organisms having many similarities to the myxomycetes. The members of the group are cosmopolitan in soils and many are **coprophilous** (dung-loving). They are probably best separated from the true slime molds by the fact that the cells of the vegetative stage (**myxamoebae**) retain their individual identity when they aggregate. Members of this group are easy to isolate, but none have been grown solely on defined media.

LIFE HISTORY

The most widely studied cellular slime mold is *Dictyostelium discoideum* (Figure 3.1) but there still remain uncertainties about its life history (Figure 3.2).

The vegetative stage is composed of individual amoeboid cells (myxamoebae) that are haploid. These cells engulf, digest, and assimilate bacteria, yeasts, and other unicellular organisms including other myxamoebae, increase in size, and multiply by cell duplication (fission) producing a large population of haploid cells. Eventually some of the myxamoebae begin to produce **acrasin** (probably a steroid hormone) that causes a chemotactic aggregation response among the other cells. The stimuli that cause the myxamoebae to produce acrasin have not been isolated, but environmental factors that suggest answers are: decrease in food supply and humidity; increase in number of haploid myxamoebae, temperature, and light exposure. Aggregation has also been said to be a function of sexual reproduction.

Aggregation does not seem to be dependent on the size of the individual cells, but during aggregation there is a reduction in their size, probably because of the loss of food vacuoles. The amoeboid cells retain their individual identity within the newly formed **pseudoplasmodium** ("slug"). In *D. discoideum*, the pseudoplasmodium may migrate for a considerable distance before the "fruiting" structure (**sorocarp**) is

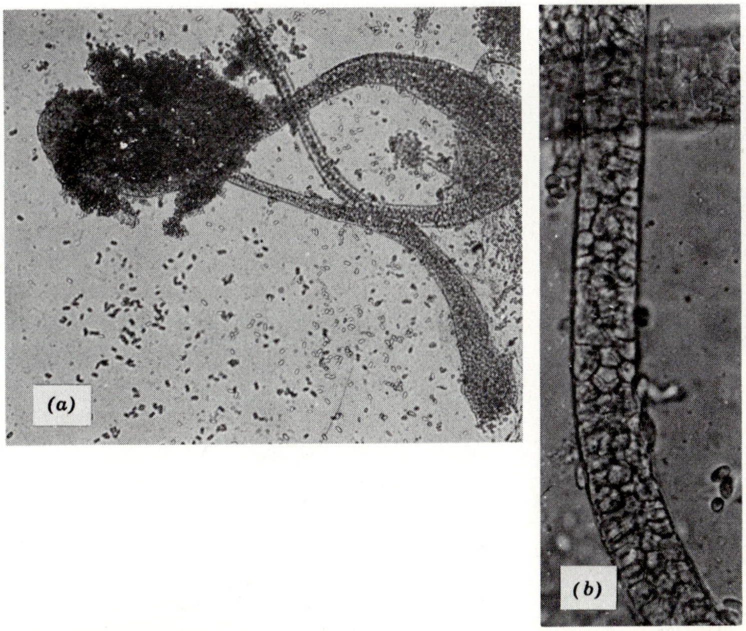

Figure 3.1
Photomicrographs of squash preparations of *Dictyostelium discoidium* (*a*) Two mature sorocarps with crushed sori, many spores and sorophore of another sorocarp present (300×). (*b*) Section of sorophore with individual cells readily apparent (1800×).

produced, but in others the sorocarp is produced at the point of pseudoplasmodial formation.

When migration of the pseudoplasmodium ceases, the anterior cells become differentiated into potential **stalk** cells and the posterior ones into potential **sporogenous** ones. As the stalk cells mature, they push down through the now flattened pseudoplasmodium and produce cellulose cell walls around themselves. The cells that were posterior to them crawl up over the outside of the newly forming stalk (**sorophore**) and increase its length. Finally, the posterior cells, which will form the sporogenous cells, crawl to the top and form a globose mass (**sorus**) of encysted cells. Each of these encysted cells is produced from one of the original amoeboid cells of the pseudoplasmodium rather than through division of a large protoplast and, upon germination, releases only one haploid, amoeboid cell. Therefore, there is not the potential for increase in numbers that takes place in organisms in which many cell duplications occur at this phase of their life history. Although sorocarps usually form, two other structures may be produced. Under conditions that are unfavorable for sorocarp formation, a myxamoeba may encyst (**microcysts**). The microcyst is surrounded by a delicate but rigid cellulose wall; the re-

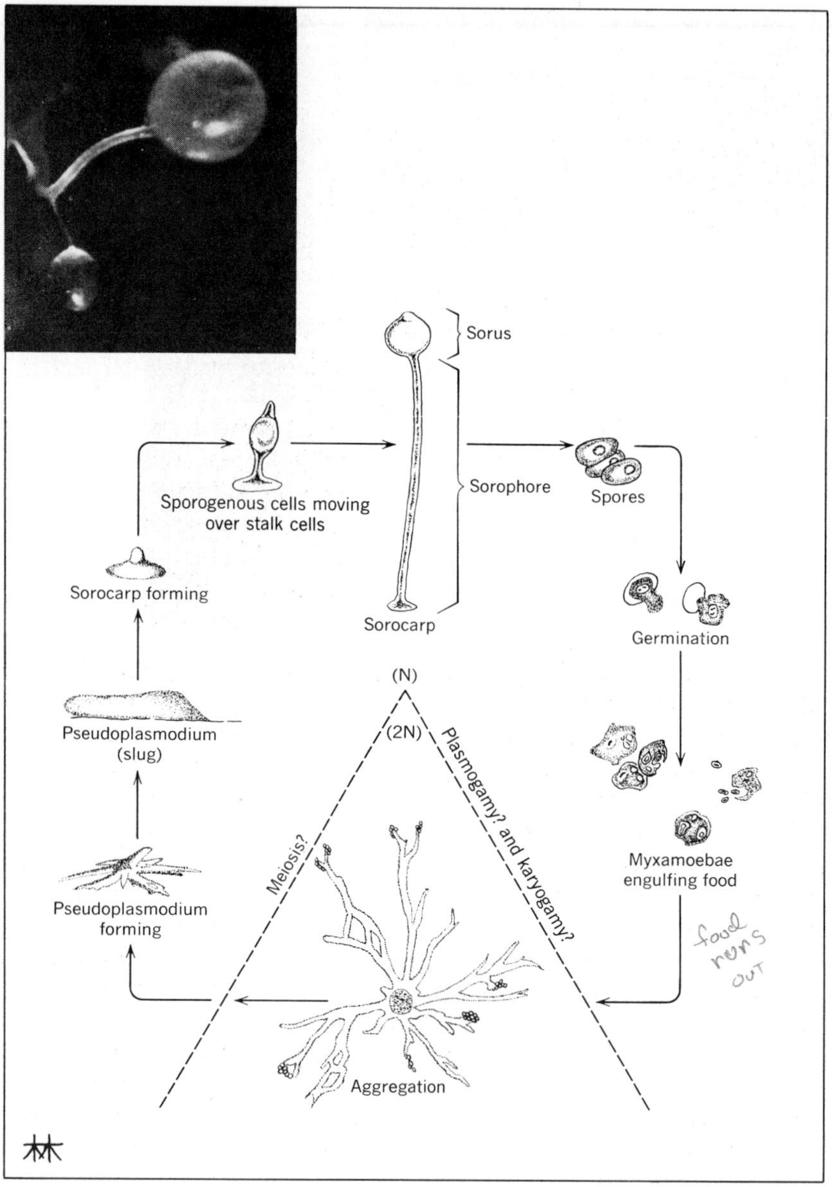

Figure 3.2
Life history of *Dictyostelium discoidium*. Areas of uncertainty indicated by question marks and dashed lines. (Redrawn from J. T. Bonner, *The Cellular Slime Molds*, Princeton, N.J.: Princeton Univ. Press.) Insert: Two mature sorocarps (40×).

turn of favorable environmental conditions allows for germination through a pore. Under other conditions, the myxamoebae aggregate into pseudoplasmodia which, instead of forming sorocarps, become divided into rounded cell masses surrounded by thick cellulose walls. These **macrocysts** are not known to function.

The formation of the sorocarp is a fascinating study in **morphogenesis.** Its morphology varies in different species and is the basis for classification. There may be a disc at the base of the stalk and the stalk may be simple or branched. There may be one or more stalks that adhere to each other along their entire length but with individual heads of spores at their tips. The stalks usually have their cells arranged in various patterns, but two genera have acellular stalks.

Asexual reproduction takes place by fission as the myxamoebae feed and grow. Sexual reproduction in the cellular slime molds is still controversial although there is evidence for heterozygosity and recombination. Some evidence indicates that **plasmogamy** (cellular fusion) takes place on the outer fringes of a population between pairs of myxamoebae of *D. discoidium.* The zygotes thus formed move to the center of the population and initiate centers of aggregation. Soon after such a center has been formed, meiosis occurs, and many more haploid cells are produced. On the other hand, studies of other species of *Dictyostelium*, as well as other genera, have produced no evidence of plasmogamy.

HUMAN RELEVANCE

The Acrasiales are important to man only as a tool in the study of morphogenesis.

the acellular slime molds | myxomycetes

4

The possession of an amoeboid vegetative stage differentiates the true slime molds (Myxomycetes) from the true fungi (Eumycotina). This creeping vegetative stage (**plasmodium**) theoretically has no limiting size and is the largest naked mass of protoplasm that is to be found in nature. It is multinucleate and feeds on dead organic material, bacteria, protozoans, and probably other small organisms as it moves inconspicuously through the rotting leaves, logs, and other debris found on the forest floor. The plasmodium also feeds on other fungal sporophores, spores, and hyphae. Various investigators have classified the slime molds with the protozoans and many invertebrate zoologists have expressed interest in them because of the plasmodium.

An equal case can be made for calling them plants because of their fructifications (**sporophores**). When conditions of temperature, pH, moisture, food supply, and light are correct, the plasmodium will clump together and form one of several possible kinds of fructifications. These bodies produce many structures that are not found in animals, but are found in many other plants. However, the deciding reason the slime molds are studied by mycologists and botanists is probably tradition. Although DeBary (1887) considered them animals as evidenced by his name Mycetozoa, they had been discussed previously as myxomycetes, the name used by Thomas H. Macbride (1899) when he monographed the group. G. W. Martin has also argued for the fungal affiliations of the slime molds and his designation of them as fungi is accepted in the most recent complete work about the true slime molds by Martin and Alexopoulous (1969). Whether they be plants, animals, or some intermediate form, they do exist, and they are fascinating. Since they are often found while searching for other fungi, and for the above reasons, they will be included here among the plant groups.

29

the acellular slime molds — myxomycetes

THE PLASMODIUM

Three major types of plasmodia have been found, but most early observations were made on the plasmodium of *Physarum polycephalum* (Figure 4.1, p. 38), a common species. This type of plasmodium (**phaneroplasmodium**) moves by means of large fan-shaped areas that possess cords of protoplasm in which rapid **cyclosis** (cytoplasmic streaming) takes place. The cyclosis is reversible and alternates, but the total movement is in one particular direction. A contractile protein, **myxomyosin,** has been isolated from the plasmodium and it is this that is probably responsible for the movement.

Plasmodia possess vacuoles that contain undigested, digested, and inert material. Digestion in the vacuoles is brought about by enzymes that are secreted into the vacuole by the cytoplasm. Not all things ingested by plasmodia can be used, and those that are not are eventually left behind with the undigested waste products. Although many plasmodia cannot be grown in pure culture, *P. polycephalum* has been so grown on chemically defined media. The plasmodium may be hyaline, white, yellow, red, or various other colors and shades. Although the colors are rather constant for a given species, variations in food supply or substrate may cause variations in plasmodium color.

Other types of plasmodia, for example, **aphanoplasmodia** and **protoplasmodia** differ from the phaneroplasmodium in size, shape, color, and ability to move, but the functions that they perform within the life history of the slime mold are similar.

The nuclei in the plasmodium are diploid, 3½ to 4 μ in diameter, and distributed at random. In a rapidly growing plasmodium, mitosis is synchronous, but as the plasmodium ages the synchrony is reduced or lost.

Under conditions of reduced water and food availability, low temperature, and other unfavorable environmental conditions, some plasmodia, particularly *P. polycephalum*, dehydrate and form a **sclerotium** containing small masses of protoplasm (**macrocysts**) from 10 to 25 μ in diameter. The sclerotium may remain in this condition for a number of years and once again become an active plasmodium when rehydrated.

FRUCTIFICATIONS (SPOROPHORES)

The part of the life history most often found in nature is the sporophore.

The **sporangiate sporophores** (Figure 4.2*a*, p. 38) are the most common of the three types of myxomycete fructifications, and possess all of the structures found in the other types. The sporangium may have a stalk **(stipe)** or it may not **(sessile).** The stalks may continue into the sporangium forming a **columella.** Sterile threads **(capillitium)** are found among the spores in the sporangium in some myxomycetes and are sometimes attached to the columella but never to the spores. The threads may be ornamented with a variety of cogs, spirals, or spines. In some sporophores, protoplasm is evacuated from strands of the plasmodium and leaves irregular threads or plates called **pseudocapillitia.** Although the pseudocapillitium often resembles the capillitium, the former is usually more massive and may have rounded ends. The pseudocapillitium may also be found as thickened strands which form a net in the **peridium.** The net often remains after the rest of the peridium has disintegrated. The sporangium wall (peridium) may be extremely fragile or somewhat persistent. A thin, shiny membranous remnant **(hypothallus)** of the plasmodium is often found subtending the fructification. If the plasmodium has accumulated lime, the lime is deposited during the formation of the sporophore. Lime may be found in amorphous masses, in crystals, or deposited in knots on the capillitium. A large number of sporangia are generally produced from a single plasmodium, and they may be loosely scattered or tightly packed upon the substrate.

The other two types of fructifications vary in general appearance from the sporangiate forms. The **aethalium** (Figure 4.2*b*, p. 38) is a type of fructification in which the entire plasmodium gives rise to a large sporophore. The rarest sporophore, the **plasmodiocarp** (Figure 4.2*a*, p. 38), arises from the plasmodium in such a manner that it appears to be a system of interconnecting tubes. Both of these sporophores are sessile, but are covered with a peridium, contain spores, and often capillitial threads.

The spores contained within the sporangium may be smooth to variously ornamented, and their colors in mass vary from black or brown to such bright colors as yellow, red, and violet.

All of the slime molds produce their spores in one of the above fructifications with the exception of the genus *Ceratiomyxa*. This genus has been described as having its spores borne on stalks originating from an extensive, threadlike hypothallus. The spore, as thus described, is probably really a small sporangium because just before germination, four small spores are produced within the "spore."

31

the acellular slime molds—myxomycetes

LIFE HISTORY

Although details of structure and function vary from species to species, the life history (Figure 4.3) of all of the slime molds is remarkably similar. *Physarum polycephalum* is a species that is extremely common, is well discussed in the literature and can be easily studied in the laboratory because of the ease with which it can be cultured through its entire life history. All of the phases of the following, generalized discussion should be easily found.

Spores from the fructification contain a single cell that possesses a single, haploid nucleus. When the spores reach a suitable environment, germination by means of amoeboid or flagellated cells occurs. The cells escape the spore case by way of a pore or crack. These **swarm cells** (if flagellated), or **myxamoebae** (if not flagellated), move about on the substrate or through thin water films. In some slime molds the spore produces a flagellated cell upon germination but the cell may lose its flagella and become amoeboid or an amoeboid cell may become flagellated. In reality, the cells released from the spores are **gametes** because plasmogamy and karyogamy occur when they meet compatible cells. There is evidence that some species are **heterothallic** while others are **homothallic.** Encystment of a gamete may occur if environmental conditions are unfavorable, but reestablishment of favorable conditions allows the encysted gamete to once again function.

Once formed by the fusion of two gametes, the zygote grows and moves about. The diploid nucleus duplicates by mitosis and the plasmodium becomes multinucleate. The zygote may also fuse by plasmogamy with other compatible zygotes. The number of plasmodia may be increased by artificially dividing the plasmodium in the laboratory.

The plasmodium, containing many diploid nuclei continues to feed and grow. When environmental conditions become proper for fructification, the plasmodium will begin to clump. Meiosis occurs either during the production of the sporophore or within the spore, thus producing haploid nuclei. If meiosis occurs during sporophore production, a spore wall forms around each haploid nucleus and its particular bit of cytoplasm. If meiosis occurs in the spore, the spore wall forms around each diploid nucleus and its bit of cytoplasm, but all of the meiotic products except one disintegrate. In either case the spores are uninucleate and haploid. Capillitium or pseudocapillitium, if typical, is formed at this time.

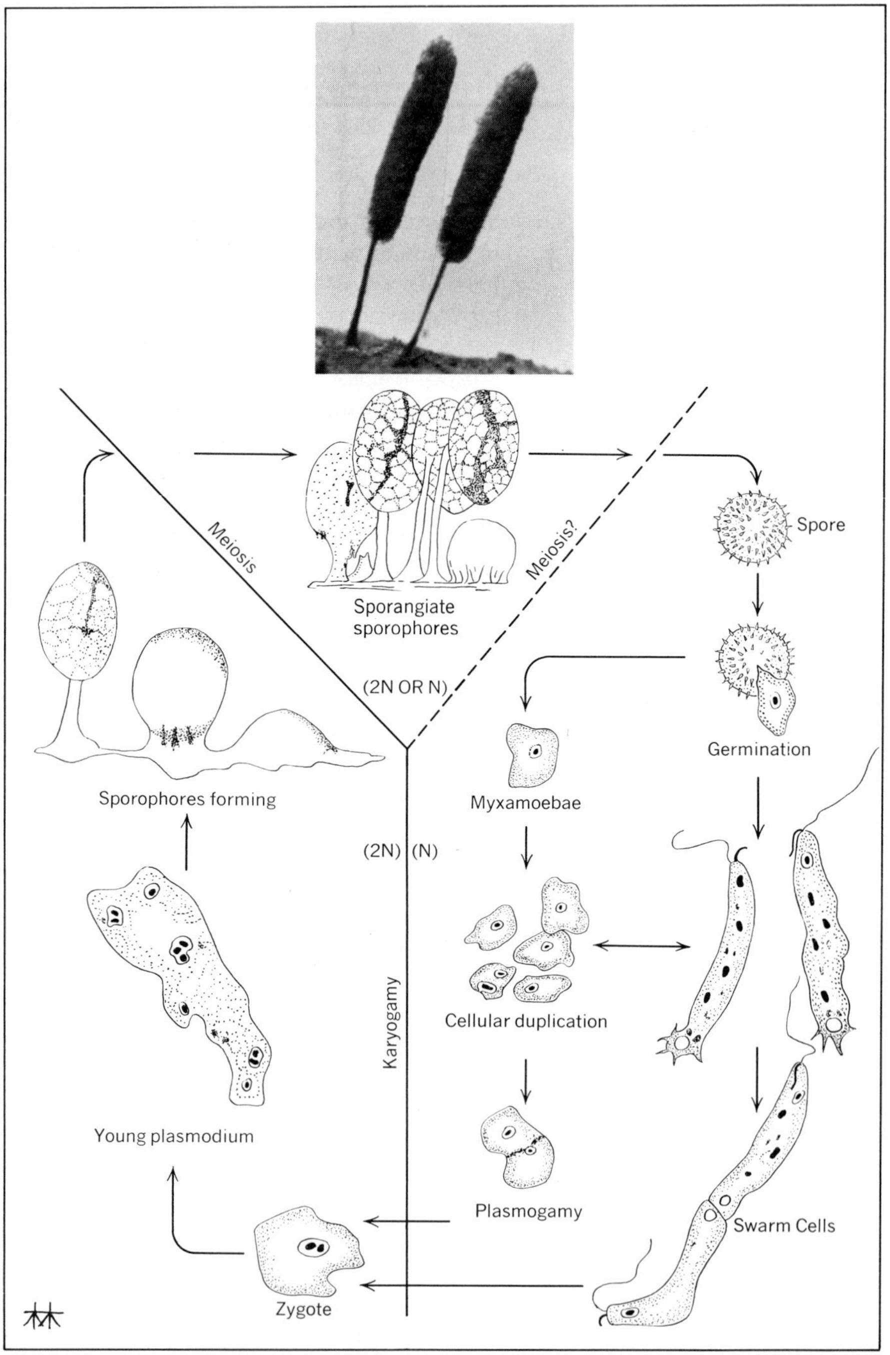

Figure 4.3
Life history of a myxomycete. (Redrawn from C. J. Alexopoulous, *Introductory Mycology*, New York: Wiley.) Mature fructifications of *Comatrichia typhoides*. Sporangia filled with spores (28×).

HUMAN RELEVANCE

The greatest importance to man of the myxomycetes is esthetic. Some of the fructifications, both in structure and color, are among the most beautiful entities to be found in either the plant or animal kingdom. In recent years, cytologists, biochemists, and geneticists have begun using the plasmodium as a tool for investigations of their particular disciplines. Because of the feeding habits of the plasmodia, myxomycetes are of some value as scavengers and decomposers.

Usually, however, the plasmodium crawls inconspicuously about in damp leaves and rotten logs, and finally "fruits." This obscure life rarely brings it to the attention of anyone who is not specifically looking for it.

It forms no disease relationships with other organisms, but the plasmodium may crawl up on grass, twigs, logs, awnings, bicycle tires, and even into basements before fruiting occurs. Such activity may cause one to think his grass or shrub is diseased, but a bit of pruning, mowing, a dry spell, or heavy rainfall will soon set his mind at ease.

algae-like fungi | phycomycetes

5

The phycomycetes are an extremely heterogeneous group of fungi whose **thalli** (vegetative bodies) range from flagellated unicells to macroscopic bodies. The extreme diversity of the members within the group has caused workers to divide them into six categories that are often assigned to class rank in taxonomic systems. Although such assignment has merit, the members of the group show characteristics that make the more generalized category a workable unit for consideration in an introductory text.

The simplest thallus in the phycomycetes is one-celled, globose to ovoid, and uninucleate. The more complex thalli are composed of many branched filaments (**hyphae**) that are collectively called a **mycelium.** The hyphae of the mycelium have no regular system of cross walls (**septa**) and are multinucleate (**coenocytic**) although an occasional septum with a large central pore is produced in older portions of the hypha. The walls contain either cellulose or chitin or sometimes both. The reproductive structures are almost always separated from the vegetative mycelium by septa.

The sexual spore of the phycomycetes is a direct result of fusion of gametes and is, therefore, a **zygote.** This zygote may immediately undergo meiosis and germinate, but the more common occurrence is the production of a heavy wall that protects the protoplast from unfavorable environmental conditions. In this case, meiosis may be delayed until just before germination occurs.

The phycomycetes do have some similarities to the filamentous green algae from which some workers see their origin. Both groups have similar vegetative bodies and methods of reproduction. These similarities have given rise to the name "algae-like" fungi, and although the phylogeny is open to question, the description is apt.

The six classes into which the members of the phycomycetes are often placed are the Chytridiomycetes, Hypochytridiomycetes, Plasmodiophoromycetes, Oömycetes, Zygomycetes, and Trichomycetes. The Hypochytridiomycetes and the Trichomycetes are small and poorly known groups at this time. The Plasmodiophoromycetes are obligate endoparasitic "slime molds" of vascular plants and contain only nine genera. The other three groups either contain a large number of species or contain species that are extremely important, or both, and these groups are considered in the following discussions and examples.

POSTERIORLY UNIFLAGELLATE PHYCOMYCETES—CHYTRIDIOMYCETES

The chytrids are small, mostly microscopic fungi and are placed in the class Chytridiomycetes by some workers and in the order Chytridiales of the class Phycomycetae by others. A generalized description of this group will introduce and define many new terms which will be used consistently in the discussions of the other phycomycetes.

VEGETATIVE STRUCTURE

The true chytrids (Chytridiales) never have a true mycelium; their vegetative thallus being primarily unicellular, although extremely variable. The simplest chytrids have their entire thallus converted into one or more reproductive structures and are said to be **holocarpic.** If only one reproductive structure is formed, the chytrid is **monocentric;** if more than one is formed, the chytrid is **polycentric.**

Rhizoids may be formed by some of the chytrids. These vegetative, assimilative (somatic) portions of the thallus may extend into the substrate where they serve as anchorage and absorptive devices for the fungus. Chytrids that produce this combination of presumably sterile and fertile thallus portions are said to be **eucarpic,** and are either monocentric or polycentric.

REPRODUCTIVE STRUCTURES

Asexual reproduction is accomplished by **zoöspores** ("animal spore") which are produced by protoplast cleavages around many nuclei in the sporangia. Release occurs by rupture of **papillae** or by the opening of a preformed "lid" (**operculum**). The zoöspores possess a single, posterior flagellum of the whiplash type. The **whiplash flagellum** somewhat resembles an old-fashioned buggy whip. The basal portion is rigid and longer than the terminal, flexible portion.

The beating of the flagellum allows the zoöspore to swim with a hopping or gliding motion, or the zoöspore may also move in an amoeboid fashion. The zoöspore contains its food reserve as an oil globule. Its nucleus possesses a crescent-shaped nuclear cap. After discharge from the sporangium, the zoöspore remains motile for a time, withdraws or absorbs its flagellum, and encysts. A short rest period is usually observed before germination which takes place by the production of a single germ tube (**monopolar germination**) after the zoöspore has settled upon its substrate.

Sexual reproduction by various methods is found within the chytridiomycetes but has not been discovered for a majority of the species. In some (e.g., *Olpidium viciae, Synchytrium endobioticum*) motile spores (**swarm cells**) produced within a sporangium may act as zoospores and develop parthenogenetically into adult organisms, or they may fuse with another swarm cell. If fusion occurs, the swarm cell should be defined as a **planogamete** (motile gamete). The method by which the swarm cell will function seems to be determined by the environment. Fusion of morphologically similar but presumably physiologically different gametes is called **isogamy.** Usually the **isogametes** that fuse are produced in different sporangia, but isogametes from the same sporangium have been observed to fuse. A thick-walled resting spore or sporangium is produced as a result of sexual reproduction, and asexual resting spores have been reported for some species. Other Chytridiomycetes (e.g., Blastocladiales) exhibit **heterogamy** in which one of the planogametes is considerably larger than the other. Thus there is a morphological as well as a presumed physiological distinction. Further morphological difference is exhibited by those Chytridiomycetes that produce specialized structures (**gametangia**) in which are produced gametes. The "male" gametes (**sperm, antherozoids**) are produced in specialized gametangia (**antheridia**), are motile, and move to the female gametangium (**oögonium**). There one sperm enters the oögonium and fertilizes the nonmotile "female" gamete (**egg**). Among the Chytridiomycetes, this type of reproduction is found only in the Monoblepharidales, and is termed **oögamy.** Other methods of sexual reproduction have been reported for members of the Chytridiomycetes, but the above three methods of planogamete union (isogamy, heterogamy, and oögamy) are probably the most common, and certainly serve as sufficient examples for the group as well as introductions to the methods of sexual reproduction (with modifications) of other plant groups.

37

algae-like fungi—
phycomycetes

LIFE HISTORY

There is no "typical" chytrid because of the wide variation found in their life histories. The "primitive" members of the group live entirely within the cells of their host (**endobiotic**). Although the early stages may be naked, a cell wall surrounds the mature thallus. The more "advanced" chytrids produce their reproductive structures on the surface of the host although their vegetative (**somatic**) structures are within the cells of the host (**epibiotic**).

An example of a chytrid life history is the life history of *Synchytrium endobioticum* (Figure 5.1) which is discussed here. This chytrid forms a disease relationship called black wart with the tuber of white potatoes. The disease has never been serious in the United States, but has caused crop losses in European countries. It is presently controlled by planting one of the several resistant varieties of potato.

In the spring, zoöspores are released from infected plant parts, swim in the soil water, and infect potato tubers in the soil by dissolving a tiny hole in the epidermal wall of the host. The flagellum remains outside the host and, after entering an epidermal cell, the spore becomes amoeboid. The fungal cell and its nucleus increase in size by assimilating materials from its host. A thick wall surrounds the enlarged parasitic cell and the structure is now known as a **prosorus.** All of this metabolic activity has caused the host cell to greatly enlarge (**hypertrophy**) and the surrounding cells to rapidly duplicate (**hyperplasia**) resulting in a tumor or wartlike mass of tissue (Figure 5.1). The prosorus is now in the lower half of the infected host cell, which is now dead.

Next the prosporus germinates by rupturing its cell wall, and its protoplast surrounded by a thin hyaline membrane pushes into the upper half of the host cell. Many mitotic duplications now occur (about 32), and thin hyaline walls divide the prosorus protoplast into multinucleate segments. Nuclear duplications then increase the nuclear number in each segment to 200 to 300.

Environmental influences cause each segment to develop into either a **sporangium** or a **gametangium;** the entire mass now being known as the **sorus.** If insufficient water is present, the "swarm" cells, when released, form **planospores** which act as gametes, are smaller than the zoöspores, and unite forming zygotes. On the other hand, zoöspores are produced if adequate water is present. This physiological difference in function can be controlled by the experimenter, but in either case the product of protoplasmic segmentation

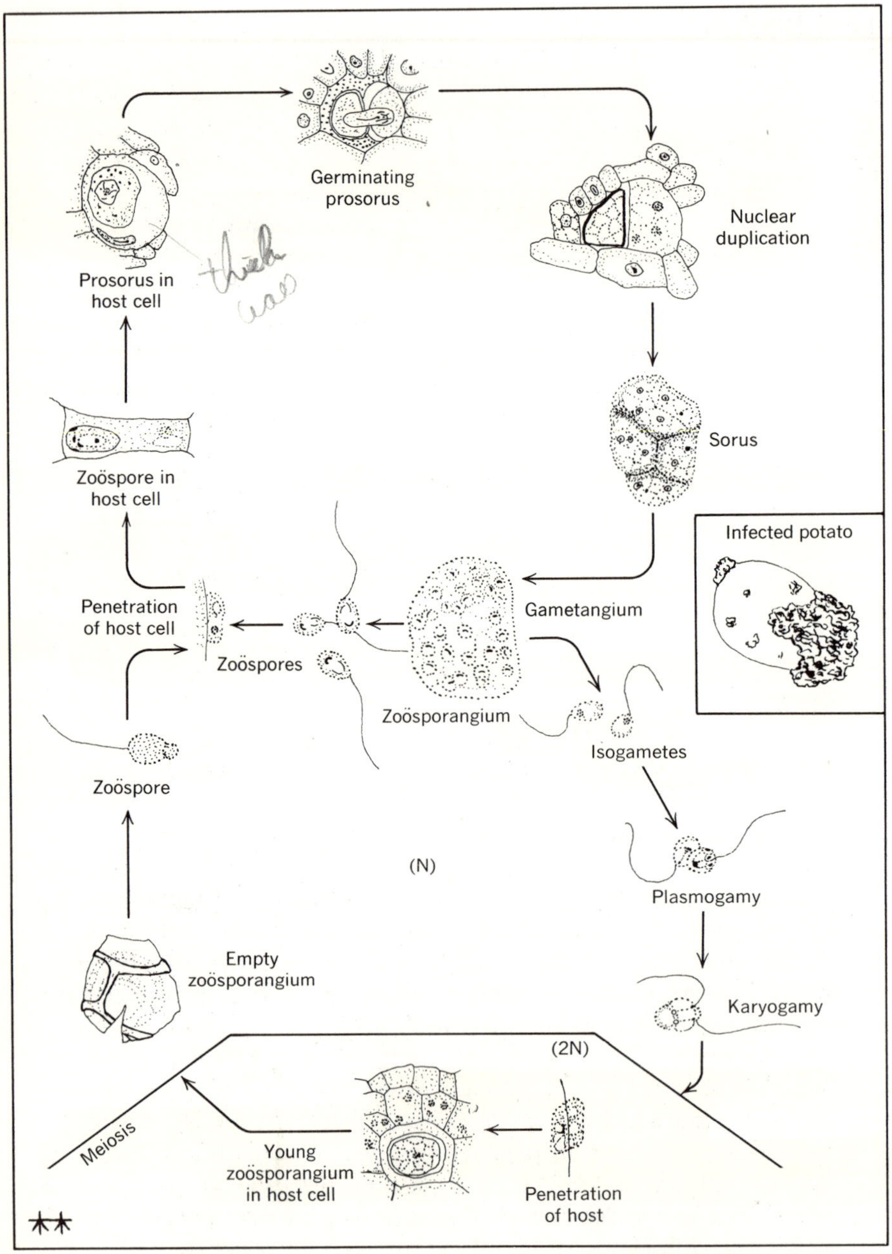

Figure 5.1
Life history of *Synchytrium endobioticum*. (Redrawn from C. J. Alexopoulous, *Introductory Mycology*, New York: Wiley.) Insert: Potato infected with *S. endobioticum* causing disease relationship known as black wart of potato.

Figure 4.1
Phaneroplasmodium of *Physarum polycephalum*. (a) in culture (b) enlargement of fan (c) digesting a mushroom (d) crawling up on sticks.

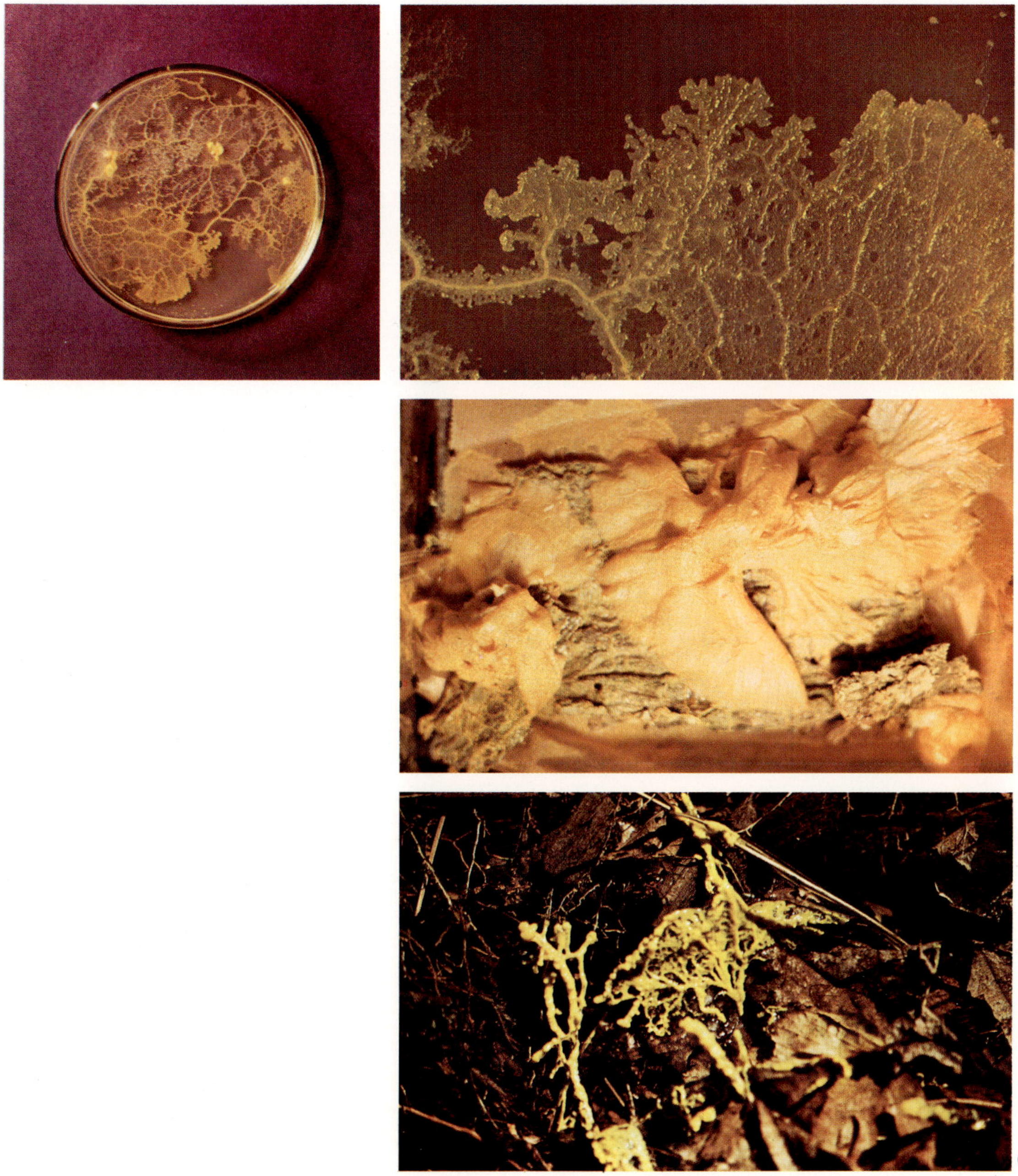

(a) (b) (c) (d)

Figure 4.2

Three types of myxomycete fructifications (sporophores). (*a*) Sporangium of *Metatrichia vesparium* with preformed end. *Comatrichia typhoides* without spores. Columella and capillitium apparent. (*b*) Aethalia of *Fuligo septica* and *Lycogala epidendrum*. (*c*) Plasmodiocarp of *Hemitrichia serpula*. Bits of capillitium expanding through broken peridium.

(*a*)

(*b*)

(*c*)

(either zoöspore or isogamete) is uninucleate. The zoöspores can reproduce the asexual part of the life history by infecting potato tubers. This, obviously, increases the soil population of the fungus.

A physiological difference between the isogametes seems to be indicated by the evidence that those produced within a particular gametangium will not fuse; however, those produced in different gametangia, but within the same sorus, will fuse. Gametic fusion takes place in a film of water on the surface of the host or in the soil. Karyogamy (nuclear fusion) follows plasmogamy (cytoplasmic fusion), the biflagellated zygote swims about for 2 to 5 minutes, settles on the epidermis of a host, and penetrates an epidermal cell. A heavy wall is produced around the parasite and the sporangium remains dormant through the winter. Meiosis presumably occurs with the resumption of increased metabolic activity during the next spring, a new "crop" of zoöspores is produced, and the cycle is completed.

The genus *Allomyces* is within the order Blastocladiales of the class Chytridiomycetes. It is of special interest and is discussed here because it is one of the "simplest" groups of organisms to have members that exhibit a well-defined alternation of generation. The species of the subgenus *Euallomyces* alternate between a haploid **gametothallus (gametophyte)** and a diploid **sporothallus (sporophyte)**. The following discussion is based on the life history of *Allomyces macrogynus* (Figure 5.2).

Allomyces has a true vegetative mycelium that is generally nonseptate except for thickened rings (pseudosepta) found at irregular intervals. The cell walls of their dichotomously branched hyphae are composed chiefly of chitin, glucosan, and ash.

A posteriorly uniflagellate, diploid zoöspore rests on the substrate, and germinates by giving rise to two germ tubes (**bipolar germination**). One of the germ tubes develops into a rhizoidal system and the other into an elaborate dichotomously branched vegetative thallus. This sporangium-producing thallus (sporothallus) produces one type of sporangium that is thin-walled and colorless, and another type that is thick-walled, resistant, and contains melanin pigments in the wall. The spores in the thin-walled sporangium are diploid and their nuclei are produced by mitotic duplications. The thick-walled sporangia are often referred to as resistant sporangia, their spore nuclei are produced by meiotic divisions and are thus haploid. In either case the spore is a zoöspore, but may be haploid

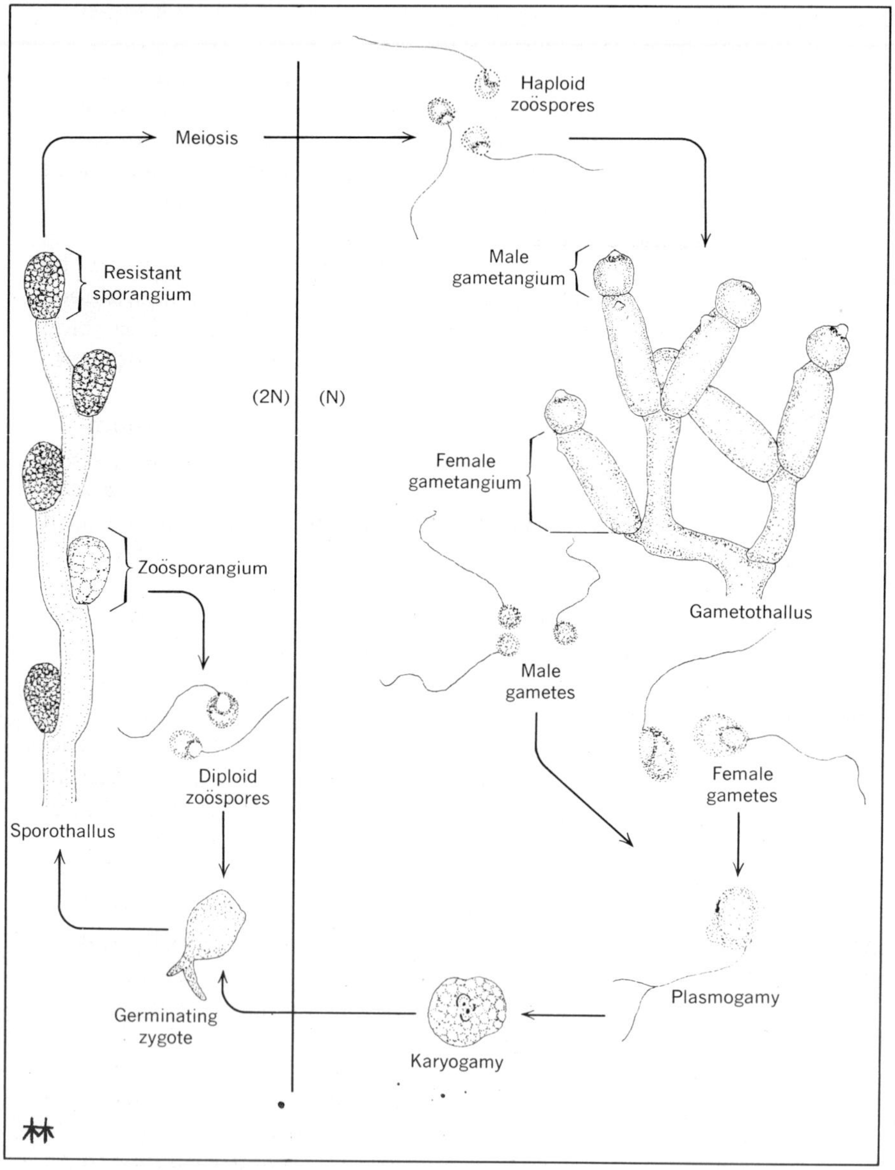

Figure 5.2
Life history of *Allomyces macrogynus*. (Redrawn from C. J. Alexopoulous, *Introductory Mycology*, New York: Wiley.)

or diploid depending on the type of sporangium in which it is produced.

The diploid zoöspores are capable of repeating the diploid cycle by germinating and producing new sporothalli. The haploid zoöspores germinate and produce a haploid thallus that is morphologically identical to the diploid thallus (**isomorphic**). This gametothallus gives rise to "male" and "female" gametangia, the colorless "female" subtending the orange "male."

The gametes are released into the water where fusion occurs. The "female" colorless gamete is larger than the "male" orange gamete, but both are morphologically similar to zoöspores typical of the Blastocladiales. Because of the difference in size, this planogamete type is referred to as an **heterogamete.**

The "female" heterogametes produce the sex hormone **sirenin** which attracts the "male" heterogametes. Karyogamy follows plasmogamy and a biflagellated zygote is produced. The zygote swims about for a short period, comes to rest, and loses its flagella. Upon its germination, a diploid sporothallus is produced, and the cycle is complete.

It is important to recognize that, in the phycomycetes, the diploid phase is usually limited to a single diploid cell that was the direct result of gamete fusion and undergoes meiosis at the time of germination. *Allomyces* is an exception to this generalization, and is a good example of alternation between a haploid and a diploid generation within the same species. Although the two thalli (n and 2n) are morphologically identical (isomorphic), their identities become readily apparent when their reproductive structures are produced. Although rare among the fungi, a well-defined alternation of generation is common among other plant groups. Therefore, it is well to introduce and understand the concept early, because it will appear again and again with many variations.

HUMAN RELEVANCE

The organisms found within the Chytridiomycetes are generally considered to be water or soil-inhabiting organisms. Several of the true chytrids are parasitic on algae, other fungi, or vascular plants. There are a few serious vascular parasites that form disease relationships with economically important crops. Some of these are *Synchytrium endobioticum* which causes potato wart, *Physoderma zeae-maydis* which causes *Physoderma* brown spot of corn, and *Urophlyctis al-*

falfae which causes crown gall of alfalfa. Most of the members of this group are, however, saprobes.

The other members of the class are chiefly saprobic water or terrestrial fungi and are economically important only because of their role in breaking down organic material into its simpler molecular constituents.

BIFLAGELLATED PHYCOMYCETES— OÖMYCETES

The zoöspores of the members of this group of fungi have two flagella, one **whiplash** and the other **tinsel.** The whiplash flagellum has been previously described for the chytrids. The tinsel flagellum is composed of a long central portion (**rachis**) with hairlike projections extending laterally along its entire length.

VEGETATIVE STRUCTURES

Most members of this group are eucarpic, but their thalli vary from unicells to many-branched, robust hyphae. Some members are holocarpic.

REPRODUCTIVE STRUCTURES

Asexual reproduction takes place by the biflagellated zoöspores that are produced within some sort of sporangium. In the advanced members of the group, the sporangium functions as an aplanospore, germinates by a germ tube, and produces a mycelium. Asexual reproduction may also take place by somatic fragmentation.

Sexual reproduction is usually accomplished by oögamy. A terminal or intercalary oögonium, containing large female gametes (**oöspheres, eggs**), is contacted by the antheridium. The oöspheres are usually uninucleate, but may be multinucleate. Male nuclei move through fertilization tubes from the antheridia to the oöspheres in the oögonium. Karyogamy takes place and a thick walled **oöspore** is formed. Both heterothallic and homothallic species are known.

LIFE HISTORY

A discussion of the life histories of two members of different orders of the group should serve to illustrate the variability that exists.

Saprolegnia is a member of the order Saprolegniales. The members of this order are often called the "water molds" because of their abundance in clear water. However, many species of the Saprolegniales inhabit soil.

The vegetative mycelium of *Saprolegnia* sp. is eucarpic, large, robust, aseptate, and variously branched. The walls contain cellulose, and cross walls delimit the fruiting structures.

The asexual phase of the life history (Figure 5.3) consists of the production of elongate **zoösporangia.** The zoöspores that are released from the zoösporangium (**primary zoöspores**) are pip-shaped with a tinsel and a whiplash flagellum. The zoöspores swim for a while, encyst, and germinate by means of a reniform (kidneyshaped) zoöspore that also has one flagellum of each type. This **secondary zoöspore** swims for a short time and encysts. It, however, germinates by means of a germ tube and produces a somatic thallus. This condition of two zoöspore types and two distinctive swimming periods is known as **diplanetism.** Asexual reproduction occurs by vegetative fragmentation and by hyphal cells that develop thick walls (**chlamydospores**), become separated from their parent hyphae and germinate by germ tubes.

The sexual phase of the life history (Figure 5.3) consists of the production of compatible oögonia and antheridia which are borne on the same thallus (homothallic) or on different thalli (heterothallic).

As the thallus becomes sexually mature, the oögonium contains from one to many large, naked nonmotile oöspheres (eggs), each of which contains one nucleus. The antheridia are much smaller than the oögonia and are multinucleate.

When fully mature, one or more antheridia become attached to an oögonium. Fertilization tubes penetrate the oögonium wall and grow to the oöspheres. Although more than one antheridium may be attached to each oögonium, fertilization is usually by only one. The fertilization tube can send branches to the various oöspheres after penetration and nuclei from the antheridium migrate to the oöspheres where karyogamy occurs. This establishes the only diploid phase of the life history. A thick wall forms around the fertilized oösphere and an oöspore is produced. The oöspores are eventually released from the disintegrating vegetative thallus. Meiosis occurs and the oöspore germinates by means of a germ tube. The haploid thallus is thus reestablished.

The genus *Albugo* is a member of the order Peronosporales. The order contains members that are aquatic, amphibious, and terrestrial. They have eucarpic, robust, freely branching, coenocytic mycelia which, in many species, develop haustoria.

44

nonvascular plant groups

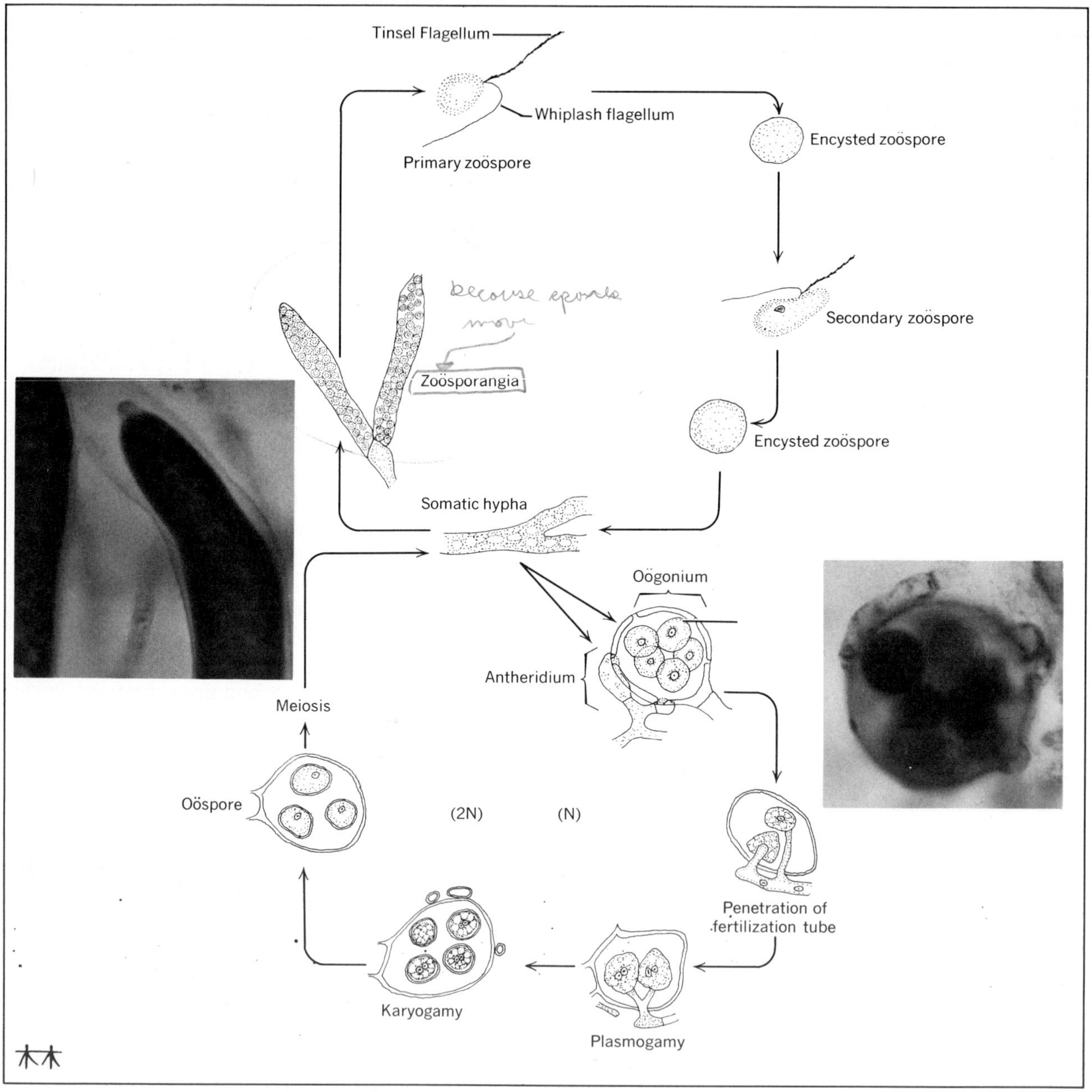

Figure 5.3
Life history of *Saprolegnia* sp. (Redrawn from C. J. Alexopoulous, *Introductory Mycology,* New York: Wiley.) Inserts: Photomicrograph of young zoösporangia with zoöspores being produced. One young sporangium produced inside wall of old sporangium (1757×). Photomicrograph of antheridia and oögonia of a homothallic species (1530×).

Albugo candida (Figures 5.4 and 5.5) belongs to the family Albuginaceae commonly known as the "white rusts." All members are obligate parasites on vascular land plants. The hyphae penetrate the epidermis of the host, grow intercellularly, and produce haustoria.

As the mycelium increases in the host, asexual reproduction is accomplished by the production of a layer of short, club-shaped **sporangiophores** just under the epidermis. The sporangiophores are tightly clustered and are produced from the tips of vegetative hyphae. As they mature, the sporangiophores cut off multinucleate sporangia at their tips. There is eventually produced a series of sporangia (often incorrectly called "conidia") that are attached in a chain with the youngest sporangium at the base. As the sporangia mature, they break off and occupy the space just under the epidermis of the host.

The accumulation of mycelium and spores creates pressure on the epidermis of the host and it eventually ruptures. The sporangia are thus released to the outside where they are disseminated by wind and water. If they reach a suitable host and have adequate moisture, they germinate either by germ tubes (in temperatures greater than 70° F) or zoöspores (in temperatures less than 70° F). The germination by germ tubes is similar to the germination pattern of true conidia and is the reason they are sometimes called conidia. The reniform, biflagellate zoöspores may swim for a while, but eventually encyst and germinate by a germ tube.

Sexual reproduction is oögamous with multinucleate oögonia and antheridia produced at the ends of vegetative hyphae within the tissues of the host. At maturity, only one nucleus in each organ is functional. In the mature oögonium, the oösphere is surrounded by a thick layer of **periplasm.** The functional nucleus migrates to the center of the oösphere and the other nuclei migrate into the periplasm. The gametangia establish contact and the antheridium produces a fertilization tube through which the functional nucleus, accompanied by some cytoplasm, passes into the oösphere (egg). Plasmogamy and karyogamy establish a diploid zygote nucleus that begins to divide. Uncertainty exists concerning just when meiosis occurs, but eventually there is produced approximately 30 haploid nuclei within the oöspore after which it enters a resting stage. The following spring the thick-walled oöspore divides into a number of uninucleate sections, each of which emerges as a single reniform,

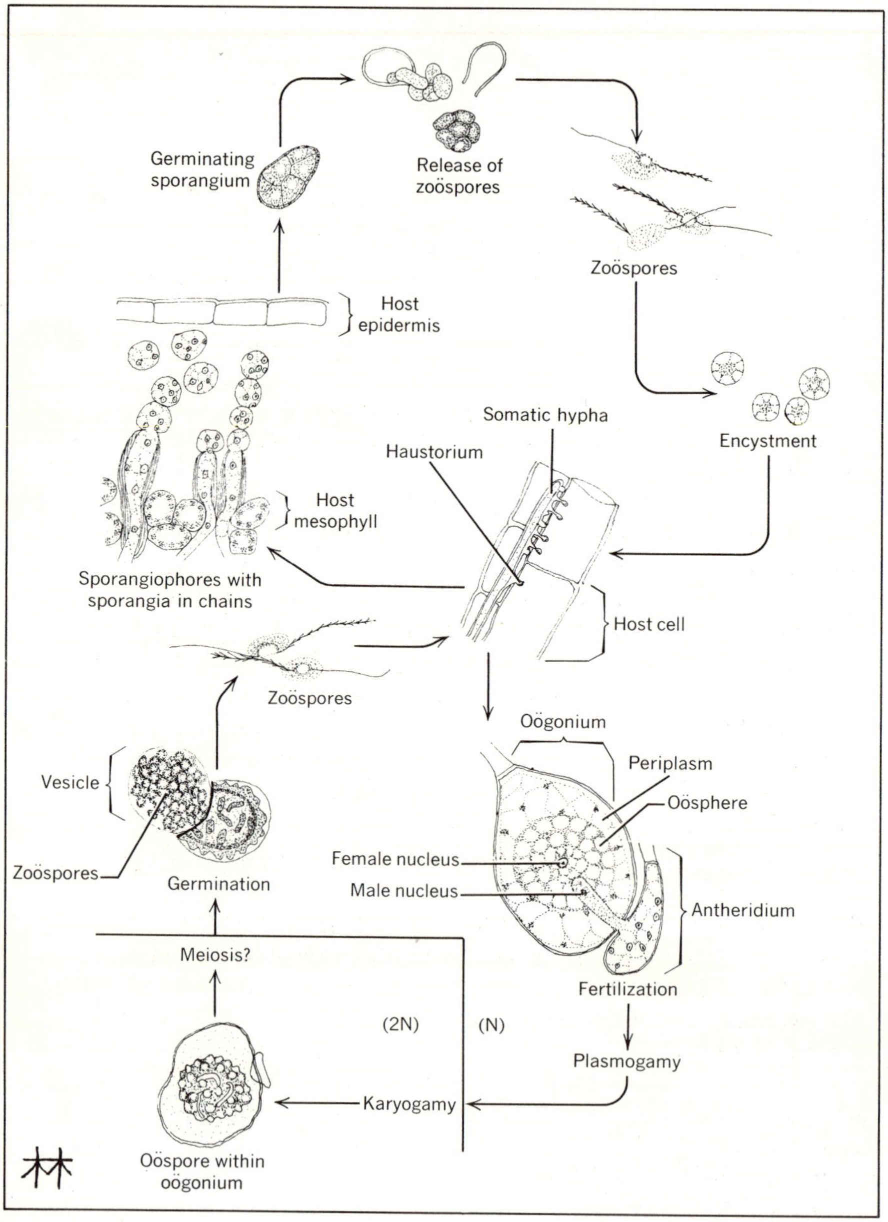

Figure 5.4
Life history of *Albugo candida*. (Redrawn from C. J. Alexopoulous, *Introductory Mycology*, New York: Wiley.)

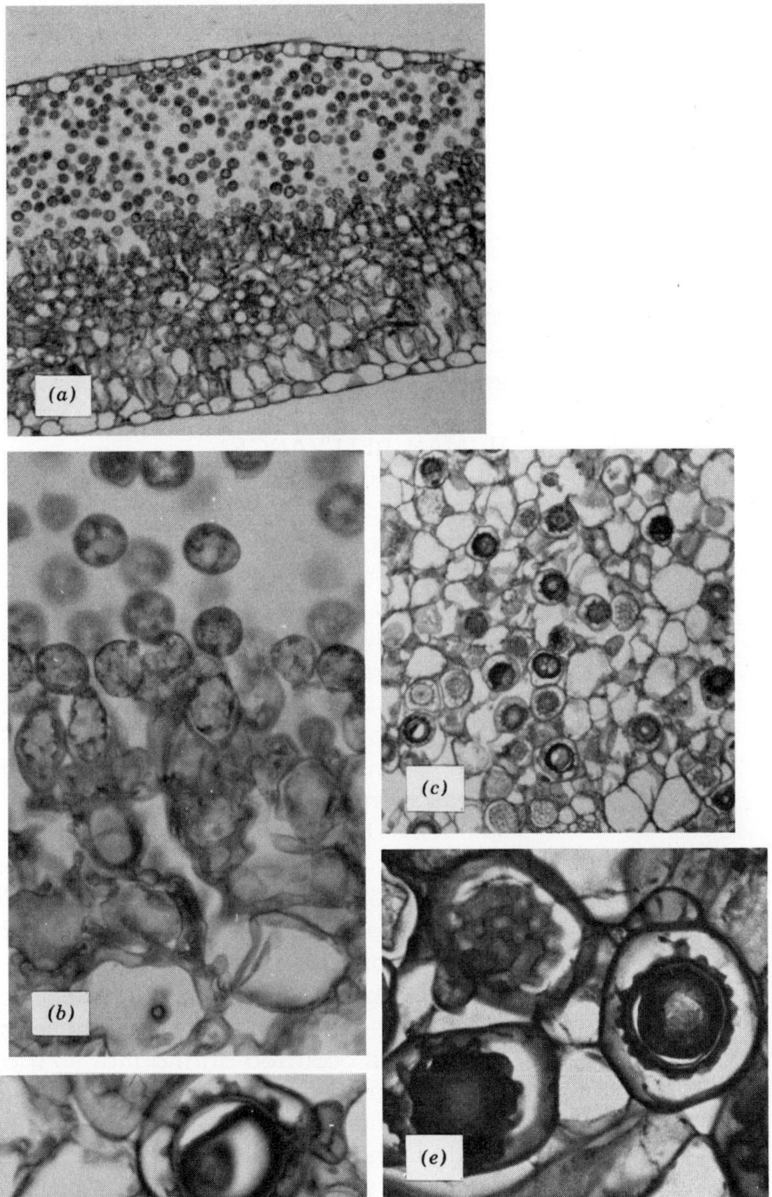

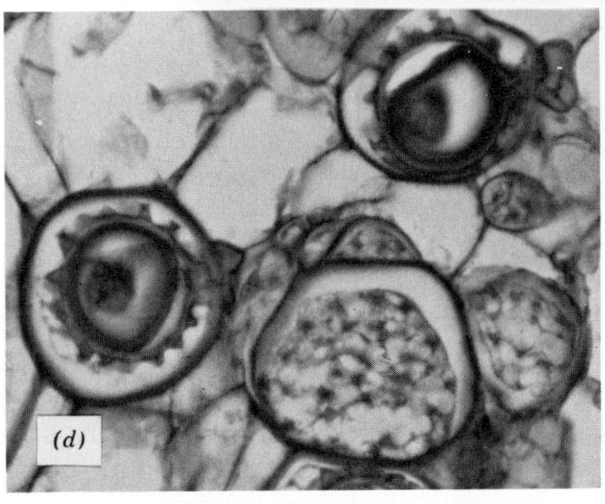

Figure 5.5
Photomicrographs of *Albugo candida*. (*a*) Section through host where sporangiophores and sporangia are produced (400×). (*b*) Same area (1350×). (*c*) Section through host with oöspores in various stages of development (370×). (*d*) Same area with antheridia, oöspheres, and sections through maturing oöspores (1800×). (*e*) Same area with reticulations on mature oöspore and thick oöspore walls (1800×).

biflagellate zoöspore. The zoöspores are extruded into a **vesicle** that is either sessile or connected to the oöspore by an exit tube. The zoöspores are released by the breaking of the vesicle wall; they swim, encyst, and germinate by germ tubes, reestablishing the haploid thallus.

HUMAN RELEVANCE

Most of the members of the Oömycetes are economically unimportant to man except, of course, as they help to fill the very important role of reducers in the environment.

Some members of the Saprolegniales attack economically important plants, and some are parasitic on fish. Others become facultative parasites on fish after the fish has been harmed in some other fashion (e.g., via scrapes or bruises).

It is within the order Peronosporales that the truly destructive members of the Oömycetes are found. *Albugo candida* cases "white rusts" of crucifers and other members of the genus cause similar diseases on other hosts (e.g., *A. occidentalis* causes white rust of spinach). Other parasitic members of this group should also be mentioned. For example, *Phytophthora infestans*, the causative organism of late blight of potatoes, and *Plasmopara viticola*, the causative organism of downy mildew of grapes, have had great influence on human activity.

Although the social and political structures in Ireland in 1845 were potentially explosive enough to cause migration, the late blight epidemics that occurred during that year and the following few years can certainly be viewed as the "triggering" motivating force. For a number of reasons, the Irish peasants had "adopted" the Andean potato to the extent that their livelihood depended directly upon a successful yearly crop. When the blight struck with such destructive force, the only choice left was to emigrate or starve. Thus was brought about the emigration of thousands of Irish to the United States during the late 1840s.

Downy mildew of grapes appeared in France for the first time in 1878. It was imported by the French from the United States on grapevines that were to be used for hybridization experiments by the wine makers. In five years it had spread to nearly every vineyard in Europe and had placed wine making in serious jeopardy in France, Portugal, and Spain. This fungus, plus the effects of *Uncinula necator*, an ascomycete, and a root-sucking aphid (*Phylloxera*) which caused a disease of grapes known as "dry leaf" disease greatly changed viniculture in Europe. As a result of the loss of wine because of the grape diseases, plus the social, political, and religious views of the time, Irish and Scotch whiskies began to be used to quench alcoholic appetites. Thus, not only did *Plasmopara viticola* influence agriculture, but it also profoundly influenced the world's libations.

Another extremely important member of the Peronosporales is the genus *Pythium*. A very large number of species have been described for this genus, and many of them cause disease relationships, known as "damping off" diseases, with economically important vascular plants. In these diseases the young seedling is often invaded at the surface of the soil by the fungus. The host tissues break down and the seedling falls over and dies. Other types of damping off result in the failure of the seed to germinate or in the death of the seedling before it emerges from the soil. The damping off diseases are most destructive during cool, wet spring weather. Members of the genus *Pythium* cause damping off of beets, celery, lettuce, tomatoes, and many other vegetable crops.

Several species of *Pythium* also cause fruit rots in muskmelon, pumpkin, and watermelon. Initial infections may occur in the field, but contact while in storage may transmit the fungus to healthy fruit. *Pythium* spp. may cause root rots in all of the cucurbits, but is important only during certain seasons. Cucumber, squash, watermelon, and muskmelon are among the plants affected by *Pythium* root rots. The roots of many other plants such as beets and sugar cane may also be affected by species of *Pythium*.

AFLAGELLATED PHYCOMYCETES—ZYGOMYCETES

The Zygomycetes is that group of Phycomycetes that produce no flagellated cells. All members of the group have gametangia that are similar to each other, and the sexual spore (zygospore) is a result of the fusion of the entire protoplasts of the gametangia.

The members of the zygomycetes most often encountered belong to the order Mucorales, although some fascinating members belong to the orders Entomophthorales and Zoöpagales.

VEGETATIVE STRUCTURE

The members of the Mucorales, the order to which this discussion will be devoted, produce extensive mycelia that grow rapidly. The young mycelium is aseptate, but septa that are solid plates of fungal tissue are irregularly produced in older filaments. Septa are regularly produced delimiting sporangia or gametangia.

Some members of the order frequently produce **rhizoids** at areas of the mycelium where it touches a firm surface. The rhizoids adhere to the substrate surface (or grow into the substrate) and firmly anchor the fungus. Absorption of minerals, carbon nutrients, and water occurs through the rhizoids as well as through the rest of the mycelium.

Rhizopus spp., the fungus most often used to demonstrate this group of Phycomycetes, produces lateral hyphae (**stolons**) from the same area that produces the rhizoids. The stolons produce rhizoids when they contact the substrate, other stolons are produced from the same area, and the fungus is effectively spread in a hyphal mat by vegetative reproduction.

REPRODUCTIVE STRUCTURES

Asexual reproduction in the Mucorales is by aplanospores (**sporangiospores**) which are produced in a sporangium that may or may not have a columella. The number of spores in the sporangium varies from one to hundreds. The sporangia that contain only a few spores are often called **sporangioles** and the monosporous sporangia are often called **conidia.** Many mycologists believe that conidia in the Mucorales evolved by reduction in the number of spores produced in the sporangium and fusion of the spore wall with the sporangium wall.

The sporangia are produced on specialized aerial hyphae (**sporangiophores**), which may be variously branched. Some species produce only sporangia that are terminal. Others produce terminal sporangia, but also produce sporangioles on lateral branches of the sporangiophore. Others produce sporangioles on a swollen apex of a sporangiophore; still others produce monosporous sporangia (conidia) in the same arrangement.

The sporangiospores differ in shape, size, color, and markings. They are formed by the production of a wall around single or multinucleate segments of the protoplast. The spore may start as a uninucleate entity, but subsequent mitoses will produce the multinucleate condition.

Sporangiospore dispersal is usually by air currents after the

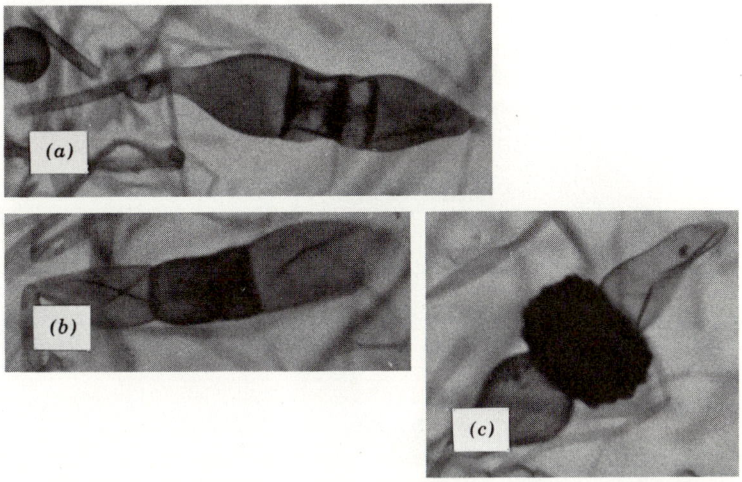

Figure 5.6
Zygospore development in *Rhizopus stolonifera*. (*a*) Gametangia from + and − strain meet (600×). (*b*) Walls between two gametangia break down. Plasmogamy occurs (600×). (c) Mature zygospore produced (550×).

breakdown of the sporangium wall. However, a fascinating method of dispersal occurs in the genus *Pilobus* (the hat thrower). The sporangiophore is positively phototropic and shoots the entire sporangium toward the light. The sporangium may be shot a vertical distance of six feet, and usually sticks to whatever it happens to hit. Other methods of asexual reproduction include the production of chlamydospores and mechanical breakage of vegetative hyphae.

Sexual reproduction is accomplished by the fusion of two similar multinucleate gametangia and the production of a heavy-walled zygospore. The gametangia are produced at the tips of compatible hyphae; they come into contact, the walls between them break down, and plasmogamy is eventually followed by karyogamy.

There are heterothallic (Figures 5.6 and 5.7) and homothallic (Figure 5.8) members of the Mucorales. It was in the former group that a mycologist first discovered the existence of a diffusible substance (e.g., a sex hormone) which caused the initiation of sexual reproduction in the fungi. The two strains that are necessary for sexual reproduction in the heterothallic members are referred to as plus (+) and minus (−).

The zygospore is heavy walled and varies greatly in the length of time (from two weeks to several months) needed for germination. The zygospore germinates by the production of a sporangiophore, at the tip of which is produced a small sporangium (sometimes called the **germ sporangium**). The genetic condition of the germinating

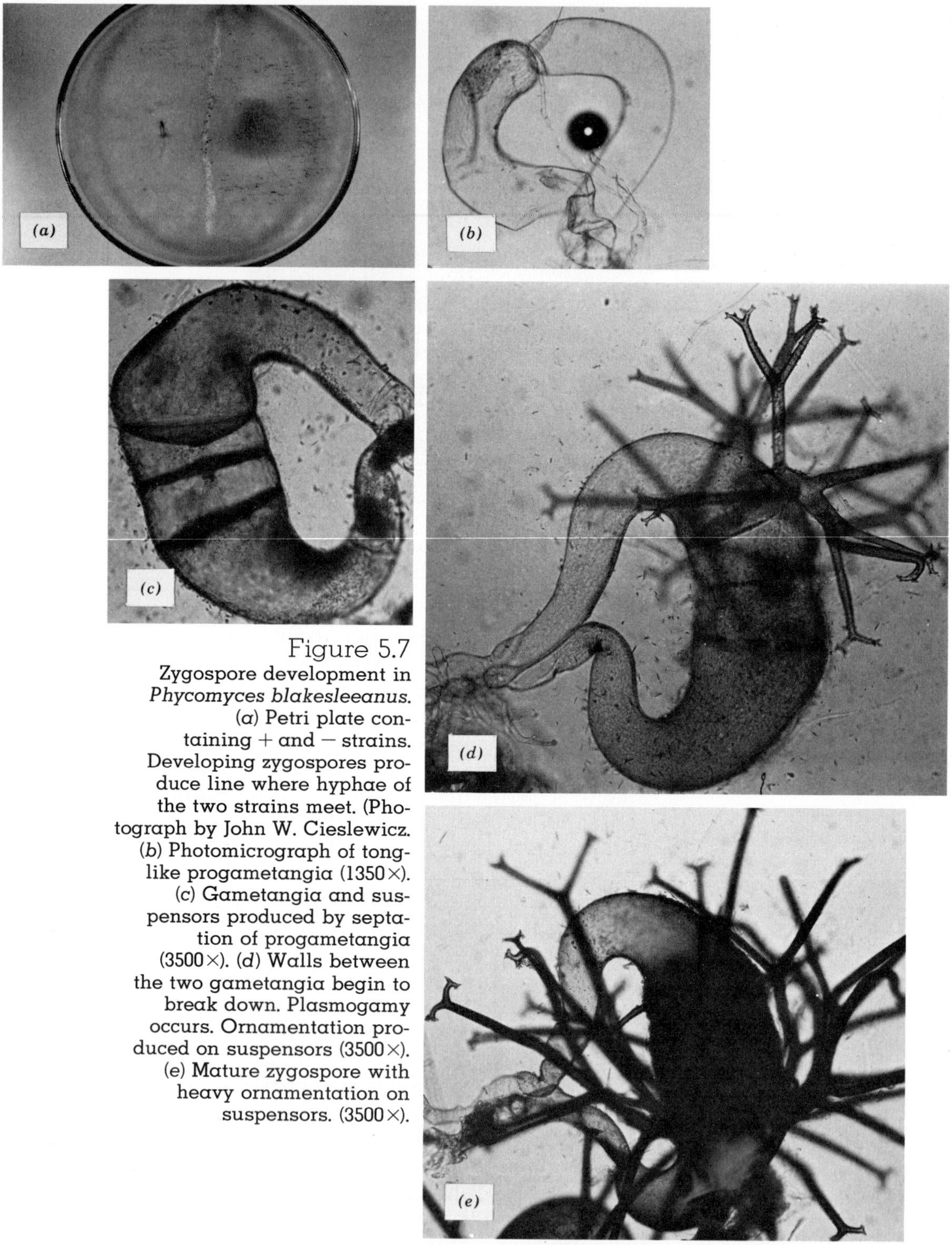

Figure 5.7
Zygospore development in *Phycomyces blakesleeanus*. (*a*) Petri plate containing + and − strains. Developing zygospores produce line where hyphae of the two strains meet. (Photograph by John W. Cieslewicz. (*b*) Photomicrograph of tong-like progametangia (1350×). (*c*) Gametangia and suspensors produced by septation of progametangia (3500×). (*d*) Walls between the two gametangia begin to break down. Plasmogamy occurs. Ornamentation produced on suspensors (3500×). (e) Mature zygospore with heavy ornamentation on suspensors. (3500×).

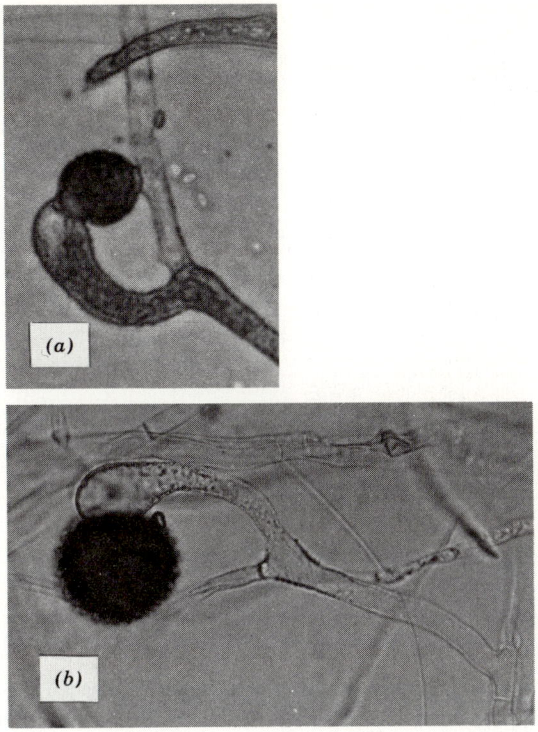

Figure 5.8
Development of zygospores in a homothallic zygomycete, *Zygorenchus* sp. (*a*) Young zygospore (1800×). (*b*) Mature zygospore (1800×).

zygospore may vary from genus to genus and, since both heterothallic and homothallic members are found in some genera, there can be variation of genetic condition within a genus.

Two examples should suffice to demonstrate this genetic variability. In *Rhizopus stolonifera* (*R. nigricans*), only some of the gametangial nuclei fuse (Figure 5.6). These fusions are accomplished by the time the zygospore wall is produced; the other nuclei abort. Meiosis is delayed until germination. At that time the sporangiospores in the sporangium may be all + or all −, or there may be a mixture of the two types. The latter condition would be expected, but the former can be explained if all of the + or all of the − nuclei disintegrated. The type of nuclei remaining appears to result by chance.

In *Phycomyces blakesleeanus* (Figure 5.7), the gametangial nuclei remain associated in groups within the zygospore throughout the dormant period. Just before germination, some of the nuclei fuse; the unfused haploid nuclei remain. At germination, meiosis occurs in some, but not all, of the diploid nuclei. Therefore, a germ sporangium can contain a mixture composed of original + and − haploid nuclei; + and − haploid nuclei that result from meiosis of the diploid nuclei; and ± diploid nuclei. The sporangiospores containing the haploid nuclei produce + or − hyphal strains. The diploid ± sporangiospores produce homothallic mycelia or heterokaryotic mycelia which behave as homothallic mycelia. The number of diploid

54

nonvascular plant groups

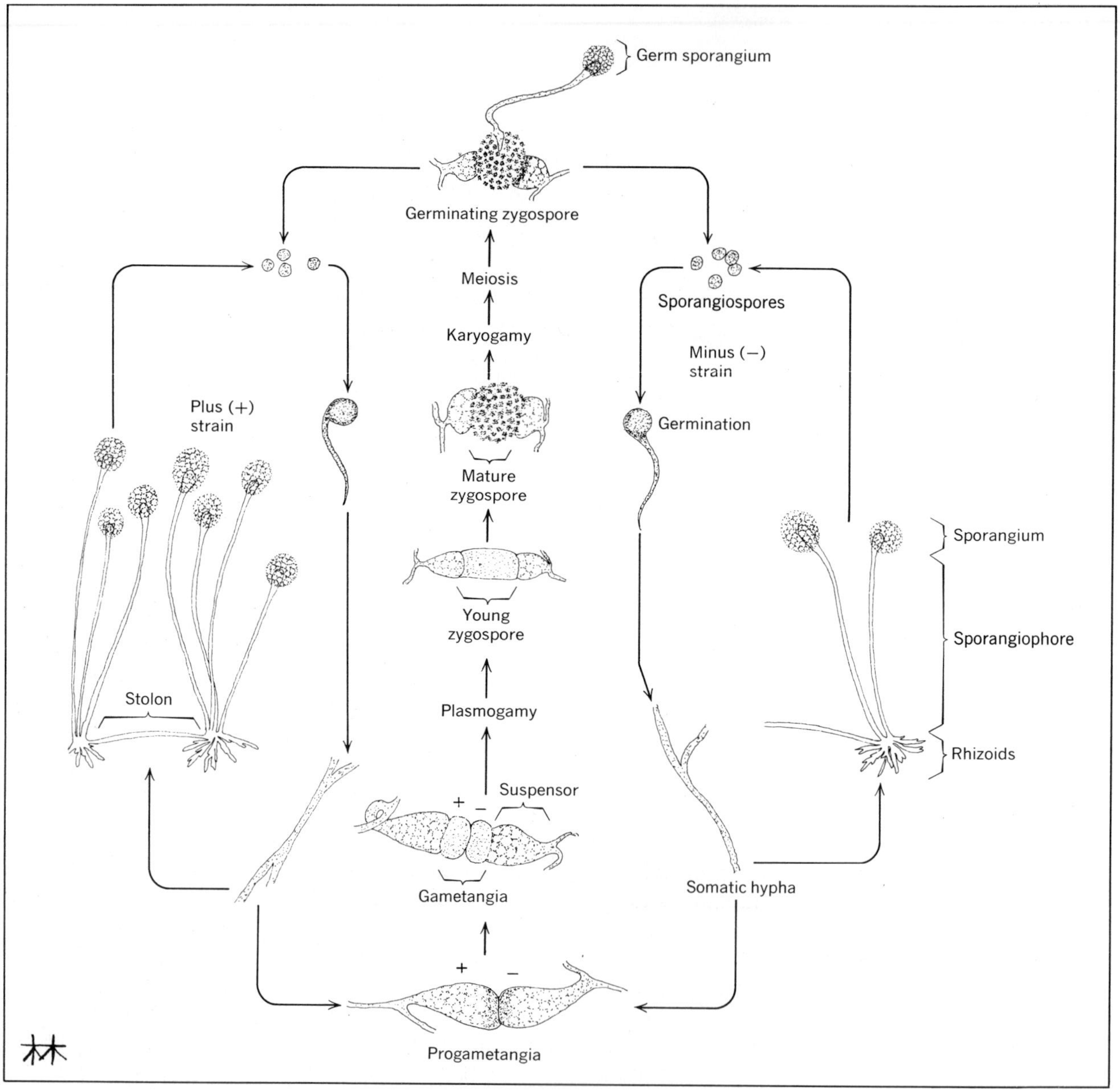

Figure 5.9
Life history of *Rhizopus stolonifera*. (Redrawn from C. J. Alexopoulous, *Introductory Mycology*, New York: Wiley.)

sporangiospores decreases with each asexual generation. Finally no more diploid spores are produced.

LIFE HISTORY

Rhizopus stolonifera, the common black mold of bread, should be familiar to most students, and its life history (Figure 5.9) will serve as an example for the Zygomycetes.

Sporangiospores are passively released when the fragile sporangium wall breaks down. A spore germinates by the production of a germ tube which soon becomes multibranched. The development of a white mycelial mat proceeds by the production of stolons at the ends of which are produced rhizoids. One or more sporangiophores are produced directly above the rhizoid area; the stolon continues to grow. As the sporangiophore reaches maturity, cytoplasm and many nuclei move into the enlarging tip. A columella forms in the center of the sporangium, and the surrounding cytoplasm is cleaved into sporangiospores.

Sexual reproduction requires the presence of + and − strains. When two opposite strains come into contact with each other, small side branches called **progametangia** are produced. Cytoplasm and many nuclei flow into the tips of the progametangia, and they enlarge. A septum forms near the tip of each, dividing it into a **suspensor** and a **gametangium.** The walls between the two gametangia dissolve and plasmogamy occurs.

Nuclei pairing and karyogamy take place as previously described, and a heavy wall is produced completing the zygospore. After a one- to three-month dormant period the zygospore may germinate. Meiosis is thought to occur at the time of germination, although unequivocal evidence is lacking.

Another zygomycete that is often used to demonstrate heterothallism is *Phycomyces blakesleeanus* (Figure 5.7). Although the nuclear conditions are different from *Rhizopus* spp., the zygospore is also produced by the fusion of the gametangia. The suspensors of this species produce elaborate outgrowths which, when mature, may tend to hide the zygospore.

Zygorhynchus sp. (Figure 5.8) is a homothallic species that produces two very unequal gametangia. At maturity, the zygospore and one suspensor are so large that they often tend to obscure the presence of the smaller suspensor.

The examination of these life histories certainly emphasizes the necessity for understanding that mitosis is strictly a nuclear phenomenon. Many mitoses (karyokineses) occur but very few septa are produced. This **coenocytic condition** allows for the flow of multinucleate cytoplasm into such structures as developing sporangia and progametangia.

HUMAN RELEVANCE

Rhizopus stolonifera has often been observed by every housewife as the common black "mold" of bread although many other fungi are equally or even more destructive. Some of the Mucorales are weak parasites on detached plant parts and can cause serious problems during the shipping and storage of foodstuffs (e.g., *R. stolonifera* on sweet potatoes). Others are parasitic on other fungi (e.g., *Syzygites grandis* on mushrooms), green plants (e.g., *Choanephora cucurbitarum* on squash), and animals including human beings (e.g., *Absidia corymbifera*, which affects the human nervous system).

Most Mucorales are, however, saprobes that live on dung and decaying plant or animals parts. Many of these are capable of synthesizing products that are of industrial value. Species of *Rhizopus* and *Mucor* have been used in the production of industrial alcohol. *Rhizpous* spp. can produce fumaric and lactic acid; *R. arrhizus* was used to accomplish the first oxygenation of progesteron in the 11-position, an important step in the synthesis of cortisone. Other members of the genus were later used to produce higher yields.

Many members of the Mucorales have been established to be important for the improvement of soil structure. They produce long-chain polysaccharide polymers that act as a "cementing" substance for the inorganic particles of clay, silt, and sand. The resulting soil aggregates greatly improve the structure of the soil. The fungi can use green manures as the raw materials necessary for the production of the polymers, and it is in this way that green manures, when plowed under, indirectly produce an improved soil structure. Other microorganisms may also use the green manure in the same way, but the filamentous fungi (including some other than the Mucorales) appear to be extremely important. Members of the group are also important as reducers in the ecosystem.

the sac fungi | ascomycetes

6

The ascomycetes are a group of fungi characterized by the structure in which sexual spores are produced. This distinctive structure is the **ascus,** and the spores produced within it are **ascospores.** The ascospore nuclei are haploid because they are produced by meiosis. Eight ascospores are usually produced within an ascus, but the number may vary from one to several hundred.

Other characteristics are shared by most members of the group. With a few exceptions (e.g., yeasts), they have a regularly septate mycelium that contains a large amount of chitin or wall material. The **septa** are produced from the wall toward the center of the hypha, and a pore is left in the center. The pore allows for a continuity of cytoplasm and nuclear migration throughout the hypha. There are no flagellated cells produced, and the asci are usually surrounded by a "fruiting" structure (**ascocarp**) that is produced from sterile hyphae. The life history of most of the ascomycetes has a sexual (**perfect**) and an asexual (**imperfect**) phase. These phases do not obligately alternate with each other, and for many members no asexual phase has been definitely associated with a sexual phase. This may mean either that the imperfect stage has been lost through evolution, that the life history of the organism has not been fully elucidated, or that there was never any imperfect stage produced during the life history of the organism. A more complete account of the asexual phase of the ascomycetes will be found in the following chapter (Fungi Imperfecti).

Three nuclear situations may exist within the hyphae. A **homokaryotic mycelium** results from the germination of a uninucleate, haploid ascospore. A homokaryotic multinucleate condition may arise by mitotic duplication of this nucleus and subsequent migration of the daughter nuclei. The **heterokaryotic condition** may arise in a variety of ways (discussed later) resulting in two paired nuclei per cell (**di-**

karyon) or a multinucleate cell in which the nuclei are not paired but are not daughter cells of the same parent cell. These various multinucleate conditions are possible because the pores in the septa allow nuclear migration.

The **heterokaryon** may arise as a result of mutation of nuclei in an originally homokaryotic mycelium, or as the result of the fusion of genetically distinct entities. Gametangia are produced by some ascomycetes. In these, **antheridia** (male) and **ascogonia** (female) produce compatible nuclei that eventually fuse. Other ascomycetes produce **spermatia** which are single cells whose nucleus eventually unites with the nucleus of the ascogonium. Because of the similarity of the sex structures in those ascomycetes that produce them to the sex structures produced by some of the red algae, some botanists have hypothesized that the former group evolved from the latter one. In still other ascomycetes, the heterokaryotic condition results from the transfer of nuclei between two different hyphae without the production of any gametangia.

The ascomycetes are divided into three subclasses by some mycologists on the basis of ascocarp production and ascus morphology. The yeasts and leaf curl fungi (Hemiascomycetidae) do not produce any ascocarp. The ascocarp producing Euascomycetes and Loculoascomycetes are separated on the basis of a combination of two characters (i.e., ascus wall; ascocarp boundary). The Euascomycetes usually produce asci with one wall (the **unitunicate ascus**). These asci are borne in an ascocarp that has a special wall produced as the result of sexual stimulation. The Loculoascomycetes usually produce asci with two walls (the **bitunicate ascus**). These asci are borne in a hole (**locule**) in a compacted mycelial mass (the **ascostroma**). There is no special wall produced as the result of sexual stimulation. Neither of the two characteristics alone will separate the two groups, but when they are combined, they form the basis for division.

YEASTS AND LEAF-CURL FUNGI—HEMIASCOMYCETES

The division of this subclass into orders and families is very controversial. As a group, however, they have a simplicity of structure. The asci are formed in a random fashion and are not enclosed in any sort of an ascocarp. There are no ascogenous hyphae or crozier bodies such as are produced by the euascomycetes or the loculoascomycetes and which will be discussed in detail with those groups.

VEGETATIVE STRUCTURES

The somatic structures of members of this group of organisms range from a well-developed mycelium as in *Dipodascus* spp. and *Taphrina* spp. to a predominantly unicellular thallus as in the yeasts. Both uninucleate and multinucleate conditions are found within the group.

REPRODUCTION

Asexual reproduction in the yeasts occurs by **budding, fission,** or a combination of the two processes. A unicellular yeast cell buds by the production of protoplasmic extensions, covered by the plasma membrane, that push out from the cell wall and form a daughter cell (Figure 6.1*a*). Successive budding may produce a series of cells that are connected in a chain (**pseudomycelium**). Fission takes place by the production of a septum through an elongated parent cell. *Taphrina* spp., a leaf curl fungus, asexually reproduces by the production of buds directly from the ascospores. These particular buds are sometimes called **blastospores,** and may either bud and produce more blastospores or germinate and form a mycelium.

Sexual reproduction in the yeasts occurs as a result of fusion between two somatic cells or two ascospores. No ascocarp is produced, but an ascus containing a varying number of ascospores is eventually formed. Heterothallic and homothallic yeasts are common. The ascospores produced by various yeasts are round to ovoid, hemispherical, Saturn-shaped, or hat-shaped (Figures 6.1*b* and 6.1*c*). The ascus breaks open (Figure 6.1*c*) and the ascospores reproduce asexually as previously described. Although *Taphrina* spp. produces a true mycelium, its method of sexual reproduction is also simple. Fusion of blastospores has been reported for some of

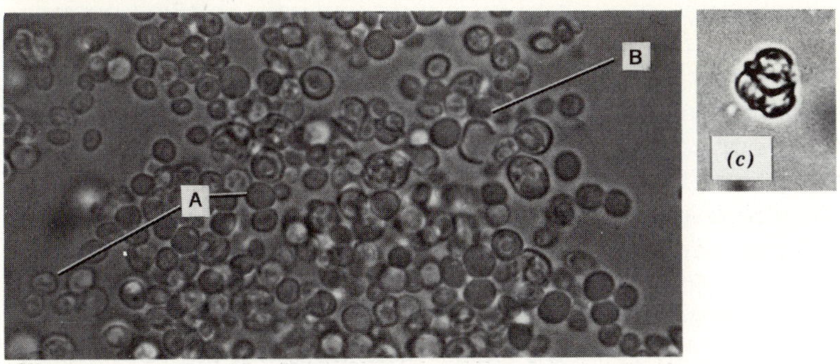

Figure 6.1
Photomicrograph of *Hansenula* sp. (*a*) Cells reproducing by budding (2500×). (*b*) Broken ascus wall with two Derby-hat shaped ascospores nearby (2500×). (*c*) Ascospores inside ascus. Three are visible; the fourth is behind the field of focus (2500×).

the *Taphrinas*, and fusion of sister nuclei within ascogenous cells has been observed for others. It is fairly widely accepted that the dikaryotic condition arises through mitotic duplication, at the time of germination, of the nucleus of the ascospore or blastospore.

LIFE HISTORY

The life histories of the yeasts and *Taphrina* spp. are discussed as examples for this group because of their tremendous importance in human affairs.

Although there are many variations, the yeasts possess three kinds of life history (Figure 6.2). One of the histories is predominantly haploid; another predominantly diploid, and a third alternates between haploid and diploid.

The yeasts that have a haploid history (i.e., *Schizosaccharomyces octosporus*) possess a very short diploid stage, the **zygote.** The zygote undergoes meiosis immediately after karyogamy, and the ascospores are produced. The somatic cells increase in number by fission, are uninucleate and haploid. Any two of these cells come together during sexual reproduction. The walls between them dissolve, and their nuclei move together in the tiny canal that is formed. Plasmogamy and karyogamy are achieved by the flow of the cytoplasm into the canal, and it subsequently widens. The zygote nucleus undergoes meiosis and the zygote cell becomes the ascus containing eight ascospores, each containing a single nucleus. After the ascus wall breaks, each ascospore matures into a somatic cell which begins to reproduce by fission.

Saccharomycodes ludwigii possesses a long diploid phase and a very short haploid phase. Two mating types have been recognized in this yeast: A and a. Two ascospores of opposite mating types fuse while still within the ascus and form a diploid cell. This cell germinates by the production of a germ tube that produces new diploid vegetative cells by budding. The diploid nuclei within somatic cells undergo meiosis and the cells become asci containing four ascospores, two A and two a.

An alternation of equal haploid and diploid generations is to be found in *Saccharomyces cerevisiae*. Both phases reproduce by budding, but fusion occurs between haploid cells. The resulting diploid cell reproduces by budding, but eventually becomes an ascus containing four haploid ascospores. Mating types A and a have been demonstrated for this yeast also, but fusion between the ascospores

Figure 6.2
Three types of yeast life histories. (Redrawn from C. J. Alexopoulous, *Introductory Mycology,* New York: Wiley.)

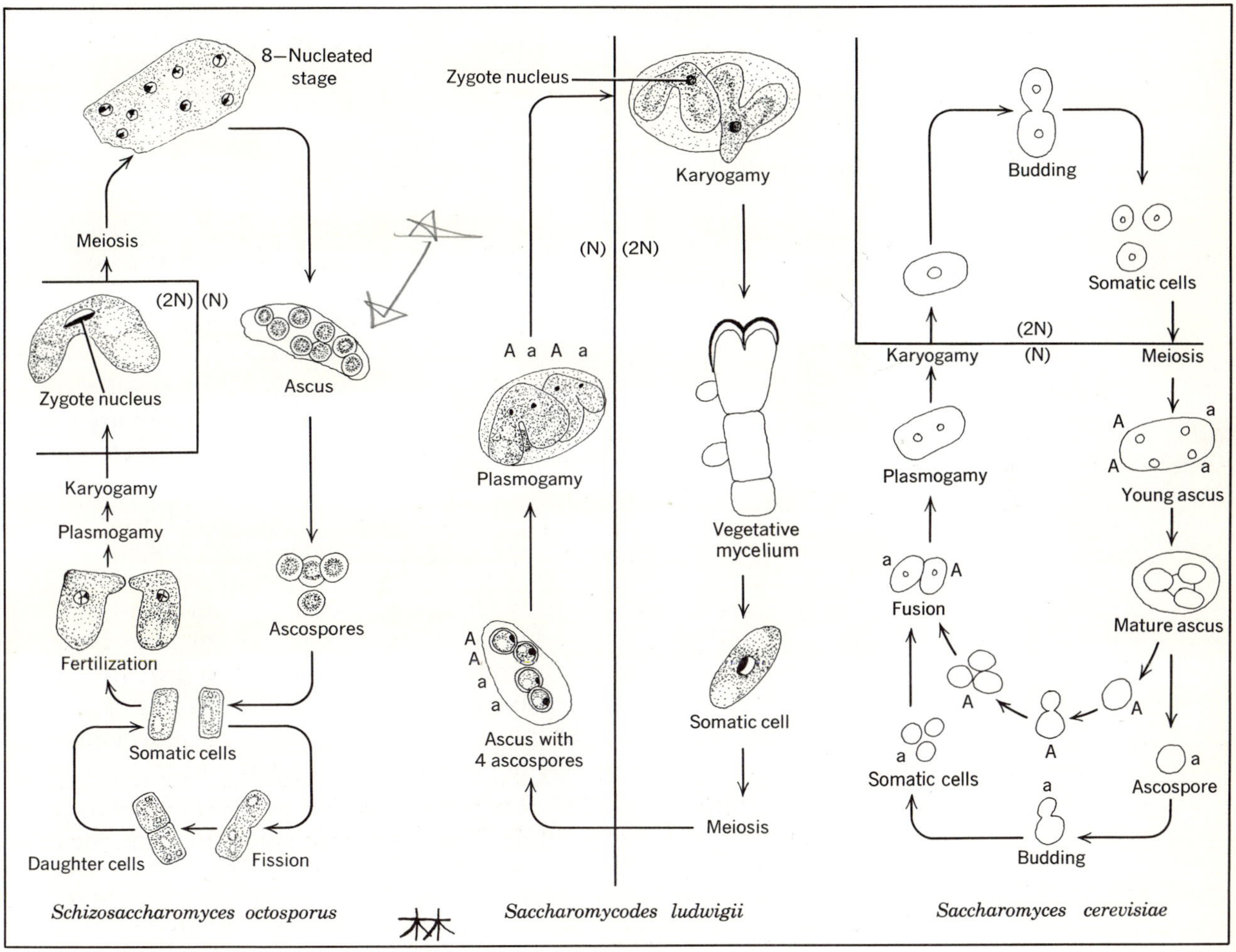

does not take place. After release from the ascus, the ascospores reproduce by budding; the resulting somatic cells are the ones that fuse.

Although, like yeasts, the ascospores reproduce by budding, the leaf curl fungus *Taphrina deformans* (Figure 6.3) does produce a true mycelium in its host. The hyphae of the mycelium are composed of binucleate cells (**dikaryon**) and grow between the cells of the host. Asci are produced in a layer, usually at the surface of the host, from subcuticular hyphal cells (**ascogenous cells**) in the following manner. The two nuclei within the ascogenous cell fuse (**karyogamy**) and the cell begins to elongate. Karyokinesis and subsequent cytokinesis produces two cells; the basal stalk cell and the ascus mother cell. Meiosis and mitosis in the ascus mother cell produce eight nuclei which mature into ascospores. The basal cell nucleus disintegrates.

Enlargement of the asci causes the cuticle of the host to rupture and a relatively uniform layer of asci (an **hymenium**) is produced on the epidermis. No ascocarp is produced and the asci are fully exposed. The ascospores may bud within the ascus or they may bud after their release. The primary blastospores may continue to bud and produce numerous secondary blastospores.

In either case, the blastospores are uninucleate and haploid, and they invade a host by the production of a germ tube which branches and grows throughout the host's tissues. At the time of germination, karyokinesis reestablishes the dikaryotic condition and subsequent karyokineses and nuclear migration maintains the dikaryon in each of the new hyphal cells. Although nuclear migration between blastospores has been reported for some species of *Taphrina*, the nuclear situation in *T. deformans* is unclear. Fusion between blastospores has not been demonstrated, but nuclear exchange could take place at the time of budding. Karyogamy between sister nuclei in the ascogenous cell has been observed. However, no gametangia are ever produced. In culture, only the yeastlike blastospores are produced by the budding of the ascospores. A mycelium is produced only in the host; not in culture.

HUMAN RELEVANCE

Yeasts are of great importance to man because of their ability to ferment sugars and are found naturally in substrata that have high sugar contents.

Figure 6.3
Life history of *Taphrina deformans*. (Redrawn from C. J. Alexopoulous, *Introductory Mycology*, New York: Wiley.)

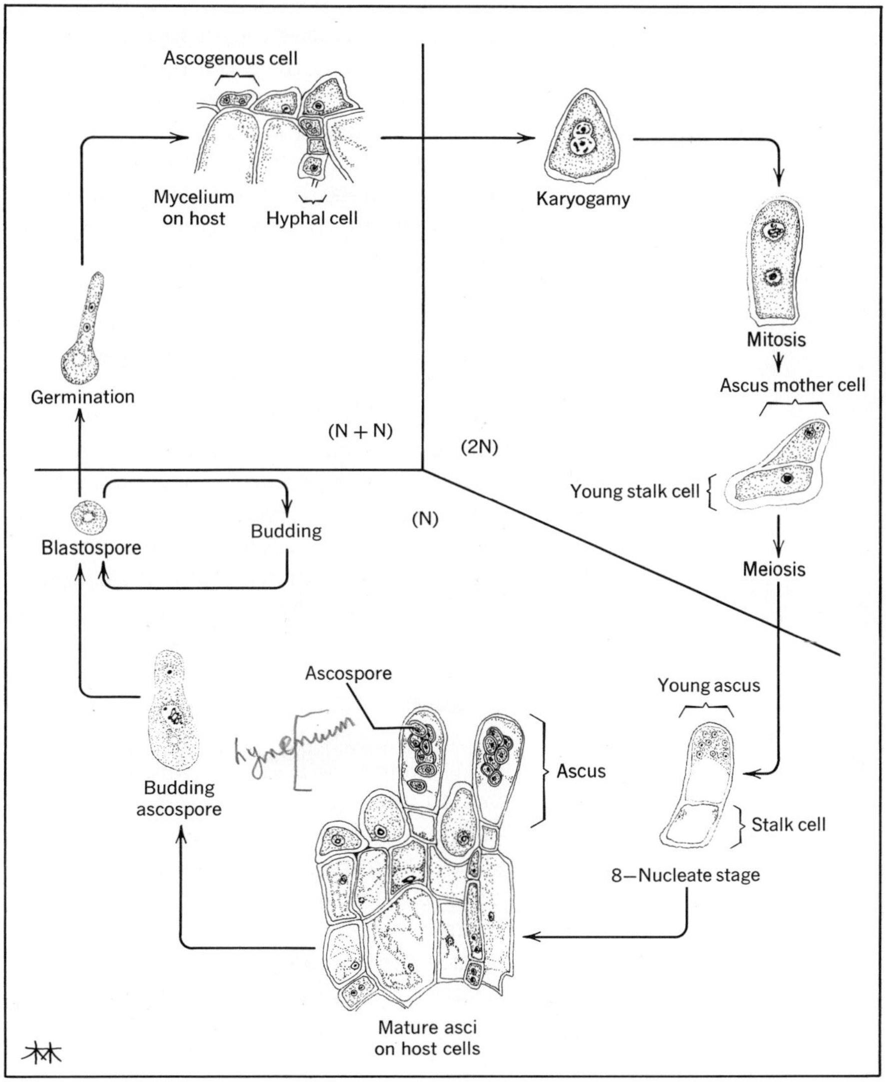

To appreciate their importance to man's welfare, one must only smell and taste newly baked bread or quaff an enjoyable alcoholic drink. Truly, Omar Khayyám depended upon yeasts when he penned "a loaf of bread, a jug of wine, and thou." The fermentation of carbohydrates by yeasts produces two products that are extremely necessary for the baking and brewing industries. The carbon dioxide that is produced is necessary for the baking industry because it causes the bread dough to "rise" by pushing the particles of dough away from each other. This creates the "light" bread of commerce. Although alcohol is produced during this fermentation, it is lost because it is vaporized during the baking process.

The alcohol produced by yeast fermentation of cereal grains is, however, used by the brewing industry for its many products. Before the alcohol can be produced, the starch in the grains must be digested to simple sugars. Although the digestion that is necessary for the production of beverage alcohol is now accomplished by the use of germinated barley that is rich in starch-digesting enzymes (malt), man's ingenuity long preceded his use of such a refined method. In his fascinating account of Captain James Cook's last voyage (1776–1780), *Seventy North to Fifty South*, Paul W. Dale described the methods by which the natives of the South Sea Islands managed the preliminary digestion. Roots of the kava plant were cut into pieces and chewed by members of both sexes. After thorough mastication, the root and the saliva of the masticator were spit into a large common bowl. Natural fermentation occurred, and a liquor was produced. After Captain Cook had observed the method of preparation, he was presented with a cup, but, as he says, "their manner of brewing having quenched my thirst, it became Omai's property." (Omai was a native of the area who accompanied Cook). Yeasts can produce, by natural fermentation, approximately 12 to 14% alcohol. When this percentage is reached (and the percentage varies with the species and strain of yeast used) the yeast cells die. Higher alcoholic percentages are obtained by distillation.

Saccharomyces cerevisiae is the yeast most commonly used in alcoholic fermentations, both industrial and beverage, and in the baking industry. Distilled beverage alcohol is produced in this country most commonly from corn, rye, wheat, and barley. Industrial alcohol is produced in greatest quantity from a cheap by-product of the sugar industry, blackstrap molasses.

Yeasts provide a means of converting cheap carbohydrates into protein without great loss of the energy of the carbohydrate. This method of protein production is far more efficient than the usual method of feeding grains to livestock. Food yeasts are also high in B vitamins and are the principal source of ergosterol which can be converted to vitamin D_2 by irradiation. Therefore, irradiated yeast is widely used as a source of vitamin D. However, yeast as a food for human beings has not become popular mainly because of its bitter flavor; when it is debittered, some of its vitamin content is lost. It has been stated that too much yeast taken at any one time may be harmful to the human digestion process. Therefore, the real value of yeasts is probably as a food supplement. For example, it has been reported that the addition of 1.5 ounces of food yeast to a loaf of bread will give the additional food value of two and a half eggs, a quarter pound of beef or two ounces of cheese.

The ability to ferment sugars also causes yeasts to create problems for man, as they often cause spoilage of sweet fruit juices and soft cheeses. At least one yeast, *Lipomyces neoformans* (form-species *Cryptococcus neoformans*) is the causative agent for cryptococcosis, a serious disease of human beings. This yeast may cause pulmonary lesions and then spread through the blood to skin (Figure 6.4),

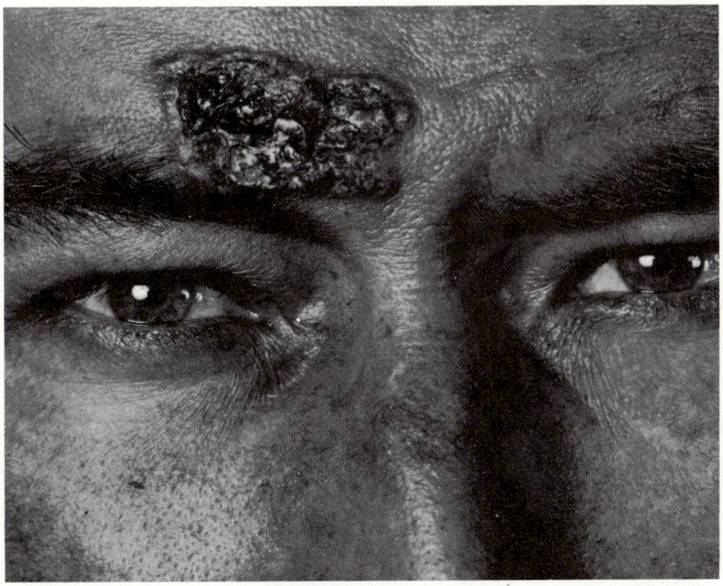

Figure 6.4
Cryptococcosis of the skin. (By permission, from *Manual of Clinical Mycology,* 2nd ed. by Norman F. Conant et al., 1954, Philadelphia: Saunders.)

bones, viscera, and central nervous sytem. The lesions in the lungs usually heal without severe damage to the host. The most dangerous manifestation of the yeast occurs when it reaches the central nervous system and causes meningeal cryptococcosis. The duration of the meningeal cryptococcosis varies from a few months to 15 to 20 years, and the prognosis is poor.

Many causative agents of human disease produce a yeastlike form when grown under certain culture conditions or in mammalian tissues (e.g., *Histoplasma* spp. and *Blastomyces* spp.). These are not yeasts, however, and their importance will be mentioned later.

The leaf curl fungi are particularly important to man because of their pathogenicity to his fruit crops. *Taphrina deformans* is the causative agent of peach leaf curl which causes considerable damage whenever peaches are grown. Other species of *Taphrina* attack plums and cherries.

Taphrina deformans seems able to invade only young tissues, but quite characteristic symptoms are soon produced. The fungus stimulates the overgrowth of the host cells, especially parenchyma cells, and because the vascular tissues are not usually invaded, a "puckering" or "blistering" results. The emerging leaves are arched and reddened. As they age, the blistered parts of the diseased leaves turn yellow and then reddish-purple.

After the rupture of the cuticle, the hymenium appears as a whitish bloom. The leaves so infected may fall, and early defoliation of a heavily infected tree may occur in the late spring. The loss of photosynthetic surface severely limits the production of fruit, and can cause the death of the tree if it continues through subsequent years.

A second crop of leaves may be produced during a particular growing season. These leaves do not become diseased because of the higher midsummer temperatures. The optimum temperature for growth of the fungus is about 20° C, and the maximum ranges between 26 and 30° C. The fungus probably overwinters as blastospores lodged in some sheltered part of the host. Blastospores that are beneath the bud scales may constitute the important inoculum for the next growing season. Although leaves are commonly invaded, some invasion of young fruits may also occur.

The disease relationship between *Taphrina deformans* and peaches is one that can be cheaply and effectively controlled. A 95 to 98% control is usually attained if the proper sprays are applied before the buds begin to swell in the spring.

FUNGI WITH UNITUNICATE ASCI—EUASCOMYCETES

The Euascomycetidae is that group of ascomycetes that usually have only one wall to their ascus (**unitunicate ascus**) and that produce an ascocarp that encloses their asci. The ascocarp is produced as a result of growth of sterile hyphae around the fertile ascogenous hyphae. Ascocarp development is initiated as a result of sexual stimulation created by the production of gametangia and ascogenous hyphae. Not all euascomycetes produce gametangia. Gametangium production is typical of the group, but no male gametangium has been found for many of its members.

The euascomycetes have been divided into three major "series" on the basis of the type of ascocarp they produce. The "series" are the Plectomycetes, the Pyrenomycetes, and the Discomycetes but they have no proper taxonomic rank, and the distinctions between them are confused by intermediate forms. However, these "series" do serve well as groups which can be discussed in an introductory treatment such as this.

SERIES: PLECTOMYCETES

This series of euascomycetes is characterized by having globose asci which arise at different levels in the ascocarp. The asci have no preformed opening and the ascospores are passively released into the ascocarp by the deliquescing of the ascus (Figure 6.5).

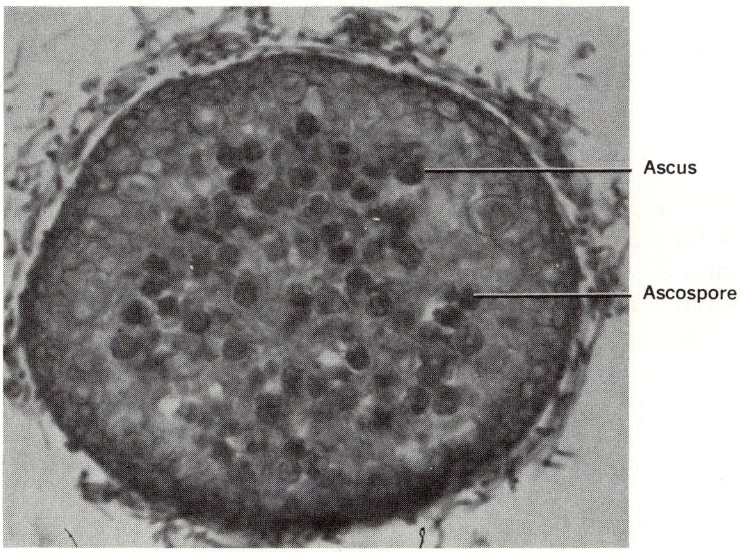

Figure 6.5
Photomicrograph of section of Cleistothecium of *Talaromyces vermiculatus*. Note the inordinate arrangement of the globose asci, each of which contains four ascospores, and the lack of any opening (1685×).

The ascocarp (a **cleistothecium**) is completely closed for most members of the "series," but it does have a preformed opening (**ostiole**) in one of the orders within the group. The wall of the cleistothecium varies from a loosely woven mass of hyphae (e.g., *Gymnoascus* spp.) to a distinct peridium composed of flattened, platelike cells that may be one to several layers thick (e.g., *Eurotium* spp., *Emericella* spp.).

The relationship between the members of the plectomycetes and members of the fungi imperfecti should be mentioned at this time. In many cases, a dual system of nomenclature for the same fungus has arisen. Many of the plectomycetes were first known by their asexual (imperfect) stages because they are of extreme economic importance to man and it was not until the life histories of these organisms began to be studied in detail that their ascomycetous affinities were demonstrated. By that time the names of the imperfect stages were well established, and it is not likely that they will be soon abandoned for the proper taxonomic designations of their perfect stages.

Therefore, we find that the plectomycetous genera of *Eurotium*, *Sartorya*, and *Emericella* have their imperfect stages within the form-genus *Aspergillus*, and the plectomycetous genera *Talaromyces* and *Carpenteles* have their imperfect stages in the form-genus *Penicillium*. There are many other such problems of recognizing proper affinities between perfect and imperfect stages, but these are perhaps the most important and best known. A further discussion of the problem will be found in the discussion of the Fungi Imperfecti in Chapter 7.

REPRODUCTION

Definite **ascogonia** and **antheridia** are delimited in some members of the plectomycetes. The antheridium often coils around the archegonium, but the archegonium may coil around the antheridium. In some of these, the antheridium appears to be nonfunctional. In others, fertilization of the ascogonium takes place by the introduction of **conidia** to the ascogonium by means of its **trichogyne** (a receptive cell, often elongated). In these types, no antheridium is produced. In still others, the asci develop from the **ascogonium** alone.

If a functional antheridium is produced, or if a conidium fertilizes the ascogonium, nuclear migration allows the male and female nuclei to pair and produce the dikaryotic condition. In some plectomycetes, two sister nuclei pair and form the dikaryon. However formed, growth of the new dikaryotic mycelium, accompanied by nuclear migrations allows for the production of an **ascogenous hyphal system.** In some, asci are produced from typical **crozier hooks** (to be discussed more completely later), but in others, asci are produced directly from cells of the ascogenous hyphal system. In one small group of plectomycetes, there is no production of an ascogenous system. Instead, the dikaryotic ascogonium (formed by the introduction of a conidial nucleus) produces chains of ascogenous cells that differentiate into asci containing eight ascospores. Both heterothallic and homothallic members are found within the group. The sterile hyphae surrounding the ascogonium are stimulated to produce the cleistothecium when dikaryotization occurs.

The asci of the group are mostly globose, and their walls are evanescent. The ascospores within the ascus have no particular arrangement, and they may remain in "spore balls" after the ascus wall deliquesces, or they may be released into the cleistothecium from which they are later released by the breaking down of the cleistothecial peridium. The ascospores of many of the plectomycetes are pulley-shaped and their walls may be variously sculptured (Figure 6.6*a*).

In one genus, *Emericella*, the ascocarp wall is composed of several layers and is then surrounded by a layer of thick-walled "hülle" cells (Figure 6.6*b*). These cells are initiated at the ends of hyphal cells, but are not known to function in any way.

Asexual reproduction in the plectomycetes occurs by means of **fragmentation, blastospores, chlamydospores,** and **conidia.** Although the first three types are important, the fourth type is the most

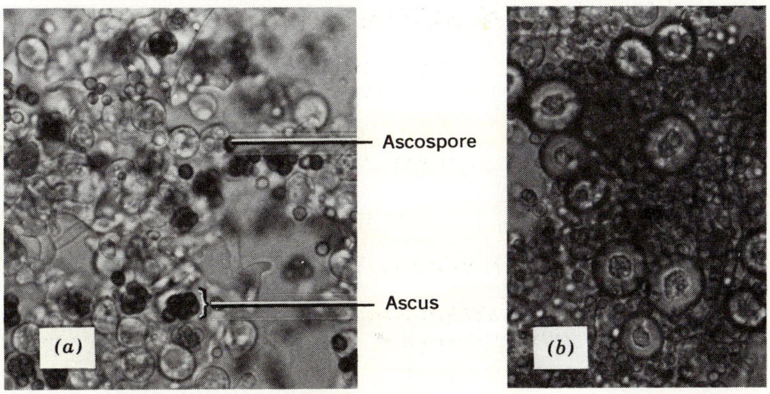

Figure 6.6
**Photomicrograph of squash mount of *Emericella nidulans*. (*a*) Asci containing four ascospores. Some asci have broken, releasing the spiny ascospores (1350×
(*b*) Hülle cells (1350×).**

common and effective method of asexual propagation. Conidial forms of the fungus are also those that are often known as Fungi Imperfecti, and have caused the taxonomic confusion that has been previously mentioned.

Within the group are to be found various methods of conidial formation. The most common method is that found in the form-genera *Aspergillus* (imperfect stages of *Eurotium*, *Sartorya*, and *Emericella*) and *Penicillium* (imperfect stages of *Talaromyces* and *Carpenteles*). In these two well-known form-genera, the conidia are produced inside the tip of a specialized tube called a **phialide** (Figure 6.7). In other plectomycetous genera (e.g., *Ceratocystis*), the conidia may be formed inside or on the surface of a hypha. The actual details of methods of conidial formation are quite varied and a complete discussion of them is outside the realm of this treatment. However, further discussion of conidia will be offered in Chapter 7 on Fungi Imperfecti.

In some members of the plectomycetes, there seems to be an absolute correlation between the ascomycetous genus and the imperfect form-genus. For example, all species of the genera *Eurotium*, *Emericella*, and *Sartorya* for which a complete life history is known have their imperfect stage in the form-genus *Aspergillus*. Furthermore, all members of the form-genus *Aspergillus* that have had their life histories traced have their perfect stage in one of those three genera.

In other members of the plectomycetes, however, the correlation is not very good. In the genus *Ceratocystis*, for example, different species may have different conidial forms, and a single species may have more than one conidial form. Such variety is not unusual, and certainly demonstrates the difficulty of pairing perfect and imperfect stages.

LIFE HISTORY

Talaromyces vermiculatus (*Penicillium vermiculatum*) will be used as an illustrative species for a life history (Figure 6.8) although some details of the sexual stages are still uncertain. Sexual reproduction occurs by the production of functional gametangia. Although plasmogamy occurs, it is questionable whether the antheridial nuclei actually move into the ascogonium.

The antheridia and ascogonia are produced on adjacent vegetative hyphae. The ascogonium is an elongated, uninucleate structure at first; later nuclear duplications produce a multinucleate condition.

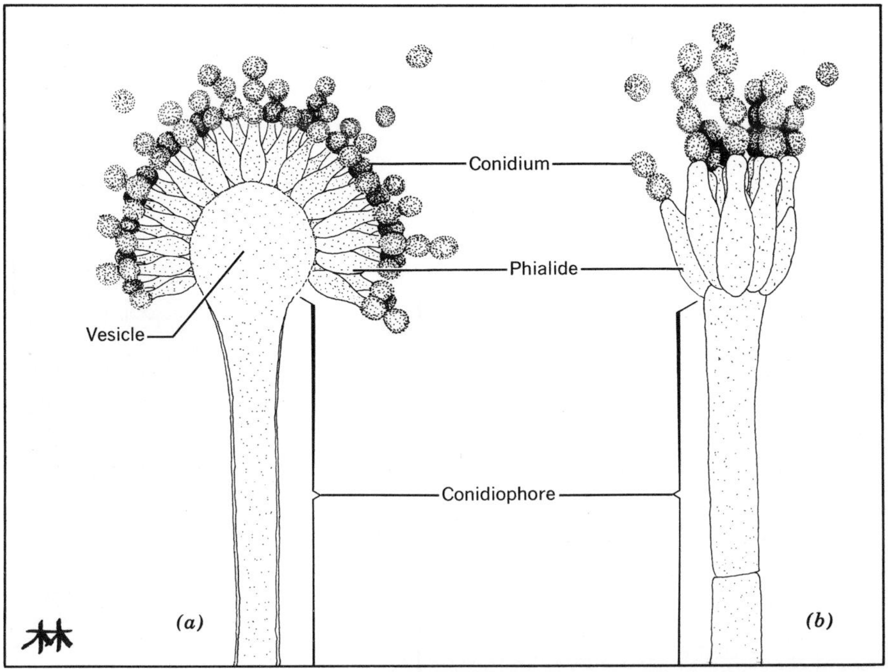

Figure 6.7

One type of conidiophore and conidia for *Aspergillus* and *Penicillium*. (*a-b*) Diagrammetic representations of *Aspergillus* and *Penicillium*. (*c*) Young conidiophores and conidia on vegetative hyphae of *Aspergillus niger* (900×). (*d*) One vesicle with young phialides of *A. niger* (1575×). (*e*) Conidiophores and conidia on vegetative hypha of *Penicillium vermiculatum* (900×). (*f*) One conidial head ("penicillus") of *P. vermiculatum* (1575×).

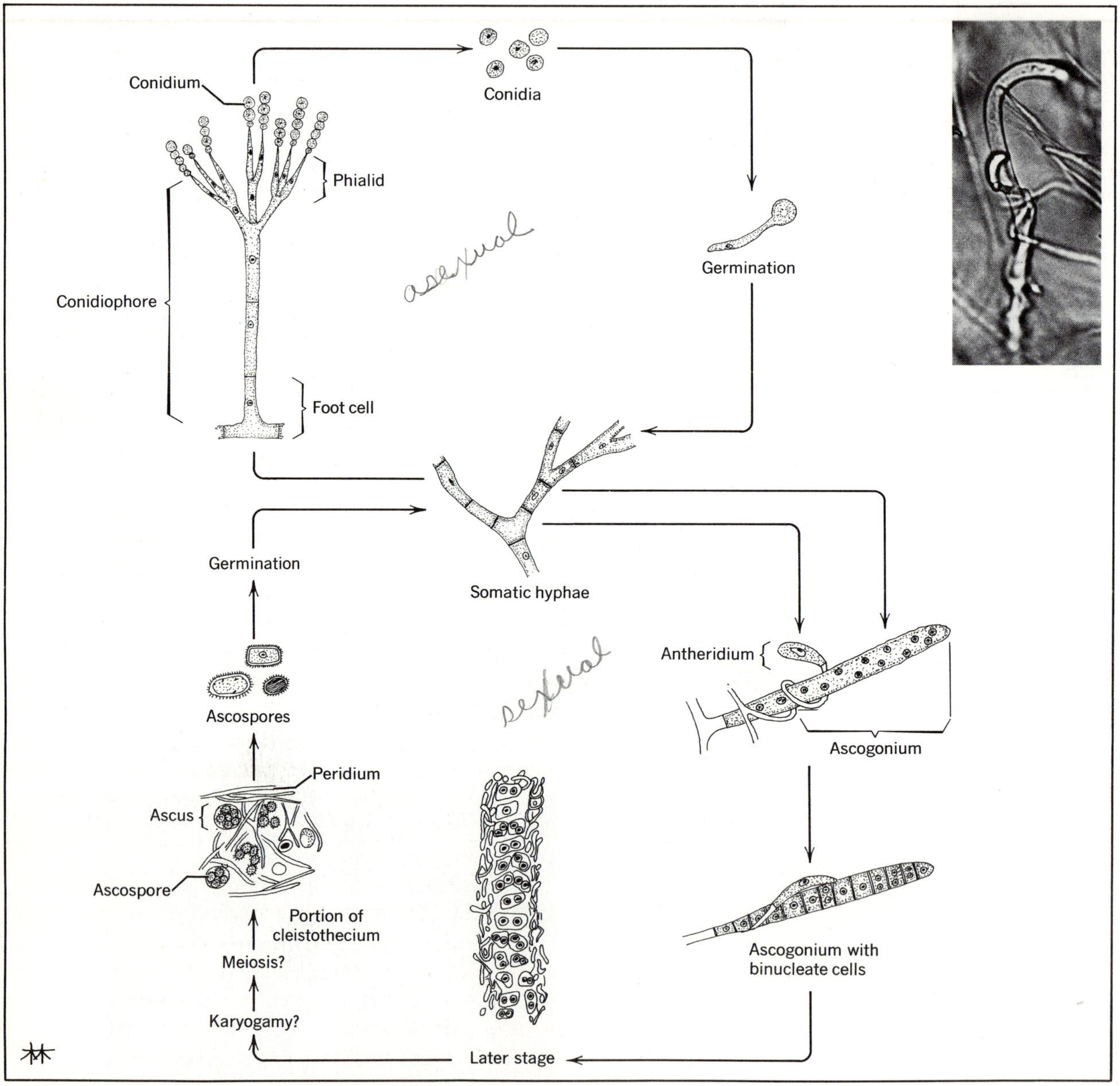
Conidia
Conidium
Phialid
Conidiophore
Foot cell
Germination
Somatic hyphae
Germination
Ascospores
Antheridium
Ascogonium
Peridium
Ascus
Ascospore
Portion of
cleistothecium
Meiosis?
Karyogamy?
Ascogonium with
binucleate cells
Later stage

Figure 6.8
Life history of *Talaromyces vermiculatus*. (Redrawn from C. J. Alexopoulous, *Introductory Mycology*, New York: Wiley.) Insert: Photomicrograph of an antheridium wrapped around an ascogonium (900×).

The uninucleate antheridium coils around the ascogonium. The antheridial and ascogonial walls dissolve where the tip of the antheridium touches the ascogonium. Thus, plasmogamy occurs, but nuclear migration has not been definitely established. Paired nuclei are observed in the ascogonium, but they may be derived from the original ascogonial nuclei without any contribution from the antheridium. Asci form from the cells of the ascogonium and are scattered throughout the cleistothecium. Each of the cells contains a dikaryon and karyogamy followed by meiosis probably occurs. The ascospores produced are haploid. As has been previously described, the walls of the ascocarp develop as a result of the stimulation of the dikaryotization. In *T. vermiculatus*, the cleistothecia are abundantly produced, and are bright yellow (Figure 6.9*a*, p. 86).

The asexual (conidial) phase of *T. vermiculatus* (which is classified as the form-species *Penicillium vermiculatum*) is produced simultaneously with the sexual (ascogenous) phase. The mycelium produces elongated, simple upright **conidiophores** (conidium holder) which branch in a symmetrical manner about one-third of the distance from the tip. This branching conidiophore is called a **penicillus** (small brush). The branches end in a group of phialides which give rise to conidia.

The conidium is produced in an elongated portion of the phialide as a result of the passage of a nucleus into the tube from the main body of the phialide. A septum is then produced posterior to the nucleus and a new section of the tip is cut off. The older cell pushes outward, and the process is continued with the result that a long chain of conidia is produced from each of the phialides. The conidia continue to derive nourishment from their parent phialide until they mature. At maturity, they lay down a secondary wall that complete severs nutritive connections with the parent.

Most of the pigmentation of the form-genera *Penicillium* and *Aspergillus* is found in the conidia. There is a great diversity in the colors, from blacks and browns to greens and yellows (Figure 6.9, p. 86). The pigmentation will vary with the substrate upon which the fungus is grown, and it is very important to use standard growth media when taxonomic considerations are being made because color is one of the major characteristics used for identification.

Ceratocystis ulmi is a heterothallic plectomycete whose ascocarps are known only from laboratory cultures. Other members of the genus are homothallic (e.g., *C. fimbriata*), but the life history of the members of the genus is fairly consistent. The mycelium produces side branches that may become coiled into an ascogonium or may produce two or three cells that act as ascogonia. The dikaryon is established by karyokinesis of a single nucleus within the cell or by the introduction of a nucleus from another hyphal cell or a conidium. In *C. fagacearum*, the ascogonium produces a specialized cell, called a **trichogyne,** which appears to act as a receptive hypha for conidia from opposite mating strains.

As a result of the stimulation caused by the ascogonium development, the adjacent sterile hyphae produce the ascocarp wall. The ascocarp is intermediate between the **cleiosthecium** and the **perithecium** (found in the next group of fungi). The ascocarp resembles a perithecium in that it possesses a preformed opening called an **ostiole.** In the *Ceratocystis* ascocarp, however, asci may form from any cell of the ascogenous hyphae, and consequently are scattered throughout the ascocarp. No true crozier bodies are produced. The asci usually contain eight ascospores that are soon released into the ascocarp as a result of the breakdown of the ephemeral ascus wall. The ascospores are extruded in a sticky mass through the neck and ostiole of the cleistothecium and may dry together in a spore "horn" called a **cirrhus.**

Conidia are produced in the genus *Ceratocystis* in various ways. The two general types are the **endoconidia,** which are produced inside a hyphae and slip out, and **exoconidia,** which are variously produced on the surface of hyphae. A structure commonly produced by the genus is called a **synnema** (pl. synnemata). This structure is formed as a result of the sticking together of a group of conidiophores in an elongated black, spore-bearing apparatus. This asexual phase has been classified in the form-genus *Graphium* of the Fungi Imperfecti (Figure 6.10). Conidia are also successively produced on the ends of short conidiophores. They are held together by a small drop of sticky liquid in a globose head. This spore form is characteristic of the form-genus *Cephalosporium*.

HUMAN RELEVANCE

The plectomycetes are extremely important to man. Sir Alexander Fleming reported in 1929 about his serendipity discovery, at St. Mary's Hospital in London, of a mold that prevented the growth of pus-producing bacteria on culture media. Although his report was ignored for 10 years, it became important when scientists began

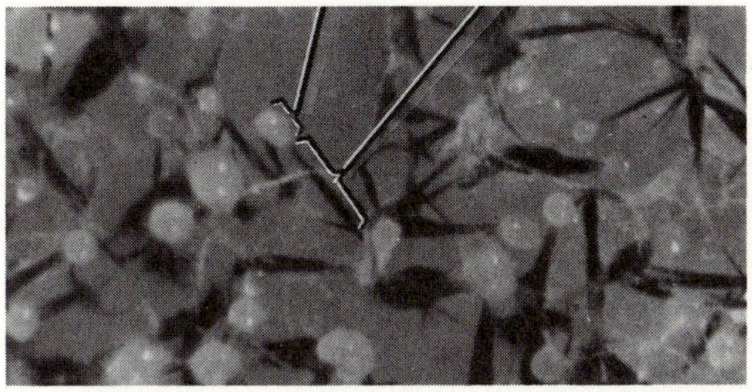

Figure 6.10
Synnemata of *Graphium* sp. (30×).

looking for effective methods of treatment for wounds and infections suffered by military personnel during World War II. The mold used by Fleming was *Penicillium notatum* (Figure 6.9c, p. 86), and the **antibiotic** finally derived is still known by the name he gave it, **penicillin.** Although a sexual phase of *P. notatum* has not been described, it is still considered to be a member of the plectomycetes because of its characteristically shaped conidiophores. Imperfect stages of other plectomycetes (even if their cleistothecial forms have not been described) are beneficial to man's health and well-being. In addition to *P. notatum*, many other species of *Penicillium* (e.g., *P. citrinum*, *P. claviforme*, *P. puberulum*, and *P. chrysogenum*) as well as many species of *Aspergillus* (e.g., *A. flavus*, *A. clavatus*, *A. fumigatus*, *A. oryzae*, and *A. spinulosum*) produce antibiotics.

Many members of this group of fungi are important in the processing of foodstuffs for human beings. The Japanese and Chinese have long used members of the *Aspergillus flavus-oryzae* group for processing many food materials; *A. oryzae* is used to make the rice wine, sake. The activity of *A. oryzae* converts the rice starch to a fermentable sugar by the **amylases** it produces when inoculated onto steamed rice. It is from the resultant moldy mixture of rice and *A. oryzae*, known as Koji, that sake is processed. It has been reported that 10% of the total rice production in Japan is converted to this national drink, so obviously the fungus is important to the well-being of the Japanese. Wheat bran has been substituted for rice in experimental work in the United States, and the bran product, Taka-Koji, has been produced. However, in the western countries, these particular filamentous fungi have been used primarily in the cheese industry.

Cheeses can be classified into a number of basic types, but only two of these types employ fungi as the ripening agents. The soft cheese known as Camembert probably originated in the northwest of France. After three to four weeks of ripening (during which many bacteria are involved) the transformation of the curd texture is brought about primarily by *Penicillium camemberti* which grows on the rind. It is this imperfect plectomycete that gives the cheese its name. Roquefort cheese, a semihard type, originated in southern France. It was originally made only from sheep's milk, but now cow's milk is also used. Gorgonzola is the Italian representative and Stilton the English representative of this cheese type. As in the Camembert cheese, many other cheeses resemble Roquefort and are worldwide in their production and usage. In Roquefort cheese, the mycelium grows throughout the curd; it is not restricted to the rind as in Camembert. The imperfect plectomycete that is the primary ripening agent is *Penicillium roqueforti*, and the blue mycelium can still be seen in the cheese at the time of its use (Figure 6.11, p. 86). *P. roqueforti* is somewhat unique because of its ability to grow under semianaerobic conditions that are not tolerated by most other fungi. Therefore this fungus is probably important to cheese making because it was able to survive in an ecological niche that could not be occupied by another fungus. Although *P. roqueforti* can grow in semianaerobic conditions, it must have some oxygen; contrary to popular belief, holes are punched in the curd rather than the holes being produced by the action of the fungus. Both fungi, of course, impart their characteristic flavors because of the enzymes that they produce.

Many members of the form-genera *Aspergillus* and *Penicillium* are used during the production of important industrial acids. Gallic acid is important in the production of various dyestuffs, permanent black ink, and pyrogallol, a photographic developer. The fermentation of gallic acid (although really a hydrolytic cleavage of gallotannin) was the first fermentation by a filamentous fungus to be investigated. The fungus used during this production is *Aspergillus niger* (Figure 6.9*f*, p. 86), and this same fungus has been found to be important in a wide range of other industrial fermentations.

Several strains of *A. niger* will synthesize citric acid from sugar; this fungus has been found superior in production to the penicillia that have also been used. Citric acid is important for use in flavoring extracts, effervescent salts, inks, and dyeing. However its major use is probably in medicine, although it may have such mundane uses as a substitute for chicle in chewing gum.

Gluconic acid is medically important because it can be used as a means of administering calcium. Calcium gluconate does not cause tissue death when injected, and is used as a calcium source during pregnancy (Figure 6.12, p. 86). Several imperfect fungi have been used in the production of this acid from glucose. These include *Penicillium luteum-purpuregenum*, *P. chrysogenum*, and, of course, *Aspergillus niger*, which seems to be preferred.

There are many other metabolic products of the *Penicillium* and *Aspergillus* form-species that are important or may become important in the future technological world. These include alcohols, esters, chlorine and sulphur compounds, and several other acids. It is also highly probable that the potential uses of these filamentous fungi have not been fully elucidated. Many of them deserve a great deal more investigation.

Although the beneficial effects of the plectomycetes on man's welfare are easily recognized, there are, unfortunately, some detrimental effects. Many of the so-called "molds" of fabrics, leather goods, radio components, photographic equipment, and foods are of the *Penicillium* and *Aspergillus* form-genera. Various kinds of fungus inhibitors have been used to retard the degradation of the cellulose and sugars that are present in such commodities, but the production of polyvinyl plastics and the maintenance of dry conditions probably solved many of the fungus decay problems—particularly in the tropics.

Deserving a special mention are those penicillia and aspergilli that cause loss of paper and food. *Aspergillus* and *Penicillium* spp. have been isolated from paper, where they cause black, brown, or colorless spots. They have also been isolated from asphalt-coated case liners and other finished paper products. If it were not for the introduction of fungicides, and methods of maintaining conditions not conducive to fungus growth (e.g., dryness), the use of paper products would indeed be limited.

One of the plectomycetes (*Byssochlamys fulva*) creates food decay problems in the food canning industry and produces a substance known to be toxic to mice. However, most of the food loss caused by this group of fungi is by their action on raw foods. Many of the blue and green "molds" found on citrus fruits, onions, tomatoes, cabbages, and breads are members of the form-genera *Penicillium* (Figure 6.13, p. 86) and *Aspergillus*. Likewise, the "moldy" condition of various leftover foods that have been stored too long by thrifty housewives who forgot about them until time to clean the refrigerator are also members of these groups.

Milk and milk products are subject to degradation by a great many organisms, but pasteurization is a relatively effective method of control for most of them. Of the fungi, however, *Aspergillus repens*, *A. flavus*, and *A. fumigatus* spores are probably the most difficult to kill, and *A. repens* has been reported to cause a "button" formation on sweetened condensed milk. The growth of the fungus is restricted by lack of oxygen in the can, but enzyme action continues even after the death of the hyphae and creates the button of curd and mycelium. Other milk products such as butter are subject to surface colonies of *Penicillium* and *Aspergillus*.

Although many species of *Penicillium* and *Aspergillus* are known to grow on fresh meat and cured bacon, they are not usually considered to be greatly harmful to the meat unless their action is accompanied by the action of various putrefactive bacteria.

Members of the form-genus *Aspergillus* can cause a group of diseases called **aspergilloses.** These disease relationships occur in man as well as cattle, sheep, pigs, rabbits, guinea pigs, cats, and captive monkeys, and frequently are the cause of pulmonary disorders of birds. *Aspergillus fumigatus* is the usual causal agent of pulmonary disease of the severe form and there may be hematogenous spread to other organs. *A. fumigatus* frequently undergoes changes in hyphal morphology *in vivo*, and these changes are usually constant.

Most infections are acquired by inhalation of spores with their subsequent germination in the lungs, but at least one infection of *A. nidulans* has been reported as being acquired with the intravenous injection of contaminated heroin. In old lesions, the hyphal branches may become so close that a fungus ball is formed. Acute infections of *A. fumigatus* may cause respiratory infections of chickens that cause mortality of up to 90%. Heaviest losses are in newly hatched chickens in which the disease is known as "brooder pneumonia." In man, the aspergillosis is respiratory in origin and is characterized by granular lesions that produce extensions into the lung parenchyma. The patient may have his sputum blood-tinged, low-grade fever, productive cough, weight loss, and toxemia. The prognosis for such a condition is very poor.

Allergic reactions may be induced in sensitized persons by *A. fumigatus*, and *A. niger* is frequently the cause of an ear disease, **otomycosis** (Figure 6.14). The fungus can grow in the external canal and produce a plug of mycelium that causes irritation and impairment of hearing.

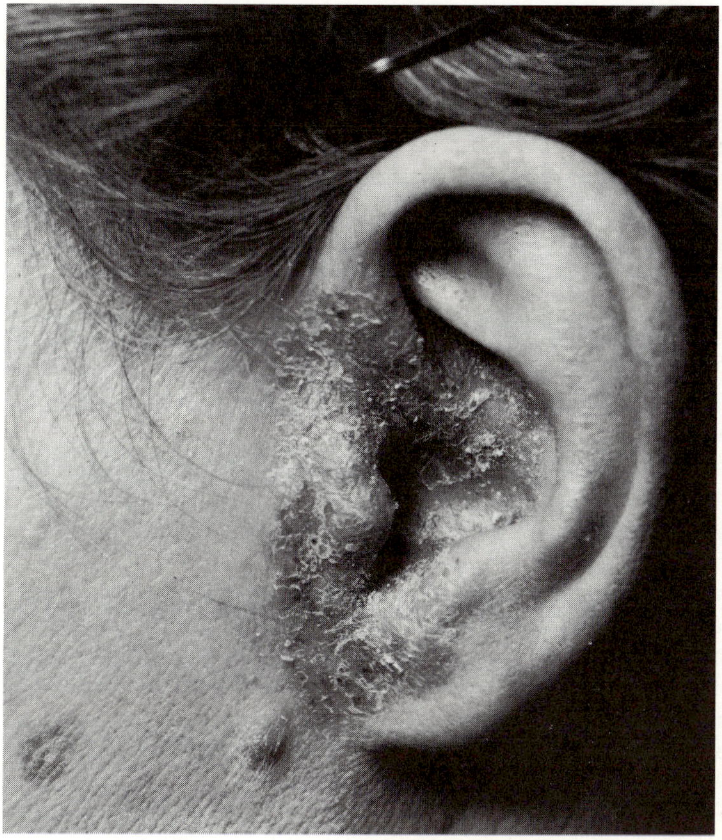

Figure 6.14
Otomycosis of ear canal and external ear. (By permission from *Manual of Clinical Mycology,* 2nd ed., by Norman F. Conant et al., 1954, Philadelphia: Saunders.)

Mycoses caused by members of the form-genus *Penicillium* (**penicilloses**) have been demonstrated in bamboo rats and experimentally in infected mice. There is great uncertainty, however, about whether it causes any disease relationships with man.

The genus *Ceratocystis* contains several species that are severe plant pathogens. Of these, probably the best known is *C. ulmi* that causes Dutch elm disease (Figure 6.15, p. 86). Cities and college campuses that, in the past, have prided themselves on planting large stands and tree-lined streets of the beautiful American elm (*Ulmus americana*) are now finding themselves plagued with the problem and expense of removing the trees (Figure 6.16, p. 86). The fungus was introduced into this country, and has practically eliminated the American elm in some areas. No satisfactory method of protection for the trees has been found, and probably the only way in which the elm will survive is through naturally or artificially selected genetic change. *C. fagacearum* is the causal agent of oak wilt, and is almost as virulent as *C. ulmi*. One of the reasons why great quantities of money must be spent for tree removal and replacement is that the same species of tree was planted close together, with no other species intermixed. This provides extremely good conditions for spread of the fungus and results in the simultaneous death of many trees. Consequently, cities find themselves in the unenviable position of having to replace large areas of mature trees with immature trees. City planners seem to gain little from the experience, however, because they tend to replace the American elm with only one species, often some species of maple. One can only predict that sometime in the future a fungus that can parasitize maples will cause epidemics, and once again the cities will be denuded. It is difficult to understand why various species of trees are not intermixed. This would result in never having all of the trees killed at one time by a disease, and would provide some security for continuity of tree-lined streets in our cities. The lesson was not learned with the introduction of Chestnut Blight (caused by *Endothia parasitica*, another ascomycete) that wiped out our American chestnut trees, and apparently is not being learned by the killing of our American elms and possibly many of our oaks. *Ceratocystis fimbriata* causes a severe disease of sweet potatoes; other species of the genus cause "blue stain" of wood. These "blue stain" fungi may occur in standing timber, in cut logs, or in curing lumber. The resulting discoloration severely reduces the value of the finished lumber.

SERIES: PYRENOMYCETES

This "series" of ascomycetes includes those members that possess a true **perithecium.** A perithecium is an ascocarp that has its asci (usually club-shaped or cylindrical) arranged in a definite layer (an **hymenium**) that lines the base and often the sides of its inner wall. The layer may be restricted to a small group of asci at the base.

The perithecium of most of the pyrenomycetes is flask-shaped with a preformed opening, called an **ostiole,** through which the ascospores escape (Figure 6.17). Some, however, possess a closed perithecium that superficially resembles the cleistothecium of the plectomycetes (Figure 6.18). However, the ascus arrangement in the closed perithecium is much more definite than the inordinate

Figure 6.17
Flask-shaped perithecium. (*a*) Photomicrograph of perithecia of *Claviceps purpurea* buried in stroma. Note escaping ascospores (1800×). (*b*) Drawing of one type of perithecium. (Redrawn from C. J. Alexopoulous, *Introductory Mycology*, New York: Wiley.)

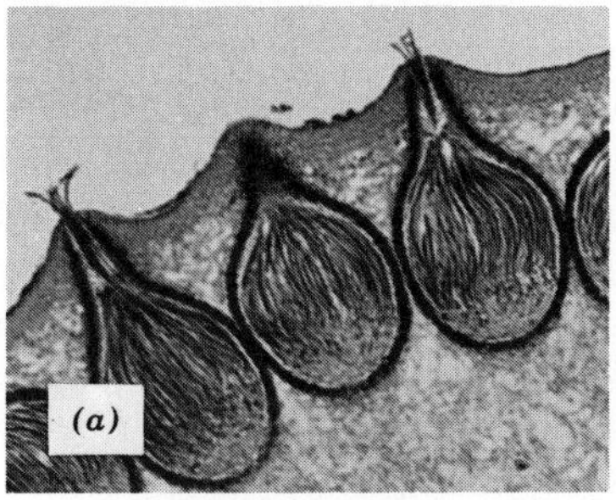

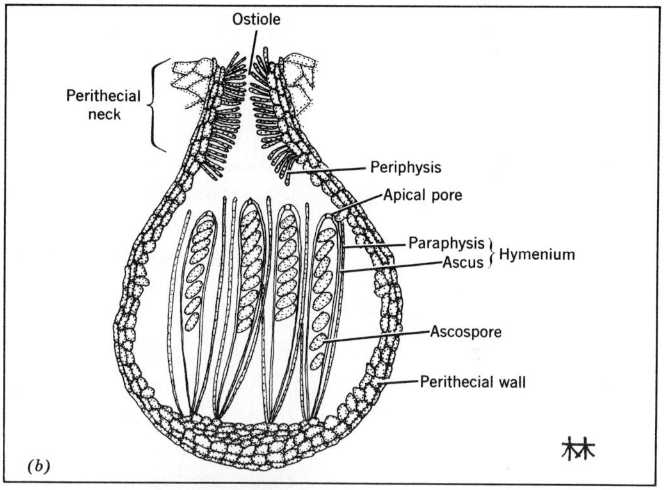

Figure 6.18
Photomicrograph of closed perithecium of *Microsphaeria alni*. Note escaping ascus with five visible ascospores (1500×).

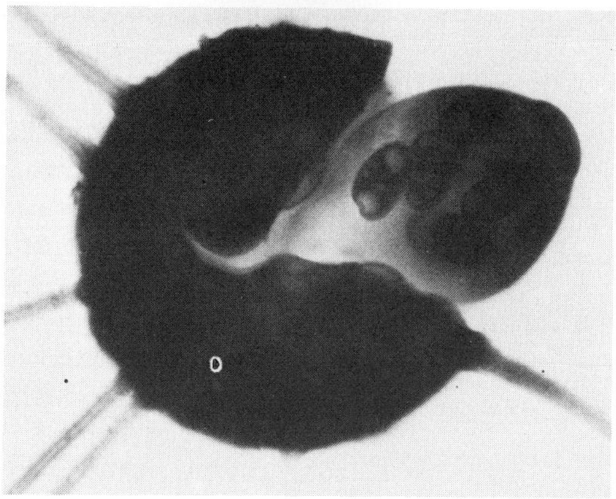

arrangement of the asci in the cleistothecium. The perithecium retains its own wall although it may be on or embedded in a vegetative structure that is known as a **stroma** (Figure 6.17*a*). The stroma may be composed completely of fungal tissue or it may be composed of a combination of fungal and substrate tissue. It varies from scanty and soft to massive and carboniferous.

The hymenium (Figure 6.17*b*) contains the asci, and also any sterile hyphal filaments that may be interspersed among the asci. These filaments are called **paraphyses** and are part of the hymenium. Similar sterile hyphae (**periphyses**) may be found lining the neck of the perithecium.

The **centrum** of the perithecium includes all of those structures that are enclosed by the perithecial wall. This would include the asci, any sterile structures present, and the inner layer of the ascocarp wall. The developmental morphology of the centrum is an extremely important consideration for an understanding of the taxonomic relationships within the pyrenomycetes. Such a consideration is outside the scope of this text, however, and descriptions of mature structures must suffice as an introduction to the group.

REPRODUCTION

Although details vary among the members of the pyrenomycetes, they all possess similar methods of sexual reproduction. Heterothallic and homothallic members are to be found within the group. Well-defined, functional ascogonia and antheridia are found in some. In others, the antheridium is formed but does not appear to function and the asci are produced as a result of the union of two ascogonial nuclei. In still others, the dikaryotic condition is produced as a result of **spermatization** of the ascogonium by the introduction of a nucleus from a conidium. Also, in some, the dikaryon is produced by plasmogamy between two vegetative hyphae, between an ascospore and a hypha or between two ascospores.

After the establishment of the dikaryon, there is usually a proliferation of the ascogenous hyphae and the production of **crozier bodies** (Figure 6.19) at their tips. The crozier body is produced as a result of elongation and bending of the terminal cell in such a way that a hook is formed. The two nuclei in the hooked cell duplicate in such a way that two of the daughter nuclei, one from each parent nucleus, lie near the bend. One of the other two nuclei is located close to the basal septum of the cell, and the other nucleus is near

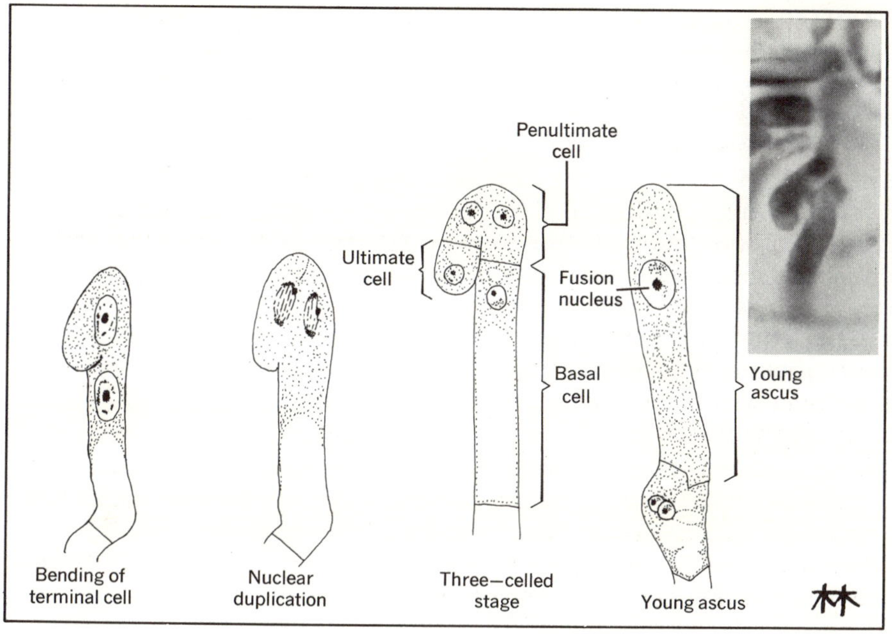

Figure 6.19
Production of a crozier body. Insert: Photomicrograph of crozier body of *Anthostoma gastrinum*. Note fusion nucleus and elongation of young ascus from penultimate cell (650×).

the tip of the cell. Two septa are then produced. This results in three cells, the **ultimate** (or terminal) cell which contains one nucleus, the **penultimate** cell which contains two nuclei (one from each of the parent nuclei), and the **basal** cell which contains one nucleus. The nuclei in the basal and ultimate cells are from different parents. The penultimate cell elongates, producing the young ascus. Its two nuclei fuse, forming the only diploid stage in the life history. As the elongation continues, the diploid nucleus migrates into the developing ascus, meiosis soon occurs, and four haploid nuclei are produced. Mitosis usually quickly follows, resulting in eight haploid nuclei. A wall surrounds each of the nuclei and a bit of its surrounding cytoplasm, and an ascospore is produced.

Some of the pyrenomycetes produce only the four ascospores that result from the meiotic division of the diploid nucleus. Commonly, however, eight ascospores are produced and, in some, successive mitotic duplications produce some large multiple of the original four.

Ascospore discharge is accomplished by a variety of methods. Some ascomycetes forcibly discharge their spores while others are discharged only by the degeneration of the ascus wall. The ascospore germinates by the formation of a germ tube and a haploid mycelium is produced. This mycelium may reproduce by asexual means until the dikaryon is reestablished by one of the methods previously described.

Asexual reproduction within the pyrenomycetes, as might be suspected, is extremely varied. All members can probably reproduce by **fragmentation.** However, most of them produce **conidia.** The conidia may be produced on differentiated, erect **conidiophores,** singly or in chains. The conidia of one group, the Erysiphales, are of this nature and give the group the common name of "powdery mildews" (Figures 6.20 and 6.32, p. 86). Other members produce conidia in specialized structures (e.g., **pycnidia** and **acervuli**) which will be described in the discussion of the Fungi Imperfecti in Chapter 7. Others produce only **microconidia** which seem to function mainly as **spermatia,** and still others seem not to have any conidial stage at all.

As in the plectomycetes, many of the asexual stages of the pyrenomycetes have been classified in the form-class Fungi Imperfecti. The same sort of taxonomic confusion exists because all of the species of a particular genus do not necessarily produce the same kind of conidia. Most of the Erysiphales have their imperfect stages in the form-genus *Oidium*, but other of the powdery mildews have been classified in other form-genera. Other common examples of classification of conidial stages of pyrenomycetes in form-genera are: *Neurospora sitophila* conidia in the form-genus *Monilia sitophila*, *Diaporthe* spp. have as their imperfect stages the form-genus *Phomopsis*, *Endothia parasitica* (the cause of chestnut blight) has its conidial stage in the form-genus *Cytospora* and the genus *Nectria* which has its conidial stages in the form-genera *Fusarium* and *Tubercularia*.

LIFE HISTORIES

The pyrenomycetes have been variously divided into orders but the life histories of the members of the orders are quite similar. Members of the orders Erysiphales, Clavicipitales, Diaporthales, Xylariales, and Hypocreales are commonly found, and are of great economic importance. The life histories of members of two orders will be discussed as examples. Discussion of members of the other orders will be deferred to the section on human relevance of the group.

The members of the Erysiphales are commonly known as the **powdery mildews** (Figure 6.20). Homothallic and heterothallic species

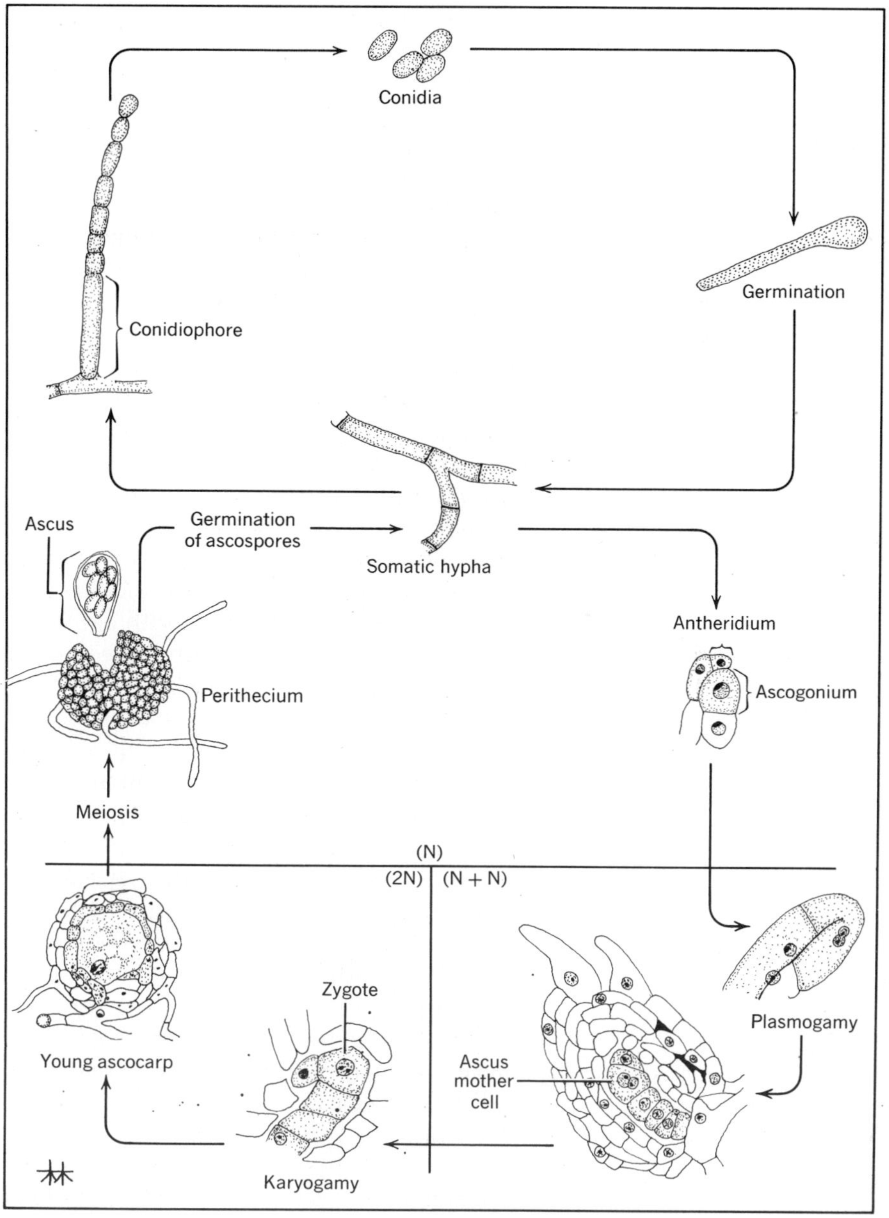

Figure 6.20
Life history of *Sphaerotheca castagnei*. (Redrawn from C. J. Alexopoulous, *Introductory Mycology*, New York: Wiley.)

are present within the group. Sexual reproduction is initiated by the formation of uninucleate ascogonia and antheridia that form as side branches from vegetative hyphae and press closely together. Although cytological details are lacking, the ascogonium may become binucleate by the migration of the antheridial nucleus through a pore. In some species, the ascus develops directly from the middle cell (penultimate cell) of a three-cell series produced by the ascogonium. In these forms, only one ascus per perithecium is produced. In other species, an ascogenous hyphal system develops from the penultimate cell. This system may branch throughout the developing perithecium resulting in the production of several asci that are arranged in a layer or basal bush.

As in other ascomycetes, the sexual stimulation results in the formation of the perithecium wall by the proliferation of the surrounding sterile hyphae. The perithecium may be regarded as unique in the pyrenomycetes because it is a completely closed structure. In this, it closely resembles the cleistothecia of the plectomycetes, but is distinguished from them because of the definite ascus arrangement. The perithecia of the North American members of this group are supplied with sterile appendages of definite form that are extremely useful in delimiting genera. The ascospores are passively released by the breaking of the perithecial and ascus walls. The ascospores are usually hyaline, unicellular, and oval. The number of ascospores per ascus varies from two to eight and is not a constant characteristic within species.

The ascospores germinate by a germ tube that produces an extensive, usually superficial mycelial mat. The colorless hyphae anchor the fungus to the host plant by penetrating the epidermal cells and producing haustoria through which the fungus gets its nourishment from the protoplast of the host. Within a few days one-celled, hyaline conidia are formed at the tips of long, hyaline conidiophores. The conidia may break off singly or remain attached in chains but they will not germinate while attached to the conidiophore. However, they germinate and repeat the asexual cycle throughout the growing season of the host plant when separated by wind and other environmental factors.

Within the order Xylariales, the genus *Neurospora* has several species that are important. They have been carefully studied by mycologists and geneticists because of the ease with which they can be grown in the laboratory (this same characteristic often makes

Figure 6.11
Roquefort cheese containing bits of the blue mycelium of *Penicillium roqueforti*.

Figure 6.13
***Penicillium* sp. growing on a lemon. (Photograph by John W. Cieslewicz).**

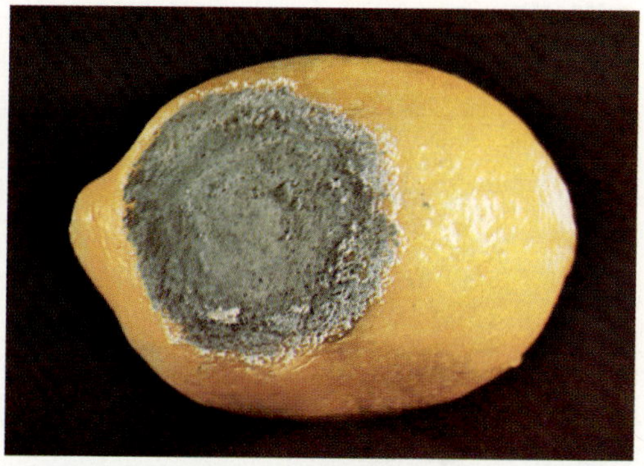

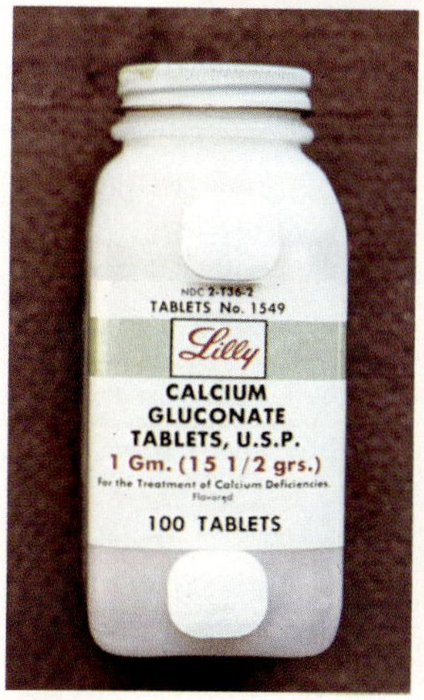

Figure 6.12
Calcium gluconate tablets.

Figure 6.15
Two American elm (*Ulmus americana*) trees infected with *Ceratocystis ulmi* causing the disease relationship, Dutch elm disease. Buffalo, N.Y.

Figure 6.22
***Rosellinia subiculata*. Perithecia on a brightly colored subiculum.**

Figure 6.23
***Hypoxylon notatum*. Perithecia in a hard, discrete stroma.**

Figure 6.24
Hypoxylon deustum. Perithecia in a hard, effuse stroma.

Figure 6.25
Daldinia concentrica. Perithecia in hard, upright stroma. Note concentric layers of stromal tissue in the longitudinal section.

Figure 6.26
Nummularia broomeiana, "nail-head fungus." Perithecia in hard, flat, upright stroma.

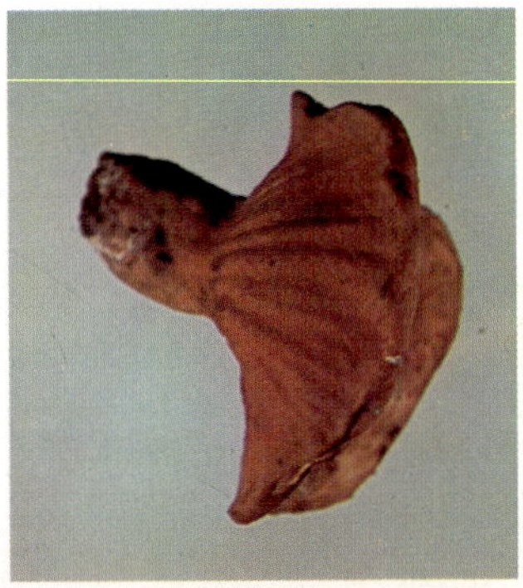

Figure 6.29
Hypomyces lactifluorum parasitic on a mushroom (probably *Lactarius* sp.). The bright orange layer is the stroma in which are buried perithecia.

Figure 6.30
Claviceps purpurea. Note the dark, hard sclerotia ("ergots") among the grains. Sample was taken from discarded "screenings" at a flour mill. (Photograph by John W. Cieslewicz.)

Figure 6.32
Microsphaera alni on lilac. The white covering is the hyphae, conidiophores, and conidia.

Figure 6.34
Sarcoscypha coccinea, "scarlet cup."

Figure 6.35
Morchella sp., "sponge or brain mushroom."

Figure 6.37
Rhytisma acerinum. Stroma on maple leaf. (Photograph by John W. Cieslewicz.)

Figure 6.39
Elaphomyces sp., a truffle, parasitized by *Cordyceps ophioglossoides*, a stipitate discomycete. (Photograph by John W. Cieslewicz.)

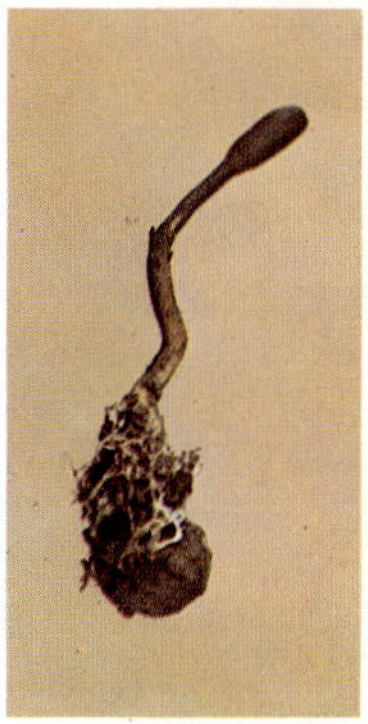

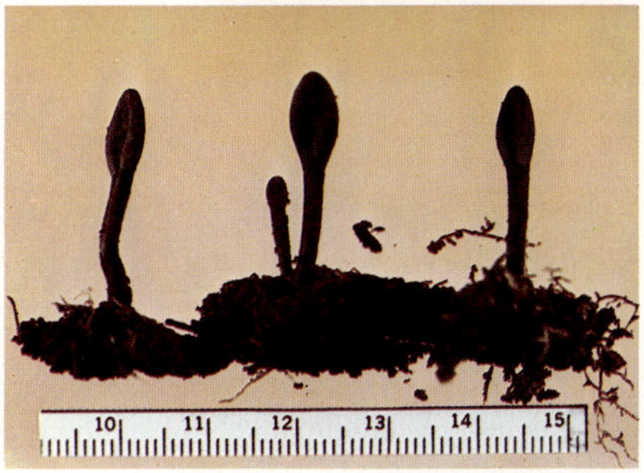

Figure 6.40
Geoglossum glabrum, "earth tongue." (Photograph by John W. Cieslewicz.)

Figure 6.41
Three species of *Leotia*. From left to right: *L. chlorocephala*, *L. stipitata*, and *L. lubrica*. (Photograph by John W. Cieslewicz.)

Figure 6.42
Scutellinia scutellata. Several apothecia growing among moss. (Photograph by John W. Cieslewicz.)

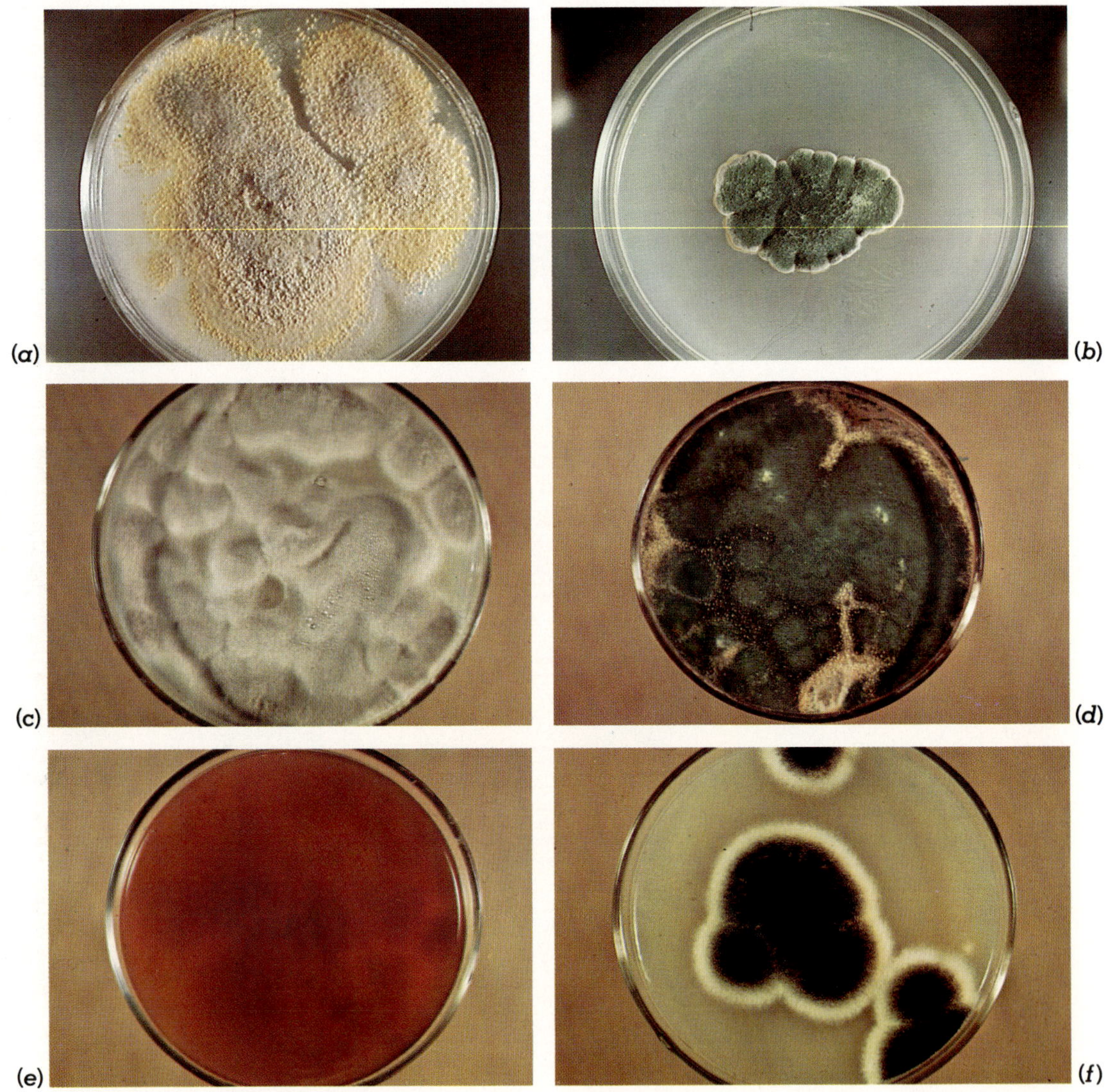

Figure 6.9

Petri plate cultures of various plectomycetes. The pigmentation is mostly in the conidia. (a) *Talaromyces vermiculatus* (cleistothecia may be seen as small dots). (b) *Emericella nidulans.* (c) *Penicillium notatum* (mutant strain). (d) *P. purpurogenum.* (e) Reverse of *P. purpurogenum.* Pigments have diffused into the agar. (f) *Aspergillus niger* (Photographs c-f by John W. Cieslewicz).

them an extreme pest).

The rather complicated life history of *Neurospora sitophila* (Figure 6.21) will serve to illustrate almost all of the possible situations which can arise in the pyrenomycete life history. *N. sitophila* is heterothallic and **hermaphroditic** (each + and − strain of the heterothallic fungus produces both male and female elements. Cross-fertilization is necessary, however, because the strains are self-sterile).

The female structure is represented by a coiled ascogonium that produces long, receptive hyphae (trichogynes). The male element is, in this genus, represented only by the conidia that are produced on conidiophores. The conidia are of two kinds, **microconidia** and **macroconidia,** and, although serving as a fertilizing agent for the ascogonium, they may also germinate and repeat indefinitely the asexual part of the life history (form-genus *Monilia*).

When a conidium (or, in some cases, other mycelial element) of the proper mating type reaches the trichogyne, the nucleus of the fertilizing element will migrate to a compatible nucleus in the ascogonium and the dikaryon will be established. Perithecial formation proceeds as previously described with the result that asci are found in a definite hymenium within the vase-shaped perithecium. Although present during development, no paraphyses are present at maturity. The perithecium is dark-colored and the asci contain eight ascospores which are brown or black and ridged. Four of the ascospores will be genetically of one strain and four of the other.

Many pyrenomycetes produce their perithecia in or on a mycelial mat. Such a mat is termed a **stroma,** and may be composed of a loosely compacted, sparse mycelium (a **subiculum**) such as found in *Rosellinia subiculata* (Figure 6.22, p. 86), or it may be composed of a tightly compacted mass of mycelium, the cell walls of which have thickened and become leathery or carbonaceous. This type of stroma may be composed of only fungal tissue and be closely affixed to its substrate (e.g., members of the genus *Hypoxylon*, Figures 6.23 and 6.24, p. 86) or it may stand upright from its substrate (e.g., members of the genera *Daldinia*, *Nummularia*, and *Xylaria*, Figures 6.25, p. 86, 6.26, p. 86, and 6.27).

Other pyrenomycetes have stroma that are submerged in their substrate and that incorporate host tissue (e.g., *Anthostoma* spp. Figure 6.28, *Diatrype* spp., *Diaporthe* spp., *Valsa* spp.). Still others have

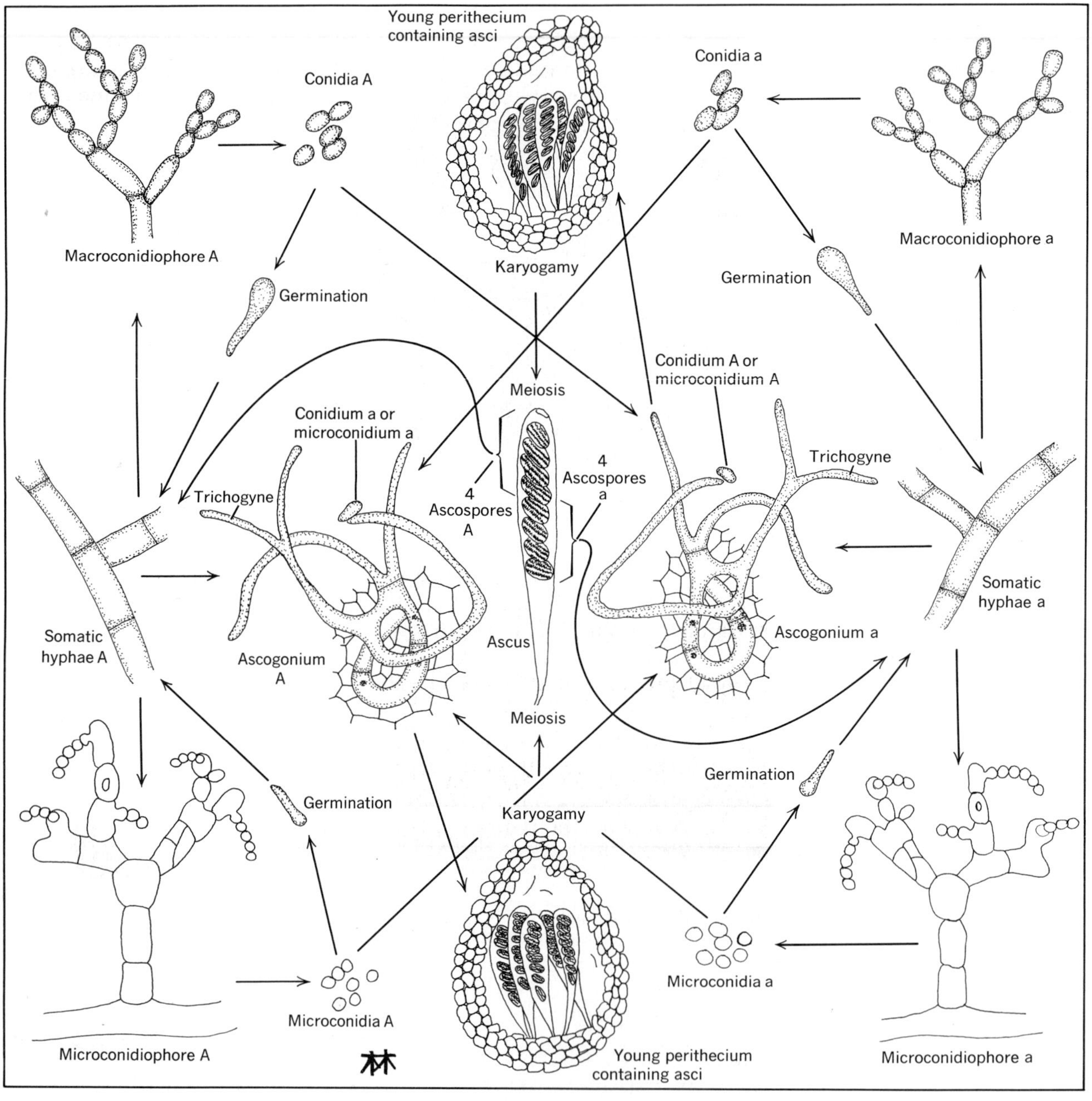
Young perithecium
containing asci
Conidia A
Conidia a
Macroconidiophore A
Macroconidiophore a
Karyogamy
Germination
Germination
Conidium A or
microconidium A
Meiosis
Conidium a or
microconidium a
Trichogyne
4
Ascospores
a
4
Ascospores
A
Trichogyne
Somatic
hyphae a
Somatic
hyphae A
Ascogonium
A
Ascus
Ascogonium a
Meiosis
Germination
Germination
Karyogamy
Microconidia a
Microconidia A
Microconidiophore A
Young perithecium
containing asci
Microconidiophore a

Figure 6.21
Life history of *Neurospora sitophila*. (Redrawn from C. J. Alexopoulous, *Introductory Mycology* New York: Wiley.)

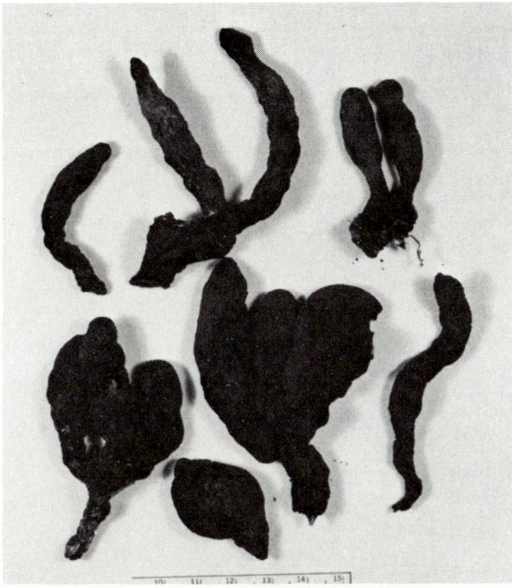

Figure 6.27
Xylaria polymorpha, "dead man's fingers." A group of stromata with variations in form.

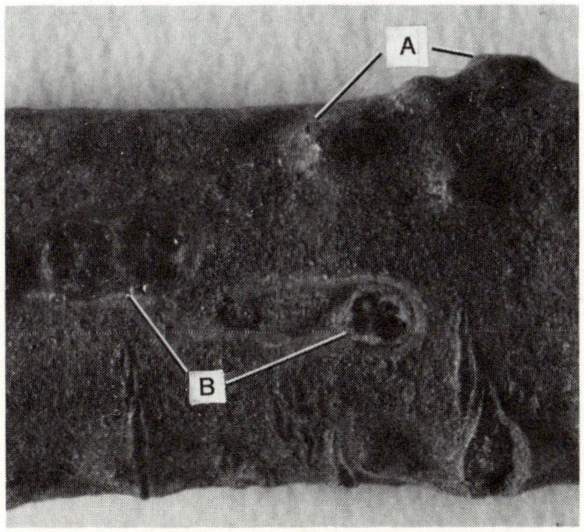

Figure 6.28
Anthostoma gastrinum. Perithecia in stroma submerged in host tissue (oak twig). (*a*) Necks of perithecia erumpent through host. (*b*) Peridermal section showing bases of perithecia.

stroma that are composed of brightly colored fungus tissue (e.g., *Hypomyces* spp., Figure 6.29, p. 86, *Hypocrea*).

Some pyrenomycetes (e.g., *Claviceps purpurea*) produce a hard, fungal structure that is capable of withstanding extremely unfavorable environmental conditions. This structure is known as a **sclerotium** (Figure 6.30, p. 86) and does not contain perithecia. However, under favorable conditions, it may resume growth and produce stromatic heads that do contain perithecia (Figure 6.31).

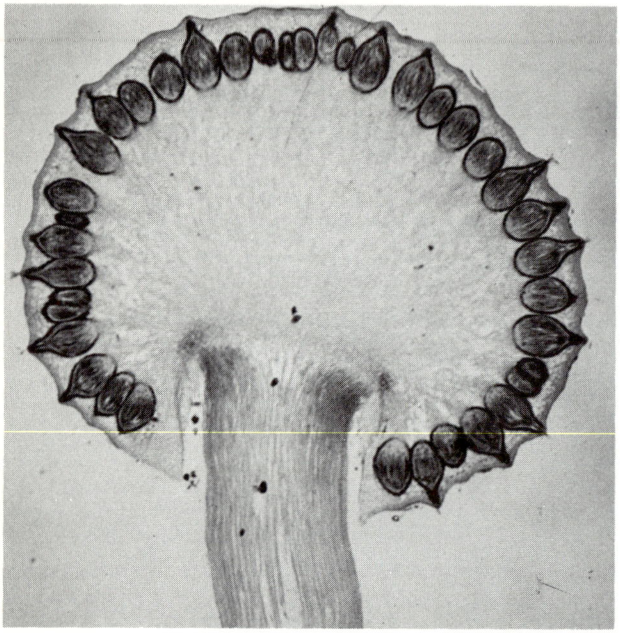

Figure 6.31
***Claviceps purpurea*.**
Stromatic head containing perithecia (350×).

HUMAN RELEVANCE

The pyrenomycetes have many members that are extremely important to man. One group is commonly known as the powdery mildews (Erysiphaceae) and causes a group of vascular plant diseases that appear as fluffy or powdery white coverings over the leaves and young stems of the host plant. The white covering is composed of a well-developed hyphal system that produces chains of conidia on specialized conidiophores (Figures 6.20 and 6.32, p. 86). Much of the mycelium is external and produces haustoria that invade the host cells and derive nutritive benefits from them. The haustoria usually invade only the epidermal cells of the host although a few species do invade the mesophyll layers of the leaf.

Although the North American members of the group are obligate parasites, they vary from a very wide to a very narrow host range, and many of them appear to be composed of a number of physiological races. Also, while some are severe pathogens that cause extensive damage (e.g., *Uncinula necator* on grape vine, *Sphaerotheca pannosa* on roses, and *Erysiphe cichoracearum* on various members of the cucumber family) others are mild and seem not to affect the host very much (e.g., *Microsphaera alni* on lilac, Figure 6.32, p. 86).

Uncinula necator was first reported to be parasitizing grapes by an English gardener, Mr. Tucker, in 1845 at Margate, Kent. At that time, plants or animals could be moved onto the European continent without any sort of quarantine restriction, and it was a short jump across the English Channel to the wine-producing areas around Versailles, France. The fungus invaded Versailles in 1848 and had traveled throughout the vineyards of Europe, North Africa, and the Mediterranean Islands within five years. It was only through the judicious use of a lime and sulfur mixture that the vineyards were rescued. The disease raged throughout the 1850s, and it was not until hybridizing with North American varieties of grape that had developed resistance to the fungus that the painstaking three sprays per season could be abandoned. The grapes imported from North America brought with them another severe grape pest, a yellow plant louse, but the losses of grapes attributable to powdery mildew were much reduced. The origin of the fungus disease has not been definitely established but it had been known to exist previously in North America and had apparently wiped out the susceptible varieties of American grapes. The imported North American grapes had resistance to the fungus probably because they were the survivors of the fungus when it was known to exist in North America.

The American chestnut tree (*Castanea dentata*) is an extremely handsome tree that reaches 100 feet in height and can have a trunk diameter of 3 to 4 feet. This beautiful contribution to our eastern forests composed, in 1904, up to 25% of the trees in the hardwood stands that ranged from Maine to Michigan and from the Great Lakes to Louisiana. However, from an estimated billion trees that produced valuable lumber, nuts, and tannins, there now remains about 160 specimens that are existing in scattered isolation throughout North America. Their decimation has created great changes in the fauna composition of the eastern forests by the reduction of food and shelter needed by deer, wild turkeys, and squirrels. The changes produced in the watersheds and the reduction of soil retention qualities have all contributed to unmeasured financial and esthetic loss.

Chestnut blight is the disease relationship that has nearly wiped out our American chestnut. The causative organism is a pyrenomycete, *Endothia parasitica*, which was introduced from eastern Asia at the beginning of this century. The first report of the disease was from the New York Zoological Park in 1904, but the disease was not considered to be serious at that time. However, one year later, the Forestry Division of the Department of Agriculture became alarmed and in 1907 set up a special laboratory to study tree diseases. State laboratories were also set up to fight the disease (e.g., the Chestnut Blight Commission of Pennsylvania was funded in excess of half a million dollars), but although the causative agent was soon discovered, its action on the host elucidated, and its method of dissemination ascertained, no method of cure, prevention or even containment was forthcoming. Consequently, 30 billion board feet of extremely durable, beautifully grained wood that was useful as interior house trim, furniture, and flooring, as well as the more mundane uses of poles, railroad ties, and coffins, was lost. Moreover, the tannin that composed 7 to 12% of the heartwood of the trees was lost to the tannic acid industry which collapsed, forcing American leather makers to obtain this necessary conditioner from other sources. Therefore, the loss of one species by an epidemic disease created far-reaching effects that have still not been completely overcome.

The government agencies abandoned their efforts to contain *Endothia parasitica* in 1915, and the American chestnut was left to its fate. Some sprouts from old trees are still to be found in the Appalachian forests, and perhaps they will provide the means of the return of the American chestnut to the American forests. Hybridizing experiments with the Chinese chestnut, which is blight resistent, has produced 50-foot trees that are free of the disease. These hybrids are sterile, but perhaps the production of a tetraploid will produce a new fertile species. The road to recovery is, however, a long and doubtful one.

Claviceps purpurea (Figures 6.30, p. 86 and 6.31) is a fungus that establishes a disease relationship with various grasses and sedges. The disease is known as **ergot,** and has had extremely far-reaching influences on the actions of human beings. The fungus invades the flowering heads of the grasses, especially rye, and completely replaces the ovary of the developing flower. The mycelium becomes compacted and hard, and produces a black **sclerotium** that hangs out of the flowering head and that is easily detached from the host plant. The sclerotium is known as an **ergot** (Figure 6.30), and it is this structure that is so very important to human affairs.

The disease has been known historically since at least the eighth and ninth centuries, and there is the possibility that Pliny the younger was describing it in some of his writings of the second century *A.D.* Ingestion of the ergots, when they are ground up with the grains of rye to produce flour, can produce a number of symptoms in human beings. The exact cause of some of these symptoms is still unknown, nor is it certain why some symptoms appear at particular times and other symptoms do not. Some of the alkaloids are vasoconstrictors and cause constriction of the distal blood vessels. This constriction can cause gangrene in the distal portions of the body. The limb becomes very swollen and inflamed, violent burning sensations alternate with sensations of extreme cold, the affected part becomes numbed, shrunken, and mummified. The victim may shed nails, toes, fingers, and even entire limbs. The progress of these symptoms may be very rapid—the first numbing and tingling in the hands and feet may occur within 24 hours after ingestion.

Another effect of ergot ingestion is a condition of extreme convulsions. These symptoms are usually coupled with a deficiency of vitamin A, and are composed of severe twitching and toxic spasms of the limbs. There are strong, permanent muscular contractions, particularly of the hands and the feet. However, the whole body may suffer violent, generalized contractions.

Although earlier writings about the disease do not say very much about it, another condition that has definitely been linked to ergotism is the hallucination caused by an alkaloid that has been fairly recently derived from the ergots. This derivative is lysergic acid diethylamide or LSD_{25}. Although it is too early to predict the final effect that the knowledge of this drug will have on human society, the immediate effect has been the mixed blessings of the promise for an aid in the treatment of schizophrenics and other mentally disturbed persons, and the spectrum of an unproductive escape for those human beings who cannot face their world as it is. Unfortunately, at the present time, one hears a great deal more about the illegal and unfortunate use of LSD by the pitiful addicts than about its proper and controlled use by those who practice the healing arts.

One of the far-reaching influences of ergot poisoning of human beings is still apparent in the power exerted by the Roman Catholic Church. The first epidemic of the disease was reported in France in 857 A.D. It had already become known as the sacred or Holy Fire, but in 1039 a Frenchman of means, Gaston de la Valloire, built a hospital for its victims in the Rhone Valley near Vienne. The hospital was dedicated to St. Anthony, and his relics were kept there. The monks who tended the "burning ones" became the Order of St. Anthony and established their hospitals throughout France. The suffering victims of the insidious disease were carefully tended in these hospitals. They received wheat bread instead of their usual rye and, because wheat is far less susceptible to *Claviceps* infection than is rye, their painful symptoms slowly alleviated. They subscribed their recovery to the intervention of the monks with God, an attitude not likely to be disputed by the monks, and thus they were further convinced of the power of the Church. The disease became known thereafter as "St. Anthony's Fire."

Although "Holy Fire," as it had been called in the Middle Ages when it was a common and dreaded disease, had been known throughout Europe for quite some time, its cause was not elucidated until 1670 in France by a provincial physician by the name of Dr. Thuillier. His observations led him to the circumstantial decision that the ergots in the fields of rye caused the ergotism in the human subjects. His contention has since proved to be correct, but he was unable to convince the farmers of the time and the disease raged on. Further discussion of this and other historical influences of ergotism are described by Carefoot and Sprott (1967).

Fuller (1968) has written an exciting medical detective story entitled *The Day of St. Anthony's Fire* in which he chronicles the events surrounding the ergot poisoning of the Village of Pont-St.-Esprit, France, and its surrounding environs. Although the poisoning occurred in 1951, litigation continued until 1965 and the cause of the poisoning, because of legal technicalities, has still not been definitely stated. However, to anyone who reads the symptoms described and who knows the effects of ergot poisoning, the cause is apparent.

The victims at Pont-St.-Esprit show the symptoms recorded from the Middle Ages (e.g., convulsions and gangrene), but they also show the frightening reactions to LSD_{25} that had just a few years previously (in 1938) been purified by Dr. Albert Hofmann in the Sandoz Laboratories in Basel, Switzerland. The refined powder was not tested until 1943, and it was at that time that its hallucenogenic power was demonstrated. The account of Dr. Hofmann's accidental "trip" is one that can be easily correlated with the accounts of the "trips" of horror encountered by the villagers of Pont-St.-Esprit. Because it has been shown in controlled experiments that LSD_{25} is dangerous in very small dosages, and that 1/100,000 of an ounce can cause unpredictable reactions, one can certainly appreciate the havoc caused when the drug is ingested in unknown quantities in bread. The bread consumed at Pont-St.-Esprit contained rye that had been parasitized by ergot. Seven people died as a result of the poisoning and over 300 others experienced up to four weeks of hallucinations that caused some of them to jump out of windows, climb bridges, cry incessantly, rage through the countryside, and bite their retaining straps until all of their front teeth were broken and bloody stumps. The hallucinations returned to some from time to time, and many of the victims were left broken and ruined. A few of the "trips" were reported to be beautiful, but the majority were frightening descents into Hell.

The active ingredients of ergot have not been fully elucidated, nor have the reasons why the concentrations of various alkaloids (e.g., LSD_{25}) vary from time to time and place to place. Some of the alkaloids have been found to be extremely useful in medicine. When properly administered, they can cause uterine contractions and thus induce childbirth. Because some are vasoconstrictors, they can cause the contraction of capillaries and reduce hemorrhaging, and have been used to treat migraine headaches. The medicinal uses of LSD_{25} are being explored in the treatment of mental illnesses, but the misuse of the powerful drug has created a severe social problem.

Although three primarily detrimental examples have been discussed at some length, and many more plant diseases are caused by other pyrenomycetes, one should not leave the group before pointing out their extreme importance as decay organisms. Many of them have enzymes capable of cellulose degradation, and hence are very important in forest ecosystems. Many others break down different types of organic matter. Their medicinal value is just beginning to be explored, and some of them are used as tools by experimental geneticists. It may be, then, that man will find a more constructive use for an otherwise extremely destructive group of fungi.

SERIES DISCOMYCETES

This "series" of ascomycetes is composed of those species that possess an open, saucer-shaped "fruiting body" called an **apothecium** that consists of an **hymenium,** an **hypothecium,** and an **excipulum** (Figure 6.33).

The hymenium is composed of the asci and paraphyses, and is exposed in an open cup as in *Sarcoscypha coccinea* (Figure 6.34, p. 86), or in large open cavities as in members of the genus *Morchella*

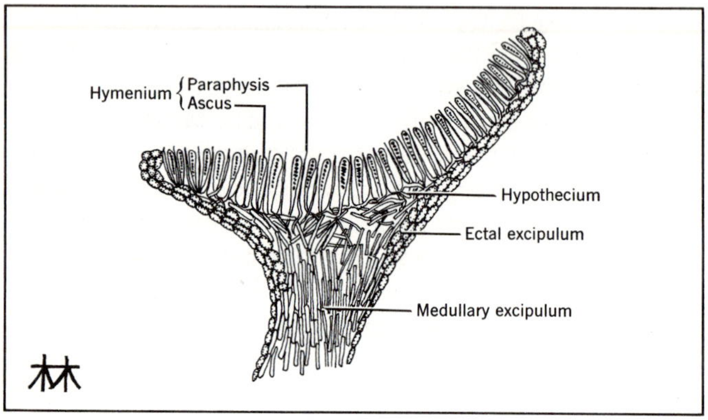

Figure 6.33
An apothecium in longitudinal section.

(Figure 6.35, p. 86). The color of the hymenium may be brown or black to brightly colored such as brilliant red, orange, or yellow. With the exception of the truffles, the hymenium is exposed at maturity although it may be enclosed during the development of the apothecium. The paraphyses vary in length and may be plain or branched. Thin tips may unite above the asci in a layer called the **epithecium.** The asci may possess a preformed lid, the **operculum,** or they may open by an **apical pore.** The discomycetes are subdivided into two large groups on the basis of this important taxonomic character (e.g., the operculate and the inoperculate discomycetes). Many of the discomycetes discharge their spores violently. Spore discharge may be easily observed by allowing a large apothecium to remain in an area of high relative humidity and then removing it to a dryer area. Often discharge can be induced by gently blowing across the surface of the hymenium. In either case, the spores will appear as a small cloud above the hymenium.

The hypothecium is the thin layer of interwoven hyphae located just below the hymenium. The rest of the tissue of the ascocarp is called the excipulum. It supports the hypothecium and the hymenium and is divided into two major tissue areas. The outer layer of cells of the excipulum is called the **ectal excipulum** and is composed of flat, platelike cells. The shape and orientation of the cells of the inner tissues of the excipulum (**medullary excipulum**) varies. These characteristics are important for the proper identification of the discomycetes, but a discussion of structural details is outside the scope of this text.

Most of the discomycetes produce their ascocarp above ground (**epigean**) but some (e.g., truffles, order Tuberales, Figure 6.39, p. 86) produce their ascocarps below the soil (**hypogean**). The hypogean discomycetes usually remain closed and release their ascospores only when the ascocarp decays or is accidentally broken. Some discomycetes are parasitic and produce their ascocarps on the stems and leaves of vascular plants. Stroma and sclerotia are also produced by some of the discomycetes, and apothecia may be produced within the stroma (e.g., *Rhytisma acerinum* Figures 6.36 and 6.37, p. 86), or on stalks that arise from the surface of the sclerotium (*Monilinia fructicola*).

REPRODUCTION

Sexual reproduction in the discomycetes is much like that described in the pyrenomycetes. Homothallic and heterothallic species are known to occur. Functional antheridia and ascogonia are produced in some members of the group but no antheridia are produced in others. Fusion between compatible strains may occur by the production of short side branches from the vegetative hyphae (e.g., *Ascobolus magnificus*), or fusion between a coiled ascogonium and a vegetative hypha may occur (e.g., *Lachnea melatoma*), or simple fusion may take place between completely undifferentiated vegetative hyphae (e.g., *Humaria granulata*). *Ascobolus stercorarius* produces a coiled ascogonium that has a long trichogyne and while the ascogonium may be fertilized by a functional antheridium, fusion between a conidium and the trichogyne may also establish the dikaryotic ascogenous hyphal system. It has been fairly well established that, at least in this ascomycete, there is a sexual hormonal system that is quite similar to the one that was discussed for the zygomycetes.

In *Monilinia fructicola*, it has been noted that spermatia (or microconidia) are produced on specialized structures. Whether or not these spermatia actually function has not been demonstrated, but it is speculated that they serve as spermatizing agents for the ascogonium.

The ascocarp is produced as a result of the sexual activity. Mature apothecia vary in size from the large cups of *Peziza repanda* (e.g., 3 to 5 inches in diameter) to the very tiny cups of members of the Helotiales (e.g., 1 mm in diameter). The apothecia may be stalked or sessile, and their height varies. The spongelike apothecium of

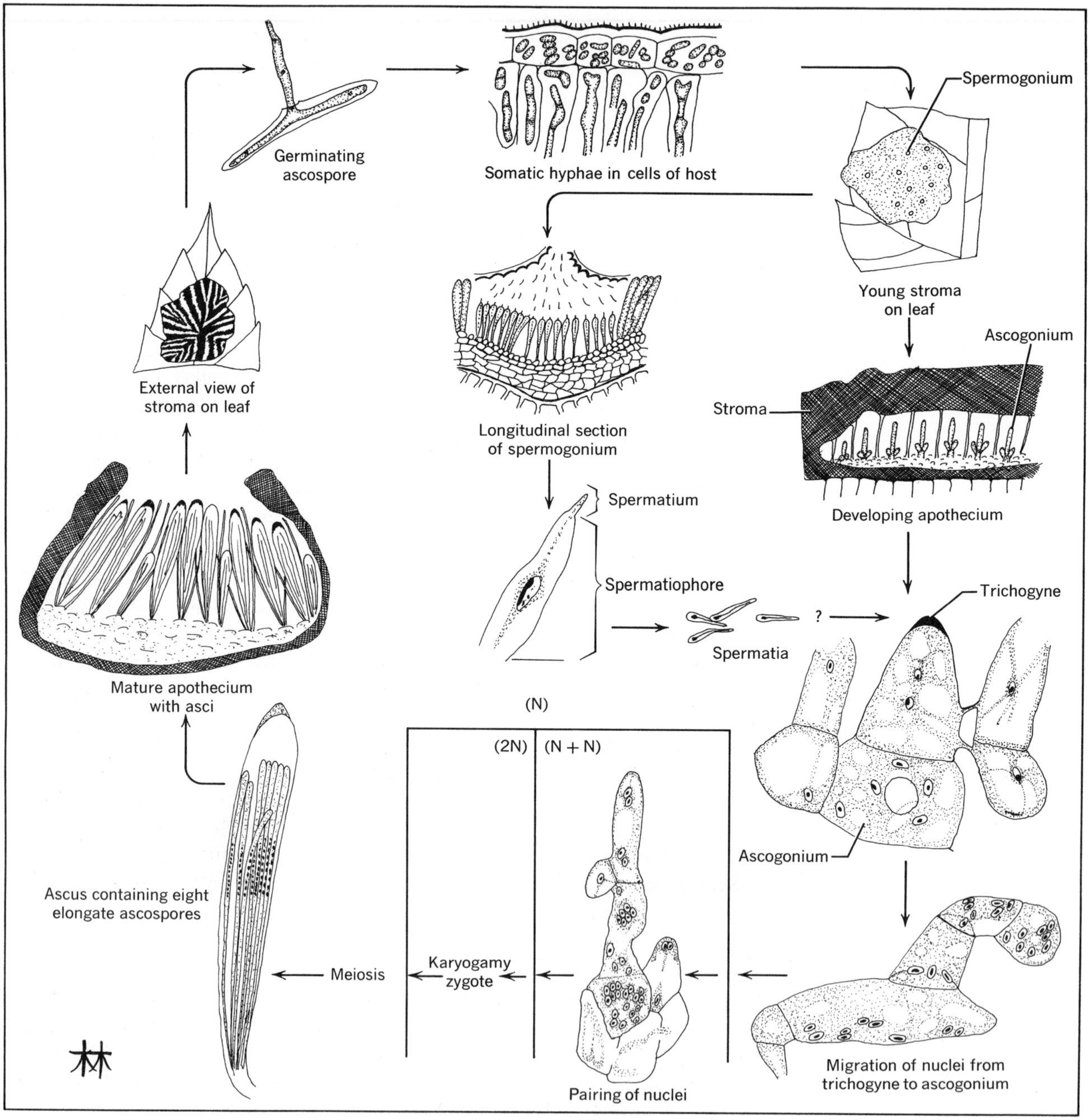
Germinating
ascospore
Somatic hyphae in cells of host
Spermogonium
Young stroma
on leaf
External view of
stroma on leaf
Longitudinal section
of spermogonium
Ascogonium
Stroma
Developing apothecium
Spermatium
Spermatiophore
Trichogyne
?
Spermatia
Mature apothecium
with asci
(N)
(2N)
(N + N)
Ascogonium
Ascus containing eight
elongate ascospores
Meiosis
Karyogamy
zygote
Migration of nuclei from
trichogyne to ascogonium
Pairing of nuclei

Figure 6.36
Life history of *Rhytisma acerinum,* "tar spot" of maple. (Redrawn from C. J. Alexopoulous, *Introductory Mycology,* New York: Wiley.)

Morchella spp. (Figure 6.35, p. 86) is as great as 5 inches tall. As previously discussed, the color of the hymenium varies greatly and the asci are usually operculate or open through a pore, although some have a longitudinal split at their tip. The ascospores within the operculate discomycetes are usually one-celled, but in the inoperculates they vary in cell number. Both groups show variation in size, color, and shape of spore. They usually germinate by the production of a germ tube, but other methods of germination have been noted.

Asexual reproduction in the discomycetes may take place by fragmentation. Conidia are produced by some members of the group but not by others. Although there are some exceptions, lack of conidia seems to be typical of the operculate discomycetes. However, many of the inoperculate discomycetes also do not produce conidia. Conidia are the major disseminating units for some of the pathogenic discomycetes (e.g., *Monilinia fructicola*) and, consequently, are quite important. As previously noted, conidia have been demonstrated to be spermatizing agents during the sexual reproduction of some members of the group.

LIFE HISTORIES

One of the important inoperculate discomycetes is the causal agent of a disease known as tar spot of maple (Figure 6.36). This fungus (*Rhytisma acerinum*) produces a discrete, hard, stroma that looks as if a bit of tar had been dropped on the leaf (Figure 6.37, p. 86). Small apothecia are produced within the stroma, and the needleshaped ascospores are released by the splitting of its surface. If the ascospores are carried by air currents to a suitable host, they germinate by means of a germ tube. The hyphae penetrate the host leaf through the stomata and derive nutrition from the mesophyll layers. The mycelium also forms a stroma within the epidermal cells, causing their disruption, resulting in the "tar spot."

Spermogonia which produce spermatia in fantastic numbers are produced within the young stroma. Ascogonia are produced within the stroma at the end of the growing season and establishment of the dikaryon takes place by the dissolution of a trichogyne septum with subsequent pairing of the trichogyne nuclei with ascogonial nuclei. The ascogenous hyphal system is produced as in other ascomycetes. It is also suspected, however, that the spermatia are capable of fertilizing the ascogonium by migration of spermatial nuclei through the trichogyne. These various possibilities are shown in Fig-

Figure 6.38
Life history of *Pyronema* sp.

ure 6.36. Overwintering of the fungus takes place in the fallen leaf litter. The apothecia mature in the early spring, and the cycle is repeated. No method of asexual reproduction has been demonstrated for *Rhytisma acerinum*.

Although the sexual process has been elucidated for relatively few of the operculate discomycetes, there have been sufficient examples to lead one to believe that some form of sexual expression precedes ascocarp formation for all of them. The genus *Pyronema* contains members for which complete life histories have been traced (Figure 6.38).

The members of this genus are usually found on soil that has been burned over or that has been steam treated (e.g., greenhouse soil). The mycelium grows on the soil as a superficial white, cottony mass. Compatible ascogonia and antheridia develop on the mycelium which develops from a single ascospore and the fungus is, therefore, homothallic. The ascogonium produces an elongated trichogyne that contacts the antheridium. Nuclear duplications in the gametangia produce approximately 100 nuclei in each sex organ. Dissolution of the walls between the antheridium and the trichogyne allow the antheridial nuclei to migrate into the trichogyne. Details of development differ with investigators, but it is generally thought that the antheridial nuclei eventually migrate into the ascogonium and eventually pair with ascogonial nuclei. Ascogenous hyphae are produced by the ascogonium, pairs of nuclei migrate into them, and ascus development proceeds as previously described for sexual reproduction in the Pyrenomycetes. The ascogenous hyphae often branch and produce clusters of asci.

As a result of this sexual stimulation, sterile hyphae grow from the cells below the ascogonium. These filaments produce a loose envelope enclosing the gametangia, several such groups may unite, and a compound apothecium up to 2 mm is produced. At maturity, the apothecium opens and the hymenium is openly exposed to the environment.

Asexual reproduction for *Pyronema* spp. can be accomplished by fragmentation, but no conidia are known to reproduce the members of the genus. Oidia have been reported for some, but the way in which they function is uncertain. Conidia have been reported for closely related genera (e.g. *Ascobolus* spp.) and various functions assigned to them, but further investigation is needed before they are completely understood.

101

the sac fungi—ascomycetes

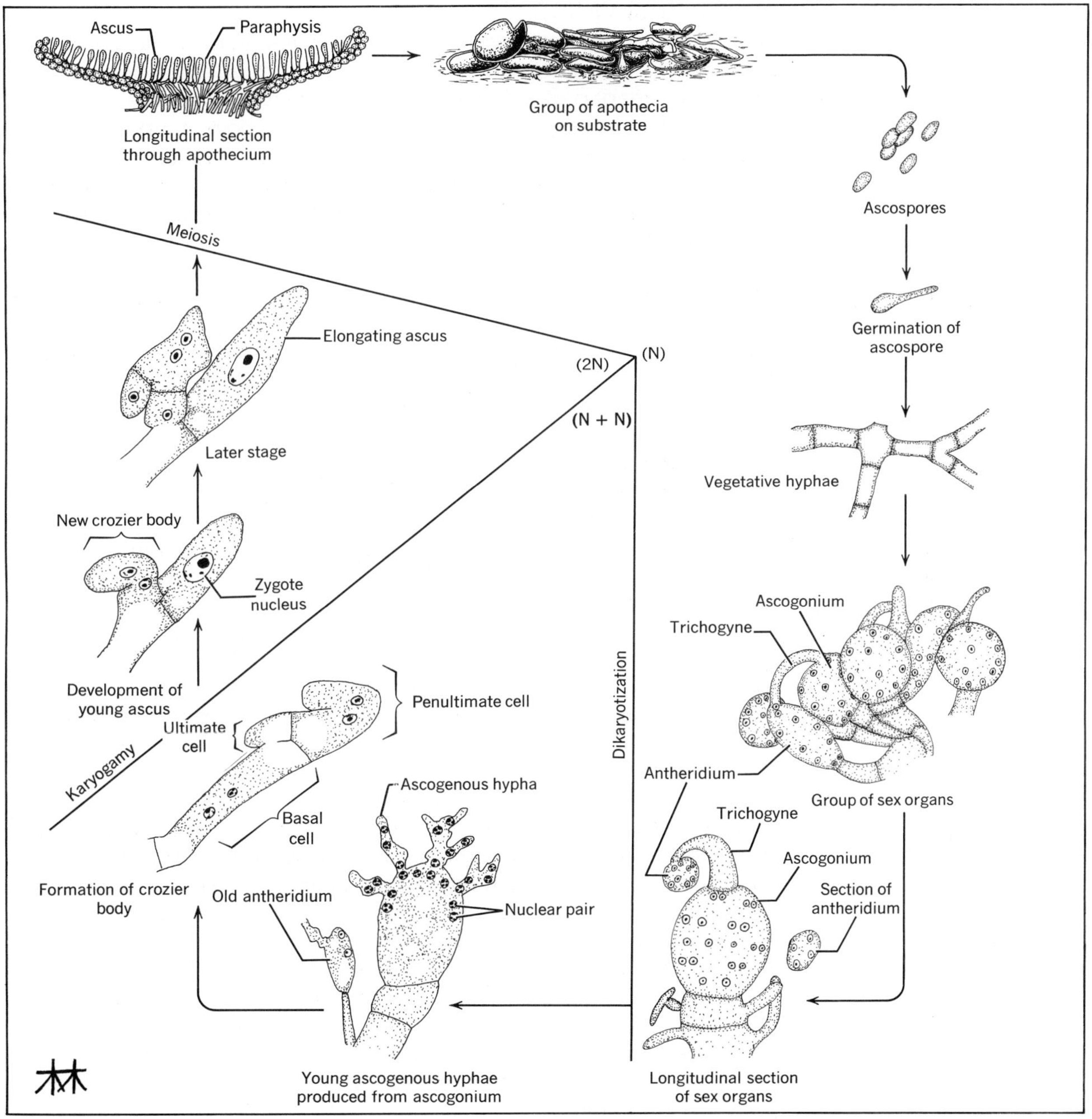

HUMAN RELEVANCE

Although most discomycetes are saprobic, several serious plant pathogens are found in the group. *Monilinia fructicola* is the causative organism of a disease known as brown rot of peaches and other stone fruits. It causes different symptoms on its host at different times during the course of its life history. It may cause a blight of the peach flowers, leaves, or twigs. It may cause cankers on the young twigs, and it causes the brown, rotting condition of the developing fruits. The affected fruits dry up and "mummify," remaining on the tree or falling to the ground.

The fungus (and hence the disease) is found throughout the world wherever stone fruits are grown and climatic factors permit. The disease has been particularly severe in the more humid eastern and southern peach-growing states; rarer in the dryer southwestern states. The disease is found in California only in the moist areas (e.g., around San Francisco) but is serious on cherries, prunes, and peaches in the humid sections of Oregon and Washington. It has never been reported from the irrigated areas of central Washington.

The disease may be found as light infections in an orchard or be so severe as to completely ruin an entire crop. The conidial stage disseminates the fungus rapidly under most environmental conditions. The effect of the fungus on the developing fruit is probably the most destructive phase of the disease. Even if there is little evidence of the presence of the disease in the orchard (because of dry conditions) the fruit may rot because of the fungus on its way to market or while being held for consumption. After the entire fruit is rotted, there is a gradual shriveling and wrinkling and the fruit is transformed from the mass of "soft rot" to a dry, shriveled mass, the "mummy." The "mummies" that remain on the trees can produce conidia as soon as the weather becomes warm the following season and infections in the orchard may start early in the growing season. The mycelium in the "mummy" is a nonseparable stroma and may produce apothecia within 1 to 3 years if it is buried in the soil on the orchard floor. The ascospores produced in the apothecia provide another source of infection for the developing fruits.

Blossom blight caused by *M. fructicola* may cause a nearly complete crop failure and fruit from unsprayed orchards of stone fruits are sometimes completely lost, either in the orchard or during shipment. It has been said that brown rot probably causes more loss to peach growers than all other peach diseases combined. The most disastrous outbreaks are to be expected during moist, humid weather when the crop begins to ripen. The disease can be controlled with proper attention to sanitary measures and cultural practices within the orchard and with an adequate spraying or dusting program.

Monilinia fructicola is one of the most destructive of the discomycetes, but many other vascular plants that are useful to man are also parasitized by cup fungi. Examples of these are: maple attacked by *Rhytisma acerinum* (Figures 6.36 and 6.37, p. 86) causes tar spot of maple, lettuce attacked by *Sclerotinia sclerotiorum* causes drop of lettuce, alfalfa attacked by *Pseudopeziza medicaginis* causes leaf spot of alfalfa, and Douglas fir attacked by *Rhabdocline pseudotsugae* causes needle blight of Douglas fir.

Many of the saprophytic cup fungi are quite beautiful. For example, *Sarcoscypha coccinea* (Figure 6.34, p. 86), is one of the earliest of the fungi to make its appearance in the spring (March through May). The hymenium of its large (2 to 3 inches) apothecium is brilliant red and gives it the common name of "scarlet cup." It is saprobic on down twigs and sometimes appears to be growing on the soil; closer investigation will reveal that it is attached to buried wood. Smaller members of this genus (*S. occidentalis* and *S. floccosa*) make their appearance in midsummer. They also have a brightly colored hymenium although they are not so spectacular as *S. coccinea* because of their smaller size.

Some of the discomycetes are among the most delectable of the fungi and are much sought after by mycophagous hunters. The genus *Morchella* (Figure 6.35, p. 86) contains several species that appear early in the spring and are delicious when properly prepared. Their apothecia are supported on a thick stalk and are extremely convoluted giving them the common name of "sponge mushroom." The color of the hymenium ranges from a grayish white to dark brown, and they range in size from less than 1 inch to 4 or 5 inches tall. This fungus is quite easy to recognize, and cannot usually be confused with any fungus that would cause gastronomic distress (with the possible exception of some members of the genus *Gyromitra*, the "false morels"). Although morels are not rare, they are found for a relatively short period of time during the season (approximately two weeks) and are unpredictable as to where they will grow. They have never been inducted to fruit in artificial culture although many mycologists and others have tried to do so.

Another group of discomycetes that provides gastronomic delight to both man and pig is the truffles (order Tuberales). Although several representatives of this order are found in North America (Figure 6.39, p. 86) none of them are suitable for food. The ones that are so desirable to epicureans are found in France and Italy. Truffle culture has been carried out in southern France for many decades under oak and beech trees where the fungus forms a part of a mycorrhizal relationship; the other part of the relationship being contributed by the roots of the tree. The truffle, which is a closed apothecium surrounded by a rind, grows from 1 to 6 inches below the soil. Although the fungus cannot be seen, it can be scented by dogs and pigs. The French prefer to use pigs because they will scent out and dig up the truffle. According to Christensen, 1951, the truffle's flavor is " . . . all-permeating, mouth-watering, tantalizing, with the promise of Elysian enjoyment, a condensation of all that is noble and good. One needs be neither a sophisticate nor a pig to enjoy truffles." Perhaps not, but he probably does need to be a true connoisseur of mushrooms to pay the price for this expensive, imported delicacy.

Many of the other discomycetes present interesting growth forms and colorations in the fungus flora. The "saddle fungi" (*Helvella* spp.) are common in heavy leaf cover in wooded areas. They have a saddle-shaped apothecium that ranges in color from pure white to cream to gray and supported by a smooth or a ribbed, fluted, heavy stalk. Although they are most abundant during the spring, some species are found throughout the summer. Their ascocarps sometimes resemble those of the morels, but humans vary in their susceptibility to the helvellic acid which they produce. Some individuals are not affected, but others are made violently ill. The "earth tongues" (*Geoglossum* spp.; *Leotia* spp.) have spatulate or rounded apothecia that range in color from black to brown to brightly colored (Figures 6.40 and 6.41, p. 86). Finally, the little (10 mm) apothecium of *Scutellinia scutellata* (Figure 6.42, p. 86) should be familiar to anyone interested in fungi. It is very commonly found from May to September on soil or wet wood in almost any wooded area. Its characteristic orange-red hymenium that is surrounded by long, brown hairs with forked bases make it easily identifiable.

One should not neglect to mention the importance of the discomycetes as decomposers of complex organic molecules. In the final analysis, this is probably the most important function they perform in the ecosystem—and thus for man himself.

FUNGI WITH BITUNICATE ASCI—LOCULOASCOMYCETES

The Loculoascomycetes is a group that is presently distinguished from the Euascomycetes by possessing a combination of two characters: a **bitunicate ascus** and an **ascostroma.** The bitunicate ascus is fairly readily identifiable because it possesses two distinct walls (Figure 6.43*a*). Usually, the outer wall is rigid and will not expand. The inner wall, however, is elastic. When the ascus reaches maturity, the outer wall breaks and the inner wall expands out of the base of the outer wall (Figure 6.43*b*). The ascospores are released through a pore in the tip of the inner ascus wall. The ascostroma is a character that is less readily distinguishable than the bitunicate ascus. Indeed, some of the ascostromatic forms are easily confused with perithecial forms (e.g., unilocular ascostroma) unless their development is followed. The ascostroma may be found on or in the host tissue before the ascogonium is distinguishable. One or more holes (**locules**) dissolve out within the ascostroma, and it is within these locules that

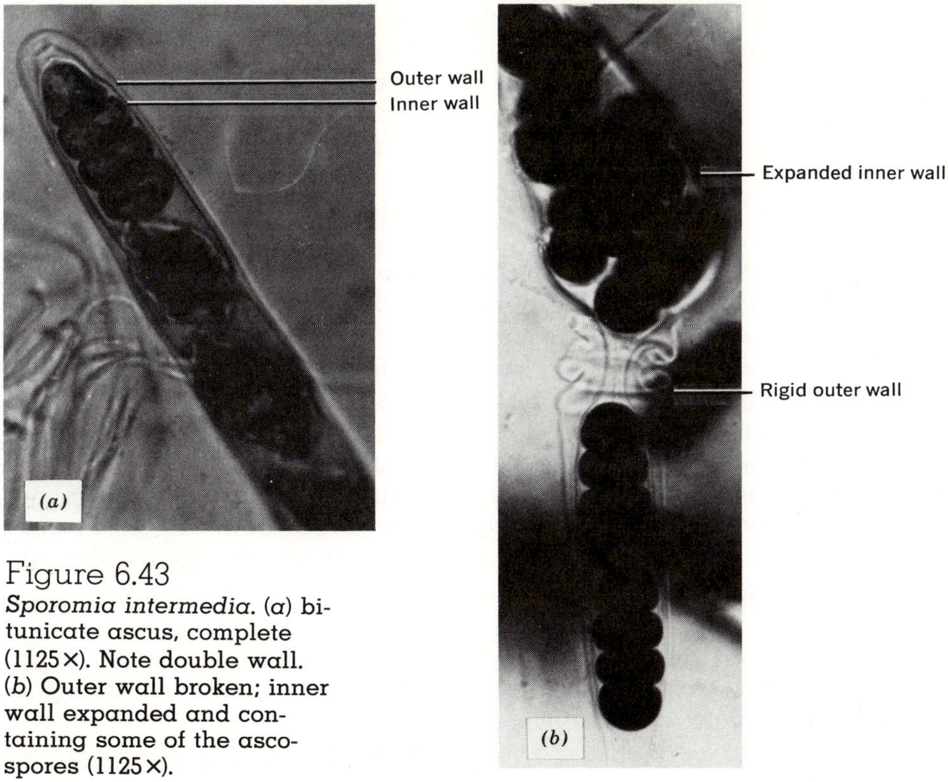

Figure 6.43
Sporomia intermedia. (*a*) bitunicate ascus, complete (1125×). Note double wall. (*b*) Outer wall broken; inner wall expanded and containing some of the ascospores (1125×).

the ascogenous hyphae and asci develop. No differentiated wall surrounds the locules in the ascostroma. This contrasts with the definite perithecial wall formed around the ascogonium in the stromatic perithecial ascomycetes. Ascostroma may have many cavities (multilocular) or only one (unilocular), and at maturity may closely resemble perithecial forms.

REPRODUCTION

Asexual reproduction occurs by the production of conidia in many of the loculoascomycetes. The conidia can repeat the asexual part of the cycle and create economic problems in such relationships as apple scab (Figure 6.44). The fungus may produce two or more types of conidia that are quite varied but many members of the group produce no conidia at all. The latter forms reproduce only by means of their ascospores.

Sexual reproduction in the loculoascomycetes is essentially the same as that which occurs in the euascomycetes. That is, there is the reestablishment of the dikaryotic hyphal system by means of gametangial contact, somatic exchange, spermatization by the production of microconidia or spermagonia, or development of the ascogonium without addition of other nuclei. Both heterothallic and homothallic members are known.

The distinguishing ascocarp of the group is the ascostroma. As previously mentioned, it is sometimes difficult to separate the loculoascomycetes from the pyrenomycetes on the basis of this character. At least three methods of development of the ascostroma have been noted, but further consideration is beyond the scope of this text. Because of the studies that link the bitunicate ascus with the ascostroma, the beginning student is advised to consider the bitunicate ascus as diagnostic for the loculoascomycetes. He is, however, cautioned against thinking that the final word on the subject has been said. The subclass is considered to contain five or six orders, but no attempt at their definition is made here.

LIFE HISTORY

Only one life history (Figure 6.44) is discussed for this group. *Venturia inaequalis* is the causative organism of a serious disease of apples known as apple scab. It also attacks many ornamental members of the genus *Malus*.

The fungus overwinters by means of ascostroma that are contained

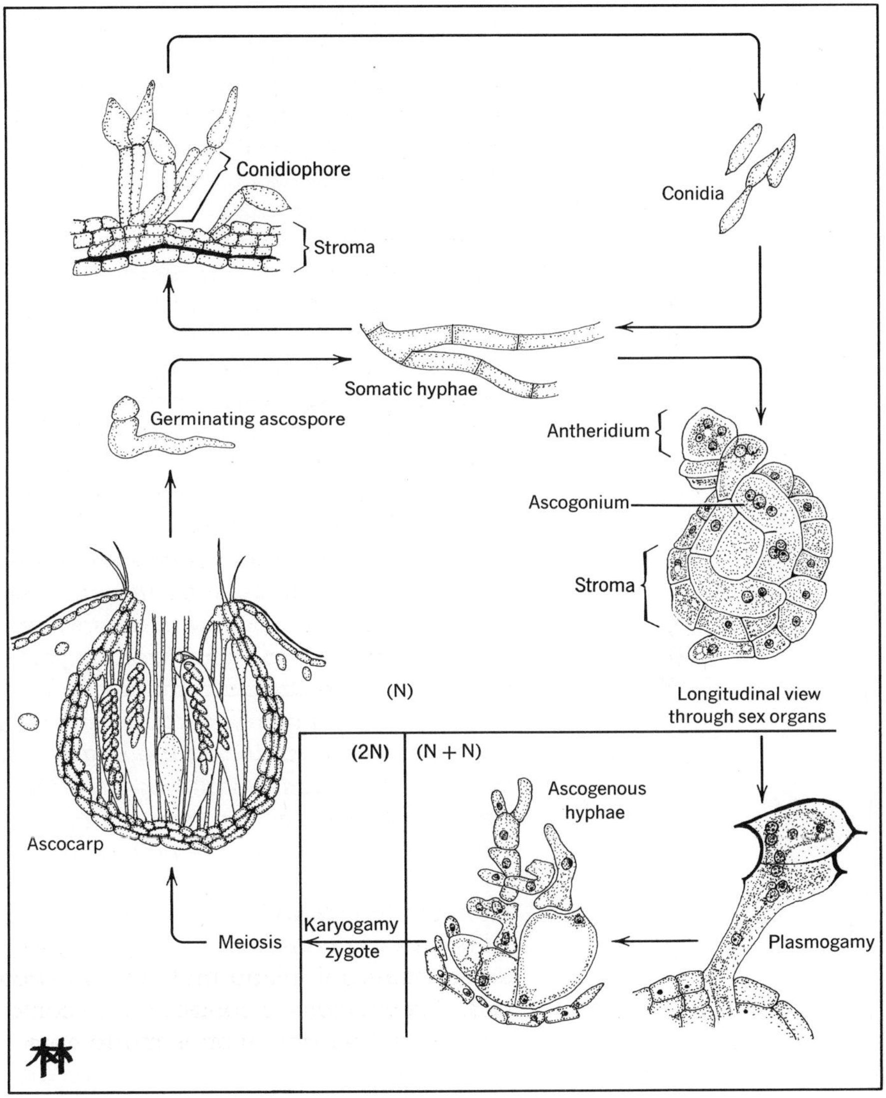

Figure 6.44
Life history of *Venturia inaequalis*. (Redrawn from C. J. Alexopoulous, *Introductory Mycology*, New York: Wiley.)

in leaves that have fallen to the ground during the previous growing season. In the spring, the ascospores are forcibly ejected from the asci. The upper cell of the two-celled ascospore is smaller than the lower (hence the species name *inaequalis*) and develops under the same environmental conditions favorable for the development of apple buds. The ascospores are passively transferred to the apple leaves by air currents.

The mycelium produces a thin stroma just under the cuticle of the host. In a few days, short conidiophores that each produce a conidium are produced on the stroma. Conidia are continuously produced at the tips of the conidiophores and are transferred to other leaves or young fruits by wind or rain. Thus the fungus is perpetuated by asexual means throughout the spring and summer.

A stroma is formed deeper within the leaf tissues later in the season by a coil of uninucleate hyphal cells. An ascogonium, consisting of a coil of multinucleate cells, differentiates within the young stroma. A trichogyne protrudes from the stromal walls and is fertilized by contact with an antheridium of the opposite mating strain. The antheridial nuclei pass through pores that dissolve in the gametangial wall, and pair with the ascogonial nuclei. Ascogenous hyphae and asci are produced as previously described for other ascomycetes. The stroma continues to increase in size. Maturation of the ascocarp continues throughout the winter and the mature ascospores are ejected during the beginning of the following growing season. The reader should be reminded that no perithecial wall is produced. The asci and ascospores develop within locules in the stroma in the manner previously discussed.

HUMAN RELEVANCE

Although the boundaries of the loculoascomycetes have not been completely defined as yet, they are known to contain several very destructive plant parasites.

Venturia inaequalis (the causative organism of apple scab) occurs throughout the world where apples are grown except where the climate is too warm or dry. It is most severe where the springs and summers are moist and cool. The apple-growing areas of northern United States, southeastern Canada, and northwestern Europe are ideal.

Crops of some apple varieties are a total loss during severe fungus years. The fungus usually attacks leaves, blossoms, and fruits, but may attack young twigs and bud scales when the proper environmental conditions exist. The fruit is most susceptible when young, and early infection causes large lesions. The older scab spots often become brown and corklike, and the fruit becomes dwarfed and misshapen. Small scab spots may develop during the storage of the fruit.

Fungicidal sprays are the most widely used control measures. The time of spraying is important because of the very early discharge of the ascospores. The fungus grows slowly near freezing temperatures and at about 85° F.; the optimum range is between 60° to 75° F. Also, the fungus thrives during rainy periods because of the increased germination of ascospores, the aid in spore dispersal by the water, and the more vigorous growth, which makes them more susceptible, of the host trees.

Although some varieties of apples are very susceptible, they are economically desirable. If proper spray programs are used, the trees bearing these varieties can be adequately protected. Some varieties of apples are less susceptible than others to particular strains of the fungus. However, an apple variety may be resistant to some strains of the fungus but not to others. Therefore, the many fungus strains complicate the problem of breeding resistant apple varieties. The same problem occurs many times with crop plants that are attacked by fungus pathogens. An extremely well-studied situation occurs with wheat and *Puccinia graminis*, the causative organism of wheat rust, and a more complete discussion of the problem of fungus strains will be offered during the account of that disease relationship.

Many other members of this group are parasitic on other vascular plants. Members of the genus *Elsinoe* attack such important commercial plants as lemons, oranges, lima beans, roses, and grasses. They produce stroma that are limited in size and development, and are called spot anthracnoses by the plant pathologist.

Several loculoascomycetes form the fungal components of lichen associations and will be discussed further in the chapter about lichens. Others are saprobic and aid in the degradation of complex carbon compounds. The number of fungi included within this group will undoubtedly be increased as more studies of the development of various ascomycetes are completed.

the imperfect fungi | fungi imperfecti

7

This group of fungi has been briefly considered in chapter 6, and the student is advised to review the introductory paragraphs of the section entitled Series: Plectomycetes before beginning this chapter.

The imperfect fungi is that group whose members, by original definition, produced no sexual stage in their life histories. They are commonly found in nature, reproduce prolifically by means of various types of asexual spores, and are of great importance to man. The original definition is no longer valid, as sexual stages have been found for many of the members of the group.

The members for which sexual stages have been discovered should properly be removed from the Fungi Imperfecti and placed in the taxon to which they belong by virtue of the sexual stage. Such changes create problems because the asexual stage has usually been known for a longer time than the sexual stage. The asexual stage may also be the one by which the fungus is mostly reproduced, and therefore the sexual stage may be found only during certain periods of the growing season or under certain specific environmental conditions. The asexual stage, then, is often the one observed and the one with which the investigator must concern himself.

There are still numerous fungi for which no sexual phase has been found. Four reasons may account for this: it exists and the investigators have just not looked in the proper place and at the appropriate time; it exists, has been described, but investigators have not made the proper associations between sexual and asexual stages; the sexual stage never evolved; or it has been lost through the process of evolution. The last two of these reasons provides man with the problems that there will probably always be a residue of organisms within this group even if all the members possessing sexual stages are removed to their proper taxonomic positions.

Therefore, one must consider this group of organisms strictly for convenience, and on purely morphological bases, with no attempt at demonstrating evolutionary relationships. The group is often labelled form-class Deuteromycetes, but this designation tends to lend a taxonomic equality to the taxa that are an attempt to express evolutionary affinity. In an attempt to further express the artificiality of the group, the designation of form-class Fungi Imperfecti will be used in this discussion.

The mycelium of the imperfect fungi is composed of branched, multicellular hyphae. The cells are usually multinucleate and the septa have perforations that allow for cytoplasmic continuity and nuclear migration. The group contains members that can utilize a variety of substrates for their nutritional requirements. Many are parasitic on plants, animals, and man himself, which brings them into direct conflict with man, but many are saprobes.

REPRODUCTION

By far, most of the imperfects for which sexual stages have been found belong to the ascomycetes. The sexual reproduction for them (as inconsistent as it appears to be to assign a sexual cycle to a group of fungi that supposedly does not have one) has already been described in the chapter about ascomycetes. Some members of the imperfects, however, have been found to possess sexual stages that belong in the basidiomycetes. These will be discussed in Chapter 8. Some fungi for which sexual stages have not been found so closely resemble basidiomycetes in their mycelial structure and other spore forms (e.g., rusts) that they have been assigned to the sexual order without ever having been assigned to the fungi imperfecti. The same situation has occurred with some of the phycomycetes. Therefore, the usage of this category (Fungi Imperfecti) has not always been consistent. If the inconsistency seems strange or confusing to the beginning student, he should remember that man is the one who is attempting to assign "pigeonholes"; the fungi do quite well without them.

Although the imperfect fungi do not possess the truly sexual cycle of plasmogamy, karyogamy, and haploidization occurring in a regular sequence and usually at a specified time, some of them derive some of the benefits of sexuality by a type of nuclear exchange known as **parasexuality.** Plasmogamy, karyogamy, and haploidization occur in a parasexual cycle, but not in a specified place or time in the life cycle. The two processes are not mutually exclusive, and fungi are known that possess both mechanisms. In parasexuality, a mycelium may become heterokaryotic by the anastomosis of vegetative hyphae that are genetically different. A nucleus or nuclei from one hypha moves into the other, multiplies, and subsequent migration through the septum pores allows the entire mycelium to become heterokaryotic. Also, nuclei that are already present in the mycelium may mutate. These two methods provide for new genetic combinations in the mycelium, providing the advantage of sexuality.

Fusion between the nuclei in an heterokaryon may occur between nuclei that are alike or unlike. Therefore, diploid nuclei that may be homozygous or heterozygous may be simultaneously present. Haploid nuclei may also be present at the same time. New gene combinations may also be produced by linkages and crossing-over during mitotic duplication of the diploid nuclei.

The various nuclear combinations may be incorporated within conidia, and thus strains of the fungus with variable characteristics can be produced. Conidia with diploid nuclei may produce a mycelium with diploid nuclei. Haploidization of the nuclei may occur during subsequent growth of the mycelium, and the diploid strains probably do not remain in existence, at least in nature, for very long. As a result of this, new haploid strains may result because of various linkages and crossing-overs that occur.

By whatever means the new strains are produced, the imperfect fungi reproduce mostly by conidia. The conidia show quite a lot of variation in such characteristics as shape, size, color and septation. The conidia are usually borne on conidiophores in various way. The conidia and conidiophores may or may not be contained in some sort of "fruiting" structure. The members of this group may also reproduce by fragmentation, and some members by budding.

The conidial characteristics, the placement of the conidia on the conidiophore, and whether the conidia and conidiophores are contained in a "fruiting" structure, are the bases on which the classification of the group is based. Further refining of this classification will probably result from studies of conidiophore structure and conidial development.

LIFE HISTORY

Talaromyces vermiculatus (*Penicillium vermiculatum*) has already been used to illustrate a life history for the plectomycetes. The asexual part of that history serves well to illustrate a reproductive cycle for an imperfect fungus, in this case the form-species *Penicillium vermiculatum*, and the reader is referred to Figure 6.8. Although *T. vermiculatus* does not produce any sort of "fruiting" structure in the asexual part of its life history, other imperfects do. It is probably fitting to discuss some of the specialized structures produced by the imperfects at this time.

The **pycnidium** (Figure 7.1) may be an urn-shaped structure that superficially resembles the perithecium of some of the pyrenomycetes. It contains conidiophores, usually short, and conidia instead of asci. The pycnidium varies in form from ostiolate to closed, and appears in various colors, shapes, and sizes. It may contain one or several cavities. Some are formed of loosely interwoven hyphae, but others are stromatic. Studies of pycnidial formation indicate that there are at least three methods by which the pycnidium develops. The pycnidium serves as the characteristic to delimit the form-order Sphaeropsidales.

The **acervulus** (Figure 7.1) is composed of a group of closely associated conidiophores that are surrounded but not covered by fungal tissue. The conidiophores arise from a hyphal mass that may

Figure 7.1
Some types of sporophores produced by members of the form-class Fungi Imperfecti.

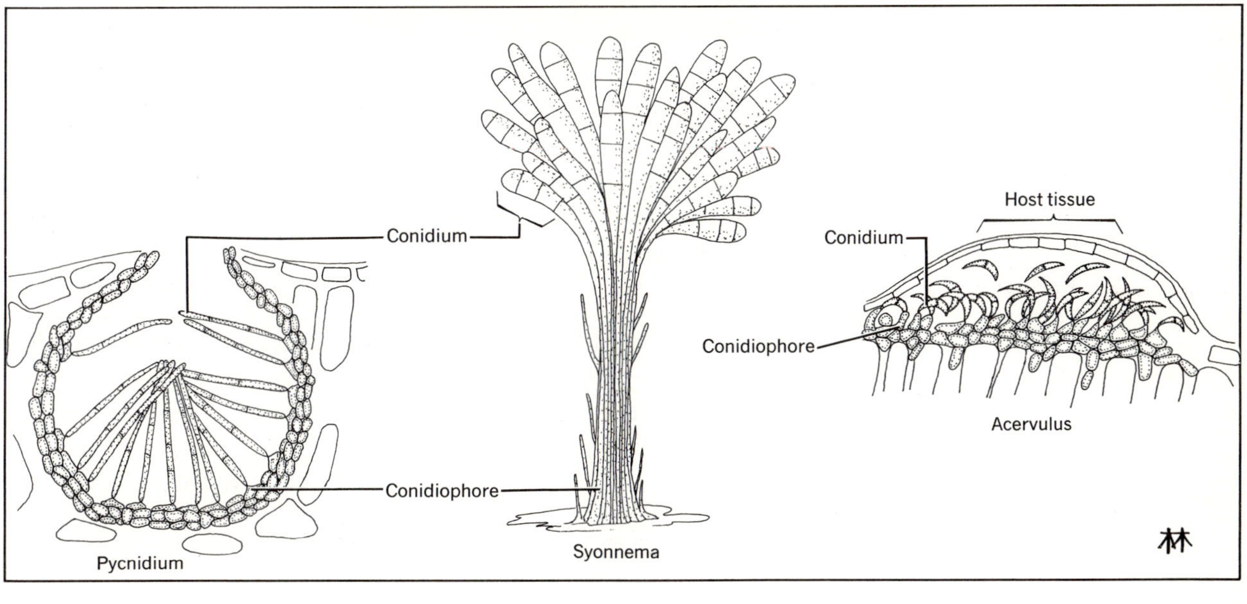

be almost stromatic. The entire structure may be either on or below the surface of the host's cuticle. The group of imperfects that produce acervuli is known as the form-order Melanconiales.

All of the other members of the Fungi Imperfecti are grouped into the form-order Moniliales. This group is often defined as all of the imperfects that do not produce pycnidia or acervuli (Figure 6.9). It contains many more members than the Sphaeropsidales or Melanconiales combined. Some members of this form-order produce their conidiophores and conidia randomly distributed over the mycelium; they are not combined in any particular fashion. The form-species *Penicillium vermiculatum* belongs to brightly colored members of this subdivision of the Moniliales. Still others produce their conidiophores and conidia in distinctive combinations. One such combination is called a **synnema** (Figures 7.1 and 6.10) and is produced by the growing together of several closely associated conidiophores, although the tips of the conidiophores often remain free. One of the methods by which the conidia of *Ceratocystis ulmi* (the causative organism of Dutch Elm disease; previously discussed) are produced is by synnemata (form-genus *Graphium*). When elm twigs suspected of containing the fungus are properly cultured in the laboratory, the synnemata can be induced, and are considered to be diagnostic for the disease. Another distinctive structure produced by some members of the Moniliales is the **sporodochium.** This structure is a pad or cushion of mycelia that may get compact enough to become stromatic, although most remain loosely interwoven. Simple or branched conidiophores are produced on the surface; the entire structure may be stalked at maturity. The form-genus *Fusarium* contains more form-species than any other in the group. It is an extremely difficult form-genus because of the great variability of its structures, although it is quite important to man because of the many disease relationships it forms with vascular plants. Sporodochia are produced by members of the form-genus *Fusarium* mostly under natural conditions; almost never in culture.

HUMAN RELEVANCE

Several important form-genera have already been discussed with the series plectomycetes where they would properly be placed by virtue of their sexual phases. These include such organisms as *Penicillium* and *Aspergillus*, both of which are extremely important to man.

However, other fungi that are found among the Fungi Imperfecti should also be considered from an economic viewpoint. Although they may possess sexual stages, the asexual stage is the one most often coming to the attention of man.

Most of the members of the form-order Sphaeropsidales are saprobic. However many serious plant parasites are to be found among the group. One of the most important form-genera is *Septoria;* several ascomycetes have their imperfect stages in this form-genus. Several form-species produce serious disease relationships on the leaves and stems of various grasses, including the cereal grains. The fungus produces leaf blotches and spots; the surrounding tissue dies and turns brown. Pycnidia develop beneath the epidermis of the diseased areas. The conidia are long, threadlike, produced in huge numbers, and wind-borne. The damage caused to the grain results mostly from the shriveled kernels and from causing the stem to fall before harvest (**lodging**). The losses caused to susceptible varieties of wheat, oats, and barley can be quite high. The best means of control are crop rotation, plowing the stubble under, and using resistant varieties.

Another *Septoria* causes a disease known as tomato leaf spot, but it also causes lesions on stems, flower parts, and sometimes on young fruits of tomatoes. The lesions vary from pinhead size to several millimeters in diameter, and may coalesce and completely destroy the leaf. Although the fungus has a narrow range of sporulation temperature, it can cause tremendous losses by defoliation in those areas where it can grow. The fungus can parasitize other members of the family to which tomatoes belong such as horse nettle, nightshade, and jimson weed. Like the ones that parasitize the grains, it also overwinters on dead plant parts. Therefore, effective control must include a good program of sanitation in the field.

Other *Septoria* attack such diverse hosts as gladiolus, wild black cherry, celery, chrysanthemum, and soybean, among others. Members of the form-genus are to be found over a wide geographic range as well as a wide host range; therefore, it is an important consideration for man.

The acervulate imperfects (form-order Melanconiales) also contain many members that are parasitic on vascular plants. These fungi produce a gradual breakdown of part of the host. Although there are usually not definite lesions produced and there is no delimiting size to the infected area, the invasion by the fungus continues until the plant part is completely brown and dead. Such a disease is called an **anthracnose,** and is characteristic of the diseases caused by members of the Melanconiales.

Collectotrichum is an important form-genus in this group. *C. lindemuthianum* causes a serious anthracnose of beans, but the use of resistant varieties have made it a much less serious consideration than it was in the 1920s before those varieties were developed. *C. fuscum* causes an anthracnose that could have implications for medicine because it attacks members of the genus from which the drug digitalis is derived (*Digitalis* spp.). Although *D. purpurea* is the principal plant used as a source of digitalis, *D. lanata* has been investigated as another source of valuable medical compounds. It has been found, however, that *D. lanata* is extremely susceptible to *C. fuscum*, and some experimental plantings have been almost a complete loss because of invasion of the fungus. The fungus causes an anthracnose of the above ground parts of the host, but conidia can be carried on the seed and cause a damping-off disease of the seedling. Conidia are spread to mature plants by splashing rain drops in the seed bed and in the field. Because of its varied modes of infection, *C. fuscum* presents quite a problem to growers of *Digitalis*. Both species of *Digitalis* are susceptible to the fungus, but *D. purpurea* is more resistant. One method by which disease-free crops of both species of *Digitalis* have been grown is by subjecting the seeds to a hot-water bath of 131° F for 15 minutes.

The form-species *Colletotrichum coffeanum* has important implications for those persons whose mornings look brighter after their second cup of coffee. This fungus attacks the coffee seedlings in the seed beds and nursery and the young leaves and twigs of older plants, and is found in all countries where coffee is grown. Strains of coffee that are resistant to *C. coffeanum* have been found, but a variety of the fungus (*C. coffeanum* var. *virulans*) appears to be particularly able to infect the coffee fruit. Fifty percent of the coffee grown at higher elevations may be destroyed by the fungus—no little consideration for caffeine lovers.

The form-order Moniliales contains by far the largest number of members of the Fungi Imperfecti. It is to this group that the form-genera *Penicillium* and *Aspergillus* belong, but many other members are also extremely important to man.

Particularly virulent plant parasites are found in the form-genera *Cercospora*, *Fusarium*, and *Helminthosporium*. *Cercospora apii* causes early blight of celery (a leaf spot), a disease that can be quite serious when the growing season is hot. Because the fungus may be seed-borne, it may cause damage in the seed beds as well as in the young plants. Growth of the celery is stunted, and the keeping quality of the plant is impaired. This same form-species has been reported as a possible human pathogen, causing severe face lesions (Figure 7.2). Other members of this same form-genus cause leaf spots of beet, lettuce, carrots, clover, cotton, pepper, coffee, and other unrelated food crops. More than 3000 members of this form-genus have been described. However, many of them were described on the basis of the host on which they were found. Further studies have indicated that several of the form-species are actually synonymous and simply have a wide host range.

The form-genus *Fusarium* is an extremely large one and taxonomically very difficult. The parasitic members may plug the vascular systems of their hosts, causing severe wilting. Several varieties of the form-species *F. oxysporum* attack such diverse hosts as gladiolus, cabbage, sweet potato, alfalfa, cucumbers, and many others. *F. oxysporum* var. *cubense* causes a catastrophic wilt of bananas that is sometimes known as Panama disease because of

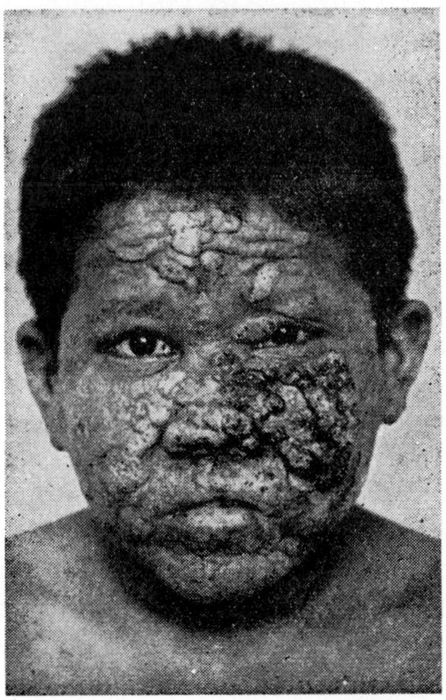

Figure 7.2
Facial lesions caused by *Cercospora apii*. (From C. J. Alexopoulous, *Introductory Mycology*, New York: Wiley.)

the severe damage done to banana crops in that country. The symptoms of the disease are rapid and dramatic. The leaves yellow and wilt almost overnight, and the whole plant soon dies. An entire banana plantation can be decimated in a few months. Methods of control are being sought, but no completely satisfactory method has yet evolved. Silting and flood followed by flooding the fields has been partially effective in some cases but the best solution is to use resistant banana varieties. The fungus is well established in the banana growing areas of the Western Hemisphere, and will probably continue to extract its toll for some time to come. There are many, many other form-species of *Fusarium* and several of them have numerous varieties that also attack vascular plants.

The form-genus *Helminthosporium* also contains many members that are destructive parasites of vascular plants. Several of them cause diseases on grasses such as corn, rice, oats, wheat, and many others. *H. maydis* parasitizes all types of corn, causing a disease relationship known as southern corn leaf blight. Although given a geographic reference in its name, the disease is found as far north as Illinois, Indiana, Missouri, and Ohio, and eastward to the Atlantic Coast. These states were hit especially hard during the growing seasons of 1970–1971 when entire corn fields were virtually wiped out by the fungus (Figure 7.3, p. 158). Abundant rainfall, heavy dews, and warm temperatures provide the ideal environmental conditions for the development of the fungus. Infection of the corn plants usually occurs as the plant reaches maturity, and losses may exceed 50% of the crop. Although the lower leaves of the corn plant are usually attacked first, progressive infection may cause the entire host to be affected. The first symptoms are small leaf spots that increase in size and may eventually coalesce, forming large lesions that can kill the entire leaf. The fungus overwinters in debris left in the field. The development of resistant strains of corn seems to be the best hope for control, although some spray programs have had limited success.

These several examples should serve to illustrate the importance of this group of fungi to man when they are placed into competition for food. However, it is among the imperfect fungi, more specifically among the form-order Moniliales, that are found most of the fungal parasites of man himself. *Penicillium* and *Aspergillus* have been discussed as the causative agents of certain diseases of man in the discussion of the plectomycetes, and *Cercospora apii* has been previously mentioned in this section.

The disease relationships that develop between animals and fungi are called **mycoses.** There are only about 12 serious mycoses of man, although there is about an equal number of less severe ones. These are caused by 50 to 60 species of fungi, most of which are free-living saprobes. Most people have probably been exposed to a fungal infection at some time or other, but the disease usually develops only under rather specific conditions such as unusual susceptibility (e.g., diabetes), exposure to a heavy concentration of spores, or exposure to the fungus over a long period of time. Some of the pathogens require deep placement in the tissues, and may develop in old wounds over a period of 10 to 20 years. However, most of them have a low ability to infect and there are few instances of transfer of the fungus from an infected to a healthy individual. One of the most obvious and convenient methods of classifying man's fungal parasites is by using the site of their attack on the host. This method produces at least four fairly easily recognizable groups. The **superficial mycoses** are those that infect the hair and outermost layers of the skin. While they may be unsightly, there is little tissue destruction, pain, or affliction. *Trichosporon beigeli* is the causative agent of white piedra, an infection of hairs that are usually found on beards and mustaches. Soft, light-colored fungal nodules are produced on the shaft of the hair, and growth of the fungus within the shaft causes the hair to weaken and easily break. The disease has been rarely reported in the United States, but it is found in the temperate zones of South America, Europe, and the Orient on man and horses. Effective treatment can usually be accomplished by removing the infected hair.

The second group is the **cutaneous mycoses** or **dermatomycoses.** These attack the hair, epidermis, nails, and other keratinized tissues. One of the most common diseases encountered within this group is ringworm that is caused by a number of different form-species of *Trichophyton*, *Microsporum*, and *Epidermophyton*. It is among this group that are found the most contagious of man's fungal parasites. "Ringworm" of the foot (e.g., "athlete's foot") is the most commonly encountered type of ringworm (Figure 7.4). The first symptoms are lesions that occur between the toes. The lesions may be accompanied by itching, and will eventually spread to the entire foot if not treated. Infections of other parts of the body may produce ringworm of the glabrous skin, beard, groin, scalp, and nails. Most of the infections are roughly circular in shape and have an advanced, raised

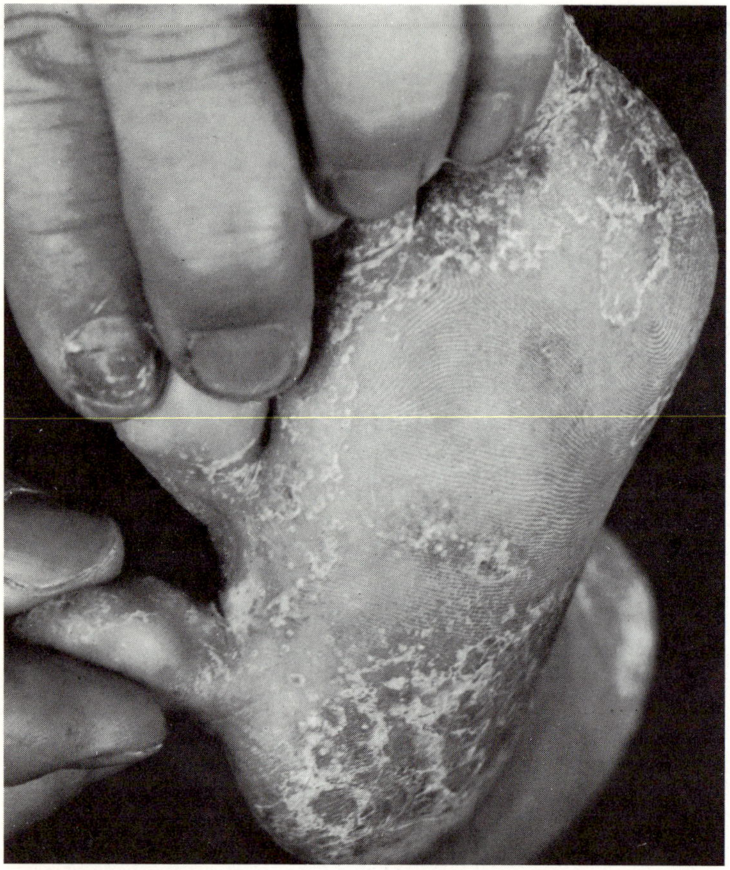

Figure 7.4
"Ringworm" (tinea pedis) of the foot. Fissure occurs between fourth and fifth toes. (From Conant, N. F. *Manual of Clinical Mycology* 3rd ed., 1971, W. B. Saunders Company.)

margin. Because of the misinterpretation of the causes of the disease and its growth pattern, the common name of "ringworm" was used. This group of diseases is now known as **tineas** (which, ironically, means "a gnawing worm moth"). Therefore, ringworm of the foot would be tinea pedis, and of the scalp, tinea capitis. Other such names designate the area of the body usually infected. The prognosis for dermatomycoses is usually very good because of the use of the antibiotic griseofulvin, which is an antibiotic derived from *Penicillium griseofulvum* among others. This antibiotic produces a barrier in the epidermal structures that cannot be penetrated by the dermatophytes. In some cases, however, the antibiotic therapy is not effective, and the more stringent methods of topical application of fungicidal agents and epilation of hair by X-ray must be used.

The third major group of human mycoses may be broadly grouped as **subcutaneous mycoses.** These affect the skin and tissues directly under the skin, although a few may rarely penetrate to the innermost tissues of the body. One of the subgroups of the subcutaneous mycoses contains the clinical entities known as **mycetomas.** There are several different form-species that can cause mycetomas, but they are all localized swellings with lesions that often contain abscesses from which drain pus that contains fungal granules. Mycetomas usually occur in the tropics or subtropics where the causative agent is introduced into the host by means of an old wound or on the tip of a thorn or other deeply penetrating object. The usual site of infection is the foot (Figure 7.5) or hand, but other areas may also be infected. Broad spectrum antibiotics are effective against some causative agents (e.g., Actinomycetes), but not against those caused by higher fungi (e.g., Fungi Imperfecti). Surgical methods, although not completely satisfactory, are used on some of the mycetomas caused by higher fungi.

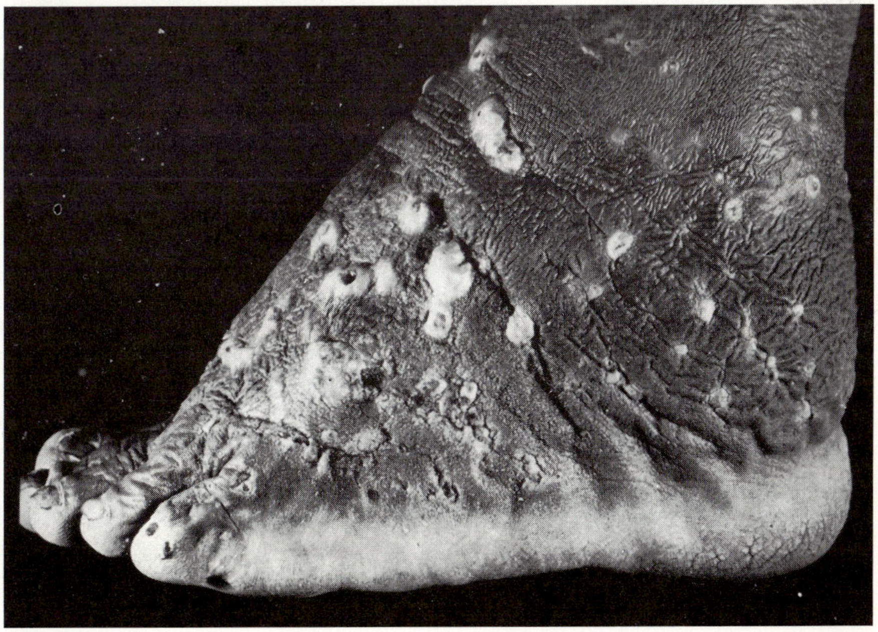

Figure 7.5
Mycetoma of the foot. (From Emmons, Benford and Utz *Medical Mycology 1970, Lea & Febiger.)*

The final group of mycoses to be mentioned is the **systemic mycoses.** These diseases may develop as cutaneous mycoses but may attack any of the internal organs. Although there is sometimes progressive infection resulting in death, the number of such cases is few, and commonly the infections are asymptomatic with a low pathogenicity level that is often self-limiting.

Histoplasmosis is an important systemic mycosis. This disease is caused by the moniliaceous fungus *Histoplasma capsulatum* which is commonly found as a soil saprobe, particularly soil that contains chicken and other bird feces. Infection usually occurs because of the inhalation or ingestion of the fungal spores from such environments. Although the disease appears all over the world, it seems to be endemic to certain areas. Such areas in the United States include the Upper Mississippi Valley.

The disease usually takes a mild, asymptomatic form that is demonstrated by the injection under the skin of a derivative (histoplasmin) of cultures of *H. capsulatum*. Sensitivity to the injection may indicate current infection or the presence of calcified lesions in the lungs that indicate previous infection. Involvement of the lungs is the most common situation, and the symptoms expressed closely resemble the symptoms expressed by tuberculosis patients. These include nonproductive cough, shortness of breath, chest pain, hoarseness, fever, muscle and joint pain, weight loss, and fatigue. In some people, the lung lesions completely heal and vanish, in others the lesions are enclosed in the calcified deposits previously mentioned (Figure 7.6). In certain areas, it has been postulated that 60 to 90% of the population has suffered exposure to *H. capsulatum*, but relatively few of the cases develop symptoms that cause them to seek medical aid.

In a very few cases (0.1%) the disease may progress to a fatal form. In this form, the symptoms are much more severe; lesions appear in the liver, kidneys, spleen, adrenal glands, lymph nodes, epithelial tissues, bone marrow, and mouth or digestive tract. Such cases often proceed to death of the patient.

There is no evidence to indicate that histoplasmosis can be transferred from person to person, but each patient is infected because of his individual exposure to the spores in his environment. The fungus is **dimorphic,** that is, the saprobic form is filamentous and produces two kinds of conidia (e.g., microconidia and macroconidia). However, when these spores or fungal fragments are introduced into the host, the fungus grows in its parasitic form with a yeast-like growth habit.

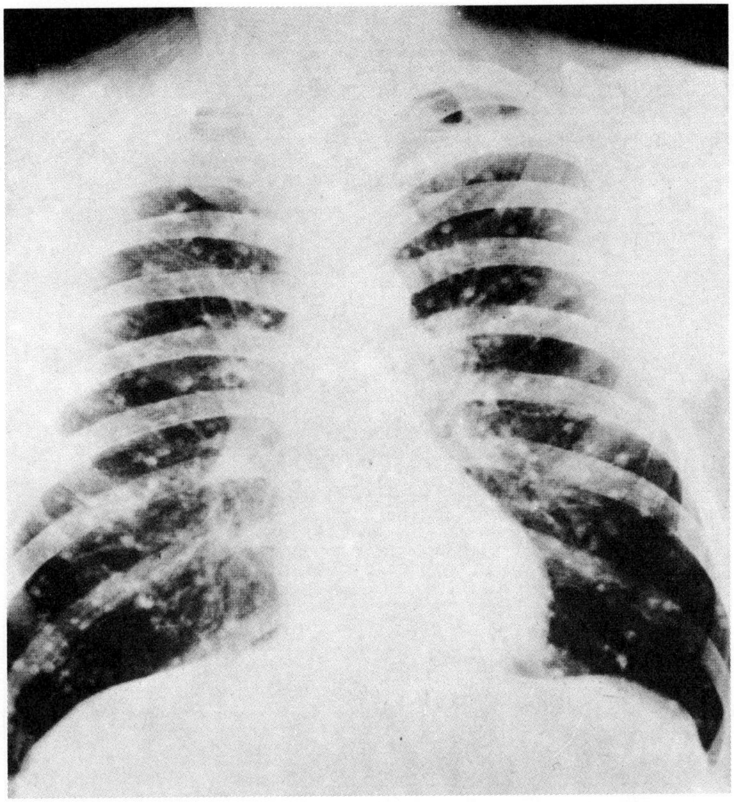

Figure 7.6
Histoplasmosis of the lungs with multiple calcified lesions present. (From Conant N. F. *Manual of Clinical Mycology* 3rd ed., 1971, W. B. Saunders Company.)

The mild, asymptomatic infections of the disease usually heal quickly with complete recovery of the patient, often without treatment. The severe, systemic forms vary in degree of severity, but the prognosis is very poor unless there is adequate treatment available. Treatment may include bed rest, proper diet, vitamins, and treatment of the symptoms. Amphotericin B, an antibiotic produced by *Streptomyces griseus*, has been effective. It is not altogether ideal, however, because it produces kidney toxins and induces chills. Other antibiotics are being tested and, hopefully, one that is more effective and less dangerous will be found.

Lest the reader be now convinced that the Fungi Imperfecti contains only villainous members, he is reminded to review the section on plectomycetes to reestablish in his thinking their importance in medicine and industry. It should also be mentioned at this time that

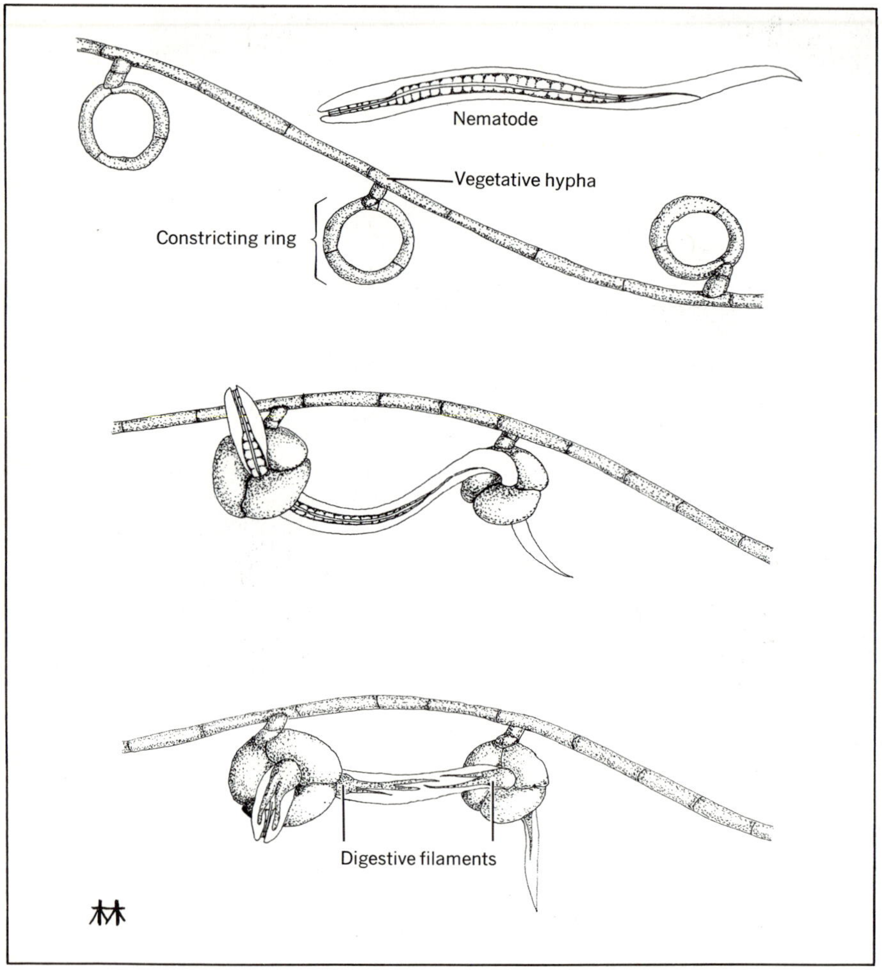

Figure 7.7
Dactylaria sp., one of the constricting ring type of nematode trapping fungi.

many of the imperfects probably also establish important mycorrhizal relationships with various vascular plants. The importance of many agarics (to be discussed later) in this relationship has long been known. Among the imperfects that are known to establish mycorrhizae are the form-genera *Rhizoctonia* with orchids, and *Phoma* with various heaths (Ericaceae).

Probably one of the most fascinating groups of fungi is the predacious nematode-trapping fungi (Figure 7.7). Although some of them are Phycomycetes and some are probably Basidiomycetes, most are imperfects. The nematode-trappers have evolved mechanisms whereby nematodes can be captured, killed, and digested. Sticky hyphal loops and branches are produced by some (e.g., members of the form-genera *Arthrobotrys* and *Dactylella*). The nematode becomes entangled in the loops or branches; other hyphae invade and digest his internal structures while he still struggles! Still others (e.g., *Dactylaria gracilis*) form rings that constrict and firmly hold the hapless worm while it is attempting to pass through. It, too, is digested while struggling to free itself. Less elaborate but equally deadly mechanisms have developed in *Harposporium*. The conidia of this fungus are quite sticky. They adhere to the worm, germinate, invade, and digest the tissues although the worm may be well away from the parent mycelium.

Because of the disease relationships many nematodes form with vascular plants, it has been suggested that the nematode-trapping fungi might be useful in disease control programs. Such has not yet been the case, however, but research into the possibility is continuing.

the club fungi | basidiomycetes

8

This final group of fungi is also a large and diverse group of organisms. The unifying characteristic is a club-shaped structure called the **basidium** (Figure 8.1). The basidium produces basidiospores that are borne externally on small projections called **sterigmata**. Because the basidiospores are products of meiosis preceded by karyogamy, they are haploid. There are usually four basidiospores produced per basidium, although this number sometimes varies.

The hyphae of the basidiomycetes are septate, although the septa possess pores that allow for cytoplasmic and nuclear migration. The movement of nuclei through the pore is limited because of a complicated arrangement of septum material that forms a "partial plug" in the pore. The cells of the mature mycelium contain two unlike nuclei (dikaryon; n+n), and are comparable to the ascogenous hyphal cells of the ascomycetes. As the mycelium grows, a specialized structure called a **clamp connection** (Figure 8.2) is formed. Not all basidiomycete mycelia produce clamp connections, but almost all mycelia in which clamp connections have been observed have proved to be basidiomycetes. The clamp connections appear to serve as a "bypass" mechanism whereby the dikaryotic condition can be maintained in successive hyphal cells, and it is thought by some mycologists to be analogous to the crozier hook of the ascomycetes. Basidiomycete mycelia penetrate the substrate, absorb nutrients, and grow. They are usually white, but brightly colored ones are also produced. Some members of the group produce hyphae that come to lie close together and that are joined by the production of an enveloping cortex. These strands of mycelia, called **rhizomorphs** (Figure 8.3, p. 158) are much stronger than the usual hyphal strands. They are often found growing under the bark of a log, or on the surface of the wood if the bark has fallen off.

The haploid basidiospore is often shot from the basidium. If

Figure 8.1
Types of basidia. (*a*) Tuning fork. Insert: Photomicrographs of two stages of development of basidium of *Calocera cornea*. (*a*′) Immature epibasidia produced. (2800×). (*a*″) Mature basidium with basidiospores (1800×). (*b*) Cruciate. Insert: Photomicrographs of two stages of development of basidium of *Tremella* sp. (*b*′) Top view showing septation of young hypobasidium (3500×). (*b*″) Lateral view with one septation and young epibasidia (3500×) (Photograph by Claire Cieslewicz). (*c*) Transversely septate. (*d*) Homobasidium. Insert: Photomicrograph of young basidia in hymenium of *Coprinus* sp. Fusion (2n) nuclei apparent (1800×).

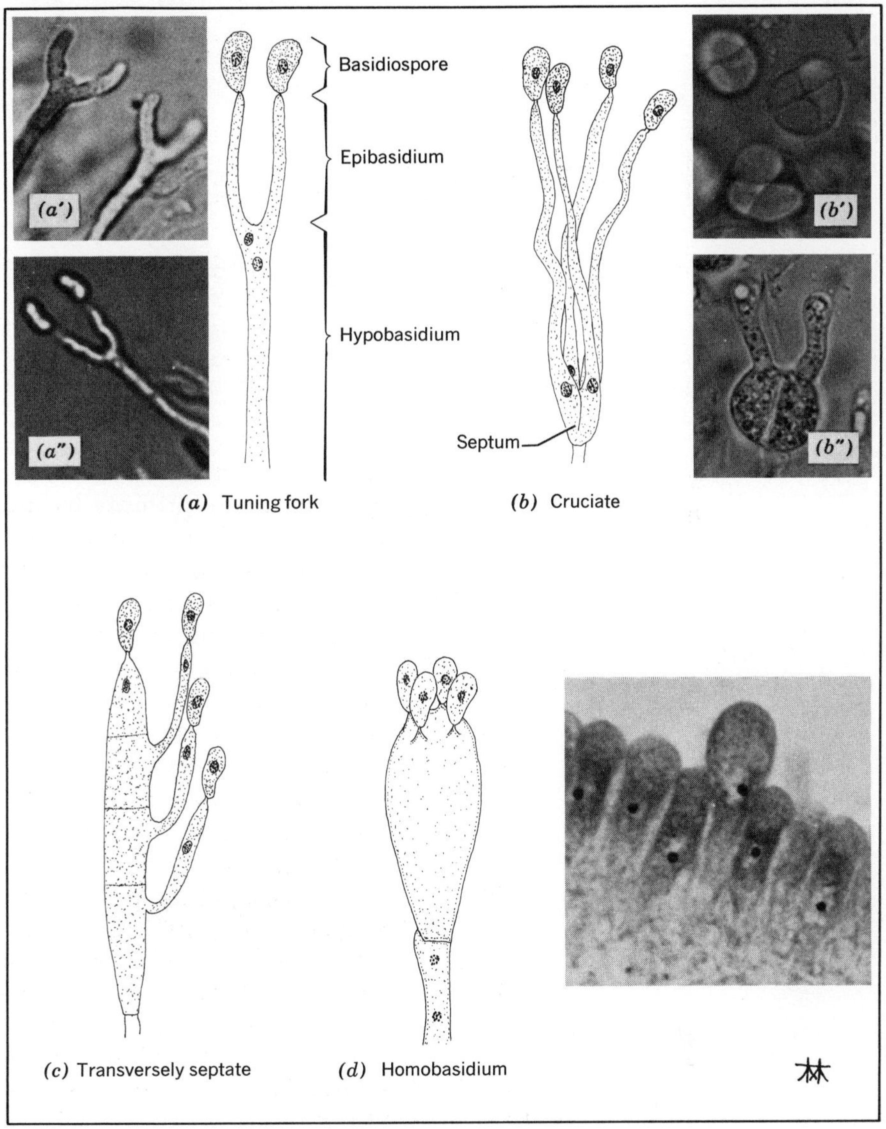

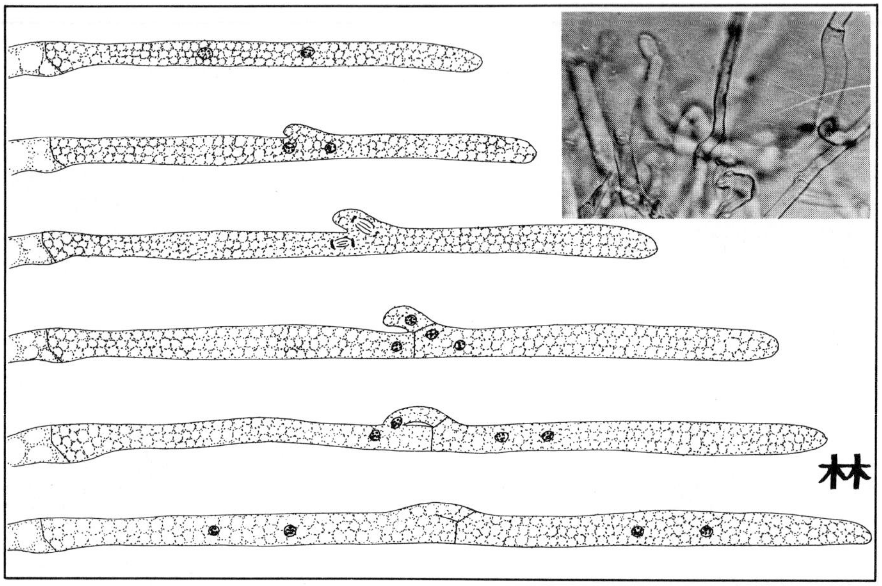

Figure 8.2
Six steps in the production of a clamp connection. Insert: Photomicrograph of clamp connections of *Crepotus* sp.

it reaches favorable environmental conditions, it germinates by producing haploid hyphae (**primary hyphae**). This hyphal system is usually short lived and does not develop further unless the dikaryotic condition is established. The dikaryotic condition may be established in a variety of ways. Two compatible primary hyphae may overlap, plasmogamy but not karyogamy occur, and a branch from this dikaryotic cell starts the new dikaryotic mycelium. Each new cell of the mycelium becomes dikaryotized by means of clamp connections. Also, two primary hyphae may carry on plasmogamy and initiate a dikaryotic cell. The differing nuclei duplicate, but instead of a new branch being formed, daughter nuclei migrate into the hypha of the opposite mycelium. Successive mitoses produce daughter nuclei that migrate through the pores and produce a new dikaryotic mycelium. In some cases, compatible basidiospores come to lie together, a pore forms between them, plasmogamy occurs, and the hyphae produced are dikaryotic from their inception. Each dikaryotic cell has the potential to produce a basidium, but specialized "fruiting bodies" are usually formed before the basidia are produced. These "fruiting bodies" are formed when environmental conditions are right for the production of the more complex tissues of which they are composed.

With the exceptions of the rusts and smuts, basidia are produced from the end cells of dikaryotic hyphae that are in the "fruiting body." As this dikaryotic cell enlarges, the two nuclei inside fuse and produce the only truly diploid stage (Figure 8.1*d*). The zygote then undergoes meiosis and four haploid nuclei are produced. During this process, four (usually) sterigmata have been produced on the top of the basidium and their tips have enlarged. Each of the four nuclei migrates to the tip of a different sterigma, and the tip matures into a basidiospore. Although four is the usual number of basidiospores, other numbers are found.

Although, with the exception of the rusts, no gametangia are produced by the basidiomycetes, it is apparent from the preceding discussion that they possess all of the evolutionary benefits of sexuality. The delayed karyogamy probably allows for a higher mutation rate than would occur if it happened immediately after plasmogamy. The production of a zygote nucleus adds the advantage of crossing-over. Spermatization is the common method of dikaryotization in the rusts, but some of the other basidiomycetes produce asexual structures (oidia) at the tips of specialized branches (oidiophores). Although the oidia may germinate and produce primary hyphae, they may also act as spermatizing agents when they unite with compatible haploid hyphae.

The form of the basidium (Figure 8.1) is used to separate the basidiomycetes into subclasses, the level of organization at which they will be discussed in this treatment. The two subclasses are the Heterobasidiomycetidae, characterized by a heterobasidium, and the Homobasidiomycetidae, characterized by a homobasidium.

The Heterobasidiomycetidae are the "primitive" basidiomycetes, and their members contain one of three possible kinds of basidia. A basal portion of the basidium, called the **hypobasidium** is produced in two of the heterobasidial types. From the hypobasidium are produced elongated tubes called **epibasidia,** and from these arise the sterigmata on which are produced the basidiospores. The hypobasidium is longitudinally septate in one kind in such a way that one epibasidium arises from each of its four parts. When viewed from its tip, the septa appear to form a cross, giving this type the name of **"cruciate" basidium** (Figure 8.1*b*).

The hypobasidium of the **"tuning fork" basidium** (Figure 8.1*a*) is not septate, but is deeply cleft into two elongate portions, the epibasidia. Because only two nuclei survive after meiosis, only two basidiospores are produced.

The third type of heterobasidium may be produced directly from a spore (**teliospore**) or vegetative hypha. It looks like a mycelium in shape, and has been described as a "promycelium." This basidium (Figure 8.1*c*) becomes transversely septate in most of the species, and basidiospores are produced from each cell resulting from the septation. In the rusts (order Uredinales), the basidia are always septate and the basidiospores are forcibly discharged from sterigmata. In the smuts (order Ustilaginales), the basidium is sometimes entire, the basidiospores are not on sterigmata, and they are not forcibly discharged. All three basidial types are found in the Tremellales, but the Ustilaginales and Uredinales do not contain any members that produce "tuning fork" or "cruciate" basidia.

The Homobasidiomycetidae are the "advanced" basidiomycetes and, although they possess a great deal of variability in their "fruiting" bodies (**basidiocarps**), they possess remarkably similar basidia (the **homobasidium**). Members of this group of fungi possess the largest basidiocarps found among the fungi, and it is in these basidiocarps that are found the basidia. The basidia are borne in differing fashions, but all are large, club-shaped, aseptate structures that usually produce four basidiospores at the ends of four sterigmata (Figure 8.1*d*). The basidiospores are forcibly ejected in those members (the hymenomycetes) that produce their basidia in exposed layers at maturity. Other members (the gasteromycetes) do not have their basidia exposed at maturity, and their spores are not forcibly ejected. The basidiospores in this group are released by the breakdown of the peridium of the basidiocarp. Both types germinate by producing a haploid primary hypha. The homobasidium is the one usually described as the "typical" basidium.

SUBCLASS: HETEROBASIDIOMYCETIDAE

The heterobasidiomycetes include those fungi commonly known as the jelly fungi (order Tremellales), the rusts (order Uredinales), and the smuts (order Ustilaginales). The types of basidia that are found in these groups have already been discussed. The basidiospores are usually violently shot from the sterigmata, and may germinate by the production of another spore (conidium). This type of repetitive germination may continue until all of the original stored food supply is exhausted or until one of the spores reaches the environmental conditions that allow it to germinate by means of a primary hypha. Dikaryotization of the primary hypha occurs in one of the manners previously discussed, and the dikaryotic mycelium produces the basidiocarp.

The jelly fungi (Tremellales) may produce basidiocarps that are jelly-like in texture, but some of them are tougher and more rigid or waxy.

Calocera cornea is a jelly fungus that looks like small jelly fingers sticking out of the wood that forms its substrate. It possesses a deeply cleft tuning-fork basidium (Figure 8.1*a*) which can usually be found in many stages of development. The color of the basidiocarp ranges from orange to yellow, depending on age.

The cruciate basidium is produced in the genera *Tremella* (Figure 8.4*b*, p. 158), *Tremellodendron*, and *Exidia* (Figure 8.4*a*, p. 158). Species of *Tremella* and *Exidia* are found on rotting wood or on soil and probably are the most common of the jelly fungi. Their basidiocarps also possess the consistency most often associated with the jelly-like texture "typical" of the Tremellales. *Tremellodendron*, however, has a tough basidiocarp of upright anastomosing branches that may be several inches in diameter. This "fruiting" structure superficially resembles a coral fungus, and the basidium must be found before identification is certain.

The transversely septate basidium is found in the group of jelly fungi that includes the one commonly known as Jew's ear (*Auricularia auricula*). The basidiocarp is ear-shaped (but with no particular resemblance to the ears of that ethnic group), large, brown to tan, and leathery to somewhat gelatinous. Ridge or cordlike areas are often produced on the lower surface.

The smuts (Ustilaginales) are an important group of heterobasidiomycetes that are obligate parasites under natural conditions, although some have been grown in culture under laboratory conditions.

The smuts produce intercellular and intracellular hyphae in their host's tissues. In some, haustoria are produced from the intercellular hyphae. The dikaryotic hyphal system penetrates throughout the host and eventually produces masses of mycelium which may or may not be covered with a membrane. The cells of this hyphal mass convert directly into unicellular dikaryotic, heavy-walled spores called **teliospores.** Although there is no well-defined basidiocarp, the hyphal mass may appear as a large, cream-colored tissue in the host (Figure 8.4*c*, p. 158). As the teliospores are produced, the mass is converted to a deep brown or black, powdery, dusty spore group that is easily disseminated by the wind. The nuclei within the teliospores fuse, providing the only diploid phase in the cycle. Although there is much variation in the pattern, the teliospore germinates with the production of a basidium which may be septate or aseptate. The basidia eventually produce basidiospores that can invade a new host plant and produce a haploid primary mycelium. Unless dikaryotization occurs, however, the primary mycelium does not become very extensive.

The rusts (Uredinales) are, in many ways, quite similar to the smuts. They, too, are obligate parasites (Figure 8.4*d*, p. 158) and very few have been cultivated successfully in the laboratory. Their vegetative mycelium grows between the cells of the host and may derive nourishment directly from the host's protoplasm although haustoria are usually produced. However, the life histories of the rusts are much more complicated than those of the smuts. The rusts regularly produce several spore forms (**polymorphic**), and may require two hosts (**heteroecious**) for the completion of their life history. They exhibit a great deal of parasite specialization, and some will attack only particular strains of a host plant (e.g., wheat rust).

REPRODUCTION

Although the heterobasidiomycetes do not produce specialized gametangia (except the rusts), all of the evolutionary advantages of sexual reproduction (i.e., genetic recombination) are present because of karyogamy between the two members of the dikaryon. The two nuclei that compose the dikaryon originally came from compatible strains of the fungus. Dikaryotization occurs by one of the methods mentioned in the introduction to this chapter, but in any case new genetic possibilities can result. The specialized gametangia found in the rusts will be discussed during the discussion of the rust life history.

The heterobasidiomycetes may produce asexual spores of various kinds. **Oidia** are asexual spores that are formed by fragmentation of a haploid or dikaryotic hyphal system in the jelly fungi. If haploid, they may dikaryotize a primary mycelium or they may germinate forming a new haploid or dikaryotic hypha. Other jelly fungi produce spores by budding.

The smuts commonly produce both uninucleate and binucleate

conidia from uninucleate and binucleate hyphal systems. Basidiospores and conidia bud repeatedly; all of these asexual spores can germinate producing a new mycelium.

The rusts usually produce asexual spores as part of a regular sequence in their life histories. The **uredospores** and **teliospores** can be considered asexual because they are derived from a dikaryotic hyphal system and their production is not preceded by any new nuclear phenomenon. The **aeciospore** and **basidiospore** may be considered sexual because dikaryotization occurs immediately previous to the production of the former, and karyogamy occurs immediately previous to the production of the latter. This will be more completely explained during the discussion of the life history.

LIFE HISTORIES

Exidia glandulosa (Figure 8.5) is a jelly fungus commonly found on decaying twigs and logs. The dikaryotic mycelium is saprobic in the substrate but eventually broader hyphae project beyond the substrate's surface and produce a cushion-shaped basidiocarp composed of intertwined hyphae. The entire fructification has a jelly-like consistency.

Some of the hyphal tips just below the surface of the basidiocarp enlarge and develop into basidia. The two nuclei in the cell fuse, forming the only diploid phase of the cycle. Karyogamy is followed by meiosis. The resulting haploid nuclei are separated by the formation of vertical septa that divide the basidium into four elongated portions. A septum forms at the base, and an epibasidium with a sterigma is then produced by each of the cells. After the epibasidium reaches the surface of the basidiocarp, the nucleus migrates from the hypobasidium (the basal portion of the basidium) through the epibasidium and sterigma, and a basidiospore is produced.

After violent discharge from the sterigma, the basidiospores may germinate by means of primary mycelia, conidia, or secondary basidiospores. The dikaryon is reestablished by somatogamy between two haploid hyphae. In some of the jelly fungi, the basidium becomes transversely septate and the basidiospores are produced from each cell of the basidium. In others, the basidium remains entire and the basidiospores are produced in a cluster at the tip.

Ustilago maydis (corn smut), a member of the Ustilaginaceae, is a common and important species (Figures 8.4*c*, p. 158 and 8.6). The teliospore is a single-celled, thick-walled, spiny, black, binucleate spore

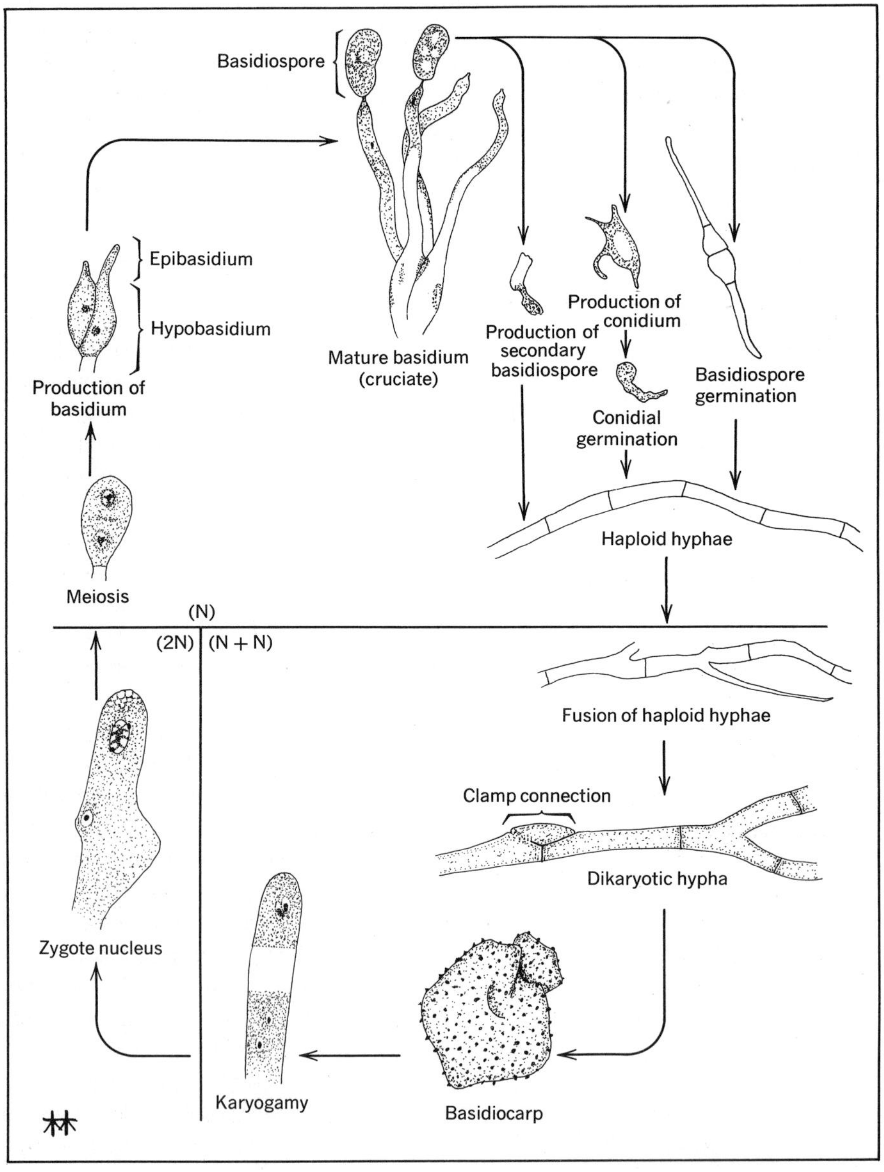

Figure 8.5
Life history of *Exidia glandulosa*. (Redrawn from C. J. Alexopoulous, *Introductory Mycology*, New York: Wiley.)

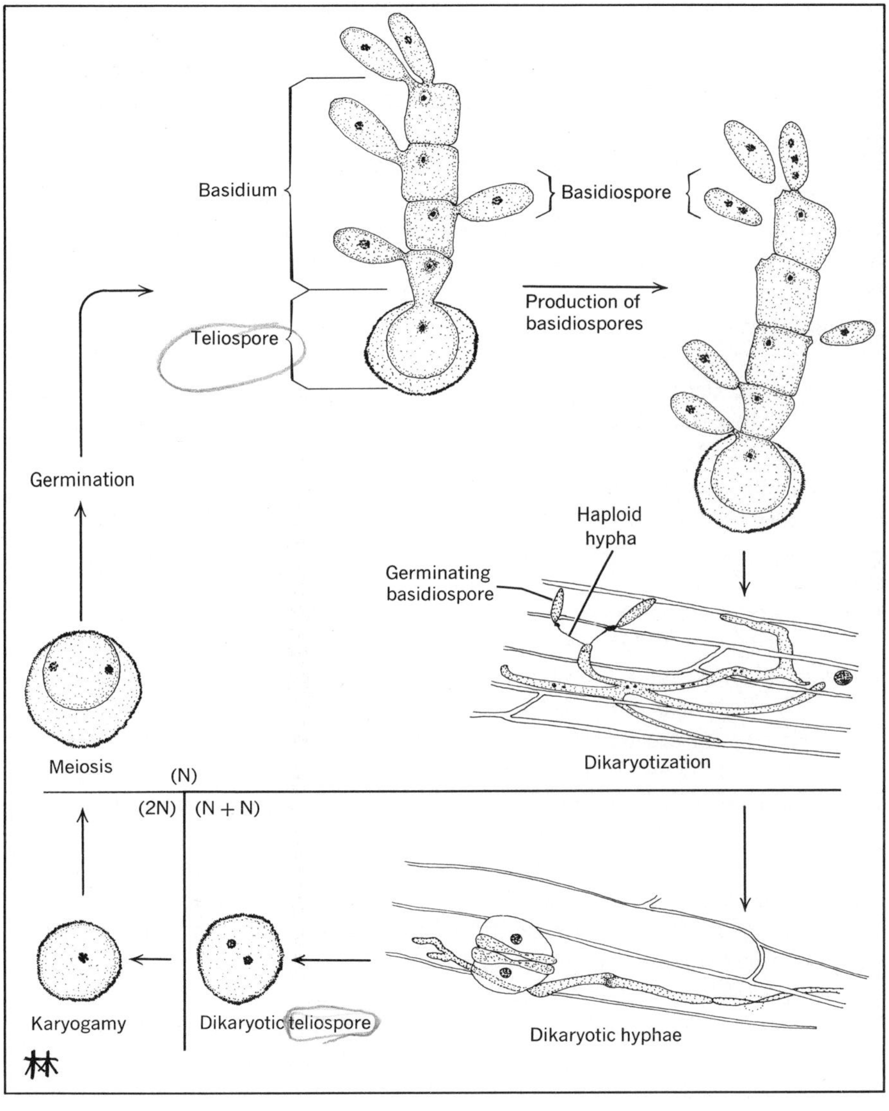

Figure 8.6
Life history of *Ustilago maydis*. (Redrawn from C. J. Alexopoulous, *Introductory Mycology*, New York: Wiley.)

that is passively released from a large gall produced on the corn plant. The teliospore may overwinter in the soil debris, and the soil may remain contaminated for several (10 to 15) years. The teliospore germinates with the production of a basidium (sometimes called a promycelium). The two nuclei fuse, producing the only diploid phase of the life history. Meiosis occurs either in the teliospore or after the diploid nucleus has migrated into the young basidium. In either case, four haploid nuclei are eventually distributed through the basidium. Transverse septa then divide the basidium into four uninucleate cells. After mitotic duplication of the haploid nuclei, one of the resulting daughter nuclei migrates to a developing basidiospore at the side of the basidial cell; the other remains in the cell. The basidiospore is produced by budding from the parent basidial cell. Another crop of basidiospores from the same basidium may be formed by the same method, and the basidiospores themselves may also bud forming a yeastlike colony.

The basidiospores reach a corn plant and produce a haploid (primary) hypha that may invade any meristematic area, particularly the floral structures. When two primary hyphae of opposite strains (mating types) meet, dissolution of the wall between them allows somatogamy and establishment of a dikaryotic mycelium. The dikaryotic mycelium establishes the nutritive relationship with the host plant and causes both enlargement (**hypertrophy**) and multiplication (**hyperplasia**) of the host's cells. This ultimately produces a greatly enlarged mass of parasite and host tissue (the **gall**). Finally the protoplasts of the cells of the dikaryotic mycelium round up and secrete heavy walls, producing new teliospores.

Although the smuts contain no specialized sex organs, they do retain sexuality in the form of the somatogamy that occurs between compatible hyphal strains. This is probably the most common method of reestablishing the dikaryon but other methods are known to exist. Certain environmental conditions may stimulate any two compatible cells to fuse (e.g., two basidiospores or buds from them may fuse or basidiospores or their buds may fuse with haploid hyphae) if they are in the proper proximity.

The life histories of some of the rusts are among the most complicated known. Some obligately require two different hosts (**heteroecious**) while others require only one host (**autoecious**). Many have four or five distinct stages (**full-cycle; macrocyclic rusts**) to their life cycle while others possess only three stages (**short-cycle; microcyclic rusts**). If a student understands the life history of a full-cycle rust, an understanding of the others offers only a slight challenge.

Although most general botany texts contain a discussion of *Puc-*

cinia graminis, it will again be discussed here (Figures 8.7 and 8.8) because it is illustrative of a full-cycle heteroecious rust, and is an extremely destructive plant parasite.

P. graminis attacks many of man's important cereal grains such as barley, rye, oats, and wheat. However, varieties of this species demonstrate specialization in that they will attack only certain species of the cereal hosts. Each variety of rust may be composed of physiological races that exhibit further specialization by parasitizing only specific races of a particular host. For example, *P. graminis* var. *tritici* parasitizes only wheat of the grain hosts, but this variety is composed of over 200 races that can parasitize various wheat varieties to a greater or lesser degree. All of the varieties and races of *P. graminis* have the American barberry (*Berberis* spp.) as their alternate host, and the fungus could as aptly be named barberry rust. However, man is much more interested in wheat than barberry; therefore he names the fungus after the host that is most important to him. The student should be aware that both hosts are necessary, and not consider them in terms of primary and secondary but simply as alternate. Man's conceit has placed the value judgment!

The terminology that has been used to describe rust life histories demonstrates more variety than the rust itself. It also shows the problems man has encountered while trying to interpret the life history of this important group of fungi. In an attempt to be consistent with the terms used for the other basidiomycetes, and in a try for simplicity, only one system of terminology will be used in this discussion. Hopefully, the student will be able to interpret other treatments when he completely understands this one.

The diagnostic spore of *P. graminis* is the two-celled, thick-walled, stalked, brown to black teliospore (Figures 8.7 and 8.8*d*). The teliospore overwinters in the wheat stubble and debris left in the field. Each cell of the teliospore is dikaryotic during its early formation but the only diploid phase in the cycle is produced by karyogamy before the teliospore overwinters. In the spring, the teliospore germinates by means of a basidium from each cell. Meiosis soon follows resulting in four haploid nuclei in the basidium. Transverse septa divide the **basidium** into four uninucleate cells. **Sterigmata** are produced on each of the cells, and the nucleus with a bit of cytoplasm squeezes through it forming a **basidiospore.**

The basidiospore is violently discharged from the basidium. It is thin-walled, and probably the most vulnerable spore in the cycle.

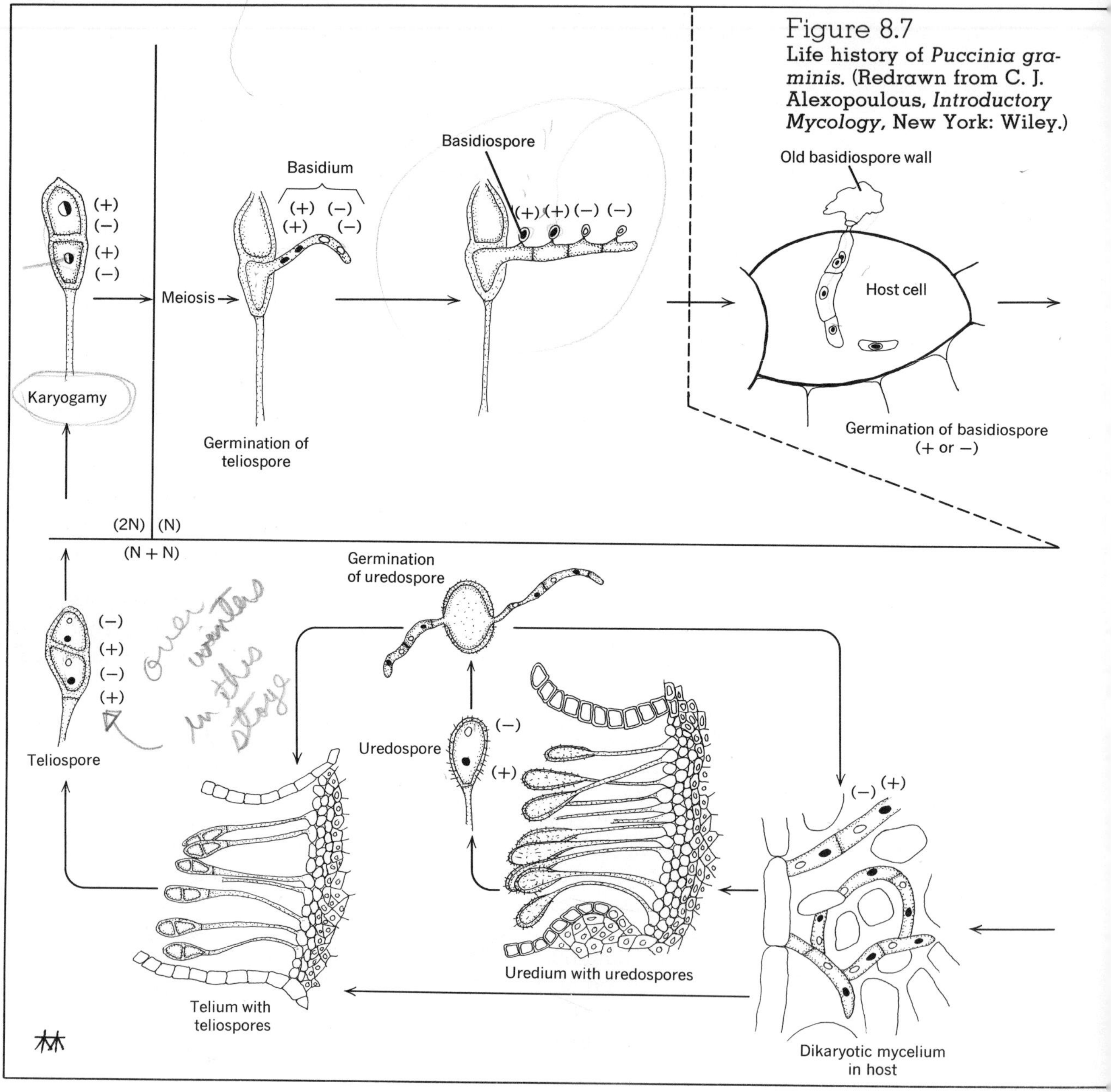

Figure 8.7
Life history of *Puccinia graminis.* (Redrawn from C. J. Alexopoulous, *Introductory Mycology,* New York: Wiley.)

the club fungi—
basidiomycetes

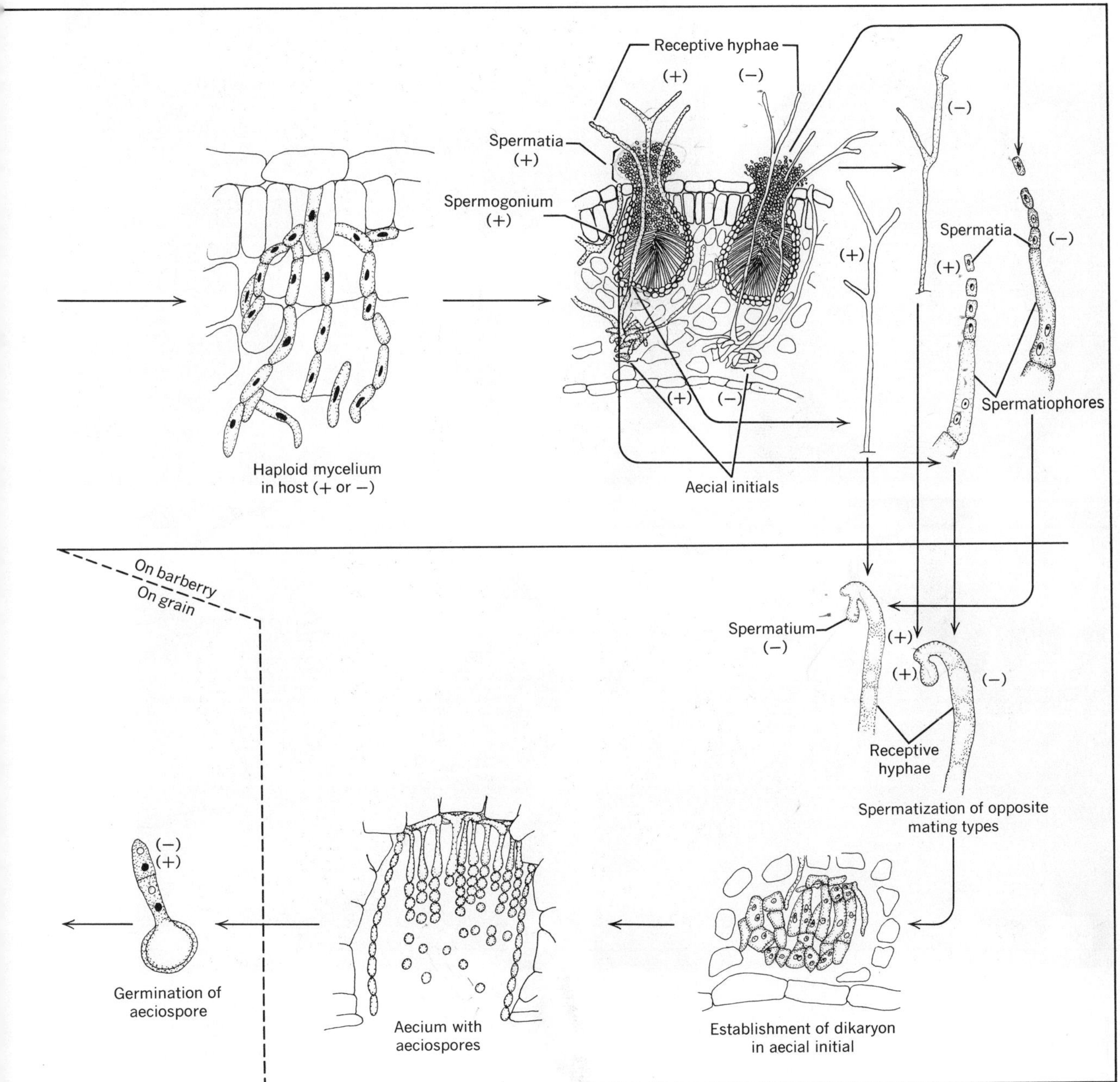

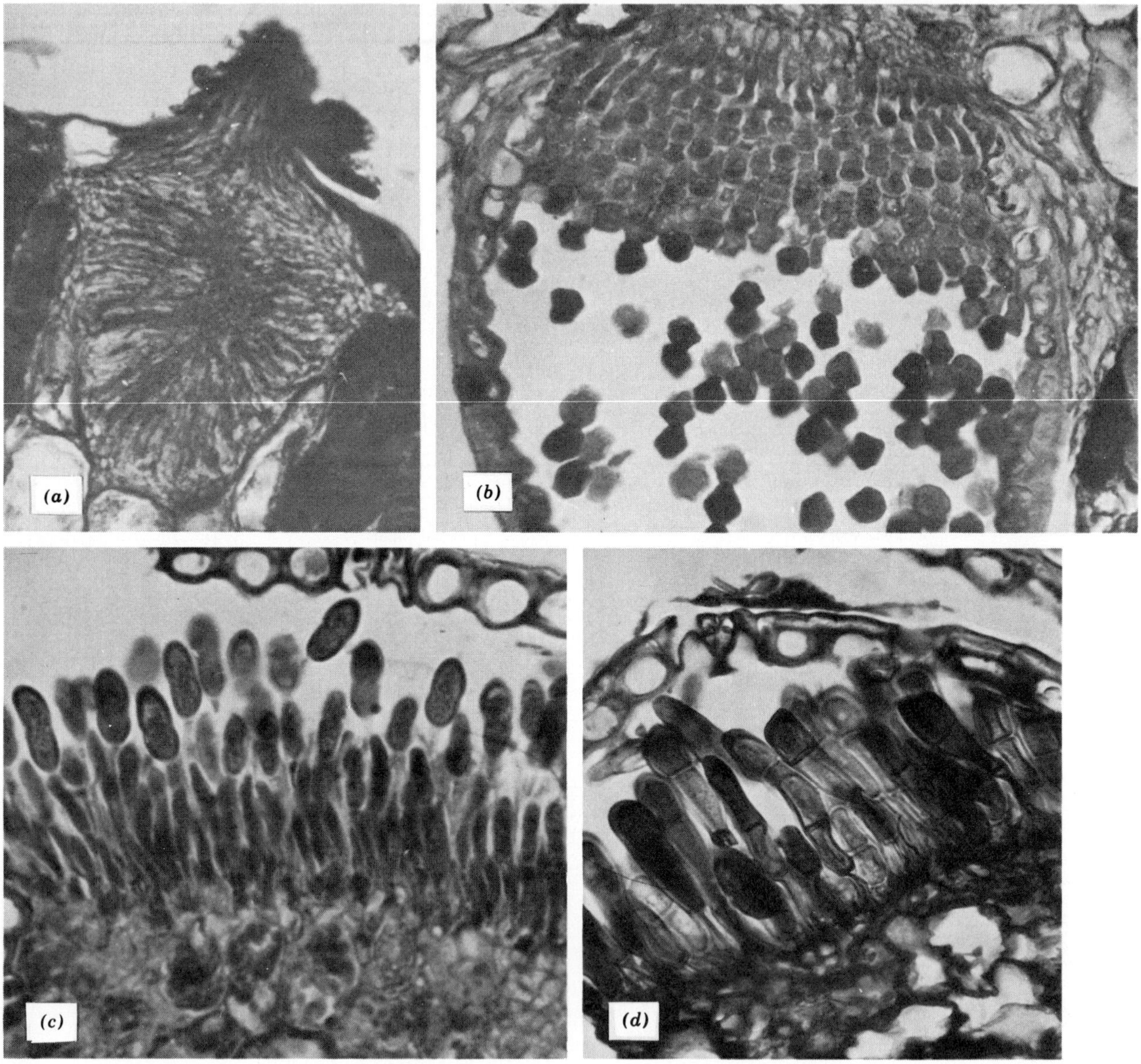

Figure 8.8
Photomicrographs of four stages of the life history of *Puccinia graminis*. (*a*) Spermagonium with spermatia (1800×). (*b*) Aecium with aeciospores (1800×). (*c*) Uredium with uredospores (1500×). (*d*) Telium with teliospores (1600×).

The basidiospores must reach a suitable host within approximately a mile or they lose their viability. However, if they reach a leaf of the American barberry, they germinate with the production of a germ tube that penetrates directly into the host. The haploid mycelium is produced intercellularly and derives nourishment from the host's cytoplasm. **Spermagonia** (Figures 8.7 and 8.8*a*) and **aecial initials** (Figure 8.7) are produced by this mycelium.

The spermagonium is a very small, flask-shaped structure. The neck extends to the outside of the upper epidermis of the host, and the inner wall is covered with **spermatiophores** that produce hundreds of uninucleate **spermatia.** The aecial initials are usually produced in the lower mesophyll and are composed of hyphal coils from which extend receptive hyphae (**trichogynes**). The receptive hyphae usually extend up through the spermagonia, although they may emerge independently in some rusts. Unless spermatization of the receptive hyphae occurs, the haploid mycelium will not live very long.

The spermatia are released in small drops of sweet nectar through the opening (**ostiole**) of the spermagonia neck. Insects that are attracted to the nectar incidentally and accidentally pick up the spermatia. They then move on to another spermagonium and passively transfer the spermatium to a receptive hypha. If the spermatium and the receptive hyphae are of opposite mating types, dissolution of the walls between them allow the spermatial nucleus to migrate into the receptive hypha (Figure 8.7). The spermatial nuclei then migrate down the receptive hyphae and dikaryotize the cells of the aecial initial. Dikaryotization can also occur by somatogamy if two opposite strains of haploid hyphae meet in the tissue of the leaf.

After dikaryotization, the aecial initial produces **aecia** (Figures 8.7 and 8.8*b*). These structures look like upside-down cups that extend out of the lower epidermis. They are surrounded by a haploid wall of hyphae (**peridium**), but the hyphae that produce the aeciospores are dikaryotic. The dikaryotic aeciospores are thick-walled, orange-yellow spores that retain their viability for some time. They can survive travel over several miles and are capable of parasitizing young wheat plants.

The aeciospore germinates by producing a dikaryotic hypha that develops into an extensive mycelium within the host's leaves or stem after penetrating through the stomates. Haustoria are formed with the host's cells and the host is severely debilitated. This mycelium produces a layer of hyphae that cut off dikaryotic **uredospores** under the epidermis (Figures 8.7 and 8.8c). The uredospore is one-celled, oval, spiny, and yellowish to reddish. Eventually the epidermis covering the spore producing area breaks open, and an elongate, reddish lesion, the **uredium** is formed. A heavily infected field of wheat appears yellowish-red and this gives the disease the common name of red stem rust. The uredospores are picked up and carried by the wind. If they reach another wheat plant, they can parasitize it. This is the only spore in the life history that can reinfect the wheat plant.

As the growing season continues into mid- and late summer, the dikaryotic mycelium in the wheat plant begins to produce thick-walled, two-celled teliospores in the same lesions that previously produced the uredospores. These brown to black spores give the name black stem rust to this stage, and the lesion is now called a **telium.**

The contemplation of the multifarious interactions between genetic and environmental forces that would allow such a complicated life history to so successfully evolve is enough to "boggle" the mind. The specialization of the various spore forms to particular hosts as well as the specialization of the many varieties and races of the species makes one question the biological axiom that specialization often leads to extinction. There are many exceptions to the axiom however, (as there are to all biological "facts") and apparently the rusts have survived because of their extremely great genetic plasticity. Not only do they possess a sexual apparatus that allows for all the evolutionary advantages of such a system but they possess several "shunt" mechanisms that also provide the advantage of genetic recombination. Examples of this are apparent in the ability of haploid mycelia to produce dikaryotic mycelia by means of somatogamy. In addition, however, it has been adequately demonstrated that somatogamy between two dikaryotic mycelia allows for the exchange of nuclei. This, of course, presents another method of genetic variability.

Although the mutation rate is extremely small, the fantastic number of spores produced allows for the production of many mutations. If even a small number of these mutations are selected for, much genetic change is introduced within the system. Because of the widespread dissemination of the spores over large geographic areas, many different environmental forces that induce mutation have the opportunity to exert an influence. The maintenance of the nuclear material for long periods in the dikaryotic condition allows for exchange of haploid nuclei for a much longer time than is usually the case. Because of all these factors, the plant breeder can stay only a few new strains ahead of the new strains of fungi that can attack the hosts.

A barberry eradication program was once thought to be the answer to the problem of control of wheat rust. While such a program did reduce the amount of wheat rust, and did practically eliminate the sexual cycle of the parasite, the rust did not disappear. Overwintering of the dikaryotic mycelium in warmer climates and the ability of the uredospore to travel over relatively long distances, somatogamy of dikaryotic mycelia, infection of winter wheat, and almost continuous wheat fields from north to south and east to west in the Western Hemisphere keep the fungus from becoming extinct.

HUMAN RELEVANCE

The heterobasidiomycetes contain some of the fungi that have been extremely important to man throughout his history. This importance has continued into the present time.

The jelly fungi are mainly of interest to mycologists who are concerned with fungal relationships. There are very few of them that come directly into conflict with man or that are of much use to man. Although the Chinese are known to eat some of the members of the genus *Tremella* (Figure 8.4*b*, p. 158), the gelatinous consistency of the fungus is considered unsavory by most others.

Some of the jelly fungi (e.g., *Septobasidium; Uredinella*) form parasitic relationships with some of the scale insects; others (e.g., *Eocronartium muscicola*) are parasitic on moss gametophytes. At least two genera are parasitic on vascular plants. *Herpobasidium lonicera* is parasitic on honeysuckles and another species of the same genus is parasitic on ferns. Members of the genus *Helicobasidium* are parasitic on the roots, crown, and stems of some trees. These relationships do not cause man any great concern, however, and most of the group is entirely saprobic. Their greatest importance, then, is their ability to reduce complex organic compounds. The other two groups of heterobasidiomycetes, the smuts and rusts, are the ones to which we must pay considerable attention.

Ustilago maydis (Figures 8.4*c*, p. 158 and 8.6) is a smut that is commonly found in almost any cornfield. Although it usually does not cause the tremendous losses that are produced by other smuts, there is a definite reduction in crop yield when it is present. Infections vary from very light to an occasional estimate of 60 to 66% loss of crop. The fields that have the higher estimates, however, have been used to grow corn for many successive years, and the incidence of corn smut has increased during the years. Rotation of crops aids in the reduction of loss caused by the smut.

The "covered" smuts or "bunts" are characterized by the fact that they invade the young plant during the development of the seedling and become systemic. They are commonly found on cereal grains, and the presence of the smut is not known until the heads of grain are examined. It is then found that the kernel has been replaced by smut spores and all that remains of the original grain is the outside covering. This replacement, of course, causes a great reduction in crop yield and value. Also, because an infected grain might not be apparent, grain growers are quite likely to sow infected grains at the same time of sowing healthy ones, insuring a new infection for the next crop. The fungus spores are also dispersed during threshing operations when the infected grains are broken open and the spores produce dense clouds that land on healthy grains and on the soil The spores may then be planted with the healthy grain. These spore clouds have been dense enough to clog machinery and to cause flash fires when ignited by sparks from machinery in the field. When conditions are favorable, the fungus spores germinate simultaneously with the grains and the mycelium invades the young plant.

One of the most destructive diseases of wheat (variously called bunt, stinking smut, or covered smut) is caused by a fungus whose effects have been noted since ancient times. The disease was not seriously investigated, however, until 1750 when the Academy of Arts and Sciences at Bordeaux, France announced that a prize would be given for the best investigation into the smutting of wheat. A Keeper of the Mint and part-time hobbiest farmer, Monsieur Mathieu Tillet entered the contest and elucidated the mechanisms by which the wheat becomes infected as well as some suggestions for control methods. He recognized that the grains were already infected when they were planted (by spores that had landed on them during harvesting operations) and postulated that soaking the seeds in something that would kill the spores would reduce the subsequent infection in the fields. He suggested soaking the seeds in everything from putrified horse urine to various types of manures. Although not eliminated, the amount of infection was reduced. His experiments were simple, direct, only partially effective, but he won the prize!

More refined experiments on wheat bunt were reported by a Swiss worker, Benedict Prevost, in 1807. He was the first worker to report that traces of copper in the water used to soak the wheat seeds kept the fungus spores from germinating. It was Tillet, however, who was honored by having his name become the generic name of the causative organism of the disease. The disease relationship is caused by *Tilletia foetida* (so specifically named because of the smell of rotting fish caused by the production of trimethyl amine) or *T. caries* (so named because of the French designation given to it by Tillet, *la carie*). Much more complete accounts of the intriguing work of these two men can be found in Christensen, 1951, and Carefoot and Sprott, 1967. M. Tillet also observed that some cereal grains were not affected by the bunt, but were attacked by a smut that caused the loss of all the grain from a cereal head, leaving only the central shaft or rachis. He called the disease relationship *le charbon;* it has now become known as "loose smut."

The loose smuts differ from the covered smuts in several ways. The teliospores of the smut fall on the open flowers of the wheat and germinate. The mycelium invades the ovary of the flower and develops simultaneously with the developing wheat seed. At the time of harvest, the infected wheat seeds appear the same as the uninfected ones. The mycelium remains dormant in the seed over the winter and is thus planted with the seed.

When the seed germinates, the embryonic plant is invaded with a systemic infection of the fungus mycelium. The mycelium continues upward with the growing tip of the host plant, and sporulates in the flowering head (**inflorescence**). The entire inflorescence (except the **rachis**) is then converted into the black teliospore mass characteristic of the disease. Flowering heads of infected cereal grains mature about two weeks previous to healthy ones. Thus there is a ready source of inoculum for the young, healthy flowers.

Because the mycelium is inside of the seed, external applications of fungicides are useless as control measures. It has been found, however, that soaking the seeds in hot water before planting will kill the mycelium. There is a delicate balance of the temperature and length of soaking time between that which will kill the mycelium and that which will kill the embryonic plant within the seed. Therefore, the hot water method of treatment has been largely abandoned for the safer course of breeding resistance into the host plant. This type of prevention is, of course, subject to the same weakness of other inbred resistances, namely that the fungus is likely to genetically adapt to the new strain. New strains, then, must be constantly developed by the plant breeders.

Many different species of *Ustilago* cause loose smuts of grains, but one of the most common is *U. tritici* that causes loose smut of wheat. *U. nuda* and *U. avenae* also cause the disease on various cereal grains. Each of these fungi has several strains, allowing for invasion of many kinds of cereal grains.

It is now proper to turn our attention to the final group of heterobasidiomycetes, the rusts. This group of fungi certainly deserves to be placed among the most important plant groups with which man has to deal, because man and rusts have long been in competition for the same food supply.

Wheat rust (Figures 8.7 and 8.8) has been one of the most destructive parasites on this important crop. Other important cereal grains are also subject to infection by various rusts, as indeed are almost all vascular plants. Man's battle with the rusts of the cereal grains has continued at least as long as the grains have been cultivated.

The influence of wheat rust on man's history and culture is a long and intriguing detective story. It is impossible to trace the entire effect in a treatment such as this, but a brief discussion will surely lead students to more complete treatments.

Wheat probably originated in the "Fertile Crescent" of Asia Minor, and wild wheat and barley are still to be found in the mountains around southern Palestine, Lebanon, Syria, and northern Iran and Iraq to the Persian Gulf. Apparently, a susceptible barberry also originated in the same area. Thus were produced the pair of hosts that were to be so important to man. Like most fungal origins, the origin of wheat rust is lost in antiquity. One can only speculate how the fungus appeared in the "Fertile Crescent," but the way in which it developed the genotype necessary to alternate between wheat and barberry is almost impossible to fathom. Indeed, after such an obligate arrangement developed, survival of so rigid a genetic system would certainly seem doomed to failure. Suffice it to say, without any attempts at explanation of mechanism (other than those used in the discussion of the life history) that the fungus-barberry-wheat symbiosis has existed for thousands of years and is still eminently successful.

Eighteen hundred years before Christ, as recorded in the Bible, Joseph, after he had been sold into Egyptian slavery by his jealous brothers, was called upon to interpret Pharaoh's dream of seven good years and seven lean years. Joseph was able to interpret the dream to mean seven years of good crops followed by seven years of famine. Whether one wishes to believe the Biblical account or not, there is good evidence from other sources to believe that famine did occur in Egypt during that particular historical period. Climatologists are also fairly certain that weather conditions were right for the widespread rust infection.

Because of his interpretation of the dreams, Joseph became extremely powerful in Egypt. His father, Jacob, sent to Egypt for grain when the rusts created the famine Joseph had predicted. Eventually, Joseph brought the Israelites to Egypt, where they lived for 400 years. At the beginning of this time, they lived as a respected, free people but toward the end they were forced into slavery. Moses led the Israelites out of Egypt into the desert where they spent 40 years. Thus, the migrations of the Hebrew people had their start as the result of famine caused by wheat rust. It is interesting to speculate, considering the present Middle East situation, upon the effect of the wheat rust that ravished the land of Palestine so many thousands of years ago. Such a simplistic explanation is not meant to suffice to explain all of the ills of that troubled part of the world, but logic seems to require that consideration of their basic causes take this part of history into account.

The Romans wrote about wheat rust as early as 700 B.C., and had been troubled by crop failures caused by it long before that time. One of their legends blamed the origin of wheat rust on a 12-year old boy. According to the legend, the boy caught a fox in his father's chicken house. The boy then tied a straw to the fox's tail, set the straw on fire, and released the animal. The god Robigus (Rubigo) was offended by such wanton sadism and thereafter punished the people of the land with the terrible disease of their principal cereal grain. The disease would cause the wheat to have the reddish-brown coloration as though it had been burned by fire as a constant reminder of the cruelty of the boy.

It was in about 700 B.C. that the Robigalia was founded. This festival was designed to propitiate the god. A yellow dog was sacrificed on April 25 every year until the Roman Empire began to decline. Certainly, then, wheat rust had a definite cultural effect on ancient Rome, and probably contributed significantly to its decline.

These two examples illustrate the effect of wheat rust on two ancient civilizations. There is no real way of knowing what reverberations we feel in our culture from these ancient happenings. We can, however, have a look at the modern effect of wheat rust on our economy.

Wheat rust created economic disaster in 1916 when it spread throughout the United States and Canada. However, even worse disaster struck in 1935. This was during the "great depression" in the United States, and was the final economic blow for many farmers. Previous poor growing conditions (drought, wind storms, hot winds) had already devastated their bank accounts and energies. Now came an epiphytotic on the wheat that cost them 100 million bushels or more than $50 million in spring wheat alone. Another 35 million bushels of winter wheat were lost. Railroads that were expecting to transport the wheat lost their revenues, and the loss of trade and income was felt throughout the country. The economy of the United States was already under much stress, and such losses could not be absorbed. Many farmers went bankrupt.

As if to administer the *coup de grace*, another wheat rust epiphytotic swept through the wheat belt of the U.S. in 1937 and in 1938 the loss in the U.S. was again about 100 million bushels. These figures are only a partial estimation of loss due to wheat rust in the U.S. and do not take into consideration the losses suffered by the rest of the world. Russia, Canada, Africa, England, Italy, and Sweden, to name only a few, are also countries that suffered severe losses. When the present world losses resulting from wheat rust are compiled, the total bill comes to an estimated staggering $500 million per year.

Why, then, if man's cultural and economic fates have been so molded by this one plant disease, hasn't man done anything to stop it? Many attempts have been made to eradicate the barberry bushes which are the alternate hosts of all of the strains of *Puccinia graminis*. The American barberry is a tenacious plant, and is not easily wiped out. It can grow in extremely remote areas and has a wide range over the Western Hemisphere. To completely eliminate it would be a task of monstrous proportions. Although barberry eradication programs have reduced the incidence of wheat rust in certain areas, they have not been entirely successful. Also, in this day of species elimination because of environmental pollution, one doubts the advisability of deliberately setting out to send still another species into the void of extinction. Many of the spore forms (see the life history) are capable of being transported by the wind and the uredospores (the only ones that can reinfect the wheat) often arrive in a field just as new young wheat plants are springing forth. This inoculum is usually provided by a more mature wheat field that is growing in a climatic area that allows for the earlier season. Thus, the rust can move from north to south and south to north throughout the Western Hemisphere by following the seasonal changes.

Chemical control has been attempted by the use of various compounds. Although limited success has been realized by using these methods, the cost of application is often prohibitive. The partial failure of each of these methods of control leads us, then, to the consideration of breeding rust-resistant strains of wheat. Such breeding programs have been successful against many of the fungi that are pathogenic to man's food plants. The rusts (including many others in addition to *P. graminis*) have such extremely great genetic plasticity because of mutations and genetic recombinations that new strains of rust evolve almost as fast as the plant breeders develop new, resistant strains of host cereals. That is, the genetic capability to infect a particular cereal strain may exist in the rust gene pool long before the host has been developed. That particular genetic character may be selected for by the development of the host, and a new genetic cereal strain then becomes obsolete. Plant breeders often consider themselves fortunate if a new cereal strain remains usable for three years before the rust strains become adapted.

Although the reader should be duly impressed by now with the importance of cereal rusts to human affairs, he should be aware that other rusts which are parasitic on vascular plants are also important.

The rust (*Hemileia vastatrix*) that is parasitic on the coffee tree created great social changes in England. Coffee is probably originally an Ethiopian plant, and its cultivation was jealously guarded by the Arabs who brought it to Yemen from Ethiopia around 1000 A.D. The roasted beans were exported by the Arabs, at a very dear price (e.g., one ounce of coffee for five pounds sterling in London), and the demand for the bean increased.

However, the Arabs were unaware that coffee had been introduced about 1600 A.D. to the hills of southern India. When the Dutch reached India, they removed some of the plants to Ceylon. From there the coffee-growing industry flourished and after about

1713 trees reached the Caribbean Islands and in South America from Brazil to Mexico.

Ceylon became a British colony in 1797 but, because of the wars with Napoleon Bonaparte, did not have the energies to maintain the coffee plantations. After Napoleon's defeat, the British, in 1825, planted new coffee trees in Ceylon. These trees flourished; by 1870 Ceylon was exporting 100 million pounds of coffee per year and had many of the characteristics of the western United States "boom" town. Because of land speculation and the huge sums of money being made from coffee, Ceylon developed a one-crop economy.

Coffee rust began to make its appearance sometime between 1850–1870, and by 1875 the disease had spread to disastrous proportions. By 1892 almost all the coffee trees had been wiped out of Ceylon.

The British are, if nothing else, an adaptable people. As they faced sure economic disaster with the coffee failure, they planted tea. Coffee had become a favorite drink throughout England during the height of the Ceylon years, but tea took its place in popularity when the coffee plantations were wiped out. Thus the drinking habits as well as the economic base changed for an entire nation because of a rust. Also, 90% of the world's coffee is now grown in the Western Hemisphere. The fungus has not yet reached the coffee plantations there, but it does exist on the west coast of Africa. If ever it is transported across the Atlantic to South America, *Hemileia vastatrix* could once again cause economic chaos.

Another important disease caused by a rust (*Cronartium ribicola*) is white pine blister rust (Figure 8.4*d*, p. 158). The rust is a full-cycle rust and is obligately parasitic on five-needle pines, especially *Pinus strobus*. Its alternate hosts are the common gooseberries and currents (*Ribes* spp.). The sexual part (spermagonia; aecial initials) of the life history occurs on the pine and the haploid mycelium may invade the needles, twigs, branches, or trunk of the pine tree. If the fungus girdles a stem, all of the host distal to the girdled area will die. The fungus can cause extremely great losses to the lumber industry. The fungus was probably introduced to the United States from Russia where it was first found in 1854. It was first found in the U.S. in 1906 in New York. A gooseberry eradication program and the breeding of resistant pines seem to be the best methods of control of the disease.

The list of rusts and their hosts is a seemingly endless one, and their relationships have had profound influences on man's culture and welfare. Any serious attempt to explain man's actions must take into consideration his food supply. Much of his food supply is directly affected by the action of fungi, particularly the smuts and rusts. Therefore, before one can really interpret the human organism, he should acquire a basic appreciation of the fungi.

SUBCLASS: HOMOBASIDIO-MYCETIDAE

This is the group of fungi with which most people come into contact because most of the fungi that are eaten directly (as opposed to those used during food preparation) are found here. Almost everyone is familiar with the mushroom that is commercially grown and available in supermarkets (Figure 8.9, p. 158), but many eaters of mushrooms (**mycophagists**) also prize highly many of the wild ones.

Although the mushrooms are certainly the best-known members of the homobasidiomycetes, there are many others of the subclass that are of equal interest if not of equal economic importance.

The homobasidiomycetes are often divided into two convenient groups, the **hymenomycetes** and the **gasteromycetes**. The first of these subgroups has its basidia exposed on an open layer (**hymenium**) before the basidiospores mature (Figures 8.10 and 8.20, p. 158); the mature basidiospores are forcibly discharged from the basidium.

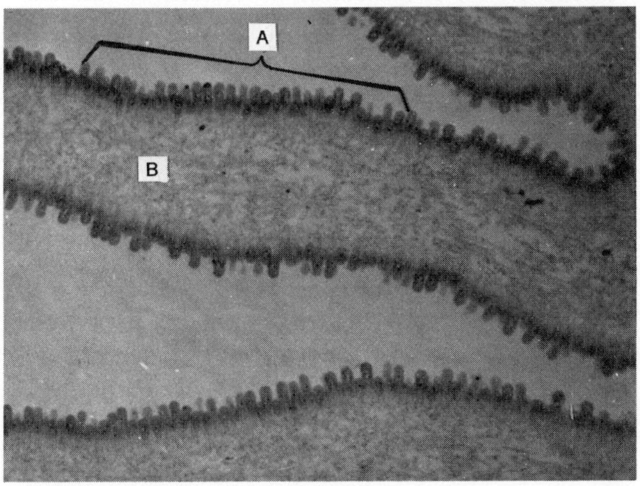

Figure 8.10
Photomicrograph of cross section of gill (lamella) of an hymenomycete. (*a*) The hymenium composed of young basidia. (*b*) The filamentous central portion of the gill (360×).

The second subgroup may or may not produce an hymenium, but the basidiocarp remains closed in either case until after the passive release of the basidiospores from the basidium (Figure 8.11, p. 158). The basidiospores reach the outside environment through the decay or mechanical breakage of the basidiocarp, although some members do have preformed openings through which the basidiospores escape.

The basidiocarps of the various families of the hymenomycetes vary greatly in their gross morphology, and it is upon these variations that the families were originally delimited. Definition of the families is now being done more and more on the basis of microscopic characteristics, and some members of the various groups have been shifted in their taxonomic positions. These taxonomic shifts do not, however, invalidate an introduction to the various types of basidiocarps.

The simplest of the hymenomycetes produce a basidiocarp that is a loose, cobwebby, and indefinite group of hyphae. The basidia are produced on the surface in an irregular fashion. Other basidiocarps found in the group are tough and leathery or hard and woody, with the basidia produced in a definite hymenium. These types of basidiocarps may be thin and crustose or rather thick. The basidiocarp may grow flat on its substrate with the hymenium exposed only on the free surface (**resupinate**), or it may be **stipitate** or **bracketlike.** The members that produce these basidiocarp types are collected in the family Thelephoraceae. Some of the common genera within the group are *Thelephora* and *Stereum* (Figure 8.12, p. 158).

The family Cantharellaceae contains members that produce stipitate, upright basidiocarps that produce their basidia on a smooth, folded hymenium that is located on the undersurface of an expanded **cap** or **pileus.** The folds or ridges often radiate from the stipe onto the pileus. The genus *Cantharellus* (Figure 8.13, p. 158) is the best known of these fungi. Some authors also place the members of the genus *Craterellus* (Figure 8.14, p. 158) here.

Many of the hymenomycetes produce club-shaped, branched, typically upright "fruiting" bodies that have a smooth hymenium that may extend over most of the surface or be restricted to some portion, usually terminal, of the basidiocarp. Because these "fruiting" bodies somewhat resemble coral, the group is commonly designated as the coral fungi (Clavariaceae). Many very beautiful fungi are found within this group. The colors of the basidiocarps may be strikingly clear white, or they may be brilliantly colored in orange, yellow, or violet. The intricate branching produced by some adds to their delicate attractiveness. The most common genus within this group is *Clavaria* (Figure 8.15, p. 158).

The toothed fungi (Hydnaceae) are those that produce their hymenium on spines or cone-shaped structures (the "teeth") that extend downward from crusts, caps, erect branches, or brackets. The "teeth" may be irregular in size and shape, or they may be even. The texture of the basidiocarp varies from leathery and fleshy to woody. Common genera within the group are *Hydnum* and *Hericium* (Figure 8.16, p. 158).

The common characteristic of the members of the Polyporaceae (Figure 8.17, p. 158) is that the hymenium lines the inner surface of pores or tubes. The basidiocarp is quite varied among the members of the group and may be resupinate (*Poria* sp.) or bracketlike (*Polyporus* sp.). The basidiocarp is usually tough, leathery, woody, papery, or corky, but some fleshy forms do occur (e.g., *Polyporus sulphureus*, Figure 8.18, p. 158). The largest brackets ("conks") produced by fungi occur in this group (e.g., *Fomes* sp.) and many of them are extremely hard and woody. Because of these brackets, the group is often commonly called the "shelf" fungi. Most of the fructifications are annual but some are perennial (*Fomes* sp.). Genera that are also in this family, other than those mentioned above, are *Ganoderma*, *Daedalia*, *Lenzites*, and *Irpex*.

All of the above families are presently placed in the order Polyporales by some authors, but the group consists of such a widely diverse collection of fungi that definition of the order is extremely difficult if not, indeed, impossible. The evolutionary relationships of the members within the group are also not very well understood. Therefore, some mycologists include these fungi in the order Agaricales by defining the Agaricales in the broadest possible sense. In either case, there are two remaining groups of hymenomycetes that should be discussed. They contain members that produce basidiocarps of two gross morphological types. Microscopic examination of the morphology of the basidiocarps is the basis for dividing them into various families (the number of families varies from five to more than sixteen, depending on the authority). The discussion here will be limited to examples that illustrate the gross macroscopic differences. Definition of families is best left to specialists studying this group.

The Agaricales have been defined as containing members that produce their **hymenia** on gills (**lamellae**), or as an inner lining of tubes. If the latter condition exists, the basidiocarp is fleshy. One can group them by referring to them as the boletes and the agarics, the boletes being the ones with hymenia lining tubes (Figure 8.19, p. 158), and the agarics being those that produce hymenia on gills (Figure 8.20, p. 158). Common examples of the boletes are found in the genera *Strobilomyces, Boletus* and *Gyrodon*. There are many common agaric genera, but good examples that are fairly easy to identify (at least to genus) are in the genera *Amanita, Coprinus, Agaricus, Russula, Lactarius,* and *Pleurotus*.

The other large group of homobasidiomycetes is the gasteromycetes, those members that produce their spores in closed basidiocarps until the spores are mature. The basidia are usually on an indefinite hymenial layer, and the spores are not violently shot from the basidium. The outer wall (peridium) of the basidiocarp may open naturally after the spores mature or it may remain closed until it decays or is broken open by an animal or some sort of environmental accident. The portion of the basidiocarp in which are produced the basidia is called the **gleba.** Several fascinating types are included in this group.

One of the largest groups of the gasteromycetes is the puffballs (Lycoperdales). These include the many types of puffballs (Lycoperdaceae) and most of the earth stars (Geastraceae). The peridium of the group is often composed of several layers of fungus tissue that surrounds an interwoven glebal area. Some of the puffballs open by a pore or split, but many of them release their spores only when the peridium has broken down because of external forces. The earth stars produce an outer peridium that peels back when mature, leaving an inner peridium that develops an apical pore. Some of the commonest members of the "puffballs" group are in the genera *Calvatia* (Figure 8.11, p. 158), *Lycoperdon* (Figure 8.21*a*, p. 158), and *Geastrum* (Figure 8.21*b*, p. 158).

The hard puffballs (Sclerodermatales) are a group that produces a hard, rindlike peridium that surrounds an extremely dark gleba. One group of earth stars that produce an extremely hard outer peridium are included in this order. Most of the common examples are in the genus *Scleroderma* (Figure 8.21*c*, p. 158). In these, the spores are released only by the decay of the peridium.

The stinkhorns (Phallales) are sometimes known as the earth's penis. The members of this group produce a membranous peridium that surrounds a moist glebal area. As the fungus matures, a long, spongy stalk pushes the glebal area through the peridium and several inches into the air. During this time, the gleba autodigests into a slimy, foetid mass that smells much like rotting flesh and thus gives the group its common name. The peridium remains at the base of the stalk as a **volva.** Spore dispersal is accomplished by carrion insects and flies that are attracted to the digesting gleba and that passively transport the spores to new environments. The most common genus is *Phallus* (Figure 8.22*a*, p. 158), but members of the genera *Dictyophora* and *Mutinus* (Figure 8.22*b*, p. 158) are also often found.

The final group of gasteromycetes that should be mentioned is the bird's nest fungi (Nidulariales). The members of this group produce mature fruiting bodies that contain a number of small, elongate, hard structures. When one looks into the fruiting body, he is reminded of eggs in a nest. These "eggs" (**peridioles**) actually represent the glebal chamber that has become divided into portions with each portion surrounded by a separate wall. Thus each peridiole contains basidia and basidiospores. The peridioles are ejected to a distance of 3 to 4 feet by raindrops that hit the "nests" which act as splash cups. A portion of the peridiole is very sticky and adheres to whatever it hits. Common genera of this group are *Crucibulum*, *Nidularia*, and *Cyathus* (Figure 8.23, p. 158).

REPRODUCTION

No specialized sex organs are produced by the homobasidiomycetes. The basidiospore germinates with the production of a primary hypha composed of uninucleate cells that contain haploid nuclei. Although some homobasidiomycetes are homothallic, the greatest majority of them are heterothallic. Therefore, two compatible primary hyphae come together, plasmogamy occurs, and a dikaryon that contains unlike, but compatible, nuclei is produced. Although this is a common method of establishing the dikaryon, other methods do exist. Some of these methods have been discussed in the introduction to this chapter, but additional mention should here be made that the primary mycelium probably has disappeared from the life history of some of the advanced forms. In these forms, a binucleate basidiospore is formed containing two unlike nuclei that migrate into the hypha produced by germination. Although plasmogamy does not occur in these forms, karyogamy followed by meiosis in the basidium provides for crossing-over. Four haploid nuclei are formed, but only two basidiospores per basidium are produced. Thus the dikaryon is established before the basidiospore leaves the basidium.

The **clamp connection** appears to be important in the maintenance of the dikaryon in successive cells of the growing hyphal system in some of the homobasidiomycetes. The clamp connection (Figure 8.2) provides a means whereby sister nuclei which are produced by simultaneous duplication of the two unlike nuclei of a dikaryotic cell are separated into two different cells. A short branch is produced on a dividing binucleate cell. One of the nuclei orients itself along the long axis of the branch. The other nucleus orients itself along the long axis of the cell and the two nuclei carry on simultaneous mitoses. The branch of the cell curves back and reconnects with the cell. A pore occurs at the point of contact. Septa form in such a manner as to separate the two daughter nuclei produced by each mitosis. However, the daughter cell produced at the end of the parent cell now contains one daughter nucleus from each of the parent nuclei. The daughter nucleus that was in the branch of the cell migrates through the bridge formed by the clamp and enters the other daughter cell of the old parent cell and since it already contained one of the daughter nuclei of the other parent nucleus, it is now dikaryotic. Thus is the dikaryotic condition maintained throughout the mycelium. The "fruiting body" (basidiocarp) arises from this dikaryotic mycelium.

As has been previously mentioned, asexually produced oidia may function either as asexual spores and produce primary hyphae, or they may serve as spermatizing agents. The homobasidiomycetes may also reproduce by other asexual means. Fragmentation of a mycelium is a common method of asexual reproduction (i.e., the "spawn" used by commercial mushroom growers), and some produce conidia. Conidial forms in the homobasidiomycetes are not nearly as diverse as those found in the ascomycetes and, consequently, most of them were never classified in the Fungi Imperfecti.

LIFE HISTORY

The homobasidiomycetes, for all their diversity of form of their basidiocarps, possess a remarkably similar life history (Figure 8.24). Most of the events in the history have already been discussed in the previous sections of this chapter but it still remains to place them in

Figure 8.24
Life history of a mushroom, a "typical" hymenomycete.

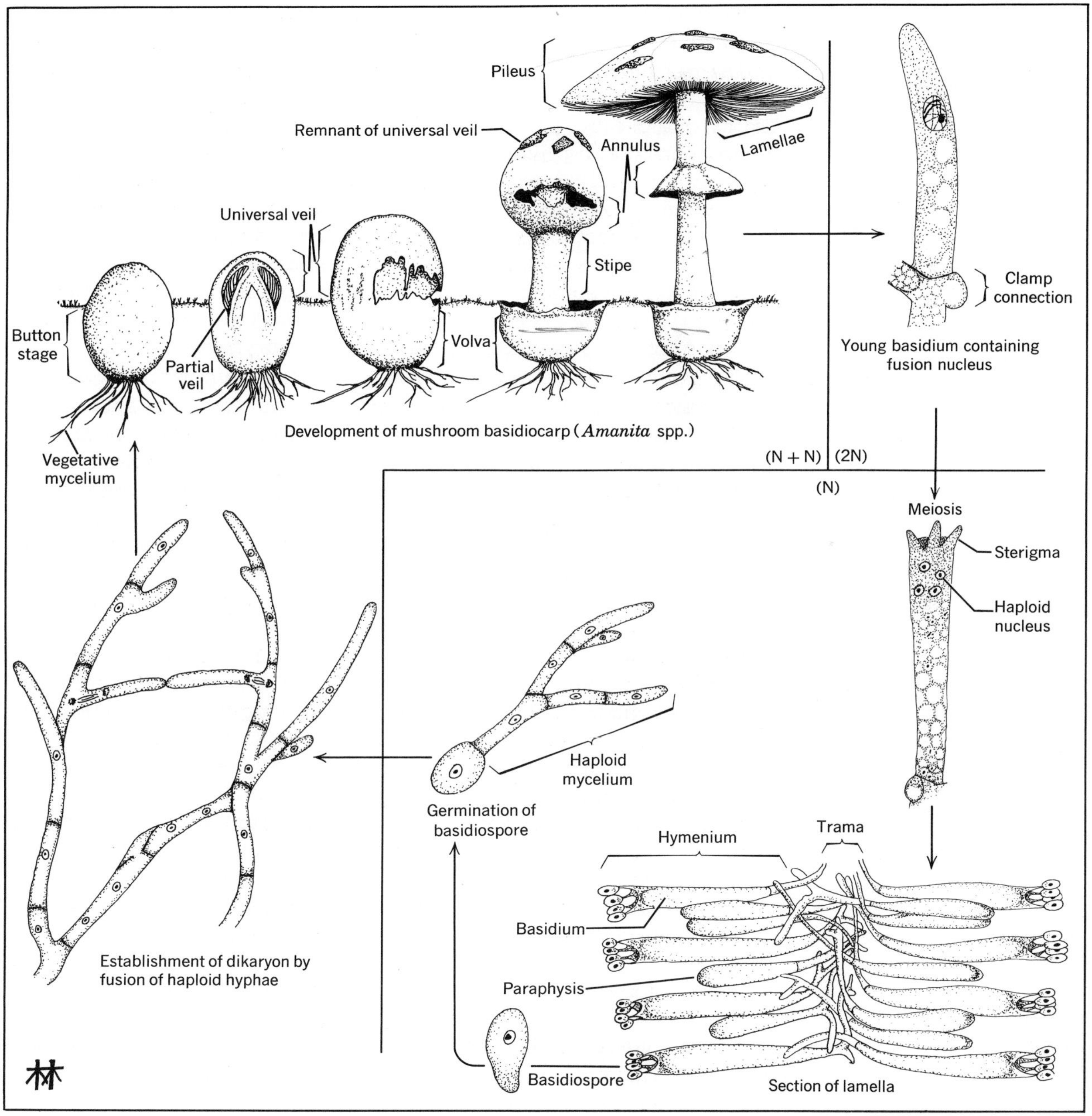

their proper relationships to each other and to add a few cogent details.

The germination of the haploid basidiospore produces a primary hypha. Nuclear exchange between two compatible primary hypha (or by one of the other methods) establishes the dikaryon. The dikaryotic secondary hyphal system (mycelium) grows throughout the substrate (usually down wood or soil). It is here that clamp connections seem to be important for some, but not all, of the homobasidiomycetes. The septal pores of the secondary hyphae are much more complex than the septal spores of the primary hyphae. Apparently there has been some evolutionary advantage of the clamp connection over the complex septal pore for the passage of nuclei from one cell to another in the maintenance of the dikaryotic condition.

The details of basidiocarp development vary somewhat among the various groups of homobasidiomycetes, but since the student is more likely to encounter and wish to identify mushrooms, the development of their basidiocarp will be used as the example here. Not all mushrooms will have the structures herein discussed, but the student should be able to understand other forms when he encounters them. The genus *Amanita* is good to use as an example, because it contains most of the structural possibilities.

The basidiocarp develops from the secondary mycelium when environmental conditions (particularly those of temperature and moisture) are favorable. The first evidence of a basidiocarp is the production of a tiny knot of hyphal cells that eventually differentiates into the so-called "button stage." A median longitudinal cut through the button (Figure 8.25 faces color inserts) reveals that the gills are already present. In some mushrooms, the edge of the cap (**pileus**) is attached to the stalk (**stipe**) by means of a membrane known as the **partial veil** (Figure 8.24 faces color inserts). The entire "button" may also be completely covered by a membranous tissue called the **universal veil** (Figures 8.24 and 8.25).

As the mushroom matures, the stipe (Figures 8.24 and 8.26 faces color inserts) increases in length and the pileus (Figures 8.24 and 8.26 faces color inserts) expands. These events cause the two veils to break. In many mushrooms the partial veil breaks in such a way that a ring (**annulus,** Figures 8.24 and 8.26 faces color inserts) remains around the stipe. The annulus may be very fragile and ephemeral, thus not easily observed, or it may be tougher and more persistent, thus easily seen. The universal veil breaks in such a way as to leave a cup (**volva,** Figures 8.24 and 8.26 faces color inserts) at the base of the stipe. The

Figure 7.3
Corn infected with *Helminthesporium maydis* causing Southern Corn Leaf Blight. (*a-b*) Lesions on "ear," leaf, and stem of corn plant, (*c*) Row of corn plants heavily infected with the fungus. (*d*) View of heavily infected cornfield.

(*a*)(*b*)

(*c*)

(*d*)

Figure 8.3
Rhizomorphs growing on a dead log.

Figure 8.4
Examples of heterobasidiomycetes. (*a*) *Exidia* sp. basidiocarp. (*b*) *Tremella reticulata* basidiocarp. (*c*) gall of *Ustilago maydis* on corn. (*d*) Aecia of *Cronartium ribicola* on white pine.

(*b*)

(*a*)

(*d*)

(*c*)

Figure 8.9
***Agaricus campestris,* the common commercial mushroom.**

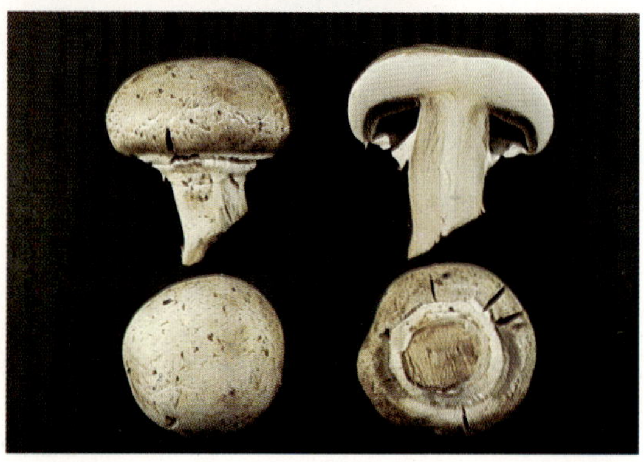

Figure 8.21
Examples of "puffballs." (*a*) *Lycoperdon pyriforme* (small puffball). Note the pore in the peridium. (*b*) *Geastrum* sp. (earthstar). Note the outer peridium that has split and peeled back. (*c*) *Scleroderma lycoperdoides* (hard puffball). Note the dark gleba in the longitudinal sections.

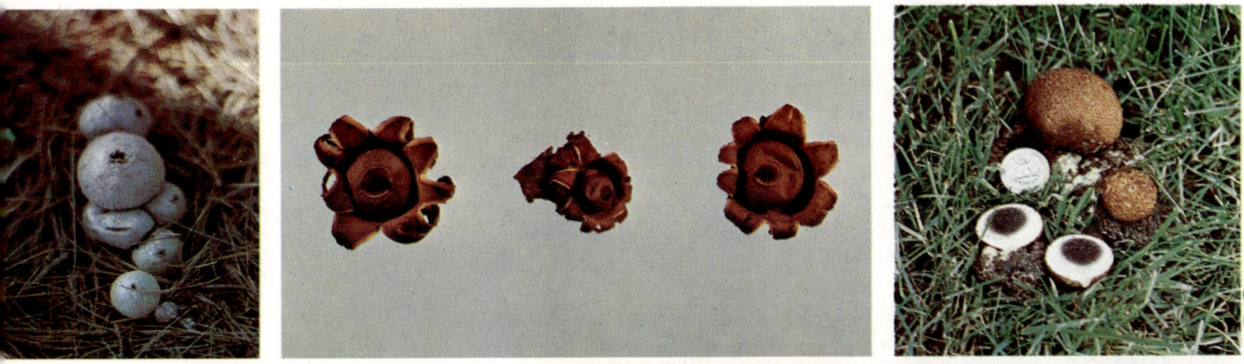

(*a*)
(*b*)
(*c*)

Figure 8.22
Examples of "stinkhorns." (*a*) *Phallus impudicus* (earth's penis). Note the "drippy" glebal area. (*b*) *Mutinus caninus* (dog's penis). Note the fly which has been attracted to the glebal area. (Photograph by Donald A. Licht.)

(*a*)
(*b*)

Figure 8.23
***Cyathus striatus* (bird's nest fungus). Note the peridoles at the base of the mature basidiocarps. The immature basidiocarp has not yet opened.**

Figure 8.16
Heracium caput-urai.

Figure 8.17
***Ganoderma tsugae*. The top surface of the basidiocarp is highly "varnished." The hymenium lines the inner surface of the thousands of tiny pores on the undersurface.**

Figure 8.18
***Polyporus sulphureus* basidiocarps on a log.**

Figure 8.19
A bolete, *Gyrodon merulioides*. Hymenia line the tubes on the undersurface of the pileus.

Figure 8.20
An agaric, *Agaricus arvensis*, the horse mushroom. Hymenia line the gills on the undersurface of the pileus.

Figure 8.11
The closed basidiocarp of the gasteromycete *Calvatia gigantia*, the giant puffball.

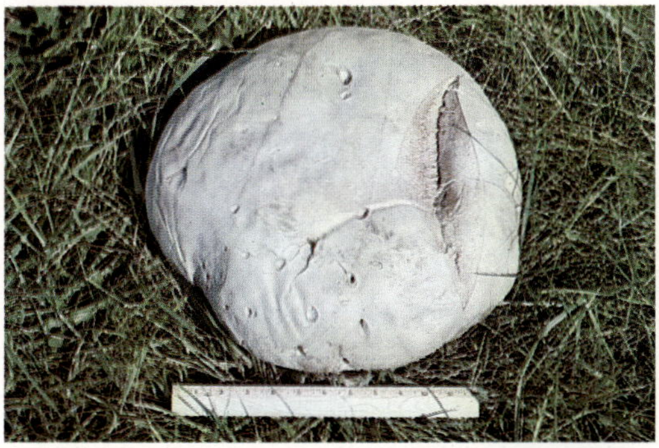

Figure 8.12
***Stereum* sp. Note the thin basidiocarp.**

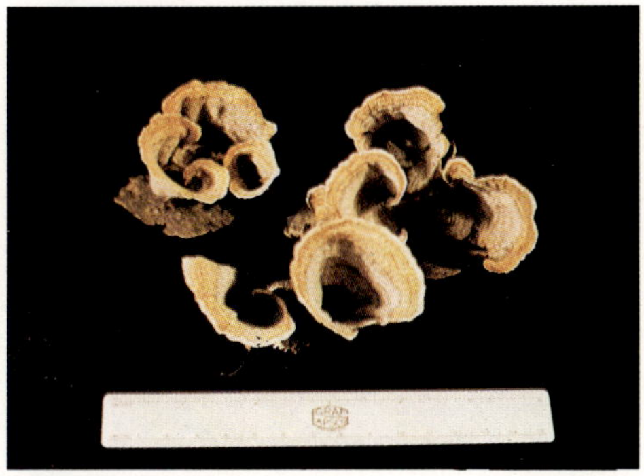

Figure 8.13
Cantharellus cibarius.

Figure 8.14
***Craterellus cornucopioides* (Photograph by John W. Cieslewicz.)**

Figure 8.15
Clavaria pyxidata.

Figure 8.25
Amanita caesarea. Median longitudinal section through "button." Both aspects of the "button" are shown.

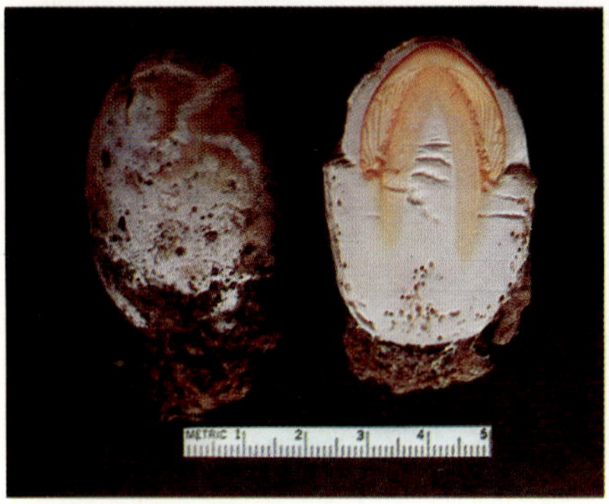

Figure 8.26
Amanita caesarea. Mature basidiocarp. (Photograph by John W. Cieslewicz.)

Figure 8.27
Pleurotus ostreatus (the oyster mushroom). Basidiocarps on a standing dead tree stem.

Figure 8.28
Amanita muscaria, the fly agaric (Photograph by John W. Cieslewicz.)

Figure 8.29
Coprinus comatus (the shaggy mane), an inky cap mushroom. The pileus is beginning to deliquesce.

Figure 8.30
Amanita virosa, the Death Angel.

Figure 8.31
"Toadstool." A toad sitting on the pileus of an *Amanita muscaria*. (Photograph by R. A. Hubbard II.)

volva may be large and shaggy or it may be closely appressed to the stipe and be difficult to distinguish. In either case, the volva may be below the surface of the soil and unless the investigator digs the mushroom (as opposed to breaking the stipe) when collecting, the volva may escape detection. Remnants of the universal veil may also persist as **scales** (Figures 8.28, p. 158 and 8.31, p. 158) on the pileus, although this is often not the case.

During this development, basidia are maturing on the lamellae that are on the undersurface of the pileus. The basidium originates as the tip cell of a dikaryotic hypha. The two nuclei of the dikaryon fuse, producing the only diploid nucleus of the life history, and the cell enlarges to the shape typical for a **homobasidium.** Four pointed tips (**sterigmata**) are produced at the end of the young basidium and basidiospores are produced at their tips. The diploid nucleus undergoes meiosis, and the four haploid nuclei migrate through the sterigmata into the basidiospores. When the basidiospores are mature, a gas bubble produced at their base provides them with the means of forcible ejection from the basidium. Under favorable environmental conditions, the haploid basidiospore germinates.

The pileus of the basidiocarp is composed of a nonlamellar portion (the **trama**) that may be quite thick or quite thin, and the **lamellae.** The trama is usually composed of filamentous hyphae, although two genera (*Russula* and *Lactarius*) also have globose cells in their trama. The lamellae have a central portion of filamentous hyphae (Figures 8.10 and 8.24) and are covered with the hymenium (Figures 8.10 and 8.24). As has been previously discussed, the hymenium is composed of fertile elements (the basidia, Figures 8.10 and 8.24) interspersed with sterile elements. In mushrooms the sterile elements may be of two kinds. One type that is difficult to distinguish from basidia because of their size and shape (and may indeed be immature basidia) are paraphyses (Figures 8.10 and 8.24). The other sterile element that may be produced is greatly enlarged, variously shaped, extends beyond the rest of the hymenial layer, is the cystidium. Some mycologists contend that cystidia are important to keep lamellae separated from each other, although such a function has not been adequately demonstrated.

HUMAN RELEVANCE

The homobasidiomycetes contain members that are important to man for a great variety of extremely diverse reasons.

They are, like most of the other fungi, extremely important in the degrading of complex organic compounds. Many of the polypores and agarics grow on down wood and, indeed, often invade dead trees while they are still standing. *Polyporus sulphureus* (Figure 8.18, p. 158) and *Pleurotus ostreatus* (Figure 8.27, p. 158) are good examples of these.

Many of the polypores parasitize standing, living trees. The trees are quite well protected from fungal invasion by their bark. However, under any environmental situation breaks in the bark are produced by insects, woodpeckers, scratching squirrels, and branches that are broken by wind, rain, snow, and ice. These breaks in the protective bark provide ideal entry for the basidiospores. The basidiospores germinate and the mycelium spreads throughout the host. Although most of the fungi cannot invade the living sapwood of the tree, they can break down the dead heartwood, and it is here that they do the most damage to their host. This group of diseases is known as heart rot and may be divided into two major groups. One group of fungi can utilize the lignins but leaves the cellulose. These are known as white rotters. Another group can break down the cellulose but leave the lignins; these are the brown rotters. Those fungi that can invade the sapwood produce diseases known as sapwood rots. Some of the wood decay fungi can destroy both sapwood and heartwood. Although the heartwood is not living, and thus technically the invading fungus may be a saprobe rather than a parasite, the latter term is used here because the loss of heartwood can weaken the tree to such an extent that it can be blown over and thus an entire living organism is killed.

A tree may live for many years with a heart rot fungus growing inside of it. However, the tree becomes less and less valuable as a timber tree as the years pass. The basidiocarps (variously called conks, shelfs, or brackets) that are periodically produced represent only a very small portion of the entire fungus body (Figure 8.16, p. 158). The fungi that invade sapwood tend to kill the tree because of the involvement with the functioning tissue. Because their mycelium does not invade the heartwood, however, the tree is often still usable for lumber.

Members of several different genera may invade a living tree, the most important being members of the genera *Polyporous, Stereum, Echinodontium, Hericium, Daedalea,* and *Fomes*. The basidiocarps of these various fungi represent varying amounts of mycelial involvement within the host. The basidiocarp of some may indicate that the entire trunk is badly decayed while the basidiocarp of others may indicate only a few feet of the trunk are decayed. Foresters and lumbermen often use knowledge of the wood decay fungi to aid in their determinations of the value of a timber stand. Although accurate estimations of timber loss caused by wood decay fungi are difficult to obtain, loss estimates have varied from 10 to 16%.

One of the mushrooms of considerable importance to forest trees is *Armillaria mellea*, the honey mushroom. It is often found growing around the base of living or dead trees, parasitizes the roots of the host and causes a root rot. The mushroom has also been reported to parasitize many plants other than trees (e.g., grapes and potatoes). This fungus is also known as the shoestring fungus because it produces hyphal strands that are grouped together and surrounded by a hard outer covering or rind. These tough strands are called **rhizomorphs** (Figure 8.3, p. 158) and are often to be found under the bark of down trees or on the surface of logs that have lost their bark. The rhizomorphs are often the means by which the fungus invades a new host. (See also the later discussion about mycorrhizae.)

Although *A. mellea* has been reported to enter directly into the healthy tissue of some of its hosts, the bark of trees seems to serve as an effective barrier to its entrance into tree hosts. The fungus is, however, a serious wound parasite, and is a facultative saprophyte in that it can continue to live on the same wood as a saprobe after the host has died. The fungus parasitizes both gymnospermous and angiospermous forest trees and is especially destructive on many types of fruit trees.

Many of the homobasidiomycetes are important as decay organisms of lumber and structural timbers. These organisms are rarely the same ones that grow on the tree while it is living, although in both cases it is the heartwood that is involved. Some of the genera that contain important wood rotters of this type are *Merulius*, *Poria*, *Polyporous*, *Stereum*, *Hericium*, *Lenzites*, and *Trametes*. Some woods are much less susceptible to fungus decay than are others. Among the very resistant conifers are juniper, bald cypress and redwood. Douglas fir is quite resistant, but some of the less resistant woods are many pines, hemlock, and spruce. The most resistant angiospermous trees include black locust, red mulberry, catalpa, and black walnut. More susceptible woods are chestnut, white oaks, and red gums. The most susceptible woods are red oaks, beech, magnolia, sycamore, basswood, and willows. Susceptibility of lumber to invasion by fungi appears to be closely related to the kinds of metabolic products such as resins and tannins that are deposited in the heartwood of the tree. Resistance to fungal invasion has been increased by man's use of various methods of drying of timber, of steam sterilization in some cases, and by the impregnation of the wood with chemical fungicides. Creosote is usually applied under pressure to timber (e.g., telephone poles and railroad ties) that is to be partially buried. Paint that contains fungicides is also often used to preserve structural lumber.

Many homobasidiomycetes form close relationships with the roots of various vascular plants. These relationships, composed of a fungal and a vascular component, are known as **mycorrhizae** (fungus-roots) and appear to be extremely beneficial to both members of the relationship (**mutualism**). Although mycorrhizae are known that have fungal components other than basidiomycetes, the relationships between members of the agarics and trees are probably the best known and most investigated. As investigation of mycorrhizae continues, the contributions of the phycomycetes, ascomycetes, and fungi imperfecti will undoubtedly prove to be extremely important, but the examples used here will be limited to mycorrhizae with homobasidiomycetous components.

Mycorrhizae occur throughout the world and in all the vascular plant groups that have been investigated. Some of the fungal components appear to be specific for various vascular plant families, but most appear to be nonspecific. Also, many species of trees may form mycorrhizal relationships with various fungi. For example, northern white pine and Norway spruce have been suspected of forming mycorrhizae with about twelve species of fungi including species of the common genera *Tricholoma*, *Lycoperdon*, and *Clitocybe*. Other mycorrhizae formers are *Boletus* (sixteen species), *Tricholoma* (ten species), *Amanita* and *Lactarius* (five species each), and *Cortinarius* (three species).

Mycorrhizae are of two general types. The **endotropic mycorrhizae** are those in which the fungal component produces hyphae that invade the young root cells (intracellular). **Ectotropic mycorrhizae** are those in which the fungal component produces hyphae that invade the root but remain outside the root cells (intercellular). The latter type is also characterized by a fungal sheath that covers the young vascular root tip. It is from this sheath that the hyphae grow between the root cells, penetrating the cortex to varying depths.

Various theories have been proposed to explain the importance of the mycorrhizal relationship. The balance between the vascular and fungal components of the relationship can be a delicate one. If the vascular component is damaged in some way, the fungal component may become parasitic and further debilitate its host. If the host dies, the fungal component may then become saprobic. Genetic changes in the fungus may also make the fungus parasitic instead of mutualistic, or it may make the fungus incapable of any type of symbiotic relationship (parasitic or mutualistic) with the vascular plant. There is also evidence to indicate that the intracellular hyphae of some endotropic mycorrhizae are digested and utilized by the cells of some vascular plants. Therefore, when one attempts to understand the mycorrhizal relationship, he is beset by a group of biological relationships that are extremely difficult to define.

Because mycorrhizal relationships are so widespread, one is led to look for a cause and effect relationship. Mycorrhizae develop more abundantly in soil where there is a deficiency of mineral nutrients, particularly nitrogens, phosphorus, potassium or calcium, and they absorb more nitrogen and phosphorus than do nonmycorrhizal roots. The fungus may contribute some growth substances to the tree. The fungus probably gets glucose from the root, and may also absorb vitamin B.

Although mechanisms are not very well understood, there are some specific examples in which benefit to vascular hosts is evident. The nonchlorophyllous vascular plant Indian pipe (*Monotropa uniflora*, Figure 26.5*a*, p. 446) has a root system that is entirely involved in a mycorrhizal relationship. Anything entering the plant must do so through the mycorrhizae, and since the plant is not autotrophic, one must assume that it derives its nutrition in some manner from the fungal component. This may happen either directly by digesting fungal mycelia or indirectly by absorbing material that had been previously digested and absorbed from the soil by the fungus. Because it is difficult to see what benefit the fungal component would derive from such a relationship, Indian pipe could be considered to be parasitic on the fungal component of its mycorrhizal relationship.

A Japanese orchid (*Gastrodia elata*) is absolutely dependent upon an otherwise severe fungal parasite (*Armillaria mellea*) for the completion of its life history. This orchid apparently will produce flowering shoots only from tubers that possess a mycorrhizal relationship with a rhizomorph of the fungus.

However, it is with research done on various conifers that the beneficial aspects of mycorrhizae can be most easily demonstrated. Several investigators have demonstrated that members of the genus *Pinus* do very poorly when planted in soil that does not contain particular fungi. The pine seedlings grown without mycorrhizae weigh less and possess smaller quantities of such essential elements as nitrogen, potassium, and phosphorus. Experiments with conifers all over the world have indicated similar results. Although experiments with angiospermous trees are not nearly so extensive, investigations on apple, oaks, beeches, birches, willows, and hickories have indicated a strong dependency on mycorrhizae before the tree does well. Although the way in which mycorrhizae perform their beneficial functions cannot be completely elucidated, there can be no doubt that mutual benefits do exist. These benefits are so important to the vascular component that vascular plant distribution may often depend upon the presence of the proper mycorrhizal-forming fungus.

The homobasidiomycetes contain those fungi most often used directly as food. Eaters of mushrooms (mycophagists) abound throughout the world, and one can find in the literature various exclamations of superiority in taste of one species over the other. Only one mushroom (*Agaricus campestris*, Figure 8.9, p. 158) is commercially grown in the United States and most of Europe. This mushroom is also the only one eaten by Americans to the exclusion of most of the common wild edible ones. Europeans, however, have long been mycophagists and enjoy a wide variety of mushrooms that they collect from the wild state. Mushrooms other than members of the genus *Agaricus* are cultivated in many oriental countries, notably Japan. Here *Lentinus edodes* and *Armillaria ponderosa* (the Japanese Shiitake and Matsutaki) are cultivated, dried, and powdered. Although mushrooms have long been an important delicacy, they seldom are used as a dietary staple. However, a friend who was in Europe during World War II reported that he learned to identify mushrooms, particularly the boletes, from a mushroom handbook stolen from a German officer. His abilities to collect and identify edible mushrooms contributed to the very sparse food supply available to him and his family.

The value of mushrooms as a food has been much discussed, and one can find many contradictory statements in the literature concerning their nutritive value. However, various chemical analyses have demonstrated that although their calorie content is low (about 28–32 calories per one-fourth pound), in many respects they are quite a good food in addition to being a flavorful condiment.

As might be expected, the various nutrient values vary with the mushroom examined but, in general, mushrooms are composed of 88 to 92 per cent water, 3 to 4 per cent protein, 0.3 to 0.5 fat, 5 to 7 per cent carbohydrates and 1 to 2 percent minerals. Apparently, some of the mushroom protein is indigestible by human beings.

Probably the best nutritive contribution made to man by mushrooms is in their vitamin content. Mushrooms contain large quantities of vitamin C, ascorbic acid, and niacin which are not usually denatured by cooking, drying, or canning.

Recipes for the preparation of mushrooms abound, and some people become emotionally involved about whether to parboil or not, whether to peel or not to peel, and even whether or not the mushroom should be washed. However, they may be used in many fashions by boiling, sauteing, stewing, frying, baking, drying, canning, and powdering. They may be eaten alone or mixed with other foods such as eggs, meat, vegetables, salads, and dairy products.

The mycophagist should exercise great care when eating a wild mushroom for the first time. Proper identification is essential for continued gastronomic enjoyment—indeed, for continued life. Contrary to popular belief, there are no easy methods of determining the poisonous from the nonpoisonous mushrooms. Such distinctions may be made only from chemical analyses, and such observations as turning silver black, easy peeling, color of gills, or any other such gross observation have absolutely no validity in the determination of the toxic properties of the mushrooms. Unfortunately, many people still use these and other superstitious beliefs to determine whether a mushroom is poisonous or not. I well remember standing in front of a bin of commercial mushrooms in 1965 in a supermarket in Richmond, Kentucky when I was approached by a sweet little gray-haired lady. She expressed concern that I was about to purchase some of the mushrooms, but in her most "motherly" manner suggested that if I were going to be so foolish as to eat such things, I should at least cook a silver dollar with them. If the dollar turned black I was to discard the whole "mess" because it was poisonous.

The only way one can be certain that he is not going to be a victim of the toxic properties of mushrooms is to be able to absolutely identify his specimen and to be aware of his own reaction to it. The former is difficult and the potential eater is warned to check his identification with someone who is very familiar with mushroom taxonomy. In November of 1967, I was asked to identify a group of mushrooms, some of which had been eaten by a man who was hospitalized and suffering from severe hallucinations. The mushrooms proved to be *Amanita muscaria* (Figures 8.28, p. 158 and 8.31, p. 158) and the patient had eaten a large quantity of them upon the advice of his automobile mechanic who assured him they were safe. The patient remained on his hallucinogenic "trip" for approximately two weeks. An interview with him after his release elucidated many of the aspects of his hallucinations, to which I will refer later, but he had learned never to trust the identifications of the novice mushroom taxonomist.

Individuals differ in their reactions to various mushrooms in the same way they differ in their reactions to other foods (e.g., strawberries). These allergic reactions vary from mild flushing of the face to strong gastrointestinal disturbances. While not usually considered dangerous to one's life, the reactions may cause extreme discomfort, and can create severe complications when the victim is subject to other health problems. Some individuals appear to be allergic to members of the genus *Coprinus* (Figure 8.29, p. 158) if alcohol is consumed with the mushroom. The reaction is not considered harmful and the symptoms of flushed face and neck, increased body heat, and increased pulse disappear after a short time. Because of individual reactions, one should eat previously untried species of mushrooms in small quantities until he is sure of his reaction.

The above discussion should not be interpreted as a dictum against eating wild mushrooms but as a warning to observe caution when doing so. Many individuals derive great personal pleasure from collecting, identifying, and eating mushrooms. The gastronomic delights afforded by such pastimes (provided one likes mushrooms to begin with) are well worth the effort required to properly identify them. Learn a few tasty ones well, and enjoy them as often as possible.

No discussion of mushrooms can be complete without mention of the ways in which they have influenced human behavior politically and mystically. Two of the best-known students of this influence (ethnomycology) are the husband and wife team of Valentina Pavlovna and R. Gordon Wasson. They have traveled widely to collect data relating to mushroom culture, but have published their findings in books suitable for perusal only by wealthy bibliophiles. Their works are extremely beautiful, but extremely expensive collector's items that are unavailable to the general public. Although a less comprehensive work, *Mushrooms and Toadstools* by John Ramsbottom is far more available and will present the mycophile with several hours of fascinating reading.

The toxic properties of certain mushrooms have been known since man began to use them as food. Indeed, the poisonous and edible varieties were separated on the basis of many fatal accidents, and the people who could differentiate between the two kinds educated others. Although many mushrooms contain toxic substances, the toxic members of the genus *Amanita* are without question responsible for most deaths caused by mushroom ingestion. Of these, *A. phalloides* and *A. virosa* (Figure 8.30, p. 158) are probably the most important. *Amanita* spp. toxins are complicated cyclopeptides that may be divided into two convenient groups, the phallotoxins and the amatoxins. The phallotoxins tend to act within 6 to 10 hours but the amatoxins are slower and act after 15 or more hours. The first action site of each toxin group is the liver where the cells of that organ are broken down. The clinical symptoms are severe abdominal pain, vomiting and diarrhea, blood and mucus in the stools, casts in urine, rapid loss of strength, and death in two to three days. Post-mortem examination reveals extensive lesions of kidney and severe necrosis and degeneration of liver parenchyma. Severe convulsions appear during the latter course of the symptoms and a very high mortality rate of 60 to 100% usually results from ingestion of these two species. Reactions to the toxins vary, but only minute quantities of mushroom are necessary to produce fatal results. Because symptoms do not appear until 6 to 15 hours or longer after ingestion, the toxins have already reached the vital organs before treatment begins.

A discussion of the toxic properties of other mushrooms is outside the scope of this text, but the details of *A. phalloides* and *A. virosa* poisoning should serve as an important warning to mushroom eaters. *A. phalloides* has also been used to further the political careers of certain ambitious types throughout history. One of the delicacies enjoyed by the ancient Roman emperors was the delectable mushroom now known as *Amanita caesarea* (Figures 8.25, p. 158 and 8.26, p. 158), a nontoxic member of the notorious genus. Claudius, who became emperor of Rome in A.D. 41, was fond of this particular species, a fondness that was to contribute to his undoing. His third wife was Messalina, the mother of his son Britannicus. When Claudius discovered how popular Messalina was with the palace guard, and the reason for her popularity, he had her executed for adultery and married his niece, Agrippina. Agrippina, by a previous marriage, was already the mother of Nero. She, as the ancient counterpart of the modern ambitious mother, wanted to secure Nero's place on the throne. After Agrippina had decided that Claudius should die, she was faced with the important problem of the proper method. Because high office was often suddenly vacated by death, the holders of the office were inclined toward suspicion. The proper poison, then, had to allow Claudius to die slowly so he would not suspect Agrippina of killing him, but he should lose his physical faculties early in the process so he would not name Britannicus as his successor. Locusta, a local "witch" who undoubtedly knew something of the properties of *A. phalloides* came to Agrippina's aid. She gave some of the mushroom to Agrippina who found a way to add it to Claudius' dish of *A. caesarea*, thus poisoning him. There is some doubt that Claudius actually died of mushroom poisoning because he expired some eight to ten hours after ingesting the mushroom. This is close to the lower limits of the toxic action.

Some accounts indicate that Agrippina and her co-conspirators became nervous after Claudius had started to react and hurried him on to divinity by getting Xenaphon the physician to finish the job with an additional poison smeared on a feather and either pushed down his throat or up his anus. In any case, Claudius became a god because of a dish of poisoned mushrooms, Nero became emperor, and Rome became a political mess.

Mushrooms have long been used in the religious rites of Central American Indians. The Wassons (1957) reported that members of three genera (*Panaeolus*, *Stopharia*, and *Psilocybe*) are the ones most often used by the Indians. They have observed in detail the elaborate ceremony involved with mushroom usage, and conclude that the mushrooms serve the Indians in their worship as the Host serves Christians in their worship. The Indians believe that their ordinary faculties are greatly enhanced and that the mushroom speaks directly to them.

The mysteries surrounding the use of *Amanita muscaria* (Figures 8.28, p. 158 and 8.31, p. 158) in religious belief are many and varied. In their first book, the Wassons (1957) discuss the effects of *A. muscaria* on the early Norsemen. Although the evidence is not undisputed, it has long been believed that these warriors went "beserk" as a result of *A. muscaria* ingestion. Illusions of great strength, sexual prowess, and mystical invulnerability to death were hallucinations common to those who either ate the mushroom directly or who drank the urine of those who had eaten it. These acts of a group of warriors would certainly tend to mystically unify them into a formidable fighting force. Although there may be dispute concerning the early "Berserkers," modern man apparently suffers from the same types of hallucinations upon eating *A. muscaria*. After the patient previously mentioned had returned far enough from his hallucinatory "trip" to be safely interviewed, I began to record his recollections. As we talked, I became increasingly impressed by the parallelism that existed between his experiences and those recorded for the early Norsemen. The same illusions of strength, sexual ability, and religious mysticism were present, but in a modern context. The patient, in his hallucination, was able to rip apart his automobile with his bare hands and was capable of beating the car into a shapeless mass of metal by using a railroad tie.

The patient's sexual delusions revolved around what he considered the unusually large size of his penis. He believed that the attending physicians were so impressed that they commented "It is so large that it is no wonder his wife doesn't want to give it to him." He also "saw" a young couple "necking" all of the time.

The patient experienced a religious phenomenon when the Blessed Virgin and Child came to sit on a curtain rod and look down upon him. They stayed there a few hours and then left. He also "saw" God on the cross moving across the ceiling. Although this patient did not eat *A. muscaria* for specific religious experiences, Graves (1970) reports such usage by various people. Indeed, one Biblical scholar (Allegro, 1970) has stated that Jesus was the personification of *A. muscaria* as it was used by the people of the Near East for hallucinogenic experiences. Therefore, Jesus was a mushroom, not a man. Allegro also discusses the erotic significance of the mushroom and maintains that the word Christian is derived from Sumerian and means "smeared with semen." As one might expect, this interpretation by Allegro has been disputed by many other Biblical scholars, but it is certainly interesting to note how often mushrooms have been important in the religious activities of peoples as geographically separated as the Near East and Central America.

A great deal more has been written about the effects of mushrooms on the culture of various peoples but space does not permit further discussion here.

The term "toadstool" is often used to describe mushrooms that are considered to be poisonous or unsavory for some other reason. The designation is an easy one to understand. When one investigates the superstitions surrounding both toads and mushrooms, he finds that they have both been associated with the devil and his evil practices. Their association with one another is probably derived from the fact that they are both often found in damp, dark, wooded areas. The implied relationship does not, however, have any basis in fact, and it is only through coincidence that a toad is ever found sitting on a mushroom (Figure 8.31, p. 158). There is, then, no such thing as a toadstool; there are only various types of mushrooms.

9

Because of various morphological similarities, the algae were at one time grouped with the fungi in a phylum named Thallophyta. Indeed, the fungi have often been thought to have evolved from the algae; later research has demonstrated that this is probably not the case.

The simplest way to distinguish the algae from the fungi is by the obvious fact that the algae contain chlorophyll and the fungi do not. Although some algae are achlorophyllous, they are derived from chlorophyllous ones. By far most algae are **autotrophic,** although some may become **heterotrophic** under certain environmental conditions. Various morphological relationships (to be discussed later) lend evidence for a close phylogenetic relationship between the bacteria and the blue-green algae. However, the blue-green algae are separated from the bacteria in that the blue-greens are photosynthetic. Although a few bacteria are also photosynthetic, they usually produce sulfur compounds as a by-product instead of oxygen as produced by the blue-greens.

Many of the unicellular algae (e.g., *Chlamydomonas*) produce flagellated vegetative bodies that may act as **gametes.** That is, they swim freely, carry on photosynthesis, and reproduce asexually. In this fashion, a large population of the organism may be quickly produced without sexual reproduction occurring. However, if these same vegetative cells encounter a compatible vegetative cell under the proper environmental conditions, they will fuse and produce a **zygote.**

With the exception of the male reproductive structure in the Charophyta, the reproductive structures in the algae are not separated into sterile and fertile layers. Most algal **gametangia** (structures in which gametes are produced) are unicellular. In those that are multicellular, each of the cells is fertile and produces a gamete. The asexual spores are also mostly produced in unicellular **sporangia;** each of the cells of multicellular sporangia produce spores. This condition is contrary to that found in the "higher" chlorophyllous plants in which the fertile cells of a reproductive structure (sexual or asexual) are produced inside a layer or jacket of sterile cells.

The algae contain no vascular tissue. They are found under environmental conditions that place them into direct contact with a water supply. Most algae are aquatic, either marine or freshwater (Figure 9.1). Those that are not aquatic may store water for a period of time, be in contact with water films that cover the structures on which they live, or establish symbiotic relationships with other plants and animals from which they derive their water. Algae have no roots, stems, and leaves. In some of the large brown algae (the kelps), a tissue that is remarkably similar to phloem has been reported. The precise definition of this tissue is still lacking.

It is difficult to distinguish between some of the simple, unicellular algae and the lower, flagellated animals. Any distinctions made between the groups are arbitrary. Those that are chlorophyllous and possess cellulose in their cell walls are usually considered to be plants, while the unicellular flagellates that do not possess these two characteristics are usually considered to be animals. Some biologists consider these organisms to be in a separate kingdom (Protista), but since other arbitrary boundaries must be set up to establish this new kingdom, its use seems only to further complicate the taxonomic difficulties. While recognizing the difficulty in establishing precise boundaries between the plants and the animals, many of those organisms that have been traditionally introduced as plants (e.g. *Euglena*, *Chlamydomonas*) will be treated in this account.

In summary, and recognizing that the boundaries are not precise, there are about five criteria that separate the "algae" into a unified group separated from the rest of the living world. These criteria are: (1) most algae are chlorophyllous; those that are not are clearly derived from chlorophyllous forms; (2) under certain conditions, vegetative organisms may act as gametes and perform sexually; (3) the reproductive structures are generally not divided into sterile and fertile layers; (4) there is no well-defined vascular tissue produced; and (5) cellulose is found in the cell wall of most members. One must remember that all organisms that are placed in the algal groups must possess one or more of these characteristics, but not all organisms that possess one or more of them are algae. Such is one of the problems that human beings face when they try to categorize the wide range of the end-products of billions of years of evolution. With

Figure 9.1
Examples of free-floating algae (phytoplankton) from Lake Ontario. Four of the fresh-water algal groups are represented in this one sample (50×). Blue-green: *Microcystis* sp. (*a*); Greens: filamentous *Mougeotia* sp. (*b*), colonial *Coelastrum* sp. (c), and *Pediastrum* spp. (*d*), desmid *Staurastrum* sp. (e); Dinoflagellate: *Ceratium* sp. (*f*); Diatoms: *Fragilaria* sp. (*g*), *Asterionella* sp. (*h*).

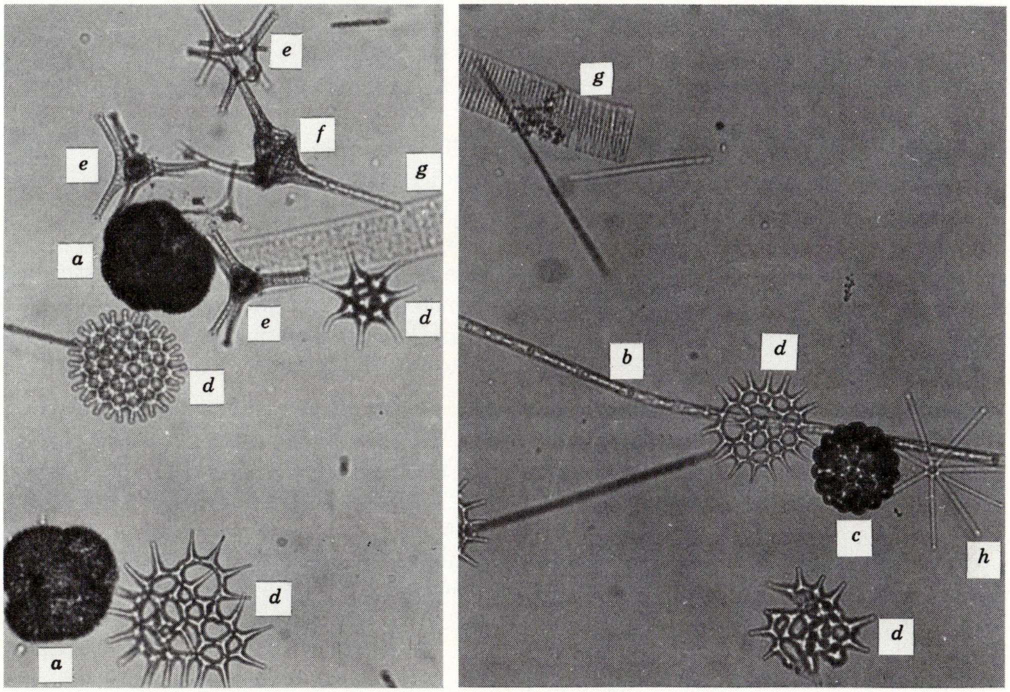

all their weaknesses, however, the categories are necessary in the study of the vast array of forms.

Algal reproduction is quite varied. Asexual reproduction is extremely common among most members and in some groups (e.g., Cyanophyta) it seems to be the only method. **Fragmentation** caused by wave action or the feeding habits of aquatic animals provide a means of population increase of the algae. Various sorts of asexual spores are also produced. Many of these are nonmotile, but others are flagellated and can swim exceedingly well. The nonmotile spores (**aplanospores**) include **akinetes, heterocysts,** and **autospores.** The motile spores are called **zoöspores.**

Sexual reproduction in the algae is by one of three basic methods, **isogamy, heterogamy (anisogamy)**, or **oögamy.** Isogamy is the type of sexual reproduction in which the two gametes are morphologically alike, although they may be physiologically different. The physiologically different gametes are often referred to as plus and minus strains. Both gametes (**isogametes**) are usually flagellated, are of the same size and shape, and are released into the water where **fertilization** occurs. An exception to this occurs in *Spirogyra*, a common green alga, where the two gametes are not flagellated and are not released into the water but unite through a specialized structure called a **conjugation tube.** This process will be discussed in greater detail in the section dealing with the green algae.

Heterogamy involves the union of gametes that are morphologically alike except for size. That is, they are both flagellated and of the same shape. One, however, is larger than the other. The larger one is usually referred to as the female gamete and the smaller one as the male gamete, distinctions that are completely arbitrary. Fertilization occurs while the gametes (**heterogametes**), after release into the water, are free-swimming. Oögamy involves the union of gametes that are morphologically as well as physiologically unlike. Again, arbitrary designations as to sex are assigned to these oögametes, with the smaller, flagellated one being designated the male (sperm or antherozoid) and the larger nonmotile form being designated the female (egg). Whereas the isogamous and heterogamous species of algae rarely produce specialized sex organs (gametangia), the oögamous species produce their sperm in antheridia and their eggs in oögonia. The "female" gamete of oögamous species is not released into the water, but remains in the oögonium until the flagellated sperm swims to it. The sperm enters through a pore in the oögonium and internal fertilization takes place. The sperm is attracted to the egg by chemical substances released into the water by the egg. Among the algae are to be found those that are sexually self-compatible (**homothallic**) as well as those that are sexually incompatible (**heterothallic**).

The mechanics that enable the sperm and egg to get together are varied among the different algal groups. Specific mechanisms are more appropriately discussed in the context of the specific group. It is possible, however, to generalize about the life-histories of the algae (with the exception of the blue-greens). These generalized life histories are diagramed in Figure 9.2. Specific life histories will be illustrated

Figure 9.2
Generalized life histories found among the algae (Redrawn from Harold C. Bold, *Morphology of Plants*, New York: Harper & Row.)

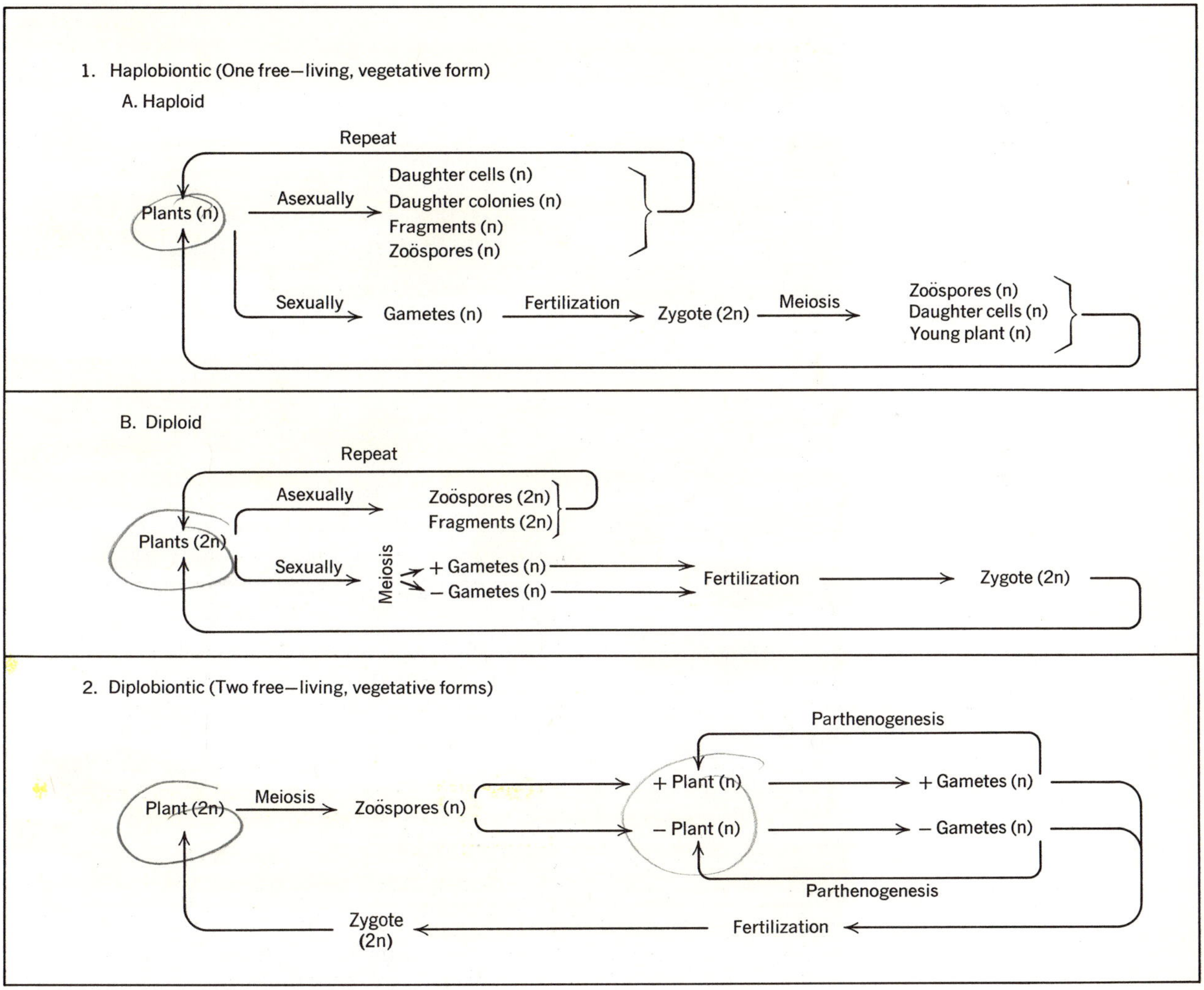

during the appropriate discussions. The algae that possess only one free-living, vegetative form in their life history are known as **haplobiontic** species. Those that have two free-living, vegetative forms in their life history are **diplobiontic.**

Haplobiontic species may be either haploid or diploid, although usually haploid. In the haplobiontic, haploid history, the vegetative organism is haploid (n) and produces gametes. The gametes unite producing the only diploid phase to the history, the zygote. When the zygote germinates meiosis occurs. Therefore the haploid number is reestablished. The germinating zygote may produce zoöspores, daughter colonies or young plants, depending on the species. The vegetative organism in the haplobiontic diploid cycle is diploid. In these algae, meiosis occurs at the time of gamete production and haploid gametes are produced, the only haploid phase to the cycle. Fertilization reestablishes the diploid number, and the zygote germinates with the production of a new diploid plant. The vegetative stage of the organism may be reproduced by asexual means in either of the above types of life histories.

The algae that have diplobiontic life histories include two kinds of free-living plants, one haploid and one diploid. The diploid plant produces haploid zoöspores. The zoöspores produce haploid plants that are plus or minus strains. The haploid plants grow to free-living entities and eventually produce gametes. The gametes unite, reestablishing the diploid number. The two plant forms may be morphologically similar (**isomorphic**) or morphologically distinct (**heteromorphic**). In some species of algae possessing this history, the gametes may directly produce new haploid plants by **parthenogenesis.** The diplobiontic life history demonstrates an alternation of haploid and diploid phases among the algae that is similar to the sporophyte and gametophyte generations that appear among the "higher" plants.

The algae are treated here in eight groups. These are the blue-green algae (Cyanophyta), the dinoflagellates (Pyrrophyta), the golden-brown algae, the yellow-green algae, and the diatoms (Chrysophyta), the brown algae (Phaeophyta), the red algae (Rhodophyta), the stoneworts (Charophyta), the euglenas (Euglenophyta), and the green algae (Chlorophyta). Particular algae have been shifted from one group to another from time to time, but these groups each have characteristics that make them fairly unique and distinct.

blue-green algae | cyanophyta

10

The blue-green algae are ubiquitous in their distribution and they occupy some habitats where one might not expect to find algae. They are found in both marine and fresh-water environments (Figure 10.1, p. 254) but many are also found on shaded bare soil, mud flats, damp rocks and in the greenhouse on damp floors and pots. They may live as **epiphytes** on both aquatic and terrestrial plants, including living on other algae. Some members live as **endophytes** in the cavities of other plants such as liverworts, water ferns, and the roots of cycads. They may also serve as the algal components of some lichens.

There are approximately 1500 to 2000 species of cyanophytes. These are divided into about 150 genera that are grouped into 3 orders. The Cyanophyta are fairly distinct as a group, but their relationship to other chlorophyllous plants is not well understood.

VEGETATIVE STRUCTURE

The blue-green algal cell, except for its pigmentation, closely resembles the cell of a bacterium. It contains no well-organized nucleus, although nuclear material (DNA) is present in the central portion of the protoplasm, the **centroplasm.** There are no **nucleoli.** There is no organization of chlorophyllous lamellae into a membrane-bounded **chloroplast.** The **lamellae** are found mostly in the outer portion of the protoplasm (the **periplasm**). There is no demonstrable **endoplasmic reticulum, mitochondria,** or **Golgi apparatus.** Various other materials are diffused throughout the protoplasts. These include ribosomes, pseudovacuoles, and polyhedral bodies. Although the protoplast of the blue-green algal cell often appears to be perfectly homogeneous through the light microscope, the electron microscope has revealed it to be quite varied in its structure.

The blue-greens contain chlorophyll and are thus able to

produce their own food. Excess photosynthate is stored as "**cyanophycean** starch." This storage product has a smaller molecular size than the starch produced by the green algae, and it does not give the typical starch reaction to potassium iodide (I_2KI). The blue-greens contain several other pigments including **carotenes** and **xanthophylls.** Red and blue pigments are present in the form of **phycoerythrin** and **phycocyanin.** The color of a specific alga in mass may range from grass-green to black to red as well as many intermediate colors as determined by the various proportions of the pigments present.

The wall surrounding the protoplast of the blue-green algal cell is apparently similar to the wall that is found in gram-positive bacterial cells. The wall may contain **mucopeptides** that also occur in the bacterial wall, but **cellulose** has also been reported. Most of the blue-green algae secrete a **gelatinous sheath** that is composed of pectic compounds. The sheath varies considerably in thickness from species to species, and may or may not contain pigments. The sheath completely surrounds the individual cells of unicellular or colonial species, but is restricted to the free surfaces of the filamentous ones. It is produced in concentric layers after cellular duplication has occurred (Figure 10.4). These layers may be evident in certain colonies, but they may become so fused in others that the stratification is obscure. In some blue-greens, the sheath is of an extremely watery consistency and may be soluble in water. The sheath can usually be demonstrated, however, by mounting the samples in a dilute solution of India ink or methylene blue.

The structure of the plant body of the blue-greens is of two basic types: **filamentous** (Figure 10.2) and **nonfilamentous** (Figure 10.3). The filamentous genera produce vegetative cells in an end-to-end arrangement which may be unbranched, falsely-branched, or branched. The filaments are enclosed in a gelatinous sheath that varies in thickness from species to species. Unbranched forms such as *Oscillatoria* spp. (Figure 10.2*a*) produce new cells in only one plane. Falsely branched ones such as *Tolypothrix* spp. (Figure 10.2*b*) produce branches as the result of pressure produced when intercalary cells duplicate. This pressure is exerted against **intercalary heterocysts** that are attached firmly to the sheath. The vegetative cells are then "shunted" in a direction away from the heterocyst, resulting in false branching. True branching is illustrated by *Hapalosiphon* spp. (Figure 10.2*c*) in which cells of the filament dupli-

Figure 10.2
Types of filamentous blue-green algae. (*a*) *Oscillatoria* sp. and *Anabaena* sp. Unbranched forms. (*b*) *Tolypothrix* sp. A falsely branched form. (*c*) *Hapalosiphon* sp. A truly branched form.

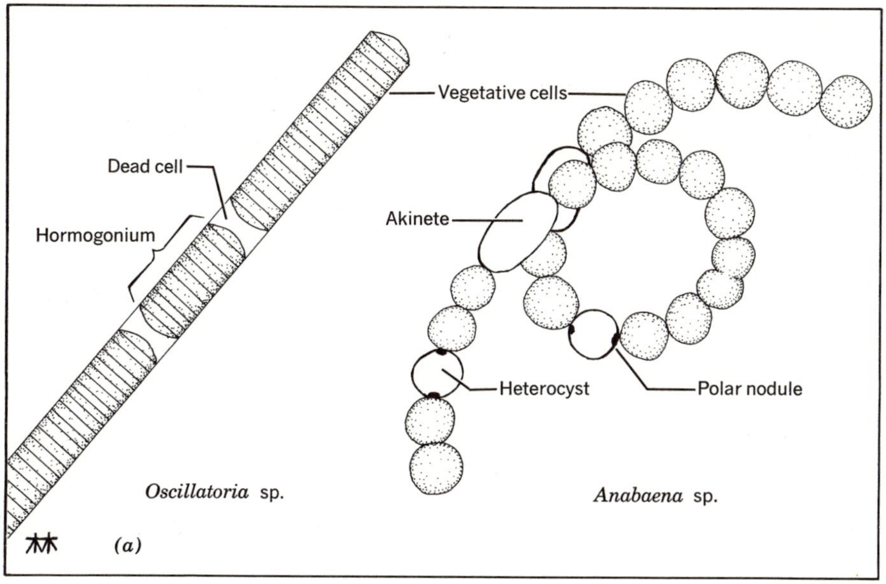

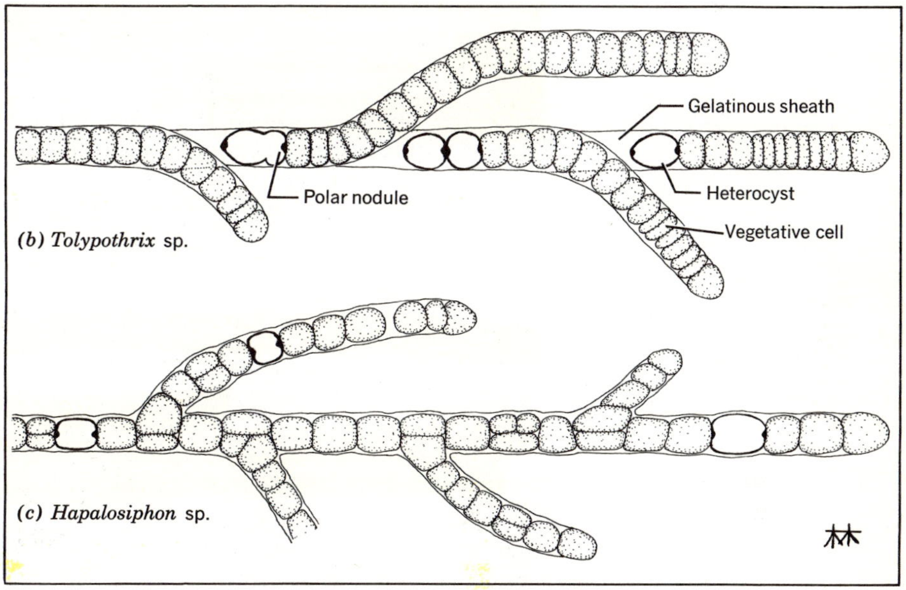

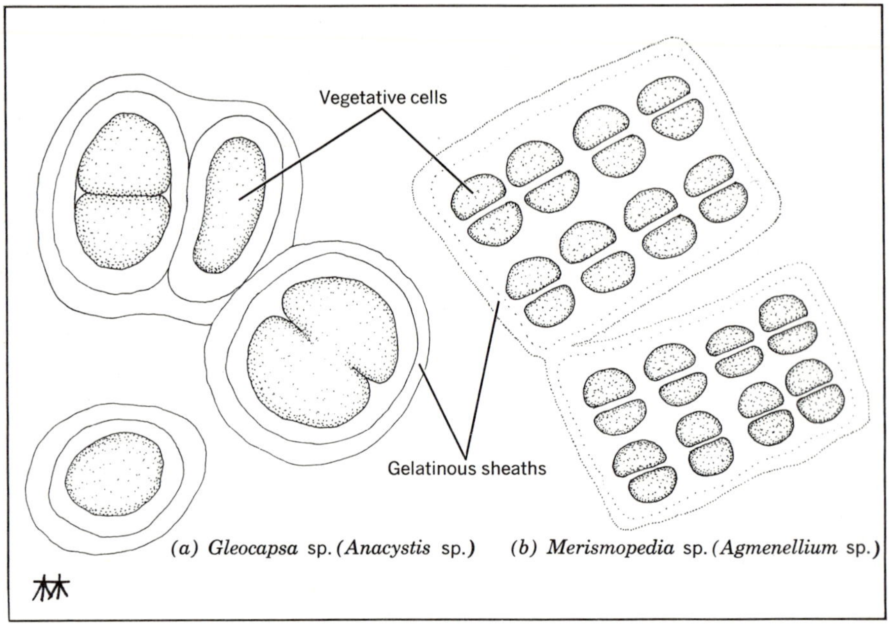

Figure 10.3
Types of nonfilamentous blue-green algae. (*a*) *Gleocapsa* sp. (*Anacystis* sp.). A single-celled blue-green alga. (*b*) *Merismopedia* sp. (*Agmenellium* sp.). A colonial blue-green alga.

cate in a plane at right angles to the main trichome.

In the filamentous blue-greens, the chains of vegetative cells are often referred to as **trichomes.** The trichome and its enclosing sheath is known as the **filament.** In some genera (e.g., *Nostoc, Anabaena*) the trichomes are enclosed in a gelatinous mass that may be oval to ellipsoid or sheetlike. These masses may be free-floating in water or occupy large areas on mudflats, fields, deserts, and other such habitats. They may also be found attached to rocks and submerged plant parts.

Several filamentous genera (e.g., *Rivularia* spp.) Produce colonies that radiate from a single, central attachment. The trichomes of these colonies are attached closely together at one end and often taper to a fine point at the other. They are often found as tiny, dark gelatinous masses on submerged plant parts.

Nonfilamentous genera may be unicelled or multicelled. Members of the genera *Chroococcus* and *Gloeocapsa* (both now in the genus *Anacystis*) are often used as examples of the unicelled forms, while *Merismopedia* spp. (*Agmenellium* spp.), *Polycystis* spp. (*Anacystis* spp.), and others have been used as examples of multicellular, or colonial forms. Unicellular forms that produce a massive sheath can be found. Duplication of the vegetative cells produce new cells that are enclosed in their individual sheaths. This process may continue until several concentric circles of sheath material may be observed in a single colony (Figure 10.4). The sheaths may anastomose in some multicellular forms to the extent that individual sheaths are no longer discernible and the colony appears to be composed of many hundreds of vegetative cells surrounded by a common sheath.

The plane of duplication determines the shape of the colony. In some, the duplication may be in only one plane but in two directions. This type produces a flat plane of cells that is illustrated by *Merismopedia* spp. (*Agmenellium* spp.). If the duplication is in two planes (thus in three directions), very regular cuboid forms may be produced. This type of duplication may be quite irregular, however, and produce an amorphous colony (e.g., *Anacystis* spp.).

The blue-greens, especially the filamentous forms, exhibit movements that still remain a puzzlement to phycologists. *Oscillatoria* spp. is commonly seen to exhibit a back-and-forth swaying (an oscillation) as well as sliding forward and backward on its entire axis. **Hormogonia** of *Oscillatoria* spp. are capable of quite good movement that superficially resembles the movement of diatoms. The trichomes

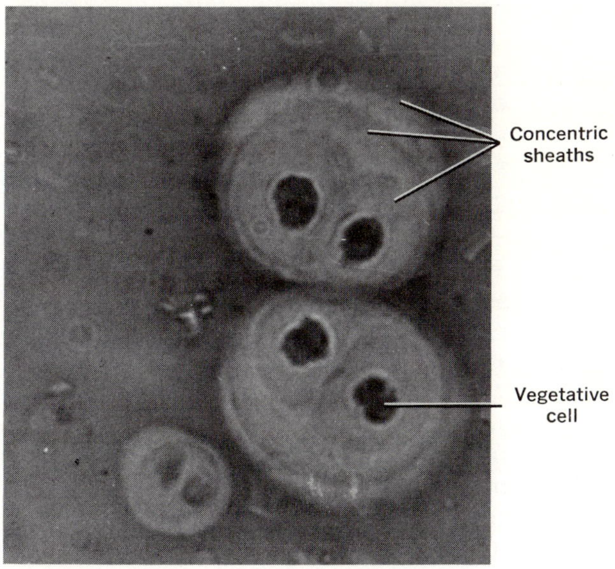

Figure 10.4
Photomicrograph of *Gleocapsa* sp. (*Anacystis* sp.) Note massive concentric gelatinous sheaths (1500×).

of *Aphanizomenon* spp. move back and forth within their bundles in an intriguing sliding motion. All attempts to explain these movements have been unsatisfactory. No flagella or cilia are present in any members of the group. Theories have been proposed to explain the motion by the secretion of slime, protoplasmic movement, osmotic changes, and other mechanisms. Each theory has so far been shown to be inadequate for one reason or another when subjected to critical laboratory analysis.

REPRODUCTION

Sexual reproduction among the blue-greens has yet to be satisfactorily demonstrated. Certainly no mechanism that is recognizable has been elucidated. There are some data, however, from experiments that have been done with species of *Cylindrospermum* that suggest that there may be some subtle method of exchange of genetic material.

Fragmentation accounts for most of the reproduction in the blue-greens. Cell duplication of individual cells in a colony or trichome is accomplished by the in-growth of the inner wall layer from the wall toward the center of the cell at a position midway in the cell. The wall continues to grow inward, apparently pinching the protoplast into two parts until it meets in the center of the cell. Because no well-organized nucleus is present, there are no mitotic figures to be seen. The nuclear material is apparently divided in the centroplasm during cell duplication. In many of the members of this group, the daughter cells remain in close proximity to one another and form a colony. However, the gelatinous envelope is often broken by various environmental factors such as wave action and each of the fragments is capable of producing a new colony. Filamentous genera are extremely subject to such mechanical breakage.

Filaments of *Oscillatoria* spp. and some other filamentous forms often break at the point where one of the intercalary cells of the trichome dies. Continuation of the breaking produces short filaments that have rounded terminal cells. These short filaments are known as **hormogonia.** They are motile and produce a new filament by continued cellular duplication.

Some cellular differentiation and specialization is noted among the blue-greens. All of the filamentous forms, with the exception of those that belong to the Oscillatoriaceae, produce a specialized cell called a **heterocyst** (Figures 10.2*a* and 10.5). A heterocyst is pro-

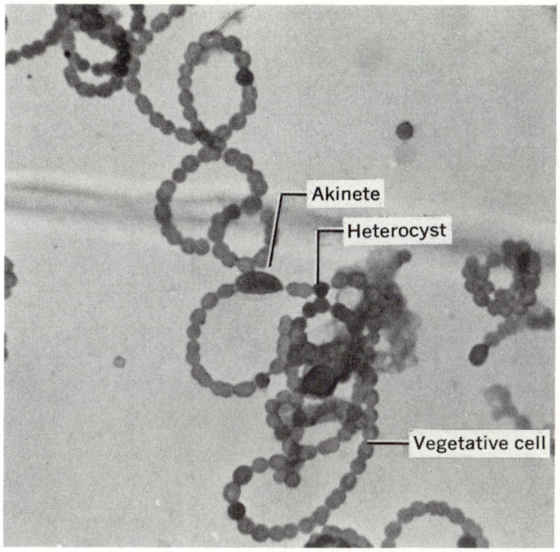

Figure 10.5
Photomicrograph of *Anabaena* sp. All of the filaments are enclosed in a gelatinous sheath (500×).

duced from a vegetative cell of the trichome and may be either terminal or intercalary in position. Although they appear to be empty cells, their protoplasmic contents are still present and metabolically active, but they have become optically homogeneous. Protoplasmic connections between the heterocyst and its adjacent cell or cells are continued for a time. However, the pores in the heterocyst through which the connections are made eventually become filled with wall material and produce **polar nodules.** The polar nodules are the only structure to be observed in the heterocyst through the light microscope. It was long thought that the connection between the heterocyst and its adjacent vegetative cells provided a weakened area in the filament that was subject to easy breakage. Indeed, many species have been observed to break at this point during hormogonium formation. However, heterocysts have been observed to germinate in some species and therefore they are known to provide a method of asexual reproduction. Although germination of heterocysts is regarded as unusual, their sporelike character cannot be denied.

With the exception of the members of the Oscillatoriaceae, most filamentous members of the blue-greens produce another type of specialized spore, the **akinete** (Figures 10.2*a* and 10.5). An akinete is a differentiated vegetative cell. The vegetative cell enlarges, increases the number of refractive granules continued in its protoplast, and produces a wall just interior to the old vegetative cellulose cell wall. This new wall fuses with the old cell wall so completely that they are indistinguishable from each other when the spore reaches maturity. An akinete may be located at any place in the trichome, and it is often associated with a heterocyst. The location and association are sometimes species specific. Although akinetes are usually solitary, two adjacent vegetative cells may differentiate in some species. The akinete is extremely resistant to unfavorable environmental conditions, and has been known to germinate after a period of 70 years.

All members of the order Chamaesiphonales produce specialized asexual spores called **endospores.** These are produced by the repeated division of the entire protoplast into a number of portions, each of which is surrounded by its own wall. When the old wall breaks, the endospores are released into the water and eventually germinate. A specialized kind of endospore is produced by the one-celled, epiphytic genus *Chamaesiphon* spp. Sometimes called **exospores,** these reproductive structures result from the unequal division of the vegetative protoplast by successive wall formation at its distal end. Each small portion of the protoplast is surrounded by a wall at maturity, and the spore is released through a terminal pore at the end of the vegetative cell. Heterocysts sometimes germinate by the production of endospores.

LIFE HISTORY

Because the blue-green algae have no clearly determined sexual phase, their life histories are simple and similar. Cellular duplication accompanied by fragmentation caused by physical or biotic environmental factors undoubtedly accounts for the production of most new plants. The fragmentation is aided in some of the filamentous genera by the production of dead cells, heterocysts and akinetes. The filament tends to fragment much easier at the location of these cells than at any other point in the trichome. The production of heterocysts, akinetes, and endospores also provides methods of continuing the life history. Although all of these spore forms are produced while the plant is growing under optimum conditions, it has been definitely demonstrated that akinetes and heterocysts do help the species to withstand periods of severe environmental stress.

HUMAN RELEVANCE

The Cyanophyta contains members that affect man in both positive and negative ways. The blue-green algal cell and the bacterial cell are similar in their simplicity and asexual reproduction. Because of this simplicity, it has been speculated that they may have been the first cellular organisms to appear on earth. It is thought by some phycologists that the blue-green cell may have been among the first autotrophic cells. From the evolutionary viewpoint, therefore, these two cell types may give evidence of the trends that took place in the internal organization of cells. Most phycologists would agree, however, that the blue-green algae probably did not provide the ancestors of the more complexly organized cells that occur in the other algal groups and the "higher" plants.

The Cyanophyta are important as nitrogen-fixers in the soil. Experiments conducted in India indicate that salty and basic soils may be greatly improved by increased nitrogen content, organic content, and water-holding capacity when blue-green algae are added to them. Blue-greens can fix enough nitrogen to support a good rice crop and still leave nitrogen in the soil after harvesting. A species of *Tolypothrix* (Figure 10.2*b*) can yield up to 1¼ tons of nitrogen per acre per year. When introduced into the soil, the alga increased the nitrogen content by 30% and the rice yield by 20%.

The blue-greens have been shown to be capable of binding the particles of prairie soils together in such a way as to aid in erosion prevention. Crusts ¼-inch thick have been found on certain sandy soils, thereby successfully preventing sheet erosion. These crusts also help to maintain a higher water content in the soils. Such action has been successfully demonstrated in Kansas, Oklahoma, Arizona, New Mexico, and Texas. Various workers have proposed that algal "blooms" be harvested from the oceans, dried, and processed as fertilizers because of their high phosphorus and nitrogen content.

The blue-greens are often the first to colonize bare rock and soil, providing the first step in bare rock succession. Because of their **mutualistic relationship** with certain fungi, the plant complex which we call a **lichen** is produced. In this relationship, the algal component probably contributes the organic "food" while the fungal component contributes the anchorage and retains the moisture. In this way, lichens are able to live in areas incapable of otherwise supporting life. Therefore, blue-greens are able to aid in reclaiming bare rock in two ways: as a single entity and as part of a plant complex.

Several blue-greens have established various types of symbiotic relationships with other organisms. In some of the plant relationships, it appears that the blue-green algal cell acts very much like a chloroplast within the cell of its symbiont. They have been found in various bryophytes as well as in gymnosperms and angiosperms. Because they are often found in nodules on these plants, there is the suggestion that they are important in providing nitrogen to them. Various relationships with animals have been described, but the exact nature of their relationships is not understood.

Several of the blue-green algae provide the various colorations in hot springs such as in Yellowstone National Park. The colors result, of course, from the pigments found in the cells and in the sheaths. Many of these same algae build rocks by depositing salts of calcium and silica within their gelatinous sheaths. The Red Sea owes its occasional red color to "blooms" of the blue-green algae, *Trichodesmium*. However, this body of water probably does not owe its name to the color but rather to a translator's error of interpreting "Sea of Reeds."

Because of the large quantities of gelatinous sheath material that is produced by most of the blue-greens, they are rarely used by man as a direct source of food. However, their unpalatability is overcome by some Asian peoples by treating the algae with sugar before consumption. A species of *Nostoc* that produces gelatinous balls, called "water plums," are eaten in areas of China and South America. They have a high protein and oil content which is usable by human beings and so are nutritious. They have also been dried and then used as a flour base. The large macrocolonies that form gelatinous sheets on damp soil may be one possible explanation of the legend of "manna from heaven" that fed the people of Israel while they were being led out of Egypt by Moses.

The planktonic species of blue-green algae provide the basis for many of the food chains or food webs that occur in aquatic ecosystems. Because of their autotrophic nature they serve well as primary producers. If the concentration of the blue-greens becomes too great, however, toxic levels may be fatal to fish and other aquatic organisms. Livestock and human beings who drink such water may also be adversely affected. The effect may be only a disagreeable flavor imparted to the water by low concentrations. However, cattle, sheep, and other livestock have been known to die as a result of drinking water containing high concentrations of *Anabaena*, *Micro-*

cystis (*Anacystis*), and *Aphanizomenon*. Experiments with *Microcystis* spp. (*Anacystis* spp.) have demonstrated that the toxin is retained within the algae cell and is thus an **endotoxin.** The toxin, when released by breakage or digestion of the cell, causes respiratory difficulty and liver damage. The algal "bloom" may be so dense as to create great oxygen demand during its respiration. This may sufficiently reduce the oxygen content of the water to the point that suffocation intensifies the toxic effects, and increases the mortality rate of the aquatic animals.

The increasing level of pollution of our fresh water streams and lakes aids in the production of algal blooms, especially some of the blue-greens. They give the water an unpleasant, "fishy" flavor and choke intake pipes and filter systems. When growing on the shore, or when washed up by wave action, they produce an odor that makes the beaches completely uninhabitable. One of the basic solutions to the algal problems is to reduce the amount of industrial waste, sewage, and especially phosphates that enter our fresh water supplies.

11

For many years the dinoflagellates were considered to be animals because of their motility and, in some cases, their ability to ingest solid food particles (**holozoic** or **phagotrophic**). Some are even parasitic or saprobic in the intestinal tracts of animals. However, later discovery of nonmotile and filamentous genera that produce motile cells that are like dinoflagellates sometime during their life histories clearly argues for the plant-like affinity of the group. Also, when present, the cell wall contains cellulose and most members of the group are completely **autotrophic.** Therefore, although many protozoologists still contend that the dinoflagellates are animals, they will be treated here as plants. The organisms are fascinating regardless of man's biased view.

Dinoflagellates occupy a variety of aquatic habitats, both marine and fresh-water. They are found in the plankton from lakes and oceans, pools, and ditches. Various species display an affinity for hard water, others for soft water. The non-flagellated forms are usually epiphytic on larger algae, and the parasitic and saprobic ones live in and on a variety of animals.

The Pyrrophyta contain approximately 125 genera and 1050 species. These are often divided into two classes, of which the Dinophyceae contains the far greater number and much more common species, and are the ones that will be considered in this chapter.

VEGETATIVE STRUCTURE

The pigments of the dinoflagellates are located in chromatids. They include chlorophylls $_a$ and $_c$, beta-carotene, and xanthophylls, three of which are unique to the group. The mixture produces a greenish-tan to yellow-brown to very dark chloroplast. The food reserves are often stored as oils. They also may contain starches that give a positive test with potassium iodide. A well-organized nucleus is present; the chromatin is produced in several beadlike strands.

Most of the genera produce asymmetrical motile cells (Figures 11.1 and 11.2) that are quite distinctive. The majority of the species are surrounded by a transverse groove that divides the cell into equal or unequal halves. The two unequal flagella of the motile cells extend from a pore at a point in the groove. One flagellum lies in the groove and undulates. It probably accounts for the spinning motion

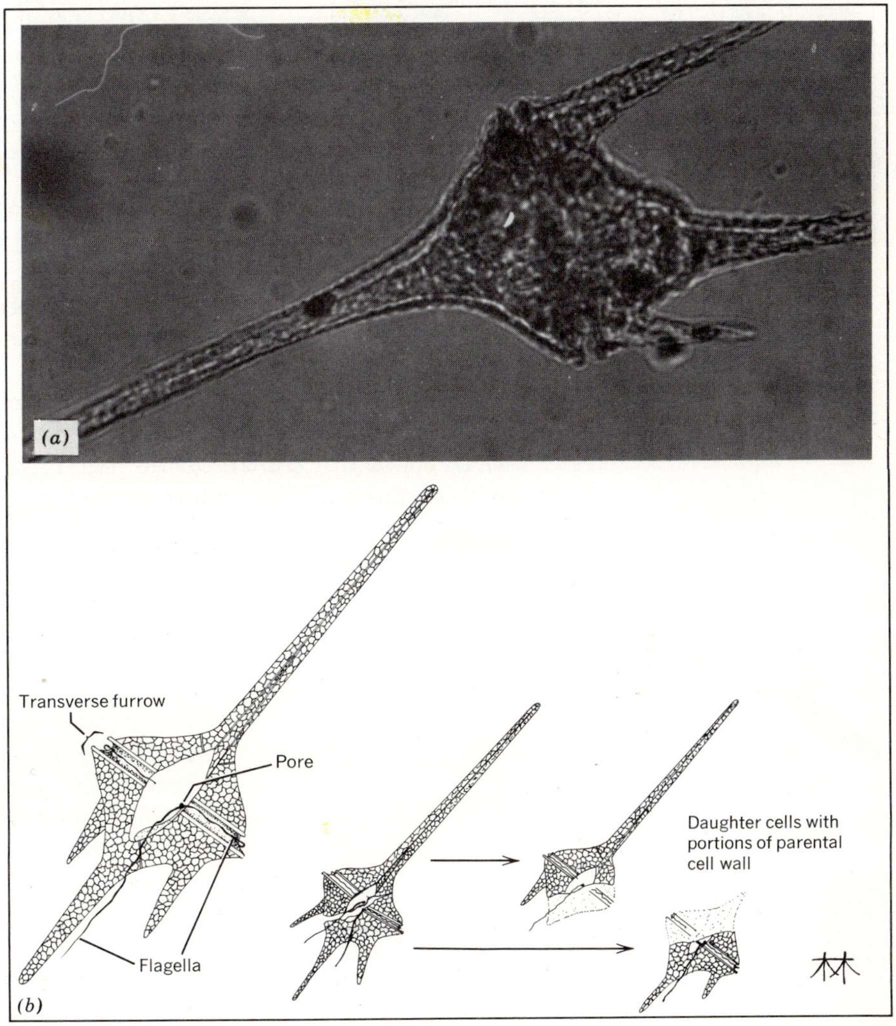

Figure 11.1
Ceratium sp., an armored dinoflagellate. (*a*) Photomicrograph of *C. hirundinella* (670×). (*b*) Vegetative cell and two daughter cells just after cell duplication.

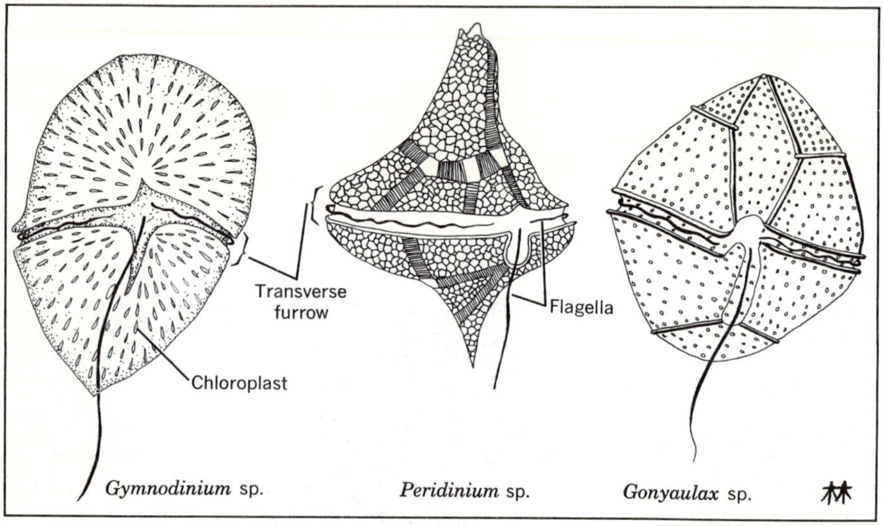

Figure 11.2
Examples of dinoflagellates.

noted in some members. The other extends outward in a position posterior to the movement of the plant and is thought to account for the forward motion.

Most of the dinoflagellates possess heavy cell walls, one layer of which is cellulose, composed of plates (e.g., *Ceratium* spp. Figure 11.1). These are known as the "armored" dinoflagellates. Those without the wall but with a rigid outer membrane that retains their shape are commonly called the "unarmored" or "naked" dinoflagellates (e.g., *Gymnodinium* spp. Figure 11.2).

REPRODUCTION

Dinoflagellates reproduce asexually by cell duplication when the motile cell is in motion or after it comes to rest. In some of the "armored" species, each daughter cell receives a predetermined portion of the parent cell wall; in others, the protoplast escapes from the old wall before dividing, and each daughter cell produces a new cell wall. Other members of the Pyrrophyta produce zoöspores, autospores, or aplanospores. Fragmentation is undoubtedly important in the reproduction of the filamentous forms.

LIFE HISTORY

The life histories of the Pyrrophyta appear to be simple and similar. Increased populations result from asexual reproduction. Although the possibilities of sexual reproduction exist, the complete life histories of most of the dinoflagellates are still unknown.

HUMAN RELEVANCE

The dinoflagellates are extremely important in the oceans and in some fresh water environments as the basis for food webs. They provide the basic food source for many of the primary consumers. Aside from this important function, most of man's interaction with the group is detrimental to him.

Many of the dinoflagellates can produce offensive odors and tastes in water. The members of the genus *Ceratium*, which can multiply very rapidly under the proper environmental conditions, produce a fishy to septic taste and smell that creates great problems in city water supplies. Along with many other algae, some dinoflagellates (e.g., *Peridinium*, Figure 11.2) clog intake filters, adding greatly to the costs of municipal water supplies.

The most spectacular of man's conflicts with dinoflagellates occurs when environmental conditions are right for sudden increases in the populations of species of *Gonyaulax* (Figure 11.2) and *Gymnodinium*. Such large populations are the cause of the massive "red tides" that kill millions of fish and other marine animals, and create chaos in the cities along the oceans where they occur. As recently as the summer of 1971, toxic conditions persisted for longer than six weeks in Tampa Bay off the coasts of Tampa and St. Petersburg, Florida. Thousands of tons of dead fish were scooped off the beaches and out of the water by hundreds of community volunteers and a fleet of seven shrimp trawlers that were hired to try to keep the fish from being washed up on shore. A 1700-foot curtain, which had been originally designed to contain oil spills, was pressed into service. Dead fish were spotted in groups that extended for six miles. The death of the marine animals is caused by neurotoxins produced by the plants. Red tides are not unique to the waters that surround the North American Continent, but have been reported from such varied locations as Peru, Japan, Australia, Africa, India, and Europe.

Research has demonstrated that red tides are somewhat predictable in most geographic areas. There is a definite association between low to moderate latitudes, warm water, upwelling, upland rainfall and upland soil that is rich in iron and phosphates. Unless most of these factors are present, red tides are not produced. Although the fish kills are large and unsightly, most of the species are trash fish such as triggerfish, parrotfish and their relatives. Therefore, most of the economic loss is borne by those dependent upon the tourist trade.

The toxins present in dinoflagellates can cause direct illness in man if he ingests shellfish that have fed on the toxic species. The toxic agent of one species of *Gonyaulax* that is eaten by clams during certain periods has been reported to be 10 times as poisonous to mice as is strychnine. The filter-feeding mussels, clams, and oysters eat the dinoflagellate and accumulate the poisons in their body tissues, apparently without harmful effects. The toxins are passed on to the next carnivore, whether it be dog, cat, rodent, or man, and causes paralysis or sometimes death. The toxin seems to be related to the curare group of poisons produced by some flowering plants.

Toxins produced by *Gymnodinium* spp. are deadly to some shellfish, however. Experimental work has suggested the toxin produced by the members of this genus interferes with the sodium exchange mechanisms that cause a deplorarization of nerve and muscle membranes.

12 the golden algae and the diatoms | chrysophyta

The chrysophytes is a group of algae that is sometimes separated into three distinct phyla. They will be treated as a single phylum here, however, because of general characteristics that seem to separate them from other algae.

The pigments are contained in plastids; the predominant pigments being carotenes and xanthophylls. These two pigment groups mask the presence of the chlorophylls. As a result, the plastids appear as some shade of golden-brown, yellow-brown, or brown. In a few instances, the plastids appear to be grass-green, and the alga seems to belong to the green algal group (Chlorophyta). Hence algae cannot be taxonomically grouped solely on the basis of color.

The chrysophytes store their food as oils and as leucosin, a carbohydrate. No starch is produced; therefore the addition of potassium iodide does not produce a positive test.

Although there is a great deal of variability in body form, most species produce cells that are composed of two overlapping halves (Figure 12.1). The cell wall is composed mostly of pectic materials, but may be heavily impregnated with **silica** (e.g., diatoms). Some cellulose has been reported in the walls of a few chrysophytes, but its presence has not been firmly established. Flagellated cells, when present, usually have two unequal flagella that are laterally inserted.

The 350 genera and over 6000 species of the chrysophytes are separated into three families; the yellow-green algae (Xanthophyceae), the golden-brown algae (Chrysophyceae), and the diatoms (Bacillariophyceae). In those treatments where the chrysophytes are divided into three phyla, these families are elevated to the phylum level. About 75% of the species are found in fresh-water habitats with the remaining 25% being marine. Some members that are not aquatic live on moist rocks or amid damp vegetation. Some are found in soil or sandy beaches. Others are found in the mud on the bottoms of aquatic habitats.

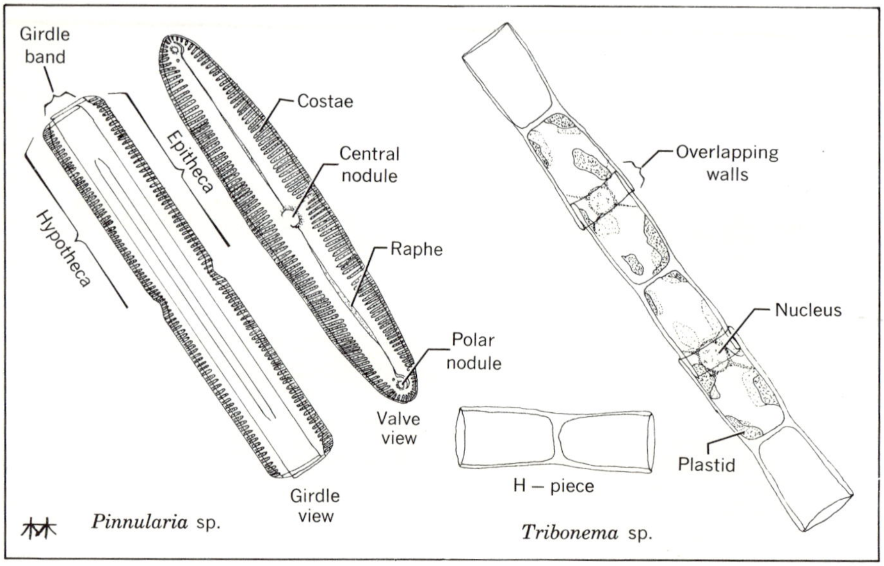

Figure 12.1
***Pinnularia* sp., girdle and valve views, *Tribonema* sp. Note overlapping walls in each case.**

VEGETATIVE STRUCTURE

Body forms of the Chrysophytes include unicells (e.g., *Botrydium*), diatoms (Figures 12.1, 12.2, 12.3, and 12.6) and filaments (*Tribonema*, Figure 12.1), *Vaucheria* (Figures 12.4 and 12.5). They remain relatively simple, although some do exhibit sparse branching (e.g., *Vaucheria*). The cells contain well-organized nuclei, plastids, and other cellular constituents. Some species have multinucleate cells while other species are completely **coenocytic** and produce cross walls only when sex organs are produced or when the filament is injured (e.g., *Vaucheria*).

Although all of the Chrysophytes do not possess walls that overlap, the characteristic is often present. The filamentous genus *Tribonema* (Figure 12.1) is composed of cells whose walls overlap in the center. When the filament disassociates, the cells break at the thin central area in a distinctive pattern. The pattern resembles an H lying on its side; these structures are called "H-pieces" (Figure 12.1). The overlapping cell walls are also apparent in the diatoms (Figure 12.1). In these chrysophytes, the walls fit over one another like the halves of a Petri dish; each half is called a **valve.** The overlapping line is often covered by a **girdle band,** but the halves are distinctive. In diatoms, the larger valve is called the **epitheca** and the smaller

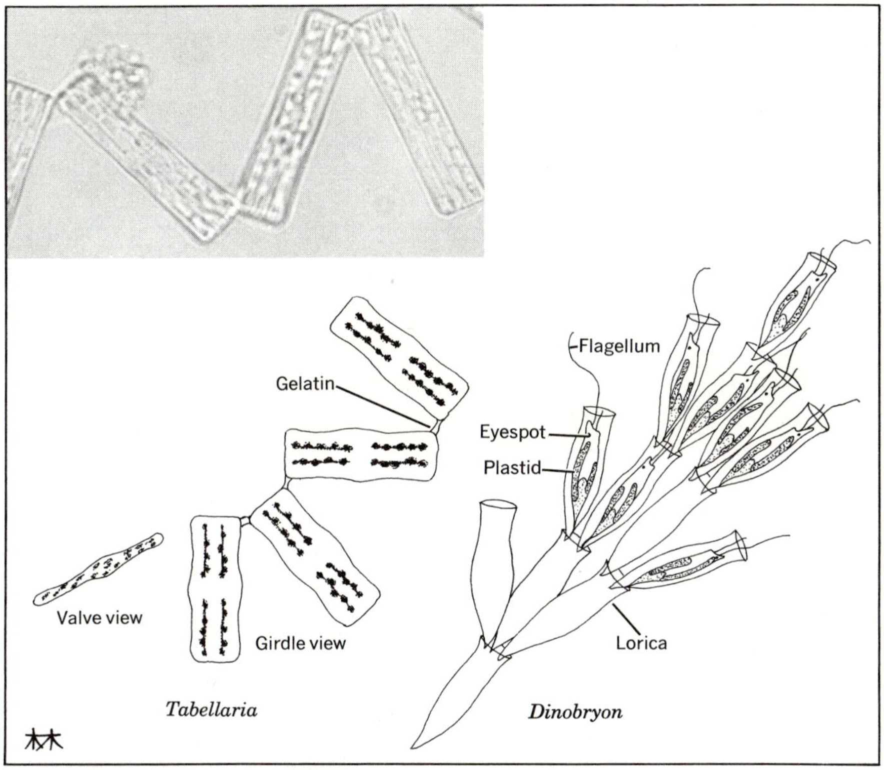

Figure 12.2
(*a*) ***Tabellaria*** **sp. Insert: photomicrograph (600×). (*b*)** ***Dinobryon*** **sp.**

Figure 12.3
A. ***Stephanodiscus*** **sp., a centric diatom. B.** ***Gomphonema*** **sp., a stalked pennate diatom.**

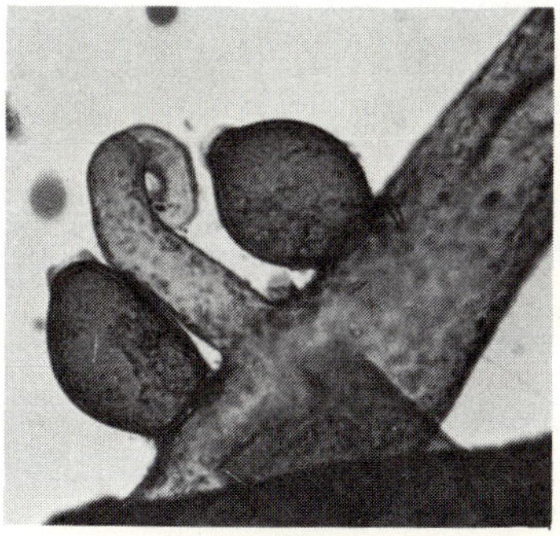

Figure 12.4
Vaucheria sessilis. **Photomicrograph of an antheridium that has discharged its sperm, between two oögonia containing zygotes (136×).**

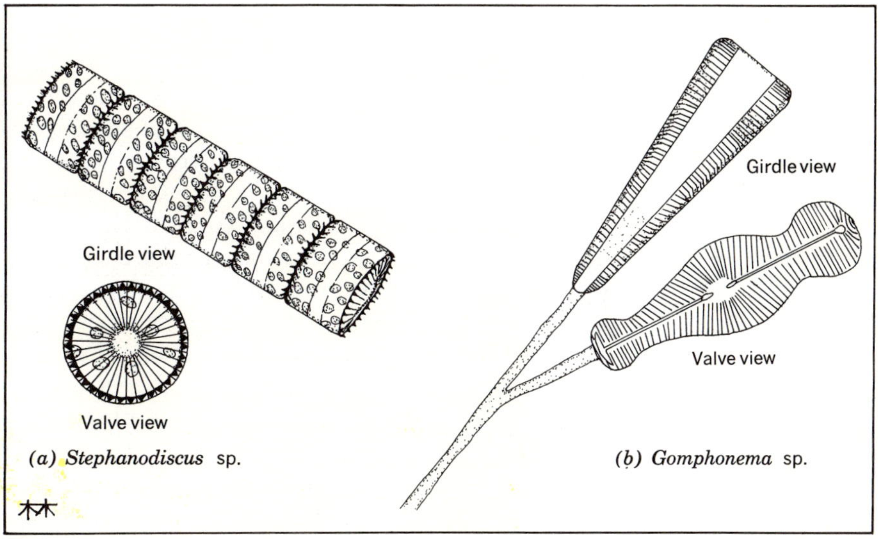

valve is the **hypotheca.** These distinctions are very important during the discussion of asexual reproduction.

Several genera of diatoms are connected at their corners by a gelatin that is secreted through pores. These connections produce beautiful arrangements of the vegetative cells. Some fresh-water genera exhibiting such arrangements are *Diatoma* spp., *Asterionella* spp. and *Tabellaria* spp. (Figure 12.2). Other diatoms are stalked (e.g., *Gomphonema* sp., Figure 12.3).

Many of the other chrysophytes do not possess overlapping walls, but have walls that are continuous. Another form is illustrated by the colonial form *Dinobryon* sp. (Figure 12.2) that produces a rigid wall-like covering, the **lorica,** which is open at one end. The protoplast lies within this lorica, is biflagellate, and possesses ameboid movement. Because of the space between the lorica and the plasma membrane, the protoplast is capable of movement within its lorica. The protoplast may move up the side of the lorica, extend from the opening, and feed on bits of solid food materials (become holozoic) although it usually produces its food by photosynthesis. The lorica is the site of colony formation during asexual reproduction.

The walls of many of the chrysophytes exhibit various types of ornamentation, but it is among the diatoms that the ornamentation reaches its fullest expression. Beautifully intricate patterns can be seen even when observing living organisms. However, when the cells are "cleared" of all protoplasm, the beauty of the silicon wall is indeed evident. The ornamentations, or markings, may be of three kinds. There may be lines that are thin "grooves" in the wall of the valve. These are called **striae.** The striae, however, may not be a groove but may be a series of small pores that are called **punctae.** One can sometimes resolve the individual punctae under careful oil immersion microscopy. There may also be ridges, called **costae,** on the inside of the valve (Figure 12.1).

Diatoms are often seen in two views (Figures 12.1 and 12.3). These are the **valve view** (when the top or "face" of a valve is seen) or the **girdle view** (when the side or edge of the diatom is observed). In the valve view, the **centric diatoms** (those with **radial symmetry**) exhibit their markings in a pattern that radiates from a central point on the valve. The **pennate diatoms** (**bilaterally symmetrical** or **asymmetrical**) exhibit their markings in a pattern that is roughly bilaterally symmetrical with respect to an axial strip in valve view. The axial strip may be perforated by a longitudinal groove that extends from one end of the diatom to the other. If such a groove is present, it is called a **raphe.** The raphe usually possesses a thickened wall about midway, the **central nodule,** and a thickened wall at each end, the **polar nodules.** An axial strip without a raphe is called a **pseudoraphe.** Sometimes the markings on the valve are so perfectly arranged that one has difficulty ascertaining the presence of a raphe. However, the presence of the nodules provides the necessary clue.

Only those pennate diatoms that possess a raphe are capable of movement. The movement may be smooth or jerky. The rate probably varies according to light, and the direction may be dependent upon the shape of the raphe, which varies from straight to sigmoid. Pennate diatoms are often observed to move back and forth along the same longitudinal axis. The cause of the movement is not completely understood, but apparently cytoplasm flows out of one polar nodule and through the V-shaped raphe to the central nodule. There is also cytoplasm flowing from the central nodule through the raphe to the other polar nodule. The resulting friction with the surrounding water or a solid substrate causes the diatom to move in a direction opposite to the direction of movement of cytoplasm.

REPRODUCTION

Large populations of chrysophytes are produced by asexual means. Many species produce flagellated zoöspores whose flagella are of unequal length. The longer flagellum is of the tinsel type and is probably responsible for the propulsion of the zoöspore. The shorter flagellum is of the whiplash type and probably acts as a rudder. Insertion of the flagella may be anterior or lateral according to the species. *Tribonema* spp. produces zoöspores that have anterior flagella. The zoöspore swims for a time, settles down and attaches to the substratum by a discoid holdfast. The flagella are lost and the zoöspore germinates with the production of a young filament. The distal portion of the filament breaks away and the plant becomes free-floating. Some species of chrysophytes produce zoöspores that may also act as gametes.

The coenocytic genus *Vaucheria* (Figure 12.5) produces a single zoösporangium by producing a cross-wall near the tip of a vegetative branch. One large zoöspore is produced within the zoösporangium and is liberated through a pore in the tip. Although the zoöspore may appear to be ciliated, it really possesses numerous pairs of slightly unequal flagella. During formation, the protoplast contracts slightly from the zoösporangium wall and the numerous nuclei move toward the surface of the protoplast. Two flagella are then produced for each nucleus. Each zoöspore is, then, a group of biflagellate zoöspores that retain their individual identity only by the production of their specific flagella. The zoöspore swims slowly for a time, loses its flagella, and germinates with the production of a new plant.

Many species of the chrysophytes produce aplanospores as well as zoöspores. Fragmentation of both filamentous and colonial forms is also important for population increases.

The unicellular diatoms possess one of the most interesting methods of asexual reproduction among the chrysophytes (Figure 12.6). Karyokinesis is followed by cytokinesis inside the old cell wall. Each of the daughter cells produces a new hypotheca by using its half of the old wall as an epitheca. Thus the old hypotheca (the smaller of the two valves) becomes the epitheca that produces a new hypotheca. This produces one new daughter diatom that is the same size as the parent diatom, and one that may be slightly smaller than the parent diatom. In theory, this practice could produce one series of offspring that would eventually be reduced to the point where they would no longer exist. Progressive reduction in size does occur among some of the diatoms, but apparently some of the daughter protoplasts increase in size before the new walls are formed and the usual size is maintained.

In those pennate diatoms whose size is progressively reduced (although probably not limited to those), a unique sexual spore, the **auxospore,** is produced (Figure 12.6). This auxospore is so named because it is a naked protoplast that increases dramatically in size soon after its formation. The auxospore may be produced by a number of mechanisms. Among these the auxospore may be produced by the conjugation of two cells, or by the fusion of two gametes from the same cell. Thus the diatoms may be either homothallic or heterothallic. After the zygote is produced, it increases greatly in size and becomes an auxospore that is larger than either parent cell. Cell walls are then produced and the new cell functions as a new vegetative cell.

The types of sexual reproduction found among the other algal groups are also present among the chrysophytes. The union of isogametes has been reported for *Tribonema* sp. Some members (i.e., *Botrydium* spp.) produce zoöspores that may act as isogametes. *Vaucheria* spp. produces very large gametangia. The antheridia and oögonia may be sessile on the main filaments (Figures 12.4 and 12.5) or they may be produced on branches from the filament. Sexual reproduction in the genus is oögamous.

LIFE HISTORY

The life history of *Vaucheria* sp. (Figure 12.5) will be described because of the ease of obtaining fresh material, and the ready availability of prepared slides. Because of their tremendous importance to man, the life histories of diatoms (Figure 12.6) will also be described.

Vaucheria spp., which is dark green in color, is often mistaken for a member of the Chlorophyta and, indeed, was until recently so classified by phycologists. After careful examination of its pigmentation, flagellation, and storage products, it was placed among the chrysophytes. The fresh-water plant is found growing as dark green, felty mats in running water, quiet pools, or on moist, undisturbed soil. The mat is composed of long, sparingly branched, coenocytic tubes. Cross-walls are rarely produced except where they separate gametangia or zoösporangia from the vegetative filament. The protoplasm in the tube is located in a layer between the wall and a large, central vacuole. Numerous discoid plastids and tiny nuclei are in this layer. Increase in the length of the tube occurs by apical growth. Asexual reproduction occurs by the method previously described in which one large zoöspore is produced within a terminal sporangium. The terrestrial species may produce thin-walled aplanospores or thick-walled akinetes. When flooded with water, however, they often produce the large zoöspores typical of the aquatic species.

The antheridia and oögonia arise as extensions from the main filament (Figures 12.4 and 12.5) or from a specialized side branch. Their protoplasm is at first continuous with the protoplasm of the main filament, but they are finally separated from the filament by the production of basal septa. Both gametangia are multinucleate when young. Just before the formation of the oögonium septum, all of the nuclei except one may either degenerate or migrate back into the main filament. Although there is some controversy as to which of these mechanisms occurs, the oögonium finally produces one large, uninucleate egg at maturity. The antheridium remains multinucleate and at maturity produces a very large number of tiny uninucleate sperm. The sperm possess two flagella that are unlike in length and mor-

Figure 12.5
Vaucheria sp. life history

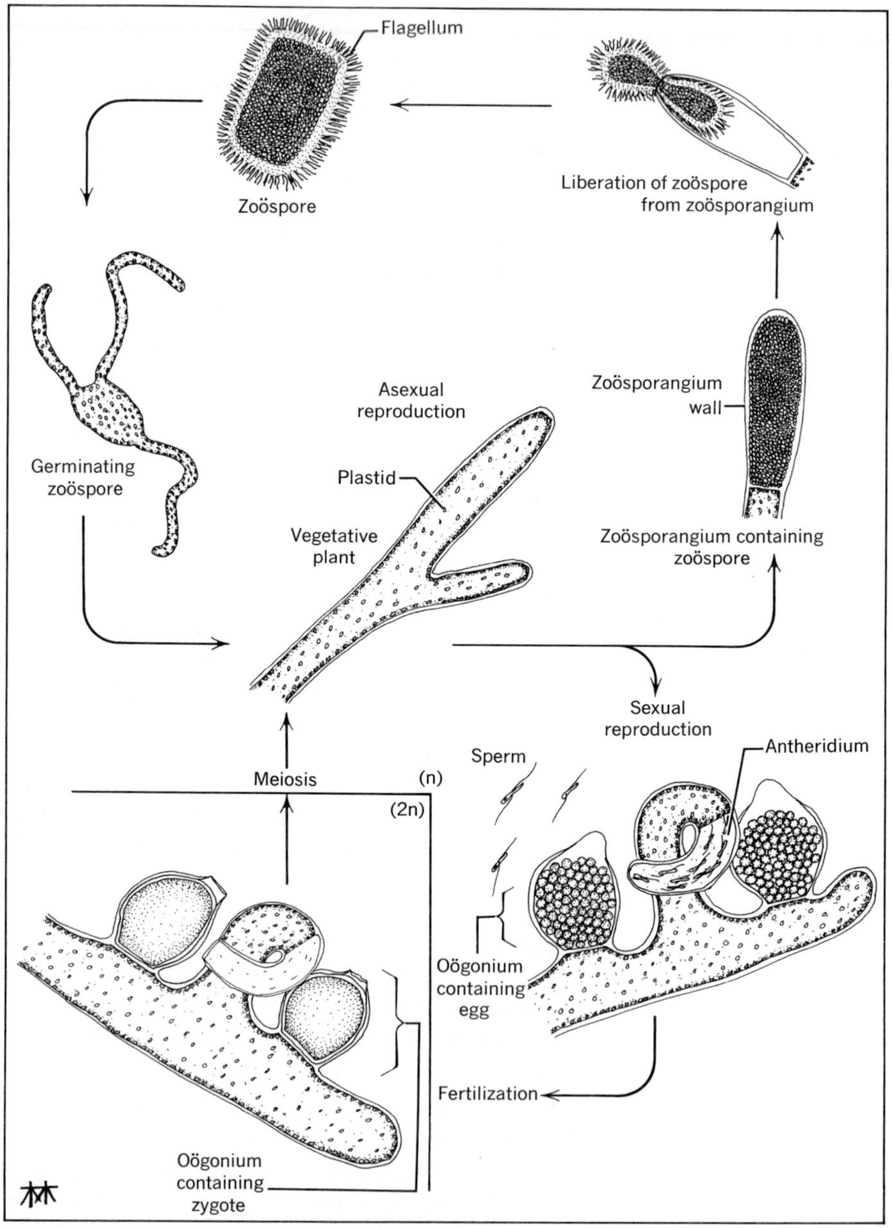

phology, and that are laterally inserted. The sperm are released from the antheridium through a terminal pore and swim to the oögonium where they enter through an apical pore and fertilize the egg. The resulting zygote soon loses its green color and becomes thick-walled. The oögonium containing the zygote often separates from the filament. After a period of dormancy, meiosis occurs in the zygote and several germ tubes are generated. The resulting new plant is haploid. The type of life history is, then, **haplobiontic, haploid.** All fresh-water species are homothallic and bear their gametangia adjacent to one another. Some of the marine species are heterothallic.

The pennate diatoms usually reproduce by the asexual method of mitosis followed by cytokinesis. As previously described, each of the valves becomes an epitheca that produces a new hypotheca (Figure 12.6). Huge diatom populations may be produced as a result of this sort of asexual reproduction.

Sexual reproduction in the pennate diatoms is by the production of **auxospores,** a specialized kind of zygote that increases greatly in size soon after its production. The vegetative cells of these diatoms are diploid and gametes are produced after meiosis in the nucleus. However, several patterns of gamete production and sexual union may occur after meiosis. Three of the four resulting nuclei may degenerate, leaving only one nucleus in a viable gamete. Compatible diatoms of this type produce only one auxospore. In other types, two of the four nuclei may degenerate, and diatoms of this type produce two auxospores. In these two kinds, the gametes are released from adjacent, compatible cells and union occurs outside of the parent cell walls. In other pennate diatoms, auxospores are produced as the result of fusion of gametes produced from the same cell. There are reports of parthenogenetic development of diploid gametes into auxospores. Homothallic and heterothallic species of pennate diatoms are known.

The centric diatoms are mostly marine, although there are some fresh-water species (e.g., *Stephanodiscus* sp., Figure 12.3). Asexual reproduction occurs as described for the pennate diatoms. Oögamy has been reported for a number of the Centrales. Tiny motile sperm cells ("microspores") are produced in vegetative cells, released by the breaking of the cell wall and swim to the egg that is contained in another cell that has not opened. The sperm, which is uniflagellate in some species, enters at the girdle band of the female cell and

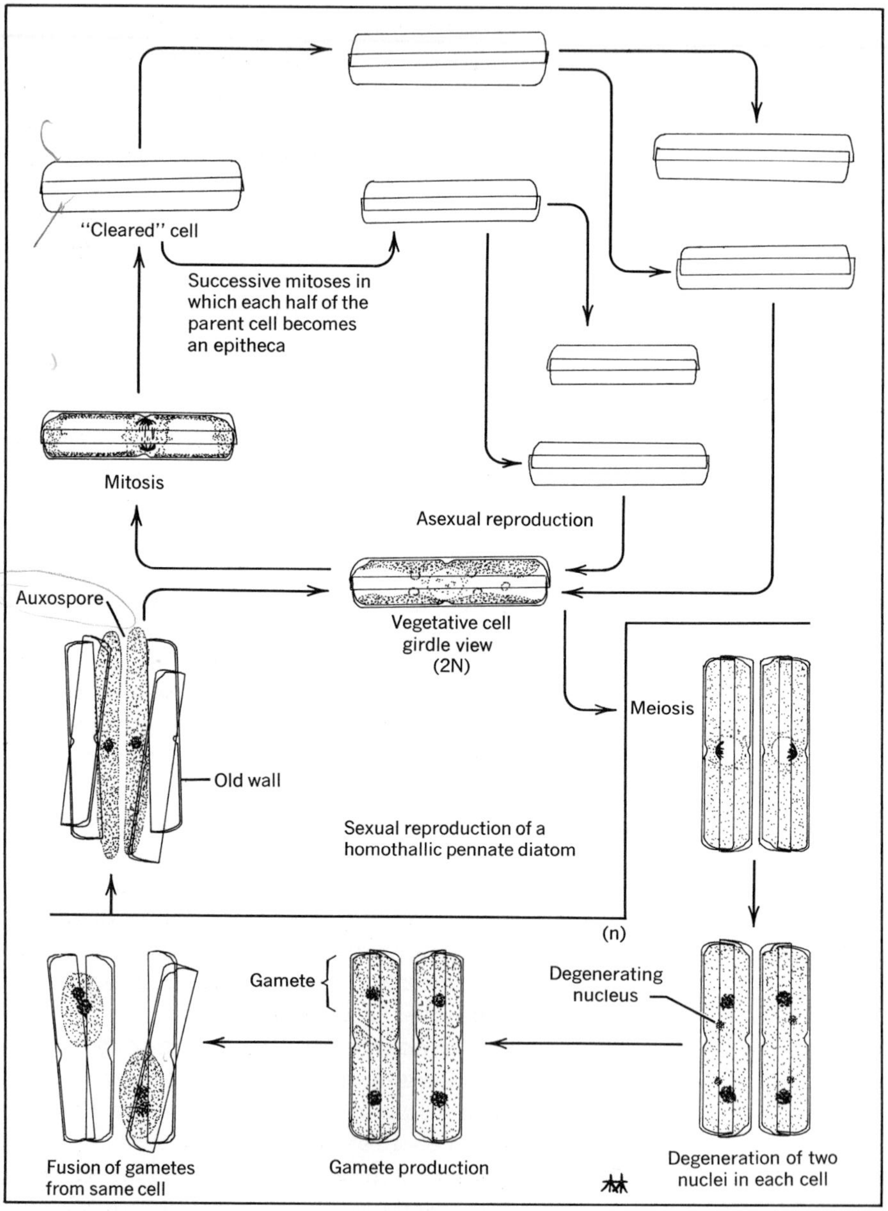

Figure 12.6
A "generalized" life history for pennate diatoms.

fertilizes the egg. The resulting zygote is released by the separation of the valves and the auxospore increases in size. Auxospore formation among the centric diatoms also occurs by the production of two gametes within the same cell. Self-fertilization occurs and an auxospore is produced. The diatoms all apparently have a **haplobiontic, diploid** life history.

HUMAN RELEVANCE

The diatoms are certainly among the most important groups of algae. Many of them are found in the phytoplankton and form the basis of food chains for aquatic animals.

However, this is one group of algae in which the fossil members from the Tertiary and Quaternary Periods are extremely important. As diatoms are deposited in both marine and fresh-water habitats, they form a greyish to white, loose to compact deposit known as **diatomite (diatomaceous earth)**. Sedimentary deposits of the fairly insoluble silaceous cell walls may form beds up to 100 meters thick. These deposits mark the bottoms of lakes or shallow seas, and beds of diatomite are still being deposited today. This material has become useful to man in a number of ways. The largest commercially valuable deposits are of marine species. Hundreds of thousands of tons are mined yearly in the United States alone. After processing, the diatomite is chemically inert and is used in many types of filtration processes. These include swimming pool water purification, clarification of antibiotics, sugar, beer, fruit juices, wines, and industrial solutions. It is used in paint to remove gloss to make flat enamels and varnishes and in highway line paint to roughen the line. It is used as a carrier for catalysts in industry and in insecticide dusts. It prevents the caking of fertilizers, reinforces rubber, toughens plastics, and increases the workability of cement. It is especially important for use as insulation for very high temperature or very low temperature processes. Polishing compounds for lenses and silver contain diatomite. At one time it was used by Nobel as an absorbent agent for nitroglycerine, and it was used to make bricks for the dome of the Cathedral of St. Sophia in Constantinople in A.D. 532.

Aside from their present commercial value, the fossil diatoms are indicators of various geologic changes. They are divided into definite marine and fresh-water genera, and can indicate previous fresh-water and marine habitats. By comparing fossil examples of *Melosira* spp. with present-day examples of the same genus, one phycologist is led to believe that he can interpret changes from a cold water, nutrient-poor flora to a warm water, higher-nutrient status, to a cooler water, high-nutrient status. Changes that have occurred in lake and drainage areas are also indicated by the study of fossil diatoms.

Certain of the diatoms as well as certain of the other chrysophytes are important as indicators of various conditions that exist in water supplies. By recognizing indicator species, one can deduce the source of a particular water sample, that is, from a hard water lake with an outlet or from a hard water lake with no outlet. Certain diatoms are cold water indicators. Many species can indicate water contaminated with industrial wastes such as copper, phenol, oil, hydrogen sulfide, iron, chromium, and sodium chloride. Others indicate conditions of high acidity.

Although man has found many constructive uses for the chrysophytes, he often finds himself in conflict with some of them, especially in his water supply. Along with members of other phyla, many chrysophytes (e.g., *Asterionella* spp., *Tabellaria* spp., *Dinobryon* spp., and *Synedra* spp.) create taste and odor problems. *Melosira* spp., *Diatoma* spp. and *Fragilaria* spp. clog intake filters. Slime is produced by such diatoms as *Cymbella* spp. and the corrosion of concrete may be caused by *Diatoma* spp.

13

With the exception of perhaps three very rare genera, the brown algae are exclusively marine. Most of the brown algae occupy the intertidal zones of colder marine waters (Figures 13.1*a*, p. 254, 13.1*d*, and 13.2*a*, p. 254). A few (e.g., *Sargassum* spp.) are distinctly warm-water plants. They are found in open waters, rocky coastlines, and marine estuaries. Many are attached to rocks, muddy bottoms, shells, or some other nonliving substrate, but many are epiphytic on other algae, including brown algae. There are a few that are endophytic but there are no known parasites or saprophytes. Some are capable of withstanding exposure to the atmosphere during low tide, but others are continually submerged. They occupy a relatively narrow band of habitats around the oceans' edges because none of them can survive above the splash zone and very few are ever found very deep in the water.

All members of the Phaeophyta are multicellular and most are macroscopic. Some are morphologically complex and have a distinctive shape. The largest algae of any group are found among the browns, with *Macrocystis* spp. (Figure 13.1*b*) reaching more than 150 feet in length.

Although chlorophyll is the photosynthetic pigment, the plastids owe their distinctive brown coloration to the predominance of **beta-carotene** and **xanthophylls.** Of the latter, **fucoxanthin** is characteristic of the Phaeophyta. The excess photosynthate is stored chiefly as a carbohydrate (**laminarin**), but there may be an accumulation of an alcohol (**mannitol**) and fat drops.

Phycologists recognize about 225 genera and 1500 species of brown algae. These are variously divided into either three classes or lumped into the one class Phaeophyceae. The former divisions occur because of the various kinds of life histories present within the group. Although the life history types will be discussed in the appropriate section of this chapter, only the class Phaeophyceae is recognized.

VEGETATIVE STRUCTURE

The protoplast of the brown algal cell is much like that found in the vascular plants. There is a well-organized nucleus with the usual structures. Mitotic figures are produced during nuclear duplication and a number of species have **centrosomes** and **astral rays** during this process. There are usually several brownish, variously shaped plastids in each cell; the color caused by the fucoxanthin. The wall of the cell is composed of an inner cellulose layer and an outer gelatinous layer that consists of **algin.**

The plant body usually has a definite form. It may be composed of a single row of cells (filamentous) or its cells may be arranged in a parenchymatous manner. Although there is little tissue specialization, many of the brown algae have recognizable **holdfasts, stipes,** and **blades** (Figures 13.1*b*, p. 254, 13.1*c*, 13.2*b*, p. 254, and 13.2*c*). These structures vary considerably in size and morphology from species to species. The plant body may be either annual or perennial. Some members have areas in which cells are actively duplicating (**meristems**). Although meristems are rarely found among the algae, there are at least three types found in this group. One of these (**meristoderm**) is found at the outermost edge of the blade of some of the Laminariales and increases its diameter. **Intercalary** meristems are those that are inserted somewhere in the length of an elongate structure. Brown algae may have intercalary meristems located at the base of a branched uniseriate filament ("hair"). This type of specialized meristem is known as **trichothallic,** and increases the length of both the filament and the blade. Other intercalary meristems that do not have any relationship to a "hair" may also be produced. Some are found at the junction of a stipe and a blade, and increase the length of each of these structures. **Apical** meristems are also produced. These may take the form of a single row of cells or as a single cell from which other vegetative cells are derived. Both types of apical meristems increase the length of the plant body.

It is among some of the brown algae (e.g., *Laminaria* spp., Figure 13.1*c*, p. 254, and *Macrocystis* spp., Figure 13.1*b*) that is found a structure that closely resembles the sieve tubes in the phloem of vascular plants. This structure is often called a **trumpet hypha** because of its shape, but it includes a **sieve plate** that separates two cells that are linked at their ends. Therefore some phycologists call these cell series sieve tubes. At maturity, the sieve tubes do not contain nuclei and it has been demonstrated that ^{14}C-labeled food materials do move through stipes containing the sieve tubes at a rate comparable to their movement through phloem in vascular plants.

Various of the large brown algae ("**kelps**") contain flotation mechanisms known as **air bladders** (Figure 13.1*b*, p. 254). The air bladders are located on the stipes or blades, and apparently allow the major photosynthetic part of the plant to float near the surface of the water. The stipe and blade of a kelp may be entire or split into a number of portions. The members of the genus *Nereocystis* have a large bladder (**pneumatocyst**) at the tip of their unbranched stipes. A large number of blades arise from the pneumatocyst and float near the water's surface.

An extremely interesting brown alga, *Postelsia palmaeformis*, the sea palm (Figure 13.2*a* and *b*), is found in the intertidal zones along the western coast of North America from British Columbia to central California. This sturdy alga is able to withstand the continuous beating of the waves because of its stout holdfast. The stipe is erect, cylindrical, hollow, and tapers toward the tip. One hundred or more blades may be attached to the tip of a mature plant. The plant may attain a height of 2 feet, and the blades a length of 10 inches. At low tide the stipes are erect and the blades hang. The appearance of the alga in that condition is one of a small palm tree; hence its specific epithet.

Fucus spp. (Figure 13.2*c*, p. 254 and *d*), is one of the most common brown algae studied in a general botany course. This plant is dichotomously branched and flattened with a holdfast that attaches it mostly to the rocks of the intertidal zone. The flattened plant body (**thallus**) possesses a prominent "midrib" and air bladders. Two kinds of apices are often found. The vegetative apex is flattened and resembles the rest of the thallus except for the lack of a "midrib." The reproductive apex is enlarged and contains several structures that will be discussed later. Because of its usual habitat among the rocks, *Fucus* spp. belongs with a group of brown algae commonly known as the **rockweeds.**

REPRODUCTION

Asexual reproduction of many of the brown algae may take place by the fragmentation of an adult or a young thallus. Longitudinal splitting of the thallus may produce two new attached thalli, but certain parts of the thallus may break away completely and form new plants while floating freely in the water. The genus *Sargassum* is the best example possessing this characteristic, and is well known because of its presence in the Gulf Stream and the Sargasso Sea. Certain browns produce specialized branches (**propagula**) that produce new plants when they become separated from the parent plant.

With the exception of the order Fucales, all of the Phaeophyta produce zoöspores. These asexual spores possess two laterally inserted flagella that are unlike. The longer flagellum projects forward and is of the tinsel type; the shorter flagellum, which is of the whiplash type, projects backward. The zoöspores may be produced in one of two types of sporangia. The first of these types contains but a single cavity and is known as a **unilocular zoösporangium** (Figures 13.3*a* and *b*, and 13.4). This sporangium is produced only on a diploid plant, but meiosis occurs during zoöspore production resulting in the zoöspores (up to 128) being haploid. These zoöspores germinate producing only haploid plants. The other type of sporangium is also produced on a diploid plant (often the same one that produces the unilocular zoösporangium). This sporangium (**the plurilocular zoösporangium**) is multicellular and produces one zoöspore in each of its cells (Figure 13.3*a*). The zoöspores produced in the plurilocular zoösporangia are diploid and germinate into diploid plants like the parent plant.

Many of the brown algae produce multicellular gametangia that have the same developmental and mature forms as are exhibited by the plurilocular zoösporangium. These particular gametangia are called **plurilocular gametangia** (Figure 13.3*c*) and each of their cells produces a gamete. Confusion about the two organs, one of which produces asexual spores and the other of which produces gametes, existed for a long time. It was then found that when the plurilocular organ was produced on a haploid thallus it was a gametangium; when produced on a diploid thallus it was a sporangium. Unless one knows the type of spore from which the thallus developed, the fate of the motile spores produced by the organ, or the chromosome number of the vegetative cells, he cannot determine whether any particular plurilocular organ is a sporangium or a gametangium. The presence of the unilocular zoösporangium indicates that the plant is diploid but if it is not present, determination may be extremely difficult.

Most of the sexual reproduction by those members that produce plurilocular gametangia is isogamous; a relatively few reproduce by heterogamy. They may be either homothallic or heterothallic. Those

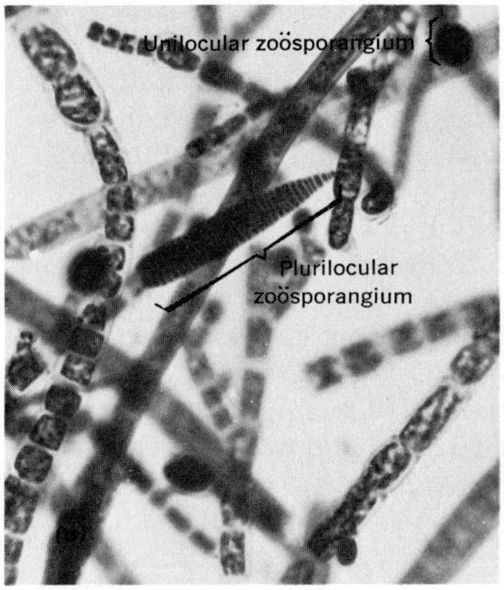

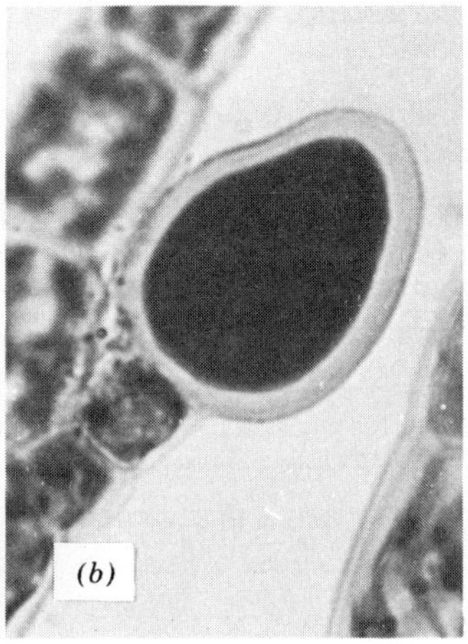

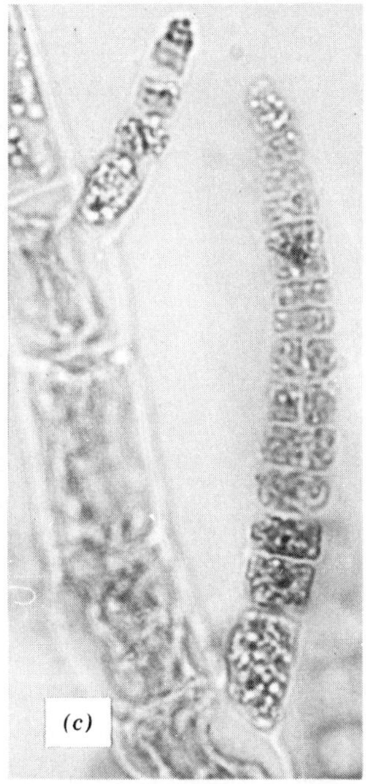

Figure 13.3
Reproductive structures of *Ectocarpus* sp. (*a*) Sporophytic plant with plurilocular and unilocular zoösporangia present (160×). (*b*) A mature unilocular zoösporangium (760×). (*c*) Gametophytic plant with immature plurilocular gametangia (580×).

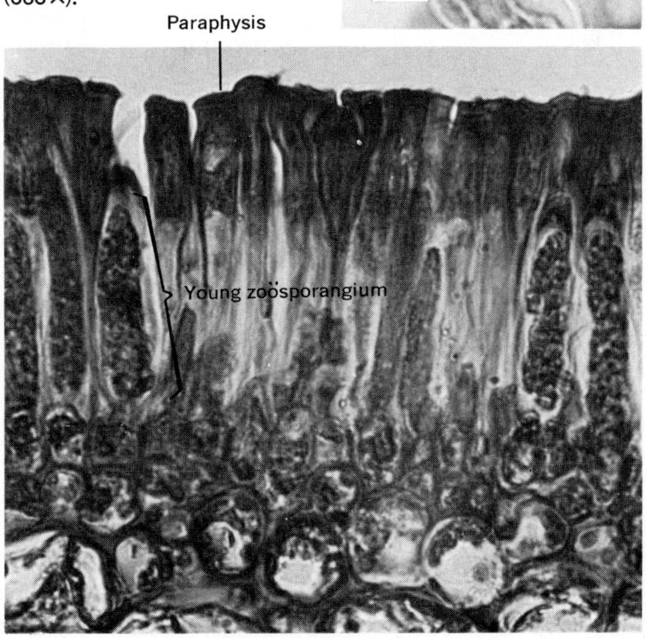

Figure 13.4
Section through sorus of sporophyte of *Laminaria* sp. (630×).

gametes that fail to fuse usually disintegrate; some develop parthenogenetically (Figure 13.6) into haploid thalli. Unfertilized female gametes of some heterogamous members regularly develop parthenogenetically. Oögamy occurs in many of the brown algae (Figure 13.8). These produce specialized gametangia (oögonia and antheridia) that produce eggs and sperm. Parthenogenetic development of gametes has also been observed in the oögamous members.

The brown alga most commonly used for classroom demonstrations is *Fucus*. Although this "rockweed" is easily collected and inexpensive to purchase from biological supply houses, it possesses a life history (Figure 13.8) that is quite different from most of the other brown algae. *Fucus* spp. (also *Sargassum* spp., a member of the same order, Fucales) does not possess a haploid thallus. Its sexual reproduction is oögamous and both sex organs may be produced on the same plant or they may be produced on different plants (Figure 13.5). In either case, the gametangia are produced in enlarged tips of the final dichotomies of the thallus (Figure 13.2*d*). These tips are called **receptacles** and contain various numbers of cavities called **conceptacles** which open to the outside environment through **ostioles** (Figure 13.5*a*). The conceptacles may contain only oögonia, only antheridia, or both, depending on the species (Figure 13.5). The gametangia are interspersed among sterile filaments (**paraphyses,** Figure 13.5*a*). The oögonia produce eight eggs (in *Sargassum* only one egg reaches maturity) but the sperms are produced in groups of 64.

Fusion of the gametes occurs after they have been released from their gametangia. The gametes may remain motile until fusion occurs (isogamous species) or one may become immotile and be fertilized by the motile one (heterogamous species). The eggs of the oögamous species are often released into the water before fertilization by the sperm, although in some species the egg is fertilized while still retained inside of the oögonium. Various methods whereby gametes are released from gametangia have developed. One of the most interesting occurs in those species of *Fucus* that have a direct correlation of gamete release to fluctuations in the tide. Low tide produces shrinkage of the receptacles and the gametangia are forced through the ostioles in slimy masses. The next high tide causes the gametangial walls to expand and dissolve, thus effecting the release of the gametes. Regardless of the method by which the gametes meet and fuse, the zygote soon germinates, producing a new diploid thallus.

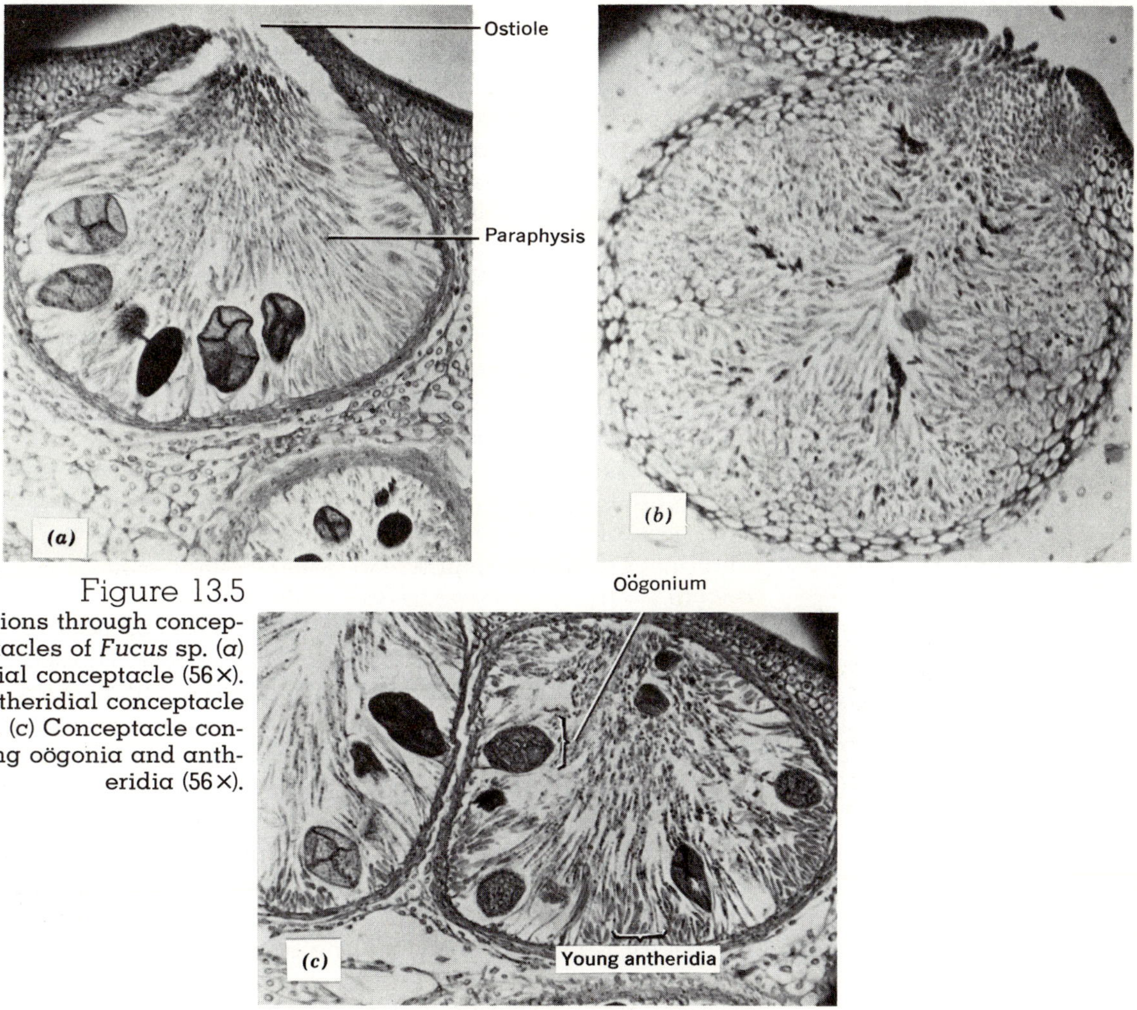

Figure 13.5
Sections through conceptacles of *Fucus* sp. (*a*) Oögonial conceptacle (56×). (*b*) Antheridial conceptacle (56×). (*c*) Conceptacle containing oögonia and antheridia (56×).

LIFE HISTORY

It should be pointed out that, with the exception of the Fucales, all of the brown algae have a definite alternation of a sexual phase with an asexual phase. That is, there is an alternation of a gametophytic generation (**gametophyte**) with a sporophytic generation (**sporophyte**). This type of life history is known as **diplobiontic.** Some of the members (e.g., *Ectocarpus* spp.) produce gametophytes and sporophytes that are morphologically indistinguishable. This condition is known as an **isomorphic alternation of generations.** Other members (e.g., *Laminaria, Postelsia*) produce gametophytes and sporophytes that are morphologically quite distinct. This condition is known as a **heteromorphic alternation of generations.** The members of the Fucales do not have an alternation of generations as the only haploid phase of their life cycle is the gametes. They, therefore, exhibit the **haplobiontic, diploid** life history.

The sporophyte and gametophyte of species of *Ectocarpus* (Figures 13.1*d*, p. 254 and 13.6) are branching filaments. The sporophyte (2n) produces both unilocular and plurilocular zoösporangia. Meiosis occurs only in the unilocular zoösporangium, and the resulting haploid zoöspore germinates producing a haploid gametophyte. It has been reported that two of the zoospores may act as gametes and unite, repeating the sporophyte. The plurilocular zoösporangium produces diploid zoöspores that germinate producing new sporophytes. The gametophyte produces only plurilocular gametangia. The haploid gametes are released, fuse, and produce a new sporophyte. Parthenogenetic development of gametes may repeat the gametophyte.

Laminaria spp. (Figure 13.1*c*, p. 254 and 13.7) possesses a macroscopic sporophyte and a microscopic gametophyte. The sporophyte produces hundreds of unilocular zoösporangia that are produced in groups (**sori**) and found on the blade. Meiosis occurs during the production of the zoöspores that are released and germinate producing gametophytes, both types of which are filamentous. The male gametophyte is irregularly branched with clusters of antheridia located at the tips of the branches. The antheridia are unicellular, and each produces a single sperm. The female gametophyte has fewer, but larger, cells than the male. The oögonia are either terminal or intercalary, and each produces one egg. At maturity the egg is extruded from the oögonium, but remains attached to it. The sperm swims to the egg and fertilization occurs. The young sporophyte begins to develop while the zygote is still attached to the gametophyte.

Fucus spp. (Figures 13.2*c* and *d*, p. 254 and 13.8) does not possess an alternation of generations; thus making its life history different from the members of the other orders. Its sex structures are produced in receptacles as has been previously described. Meiosis occurs during the production of gametes that unite forming a zygote that germinates producing a new sporophyte.

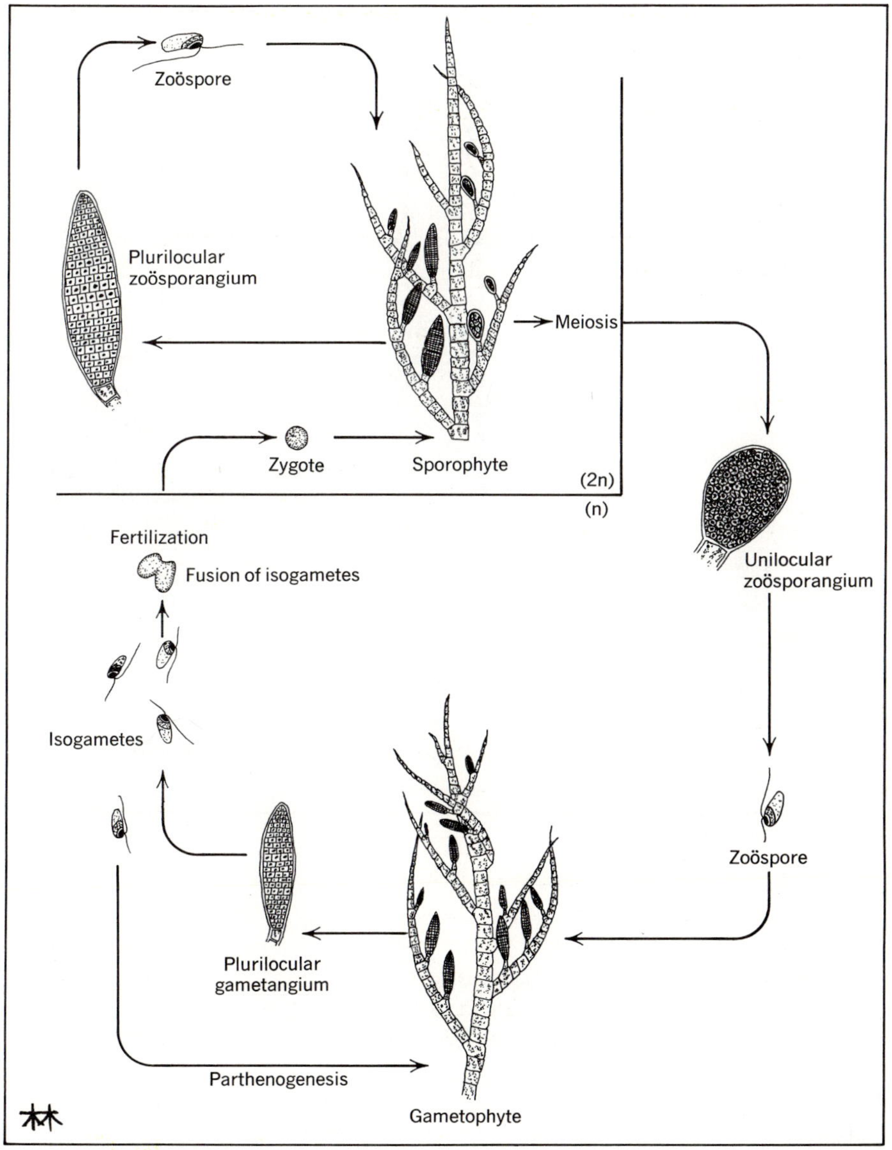

Figure 13.6
Life history of *Ectocarpus* sp. A diplobiontic, isomorphic life history.

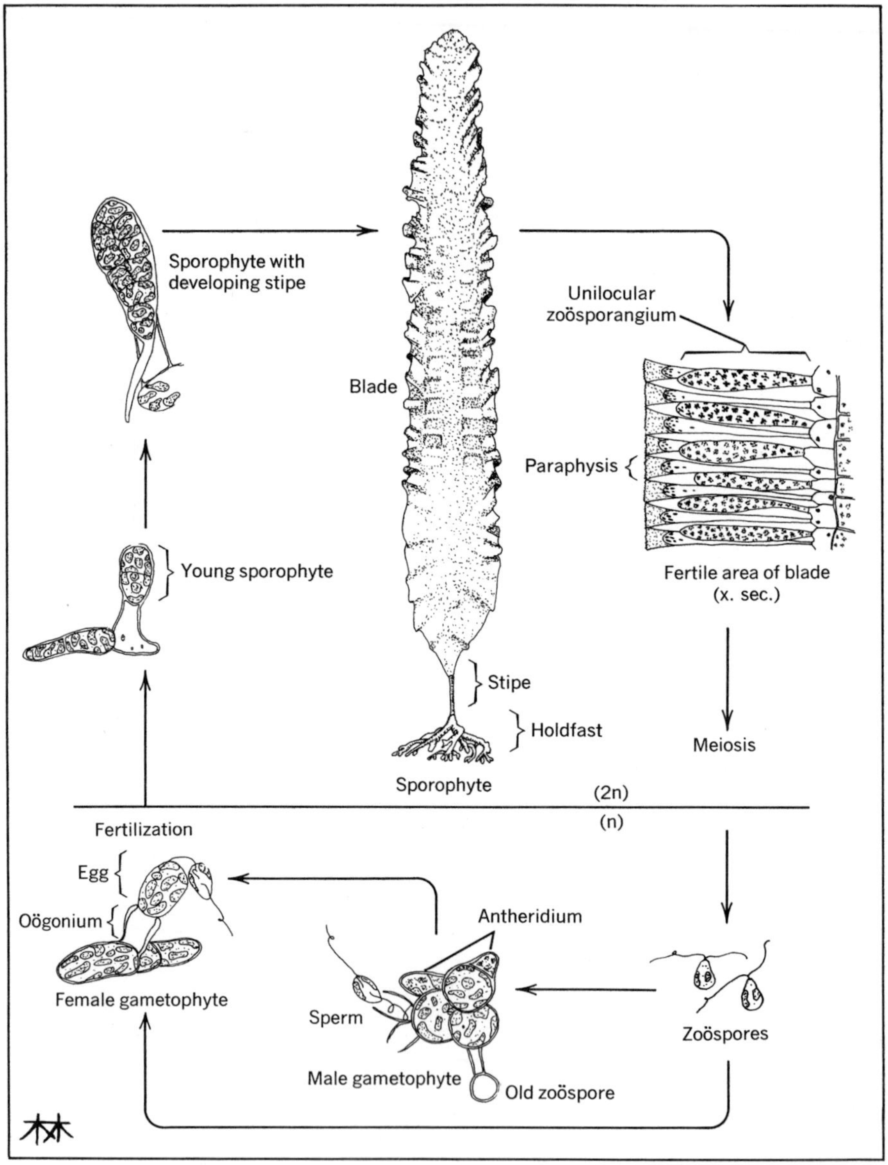

Figure 13.7
Life history of *Laminaria* sp. A diplobiontic, heteromorphic life history.

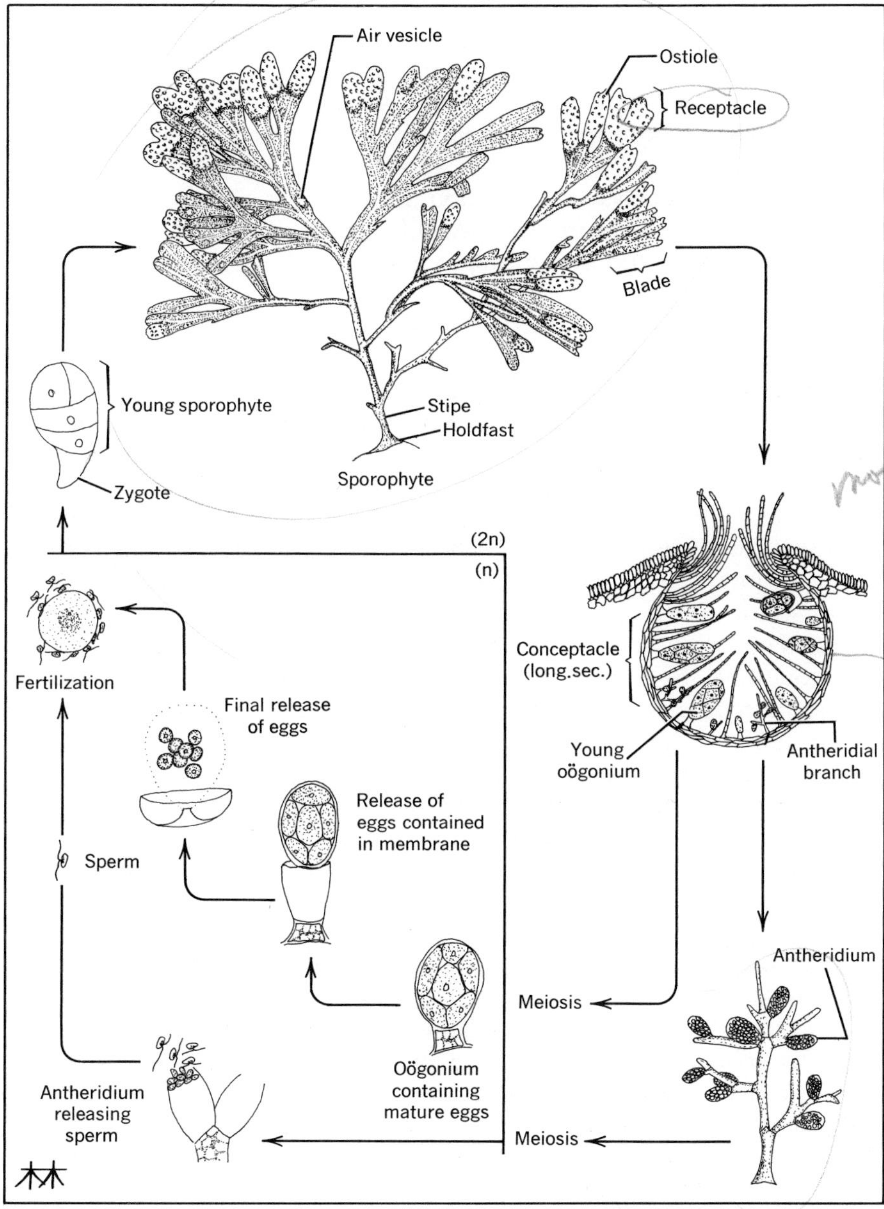

Figure 13.8
Life history of *Fucus* sp. A haplobiontic, diploid life history. The one illustrated is bisexual (both types of sex organs in the same conceptacle).

HUMAN RELEVANCE

Aside from their important position as primary producers in the marine environment, the brown algae are of great direct importance to man. Kelps that were washed ashore during storms and the intertidal rockweeds were extremely important as sources of potassium and iodine, and the "algae rights" along the beaches were hotly disputed by various processors. Although no longer so important in this country because of natural deposits of these elements, the ash obtained by burning the brown algae is still used in many other places in the world.

Brown algae (e.g., *Laminaria, Alaria*) is used as a direct food source by many Oriental peoples. Kombu is a product of various browns and is eaten in Japan as a cooked vegetable, or boiled with fish or other meats. Kombu is processed by boiling the blades in fresh water for a few minutes and then pressing them together in large flat plates that are dried and shredded. The "fishy" flavor is unpleasant to most Western palates. The browns are used as food for domestic animals such as hogs, horses, cattle, sheep, and poultry in the Scandinavian countries, Germany, and France. They are used as a vitamin source in various processed pet foods.

Fucus spp. has been used as a red dye for textiles, and was used as a rouge by the ladies of ancient Rome. Other brown algae are important medicinally as components of greaseless lubricating jellies and ointments. Iron alginate is used to increase the amount of hemoglobin in the blood. Long before the cause of goiter was known, the Chinese were using the brown algae as a cure for this disease. Because many of the algae contain up to 5% iodine, the cure worked. During the 17th century, kelp ash was used to glaze ceramics. *Laminaria* spp. has been used by the Chinese and others to produce a tough paper. Derivatives from this same alga have been used as binders in cartridge primers and in the firing mechanisms of marine mines.

Undoubtedly the most important uses of the brown algae comes from the colloidal gel **algin** that is found in their middle lamellae and primary cell walls. The amount of algin varies from 14 to 40% in these structures, and may comprise up to 10% of the net weight of the plant. The amount of algin present varies with the season with more algin being present during the winter than in the summer. Many of the algae are difficult to harvest because of their habitat (e.g., *Laminaria* spp. on subtidal rocks). However, the giant kelps (e.g., *Macrocystis* spp.) are easier, and thus cheaper, to harvest because they grow on long stipes in deep water. The stipes of *Macrocystis* spp. are cut by blades that extend three feet into the water from barges. Thus the large floating plant blades are released and can be pulled on board a barge by a chain elevator. A five-man crew can harvest 300 tons of kelp a day in this fashion. The blades of *Macrocystis* have been reported to regenerate at the rate of 10 to 15 feet per week.

Alginic acid and its salts are derived from the plants by various industrial methods. These products are extremely important in industry and food production because they are nontoxic, highly viscous and good gel formers. When added to ice cream, algin prevents the formation of coarse ice crystals. It is used as a stabilizer and moisture retainer in many products such as icings, candy, marshmallows, cream, mayonnaise, and cheese. It is important in various jellies. Algin is extremely important in the cosmetics industry because of its high viscosity (37 times greater than gum arabic). It is thus used in hand and setting lotions, shaving creams, suntan lotions, soaps, and hair creams.

The rubber industry uses algin as a creaming and stabilizing agent in the processing of natural and synthetic rubber latex. The pigments in paint are kept in suspension and brush marks are reduced when algin is added to paint. Surface films on papers, and glazes on ceramics also contain algin. Textile yarns, sausage casings and medical gauze are made from alginate fibers. Thus the brown algae are extremely important to man's health and well-being.

the red algae | rhodophyta

14

The red algae are so named because the red pigment **phycoerythrin** often covers the presence of the other pigments that are present. Phycocyanin, a blue pigment, is also often present and may give the plant a deep blue-green or even blackish color. These two pigments absorb light energy that is transferred to chlorophyll$_a$ (and sometimes$_d$). The accessory pigments may have a direct role in photosynthesis in the red algae. The storage product for the red algae is **floridean starch,** a complex carbohydrate that gives a red color when tested with I_2KI.

The red algae are further characterized by a type of sexual reproduction that is unique in the algae. This will be discussed more fully later, but it should be pointed out here that there are certain parallels with the sexual reproduction found in the ascomycetes. There are no flagellated reproductive bodies present, and the sperm is passively transported to the female sex organ.

Like the brown algae, most red algae are found in marine habitats. There are more red than brown algae that occupy fresh-water, but the number is small. The most common fresh-water genus is probably *Batrachospermum*, a red alga that is gray-green to blue-green to olive in color. It occurs in branched masses on stones and sticks in flowing water and quiet pools.

Although members of the Rhodophyta are found in all marine environments, many more are warm-water than cold-water species. They usually are found as sessile organisms as epiphytes on other algae, or on rocks and other nonliving substrates in the littoral or sublittoral zones. Some red algae are parasitic on other red algae that are closely related to them, and are mostly host specific. Death usually occurs soon after a thallus becomes detached from its substrate. Many of the red algae grow at great depths (about 250 to 300 feet) in clear, warm tropical waters.

All of the red algae are included in one class, the Rhodophyceae. There are presently about 400 genera and some 3500 species known.

VEGETATIVE STRUCTURE

The vegetative cells of most of the red algae are uninucleate with a small central vacuole. Some species have multinucleate cells with very large vacuoles. The vegetative nuclei are usually small, but they contain the structures usually found in the eukaryotic nucleus. The cell wall is composed of cellulose and pectic compounds. The entire plant may be covered with a gelatinous material that has been variously classified and has been put to use by man in many ways to be discussed later. The walls between two adjoining vegetative cells in most red algae have pit connections that probably allow for protoplasmic continuity between cells. The floridean starch is usually found in small granules scattered within the cytoplasm. Although pyrenoids are sometimes present, the starch granules are usually more closely associated with the nucleus than with the pyrenoid bodies.

The plant body of the red algae is basically filamentous (e.g., *Polysiphonia* spp. (Figures 14.1*a*, p. 254 and 14.4). The filament is initiated by a single apical cell. Many species of red algae are membranous (e.g. *Porphyra* spp. Figure 14.1*b*, p. 254) but their filamentous nature can be demonstrated because the development of their thalli is initiated by one or more apical cells. Most of the species are highly branched. A few species are unicellular or colonial. One group of red algae has a thallus that is heavily encrusted with calcium carbonate. This group is commonly known as the coralline algae and are usually found attached to rocks although some are epiphytic. Most red algal thalli are only a few inches in length, but some may get to be six feet long.

REPRODUCTION

The red algae rarely reproduce by vegetative fragmentation, but there are various kinds of asexual spores produced. The most common asexual spore produced by the "lower" (e.g. *Bangia*) red algae is the **monospore** (**neutral spore**). The monospores are produced in special cells that enlarge forming **monosporangia** and are released by the breakdown of the sporangium wall. The ameboid spores eventually settle to the substrate and germinate forming a new plant.

The most common asexual spore among the "higher groups" is the **tetraspore.** Tetraspores are produced in **tetrasporangia** (Figure 14.2*a*) on a diploid, free-living phase of some of the red algae. Meiosis occurs in the tetrasporangia and the tetraspores are haploid. They germinate producing a new haploid phase.

Sexual reproduction among the red algae is quite unlike sexual reproduction in any other algal group, and a special terminology has evolved to emphasize that difference. The male sex organ is one-celled and produces a single, slightly ameboid sperm. The organ is called a **spermatangium** (Figure 14.2*b*) but is analogous to the antheridium of other algae. The female sex organ is called a **carpogonium** and is really an oögonium with a long, sometimes twisted protruberance called a **trichogyne** (recall the ascogonium of some of the Ascomycetes). Sperm are released in large numbers into the water. They are passively transported by water currents to a trichogyne where they lodge. The nucleus of the sperm then divides and the walls dissolve at the point of contact. One of the sperm nuclei then migrates from the sperm into the trichogyne and finally down to the bottom of the carpogonium where it fertilizes the egg nucleus.

Postfertilization activity in the red algae may be either simple or quite complex. Specific examples of life histories will be discussed in the next section, but the beginning student should be introduced to some unique structures before that time. In the simpler red algae (e.g., *Nemalion* spp.) the zygote nucleus undergoes meiosis and then successive mitoses. Cellular duplication follows with the ultimate production of a group of short filaments. The cells of these filaments mature as **haploid carposporangia** and their protoplasts are finally released as **haploid carpospores.** The carpospores germinate and become gametophytic plants.

More complicated life histories are present in other red algae (e.g., *Polysiphonia* spp. and *Griffithsia* spp.) however. There have evolved other plant phases between fertilization and gametophyte development in these algae. These phases arise as a result of delayed meiosis, the migration of the zygote nucleus, and the addition of various kinds of accessory cells. The first of these phases is the **carposporophyte** which is a diploid phase that develops from the base of the carpogonium after fertilization. A series of cellular duplications produces filaments that are densely packed with protoplasm that probably nourishes the developing carposporophyte. In this case, the

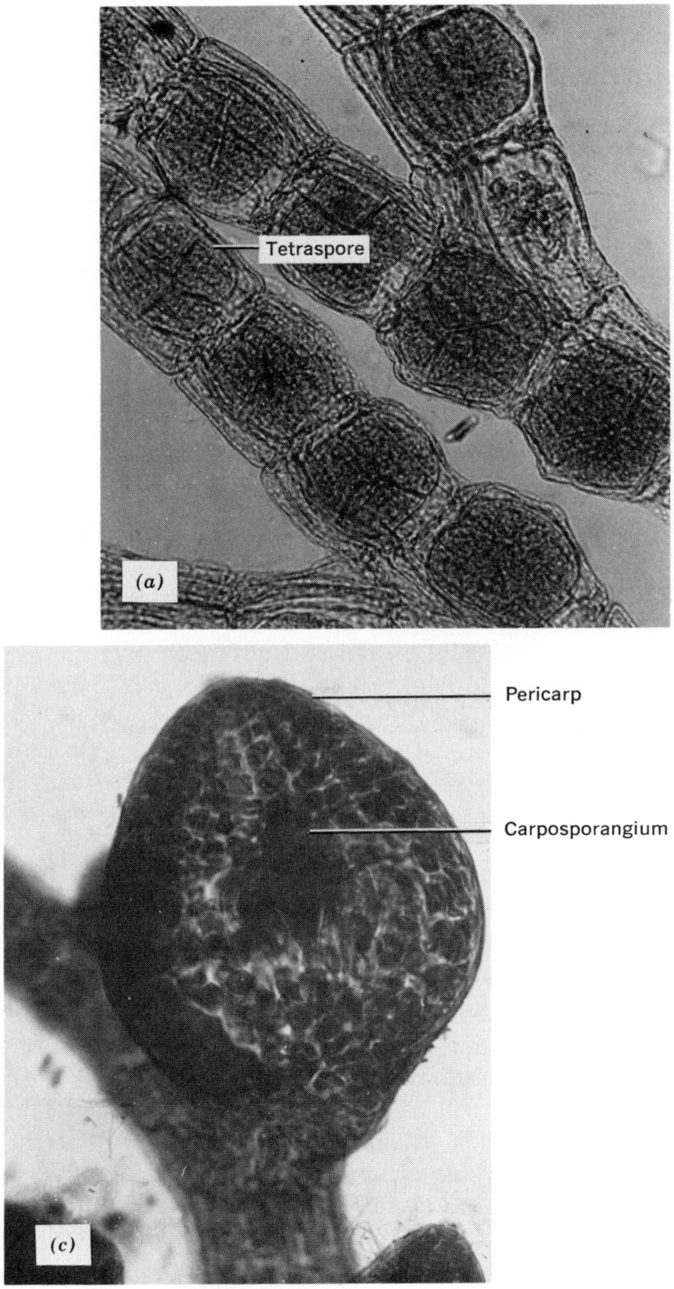

Figure 14.2
Reproductive structures of *Polysiphonia* spp. (140×). (*a*) Tetrasporangia, each of which contains four tetraspores, on a tetrasporophyte. (*b*) Male gametophyte with spermatangial branches. (*c*) Female gametophyte with pericarp enclosing young carposporangia produced by the carposporophyte.

carposporophyte is diploid and produces **diploid carposporangia** and **carpospores.** In contrast, the group of filaments in the previously discussed carposporangia (as in *Nemalion* spp.) are haploid, but they are also referred to as carposporophytes by some botanists. The diploid carpospores are liberated from the carposporangia (Figure 14.4) and are carried about by the water currents. They eventually germinate and produce still another phase to the life history, the tetrasporophyte.

The **tetrasporophyte** (Figure 14.2*a*), since it develops from the diploid carpospore, is also diploid. In form, it resembles the gametophyte phase, but when it reaches maturity it produces **tetrasporangia** (Figure 14.2*a*) rather than gametangia. The tetrasporangia are produced on short stalk cells and meiosis occurs early in their development. Four **haploid tetraspores** are produced from each tetrasporangium and, when released, they germinate into gametophytic plants.

The previous discussion of sexual reproduction among the red algae is vastly simplified. Many intermediate nuclear migrations and cellular duplications occur during the production of carposporophytes and tetrasporophytes. In summary of the reproduction of this group, however, it should be pointed out that the carpospore is the only sexual spore produced by the red algae. It may be haploid and produced rather directly from the zygote or it may be diploid and produced after the interpolation of a carposporophyte phase. If haploid, it will germinate directly into a new gametophyte. If diploid, it will germinate into an asexual diploid plant, the tetrasporophyte.

LIFE HISTORIES

The life histories of most of the red algae are not very well known. For example, the life history of members of the genus *Porphyra* are known to have a **diplobiontic, heteromorphic** life history, but the time at which meiosis occurs is still open to question. Some phycologists believe that meiosis may occur at different times during the cycle. Two examples will be used in this treatment. The first is *Nemalion* spp. (Figure 14.3), which represents the more "primitive" or simplified type of history. The second is *Polysiphonia* spp. (Figure 14.4) which represents the more "advanced" or complicated type.

The gametophytic thallus of *Nemalion* spp. is wormlike in appearance and reddish-brown in color. It usually clings to rocks in the in-

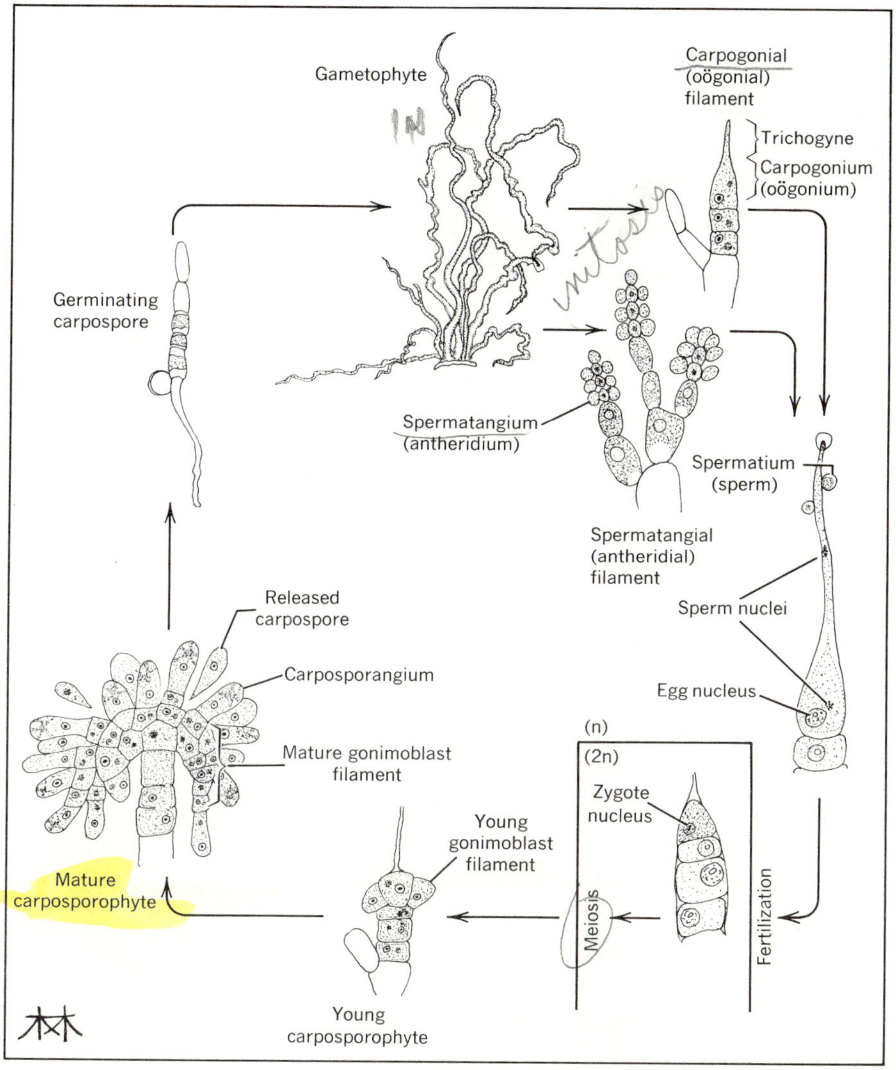

Figure 14.3
Life history of *Nemalion* sp.

tertidal zones. *Nemalion* is homothallic, but sometimes the spermatangial branches, which are located at the tips of the vegetative filaments, appear before the carpogonial filaments (**protandry**). The carpogonial filaments are produced in an intercalary position. The apical cell of a carpogonial filament becomes the carpogonium. After the sperm has been released from the spermatangium and

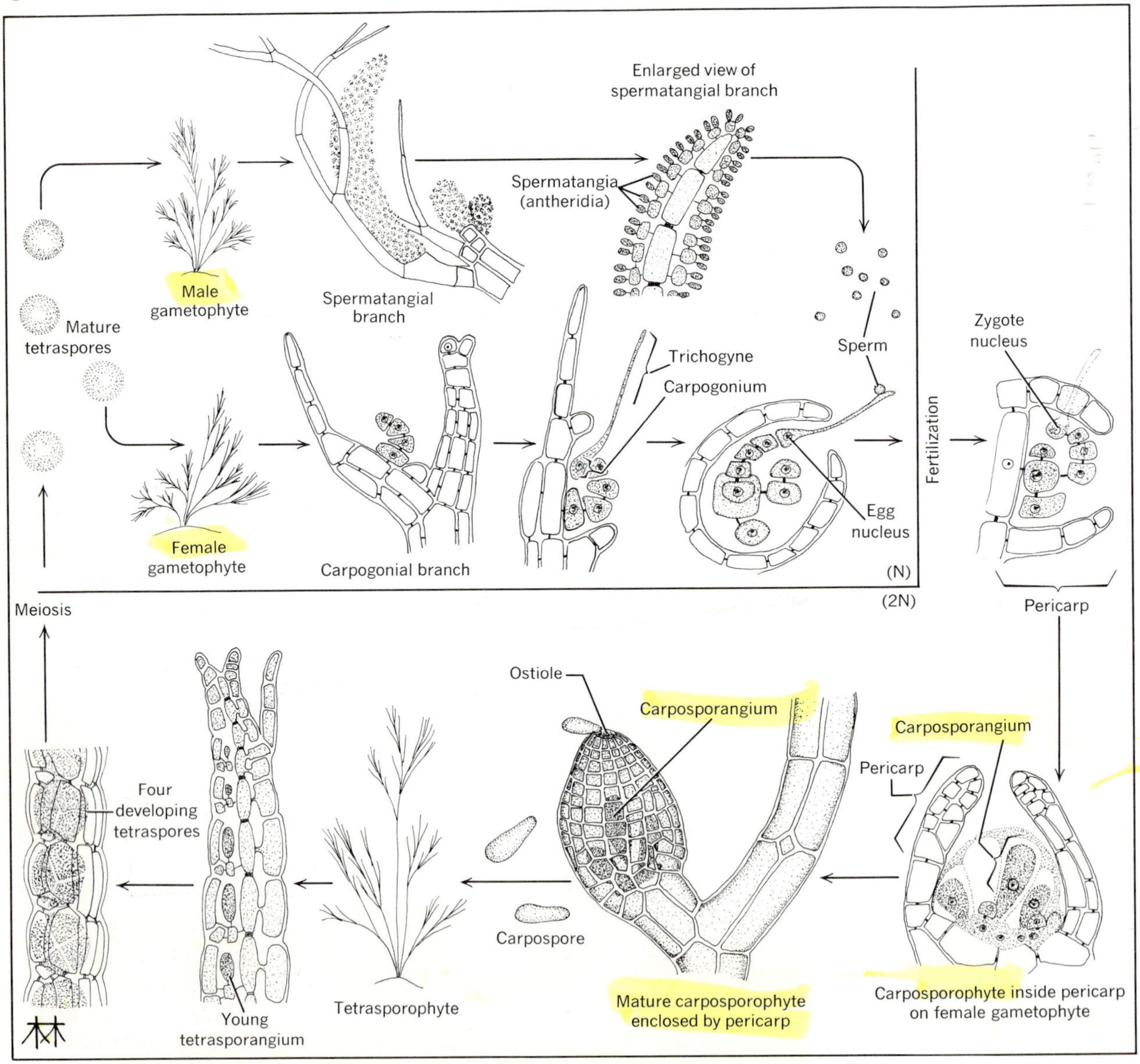

Figure 14.4
Life history of *Polysiphonia* sp.

become lodged on the trichogyne, mitosis of the sperm nucleus occurs. The walls between the sperm and the trichogyne break down and one nucleus migrates to the base of the carpogonium where it fuses with the carpogonial nucleus producing the only diploid phase in the life history.

The zygote nucleus increases in size and divides meiotically. Several filaments **(gonimoblast filaments)** are initiated from the carpogonium and the terminal cell of each develops into a carposporangium. The gonimoblast filaments and carposporangia collectively are recognized as the carposporophyte by some botanists. The carpospores are released through the distal end of the carposporangium and germinate by the production of a prostrate protonemalike stage that is thought to overwinter. The erect portion of the gametophyte is produced during the next spring. The life cycle of *Nemalion* is, therefore, **haplobiontic and haploid.**

Polysiphonia spp. is heterothallic and **triphasic.** That is, the gametophytes are self-sterile, and there are three phases to the life history (at least according to many botanists). The gametophyte is a delicately branching filamentous plant that is usually attached to rocks or to large algae (often kelps). The gametangia are produced at the tips of branches. Each gametophyte is unisexual. Sperm is released by the rupture of the spermatangial wall and is passively carried to the trichogyne where it lodges. After the walls dissolve, fertilization occurs as has been described for *Nemalion*.

The details of postfertilization activity are rather complicated. Through an involved series of cellular duplications, fusions, and nuclear migrations there are produced a number of gonimoblast filaments, the terminal cells of which become carposporangia. The carposporangia along with the accessory structures that have been produced are usually known as the carposporophyte. Since meiosis has not yet occurred, all of these structures are still diploid. Each carposporangium produces a single, diploid carpospore. Before and during fertilization, certain cells at the base of the carpogonial branch have been duplicating in such a manner that there is finally produced a sterile covering, called the **pericarp,** that encloses the entire carposporophyte. An opening (ostiole) in the pericarp allows the mature carpospores to be released into the water after they leave the carposporangium.

When the carpospores settle down to a substrate they germinate and a diploid plant that looks very much like the gametophyte is produced. Instead of producing gametangia, however, this plant produces tiers of tetrasporangia. Meiosis occurs during the production of the four tetraspores in each of the tetrasporangia, and the tetraspores are, therefore, haploid. The tetraspores are released by the splitting of the tetrasporangial wall, and settle to their substrate. They then germinate and produce the new gametophytic generation.

The life history, as described here, of *Polysiphonia* spp. is **diplobiontic and triphasic.** The gametophyte and tetrasporophyte phases are isomorphic. The carposporophyte is recognized, but it is attached to the female gametophyte. Not all of the red algae that are diplobiontic are isomorphic. For example, *Porphyra* has a sporophyte (the *Conchocelis* phase) that is quite unlike the gametophyte. Therefore, although the nuclear cycle for *Porphyra* has not been completely elucidated, the life history is known to be **diplobiontic but heteromorphic.**

HUMAN RELEVANCE

The red algae, as would be expected of any photosynthetic plant, occupy important places in the environment as trappers of radiant energy. They are also important directly to man's economy in a number of ways.

The most important product of red algae that man uses is probably agar. This colloid is usually obtained from the genera *Gelidium* and *Gracilaria* although other genera are sometimes used. The algae are dried and washed. They are then boiled, the extract filtered and allowed to gel. The gel is then frozen for about two days at 14°F. The water is removed after the gel begins to thaw. The dry material is then bleached and converted to flakes or powder for commercial sale and use.

Agar is extremely important for research purposes. It can be melted in water of 90° to 100° C and will gel at about 35° C. A 1 to 2% solution of agar and water, with the addition of various nutrients, can be used to grow and maintain cultures of fungi and bacteria that are used for study and experimentation.

Aside from its scientific value, however, agar has many industrial uses. It is used in many of the same ways as the alginates. That is, it prevents disintegration of canned fish and meats while they are cooking, it prevents the cocoa in chocolate milk from settling out, it is important for preventing drying in bakery goods, and it stabilizes sherbets, mayonnaise, icings, and candies. It is also a very important component of laxatives because it stimulates peristalsis. Funori is a glue and sizing agent that is produced from several genera of red algae. The agar used in the United States per year for all purposes amounts to hundreds of thousands of pounds.

Several of the red algae are used as a direct food source by man. One of these, *Porphyra* spp. is regularly cultivated on "*Porphyra* farms" in Japan and the Philippines. The alga is cultivated by sinking bundles of bamboo or other wood or nets of hemp or coconut fibers or lattice-works of wooden slats into the mud in shallow water during the autumn months. Spores settle on these structures and develop into germlings. The substrates are then often moved to the mouths of fresh-water rivers where the water is less saline. The lower salinity favors the development of the gametophyte which is harvested in January and converted into "Amanori" or Asakusa-nori. These products may be eaten directly in soups or with fish and rice, or they may be mixed with oatmeal and fried in butter. Because of the popularity of this alga in the Orient, *Porphyra* farms yield millions of dollars worth of produce each year.

Irish moss (*Chondrus crispus*) has an important polysaccharide cell wall component known as **carrageenin.** This gel has different properties from either agar or algin, and was widely used at one time in the preparation of a sweet, molded, jellylike dessert known as blancmange. Because the carrageenin was mixed with milk to produce the dessert, it was also used as an important part of invalid diets. Members of the genera *Gigartina* and *Iridophycus* have been used in a similar way. *Chondrus crispus* grows in great quantities along the shores of Massachusetts, Nova Scotia, France, and Ireland where it is harvested with long wooden rakes. The harvest in Nova Scotia alone in 1958 was about 17.1 million pounds. The value of such a harvest reflects the additional uses that have been found for it in the food, pharmaceutical, leather, textile, and brewing industries where it is used as an emulsifying and stabilizing agent.

The coralline algae are those that have cell walls heavily impregnated with calcium carbonate. Because of these deposits, the algae are sometimes ground up and used as a calcium source for fertilizer along coastal areas. More importantly, however, these algae contribute to the building of salt-water reefs. In some waters, the coralline algae are probably as important as, if not more important than, coral.

15

The stoneworts have often been included as a class in the green algae. They do, however, possess many morphological characteristics that justify separating them from the green algae and placing them in a phylum of their own.

Like the green algae, the stoneworts (e.g. *Chara* spp. and *Nitella* spp.) are green because they possess more chlorophylls $_{a\ and\ b}$ than carotenes or xanthophylls. Their color is sometimes obscured because certain species in some habitats are covered with a layer of calcium carbonate that precipitates from the water. The stoneworts are usually found growing upon the muddy or sandy bottoms of fresh water pools, although some occupy brackish waters. They often produce underwater "meadows" that may be quite extensive; they thrive best in hard, clear water.

The phylum Charophyta contains only one class (Charophyceae), one order (Charales), and one family (Characeae). There are about six genera and 250 species that are presently living. Unlike most other algae, there is a fossil record of the stoneworts because of the lime that they deposit. Female structures, encrusted with calcium carbonate, have been found dating as far back as the Devonian era. There are some species that are known only from fossil deposits.

VEGETATIVE STRUCTURE

The vegetative body of the stoneworts is an erect, branched axis that has many macroscopic characteristics. The axis and its branches are divided into **nodes** and **internodes** (Figure 15.2). The nodes are the areas at which is produced a whorl of branches (sometimes incorrectly called leaves). The unbranched sections between the nodes are the internodes. The branches produced in whorls at the nodes grow only to a particular length and are thus said to possess determinate growth. The indeterminate branches of the axis arise at the nodes in the axis of the determinate branches. The plant axis is attached to the substrate by a series of branching filaments called **rhizoids.**

The vegetative body is produced from derivatives of an apical cell that duplicates regularly in a manner similar to some of the vascular plants. Nodal and internodal cells are ultimately produced and the nodal cell undergoes elongation until it is several times longer than wide. This cell may remain naked as in *Nitella* spp. or may be covered by a layer of cells known as **corticating cells** (most species of *Chara*). Half of the corticating cells originate from the node above the internode, and half from the node below. The two groups of cells grow toward each other and meet in the middle of the internode.

Young cells near the apex of branches are uninucleate and have no conspicuous vacuoles. The more mature cells develop large central vacuoles and are multinucleate. Many small, ellipsoid chloroplasts are contained in the cytoplasm.

REPRODUCTION

No asexual spores are produced by members of the Charophyta, but many of them produce asexual reproductive bodies of a vegetative nature. These bodies consist of groups of cells that are filled with starch (**amylum stars**). Filamentous outgrowths from nodes and bulbils from rhizoids also effect vegetative reproduction.

Sexual reproduction within the group is oögamous (Figures 15.1 and 15.2). The gametes are produced in complex antheridia (**globules**) and oögonia (**nucules**). Most species are homothallic; a few are heterothallic. The gametangia are produced on the determinate branches at the nodes, and the homothallic species produce both at the same node.

According to one interpretation, the male reproductive structure is the exception to the generalization that algal gametangia have no division into sterile and fertile layers. This complicated structure is composed of chains of single-celled filaments that are attached by various sterile cells to the outermost layer of eight cells which covers the entire structure. These outer cells are large, epidermallike, shield-shaped cells that are orange-red at maturity. They produce the sterile cells which finally produce the fertile ones. Each cell of the fertile filament produces a single sperm and it is this specialized cell that is called the antheridium by many botanists. If the fertile filaments of cells are antheridial filaments that are composed of strands of single-celled antheridia, then the male gametangium of

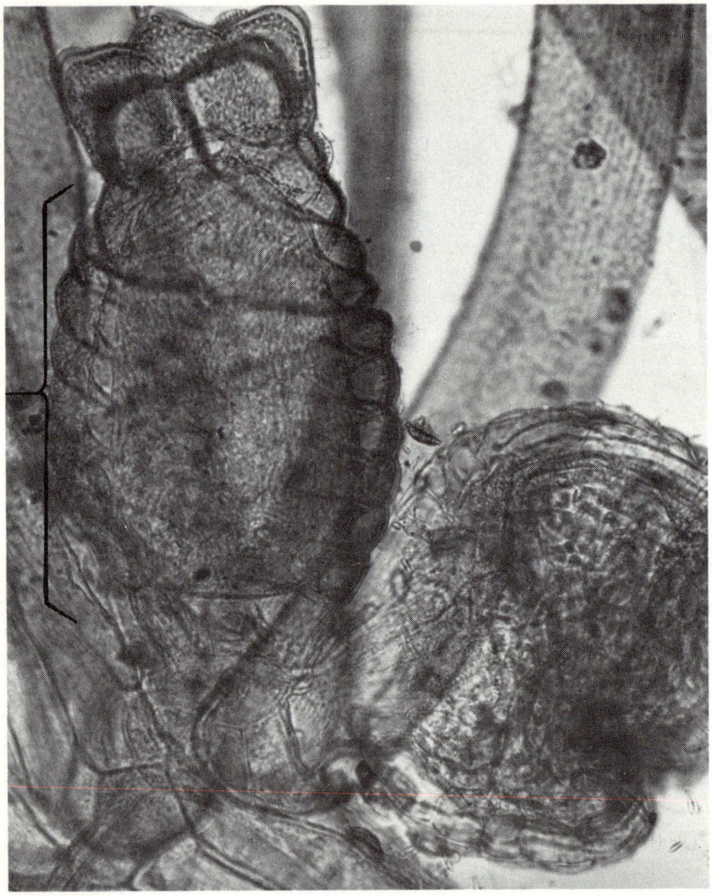

Figure 15.1
Sexual reproductive organs of *Chara* sp. (160×).

the Charophyta, like those of other algae, is also single-celled. The evolutionary and taxonomic arguments regarding the true nature of the male sex organ of the stoneworts is beyond the scope of this treatment, but they are of concern to some phycologists. In either case, the entire structure is attached to the plant at a node by means of a large basal cell, the **pedicel.**

The female gametangium is also attached to the node by a pedicel cell. The pedicel cell subtends the elongate oögonium which contains a single egg. Five spirally elongate sterile cells (**tube cells**) grow up from the base of the oögonium and form a sheath around it. The tips of these cells are cut off in one tier of five cells (*Chara*

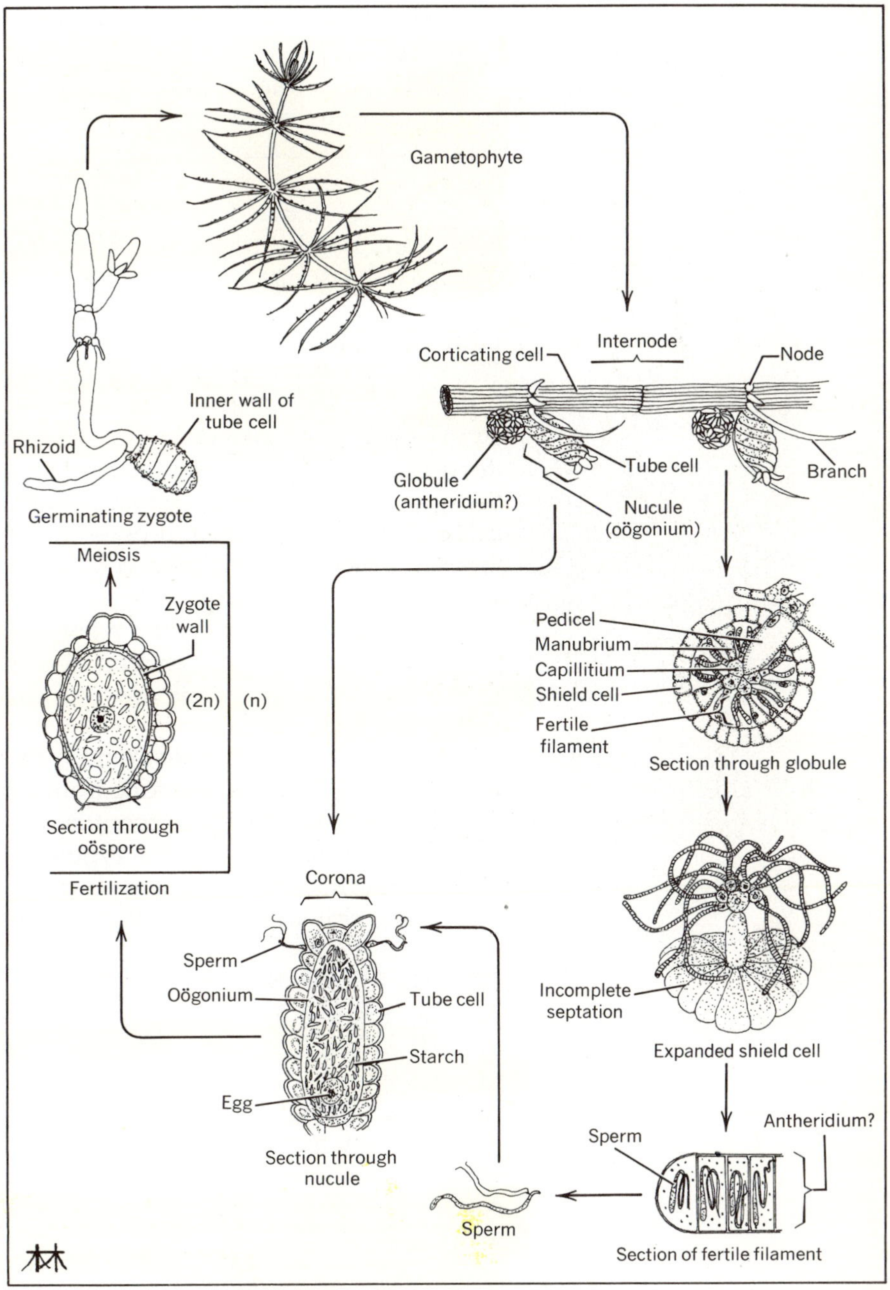

Figure 15.2
Chara sp. Life history.

spp.) or two tiers of five cells each (*Nitella* spp.). These tips form a crown (**corona**) over the distal end of the oögonium.

The sperm is released from its cell in the antheridial filament through a pore. It escapes from the antheridium because of a slight separation of the shield cells. At maturity, the tube cells separate slightly from one another just below the corona. The sperm cells enter through these openings and penetrate the wall of the oögonium. Fertilization is accomplished when the egg and one sperm nucleus fuse.

LIFE HISTORY

The members of the Charophyta possess a **haplobiontic, haploid** life history (Figure 15.2). Their sex cells are produced in specialized structures, the nature of which is disputed. Some botanists consider the gametangia to be single-celled structures that are comparable to the gametangia of other algae. Other botanists consider them to be multicellular structures that possess certain sterile layers. The latter interpretation would suggest a closer affinity to the mosses and liverworts than to the algae.

Sexual reproduction occurs when the sperm swims to the oögonium and fertilizes the egg. The zygote is the only diploid phase of the life history, and it develops a thickened wall soon after fertilization occurs. The zygote with its wall is called an **oöspore.** The inner walls of the tube cells remain as spiral markings on the oöspore wall. The oöspore falls from the plant and remains dormant on the bottom for weeks or months.

Just before germination the zygote nucleus is thought to undergo meiosis. Germination produces a green filament that finally produces the thallus and rhizoids of the mature plant.

HUMAN RELEVANCE

The members of the Charophyta are of very little importance to man. They are, of course, primary producers because they carry on photosynthesis.

They sometimes block waterways because of their extensive growth. They may hasten the filling-in of ponds and streams because of the precipitation of calcium carbonate. As the plants die, settle to the bottom and decay, a considerable amount of crumbly soil consisting of a mixture of clay, sand, and the calcium carbonate may be produced. Although this mixture (called **marl**) may eventually fill the pond, it has been put to some use as fertilizer and in the making of cement blocks and bricks. Some of the Indian tribes used species of *Chara* and *Nitella* to scour their cooking utensils.

Studies of *Chara* spp. have been used to determine the minimum light needed for photosynthesis and to illustrate one mechanism of rhizoidal response to gravity. In the former case, samples taken from the bottom depth of 450 feet of a very clear lake contained specimens of *Chara* spp. At this depth, the light intensity is only about 0.1 percent of full sunlight. Because algae are rarely found below 250 feet, this is probably about the lower limits of light requirement for algae. When *Chara* spp. rhizoids are placed on their sides, they will show a positive geotrophic response within 25 minutes. This response is attributed to cell structures (**statoliths**) which move to the lower side of the rhizoid in response to gravity. Their movement apparently causes a change in the distribution of wall material to the growing tip. Because more wall material is deposited in the upper wall, the tip moves toward the gravitational pull.

the euglena | euglenophyta

16

The euglenas are organisms about which much discussion has revolved as to whether they are plants or animals. They possess characteristics of both major divisions of living organisms. Since they contain chlorophylls $_{a \text{ and } b}$ and carry on photosynthesis they will be discussed in this treatment. Although they are photosynthetic, few are wholly autotrophic. Several chlorophyllous members must obtain vitamin B_{12} from their environment and many colorless members are saprophytic or parasitic. Some possess the ability to produce chloroplasts and photosynthesize while exposed to light, but lose their chlorophyll and assume a saprophytic existence when kept in the dark. Their animallike characteristics include their motility, their light receptors, and their gullet. However, none of these characteristics is entirely diagnostic for either plants or animals.

The euglenas are found in small fresh-water pools that contain a lot of organic matter. Some are found on mud banks of rivers, ponds, and estuaries or along salt water marshes. Some even grow on snow. They may be present in such numbers that they color their habitat.

The euglenas produce chloroplasts that contain chlorophylls $_{a \text{ and } b}$ beta-carotene and several xanthophylls. They store their excess photosynthate as an insoluble carbohydrate (**paramylum**) that is similar to starch but does not give the starch test with potassium iodide. Some fats are also stored.

The group is a small one possessing only about 25 genera and 350 to 400 species. They are usually combined into one class, the Euglenophyceae and the order, the Euglenales. The family Euglenaceae includes all of the chlorophyllous genera, although two other families have been erected for the achlorophyllous ones. Almost all of these are fresh-water organisms.

VEGETATIVE STRUCTURE

With the single exception of the genus *Colacium* (Figure 16.2), all members of the Euglenophyta are single-celled and solitary. *Colacium* spp. is attached by gelatinous stalks to its substrate which is often copepods, rotifers, or zoöplankton. It may be solitary or the stalks may be united producing an amorphous colony. Other members produce a "Palmella" condition during certain reproductive stages; these will be discussed later in more detail.

The cell boundary of the members of the Euglenophyta is a differentiated plasma membrane called the **periplast.** The periplast is often **striate** or **punctate** and may be so rigid (e.g., *Phacus*, Figure 16.2) that the cells have a definite shape, or so flexible (e.g. *Euglena*, Figure 16.1) that the cell shape constantly changes. No cellulose wall is present. Some genera (e.g. *Trachelomonas*, Figure 16.2) produce a firm polysaccharide layer external to their periplast. This layer is the **lorica** and, although colorless when first produced, it usually becomes impregnated with iron compounds that color it yellow to dark brown. The lorica always has an opening at the anterior end through which the flagellum projects.

The anterior end of the *Euglena* cell (Figure 16.1) is differentiated into an elongate flask-shaped structure called the **gullet.** The neck of the gullet is the **cytopharynx,** and the bulblike posterior portion is

Figure 16.1
***Euglena* sp. Structure of vegetative cell.**

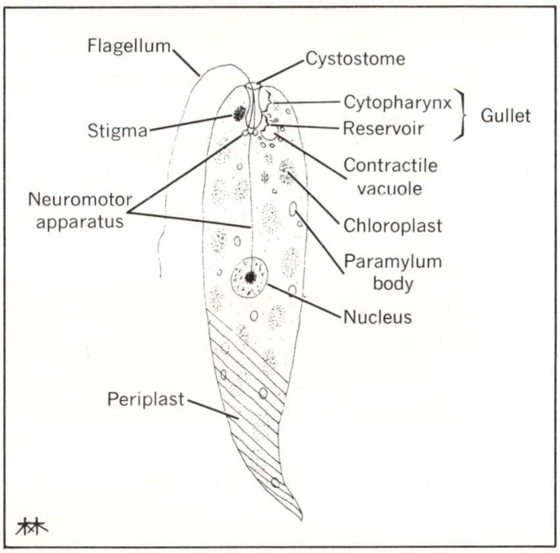

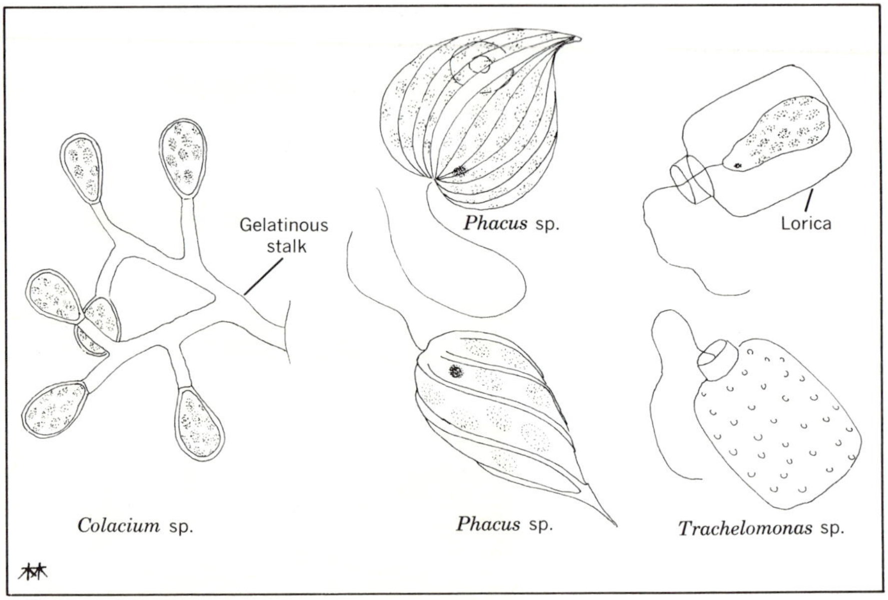

Figure 16.2
Examples of members of the Euglenophyta

the **reservoir.** One or more **contractile vacuoles** are adjacent to the reservoir. The flagellum or flagella are attached at the base of the reservoir and are of the tinsel type. The base of the flagella are inserted into a **neuromotor apparatus** (Figure 16.1) much like that which will be discussed in the chapter about green algae.

All of the chlorophyllous forms, and some of the achlorophyllous ones, possess an anterior pigmented eyespot or **stigma.** This structure absorbs light and allows the organism to produce a positive phototactic response. All members of the group possess one large nucleus.

REPRODUCTION

Reproduction in the Euglenophyta is by mitosis followed by cell duplication. The duplication may take place while the organism is active, or the cells may withdraw their flagella, secrete a thick, encysting layer and slime, and remain nonmotile for a while. Cellular duplication often occurs during this time. The daughter cells may become surrounded by a gelatinous sheath and divide again before escaping from the sheath. Continuation of this process produces temporary **palmelloid colonies** that can revert to the motile condition at any time. The immotile colonies or encysted cells may accumulate a combination of pigments that turn them blood red. These pigments are collectively known as **hematochrome,** and the encysted organisms with their pigments may produce a red surface film on ponds and lakes (Figure 16.3, p. 254). In those species that produce them, one of the daughter cells remains in the old lorica, but the other escapes and secretes a new lorica. Duplication of the motile cells is longitudinal and begins at the anterior end.

Sexual reproduction has not been well established for the group. Although reports of gametic union of vegetative cells have been made for some euglenoids, they have not been fully verified.

LIFE HISTORY

The Euglenophyta apparently possess a haplobiontic type of life history. Meiosis and karogamy have not been established for the group. Whether the vegetative cells are haploid or diploid is still not known. Therefore, the organisms that are assigned to this phylum seem to be quite successful by reproducing only asexually.

HUMAN RELEVANCE

In large quantities they come into conflict with man when they give his water a fishy odor and a fishy-sweet taste. Certain lorica-producing genera such as *Trachelomonas* may cause some filter-clogging. They may also produce slime in the water and cause discoloration. Certain ones have been known to cause corrosion of concrete.

The Euglenophyta is of some use to man as an indicator of the biochemical state of certain bodies of fresh water—they may indicate pollution and/or eutrophication in fresh water ponds and small lakes. *Euglena* sp., *Phacus* sp., and *Trachelomonas* sp. are often found in hard-water lakes with no outlets. Several species of *Euglena* are indicators of conditions of high acidity. *Trachelomonas* sp., because it accumulates iron in its lorica is a good indicator of the presence of iron in industrial wastes, and a species of *Euglena* indicates chromium.

This chapter has concentrated only on those Euglenophyta that are photosynthetic. The parasitic and saprophytic members are not included, but they also occupy important ecological positions.

the green algae | chlorophyta

17

The members of the Chlorophyta are called the "green algae" because the chlorophylls contained in their chloroplasts generally mask the other pigments that are present. Both chlorophylls$_{a \text{ and } b}$ are present as are α and b carotenes and several xanthophylls. All of the pigments of the green algae are restricted to chloroplasts that exhibit a wide variety of size and shape. The chloroplast characteristics are often important taxonomic considerations. Most members of the group produce **pyrenoids** (Figures 17.8 and 17.14) in their chloroplasts. These bodies are the centers of starch formation within the cells. The **starch** is the method of storage of the excess photosynthate and indicates a relationship between the green algae, the Charophyta, and the green land plants. The latter two groups also store their excess photosynthate in the same manner.

Members of the Chlorophyta occupy an extremely wide range of habitats. Although only about 10% of the species are marine, many of them occupy definite geographical locations determined primarily by water temperature, light intensity, and nutrient availability. Most of the marine species live in shallow waters along the coastlines and are exposed during low tides. Other species are to be found submerged in 300 feet of water.

The other 90% of the species are either fresh-water or terrestrial species, and most are cosmopolitan. Many of these species grow submerged in shallow pools or deep ponds. Many are free-floating or free-swimming (**planktonic**). Both marine and fresh-water species use a variety of substrates: mud, sand, rocks, logs, shells, the backs of turtles, and even other algae and aquatic vascular plants (epiphytic). Some members are endozoic in the cells of protozoans, coelenterates, and sponges. Terrestrial species are found growing in or on soil and on the bark of trees. Some green algae have been recovered from airborne dust and some have been found growing on snow or ice. The group could, then, be called ubiquitous in its distribution.

In the past many botanists included the members of the Charophyta within the Chlorophyta. However, there appear to be enough characteristics unique to the Charophyta to distinguish them as a separate phylum from the Chlorophyta and that is how they appear in this treatment. Therefore, as discussed here, the Chlorophyta contains only one class, the Chlorophyceae, which contains approximately 425 genera and 6500 species. The numbers of recognized orders and families varies according to the opinions of the various phycologists. The arguments concerning the various treatments are not of concern here.

VEGETATIVE STRUCTURE

The majority of the green algae possess vegetative cells that produce a cellulose cell wall. The exceptions produce an outermost protoplast layer that is rigid and has a definite shape. The cell wall is surrounded by a pectose layer that varies in thickness from species to species. The sheath may also vary in thickness during the growing season.

The protoplast of the cell is well organized with an endoplasmic reticulum, Golgi bodies and mitochondria. The nucleus is eukaryotic. Large central vacuoles and smaller contractile ones are also often present. All of the cells of the members of one order (Volvocales) are flagellated (Figures 17.1 and 17.2) as are the reproductive cells of most of the other green algae. The flagellated green algae cells, and the flagellated cells of many other plants, possess a **neuromotor apparatus** (Figure 17.1*b*).

To understand the motility of these cells requires an introduction to the neuromotor apparatus. A member of the genus *Chlamydomonas* (Figure 17.1) will be used as an example, but with minor modifications it will serve as a description of the neuromotor apparatus of other flagellated cells (e.g. *Euglena* spp., Figure 16.1).

The flagella of *Chlamydomonas* spp. are approximately 25 microns long. They possess a central filament that is surrounded by a cytoplasmic sheath over most of its length. The central filament consists of two fibrils that are surrounded by nine other fibrils (Figure 17.1*c*). The tip of the central filament extends for a short distance beyond the end of the sheath. This is known as the whiplash type of flagellum, and its construction is very consistent throughout the

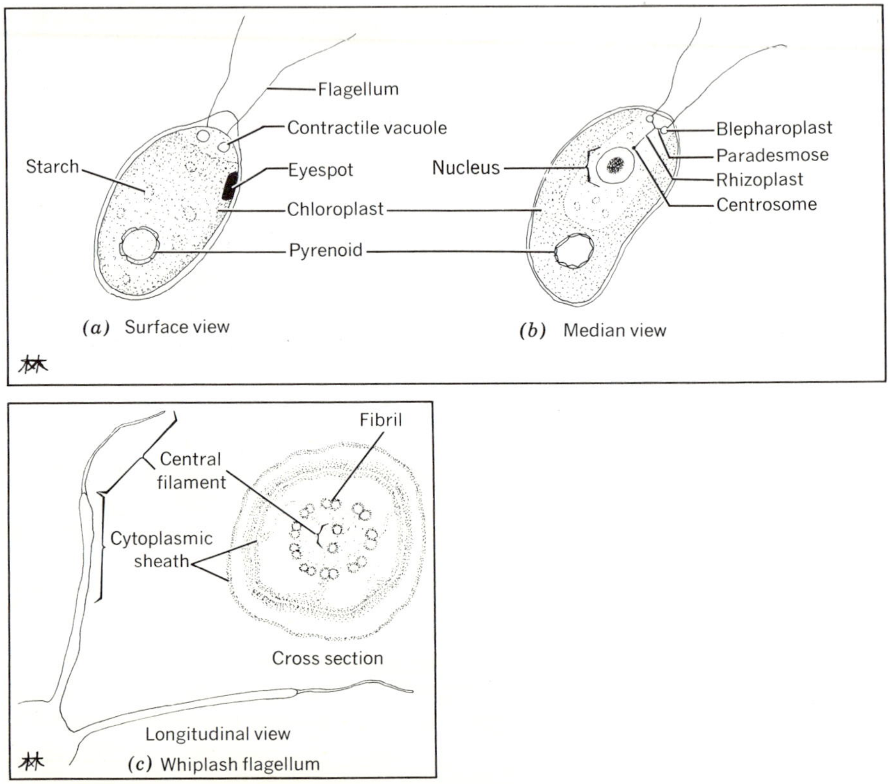

Figure 17.1
Chlamydomonas sp. (*a*) Vegetative cell. (*b*) Neuromotor apparatus. (*c*) Cross section of a whiplash flagellum.

plant and animal kingdoms. It appears to be the only flagellum type found in the Chlorophyta. Other plant groups with flagellated cells may also have flagella with minute lateral appendages arranged in one or two rows. These are known as the tinsel type.

Each flagellum is inserted at the anterior end of the cell and extends to a granule (**blepharoplast**) just inside the cell wall. The granules are connected by a transverse fiber (**paradesmose**). A descending fiber (**rhizoplast**) extends from the transverse fiber to a **centrosome** that may be just inside or outside of the nucleus (Figure 17.1*b*). Therefore, the action of the flagella appears to be directly related to nuclear activity and stimulated through a primitive nervous system.

Chlamydomonas spp. and many other unicellular genera possess a red eyespot that is a **photoreceptor** (Figure 17.1*a*). Its structure suggests that it may function as a primitive lens, and it appears to aid in flagellar control during phototactic response. It is usually found at the anterior end of the cell, but it may be median or even posterior.

The members of the Chlorophyta may be unicellular as in *Chlamydomonas* spp. or multicellular. The unicellular forms may be motile or nonmotile. Some members are colonial and may also be motile or nonmotile. Other species are filamentous, membranous, or tubular. Although most of the cells are uninucleate, some members possess many nuclei per cell.

A unicellular genus that is much studied is *Chlamydomonas* spp. (Figures 17.1 and 17.9). It is readily found in many natural environments or bought from biological supply houses. Members of the genus are motile, because they possess two flagella and a neuromotor apparatus during most of their life. At times, however, they may settle on a moist substrate, lose their flagella, and secrete a thick mucilage layer. The cells continue to duplicate vegetatively, and a large colony is sometimes produced. After a time, the individual cells of the colony develop new flagella and again become motile. This nonmotile stage resembles the genus *Palmella* spp. and has been called the palmella stage of *Chlamydomonas* spp. The palmella stage has been thought to be a response to unfavorable environmental conditions, but research has demonstrated that it occurs during optimum growing conditions for the genus, and must therefore be regarded as a part of the life history (Figure 17.9).

The cell structure has been previously described in this section. Most species contain large, distinctive, parietal, cup-shaped chloroplasts that obscure most of the other cellular detail (Figure 17.1*a*). Some species may possess band or H-shaped chloroplasts. Most species contain but a single pyrenoid, but others may produce several. Pyrenoids are important as sites where glucose is converted to starch. The eyespot is usually evident. There is always a cellulose cell wall present, and some species produce an external pectic layer. Contractile vacuoles are usually found at the base of the flagella. All species are uninucleate.

Chlamydomonas spp. is a member of the order Volvocales which is the only order of the Chlorophyta whose members all possess vegetative cells that are flagellated and actively motile. Many of the other members of the order are multicellular, colonial forms. Each of the vegetative cells is very much like *Chlamydomonas* spp., and it is thought by some phycologists that the colonial types probably origi-

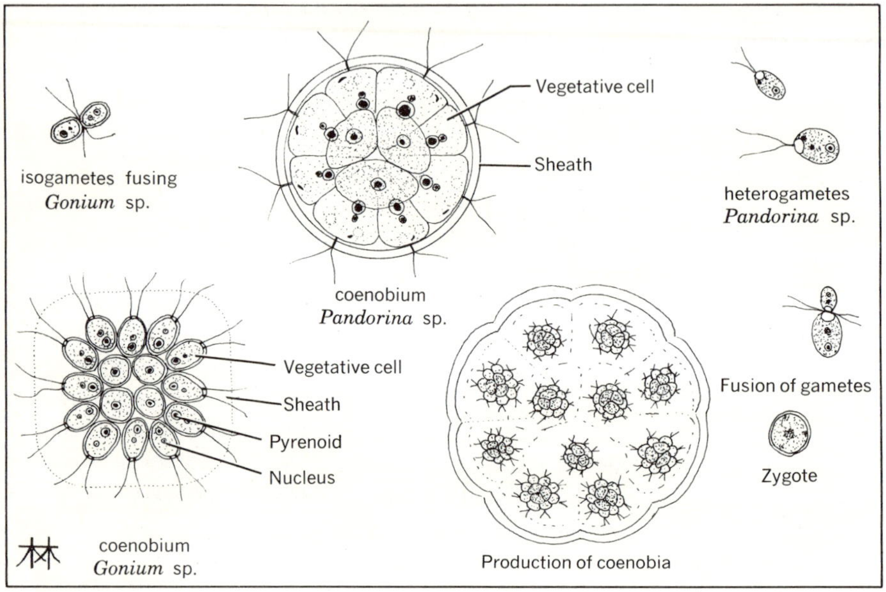

Figure 17.2
***Gonium* sp. and *Pandorina* sp. Two colonial members of the Volvocales.**

nated from the tendency of *Chlamydomonas*-like forms to remain closely associated following asexual reproduction.

All colonial members produce colonies with a definite number (a multiple of 2) of vegetative cells arranged in a definite manner (**coenobia**). For example, members of the genus *Gonium* (Figure 17.2) possess 4 to 32 or 64 cells arranged in a flat plate surrounded by a sheath. The colony tumbles over and over through the water. *Pandorina* spp. (Figure 17.2) may possess 8 to 32 cells (usually 16), also in a sheath, but the colony is spherical or oval. The individual cells are slightly pyriform, but the colony tumbles or rolls through the water. The colonial form that possesses the largest number of vegetative cells is *Volvox* spp. (Figure 17.10). Estimates of the number of vegetative cells run to 500,000, and the colonies may be as large as 1 mm in diameter. Such a colony is visible to the unaided eye, and thus the genus has been known for several hundred years. The cells are contained in a sheath, but in some species they are connected by thin protoplasmic extensions. Daughter colonies are often found contained in the parent colony as it rolls gracefully through the water.

Several genera of green algae are unicellular, nonmotile forms.

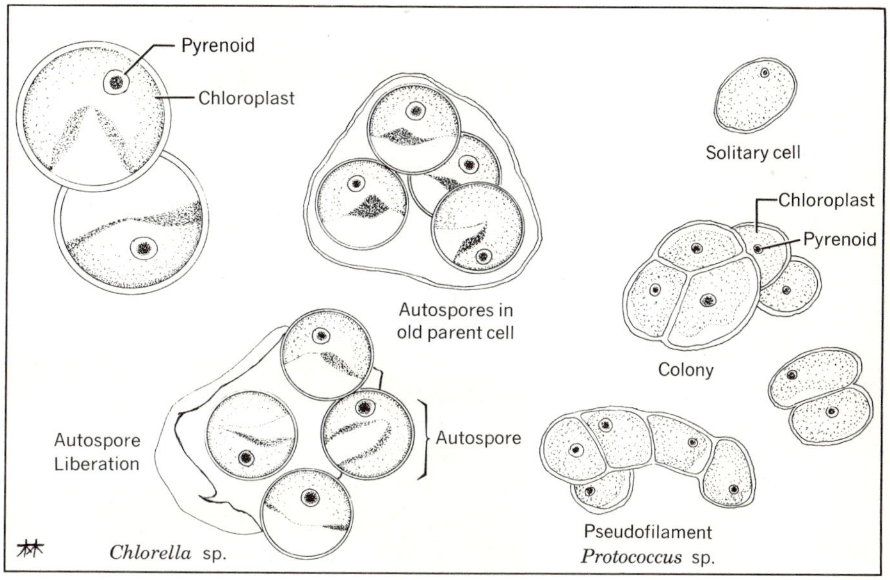

Figure 17.3
***Chlorella* sp. and *Protococcus* sp. Two nonmotile unicellular green algae.**

Some of these forms produce flagellated cells during some portion of their life history. The flagellated cells may be either zoöspores, gametes, or both. Two of the most common nonmotile, unicellular greens are *Chlorella* spp. and *Protococcus* spp. (Figure 17.3). Neither produces any flagellated cells during its life history.

Chlorella spp. is found in salt and fresh water as well as in soil. It may also be endozoic in protozoans, *Hydra* spp., and sponges. Its cells are small and oval or spherical. The chloroplast is thin and cup- or platelike. It may contain pyrenoids of different sizes, shapes, and numbers, or it may not contain any pyrenoids at all. Its minute nucleus is buried in colorless central cytoplasm. *Protococcus* spp. is an organism that has undergone much nomenclatorial change (e.g. *Pleurococcus*, *Desmococcus*). The organism (regardless of the assigned name) grows on moist substrates and is particularly apparent when growing on tree trunks, often on the north exposure. It was this organism, not moss, that was often used to determine direction by the Indians and pioneers, and during the present day by Boy Scouts (Figure 17.7*a*, p. 254). Solitary cells of *Protococcus* spp. are oval to spherical and develop a rather thick wall. Cell duplication produces two cells that may separate or remain united. Further duplications may result in larger colonies that may have a tendency to produce short pseudofilaments. The cells are usually sharply angled because of mutual compression. The cells have a single parietal chloroplast that may or may not have a pyrenoid. Some phycologists separate those members without a pyrenoid into another genus (*Aptococcus*).

The nonmotile colonial forms also have members that produce motile reproductive cells and members that do not. Of the former group, *Pediastrum* spp. and *Hydrodictyon* spp. (Figure 17.4*a*) are two of the most beautiful algae among the fresh-water flora. *Coelastrum* spp. and *Scenedesmus* spp. (Figure 17.4*b*) are common, and quite pretty, members of the latter group.

Pediastrum spp. and *Hydrodictyon* spp. both produce coenobic colonies. That is, the number of cells in their colonies is fixed at the time of the colonies' origin and no cells are ever added to the colony. Both genera are found in the bottoms of lakes, in the **phytoplankton,** and in marshy or swampy areas. *Pediastrum* spp. produces its cells in a flat, circular plate with the peripheral cells usually being quite different in shape from the cells in the interior of the colony. The genus is easily identified because of the overall similarity of its many species. A colony may consist of from 4 to 128 cells, although dead cells or spaces where cells are missing are often observed. Young cells contain a single parietal chloroplast with one pyrenoid, and a single nucleus. These conditions are usually altered as the cell ages.

Hydrodictyon spp. is commonly and aptly called "water net." The cells of the plant are long, cylindrical, and joined together in such a way that they form a hexagon. Many hexagons are united in such a way that a net is produced. The entire colony is cylindrical and may reach up to two feet in length.

The young cells of *Hydrodictyon* spp. are uninucleate, but as they enlarge, karyokinesis produces many hundreds, or even thousands, of nuclei. The cytoplasm is eventually pushed to the peripheral part of the cell by a large central vacuole. The cytoplasm eventually divides into individual, uninucleate bits that mature into zoöspores. The zoöspores finally arrange themselves into the pattern possessed by the parent colony, and a coenobium or **autocolony** is formed.

Scenedesmus is a genus that contains a number of species that vary in cell number from 4 to 32. The individual cells may be oval to crescent-shaped according to species. All of the cells of the coenobium may be alike, or the outside cells may differ by possessing spines of various numbers and shapes. The cells may be in one or two rows. Each of the cells is uninucleate and possesses a large parietal chloroplast that contains one pyrenoid. Each of the vegetative cells is capable of producing an autocolony that ruptures the parent cell wall and increases to the size of the parent colony. Although zoöspores have never been reported among naturally occurring *Scenedesmus* spp., some species have been reported to produce them when grown under laboratory conditions. *Scenedesmus* spp. is a common alga in most fresh-water habitats.

The coenobium of *Coelastrum* spp. may possess as many as 128 cells that may be globose to pyramidal in shape. The cells are either closely compressed or connected by processes forming a hollow, spherical colony that is commonly found in the phytoplankton. Autocolonies are produced within the parent cells. The colonies are released by the breaking of the parent cell wall.

Another of the major body forms of the green algae is filamentous. If the cellular duplication is restricted to one direction, the filament is unbranched, but if cellular duplication occurs in certain cells of the filament in a different plane, branching occurs. There are many filamentous unbranched and branched species. Most of the filamentous algae produce zoöspores, but one major order (the Zygnematales) produces motile spores that are not flagellated but move in an amoeboid fashion.

One of the most interesting, unbranched filamentous forms is *Oedogonium* spp. (Figure 17.5). This alga has a unique type of sexual reproduction, which will be discussed later, but it also has a distinctive type of vegetative cellular duplication that is either terminal or intercalary (Figure 17.5*a*). During karyokinesis a ring of wall material encircles the cell just below its distal end. The ring increases in thickness and a small groove is formed in the portion that adjoins the lateral wall. After karyokinesis is completed, transverse cytokinesis of the protoplast occurs in the middle of the cell. Each daughter protoplast elongates to about the size of the parent protoplast. This action pushes the ring distally. The ring persists as an apical cap on the cell. A new cell wall is then produced between the two daughter protoplasts. All of the cells in the filament do not carry on cellular duplication; thus only a few of them possess apical caps. The number of the rings in an apical cap indicates the number of cellular duplications that have taken place in that cell.

Oedogonium spp. is a submerged aquatic alga that is usually attached. Cattail and rush stalks are common substrates. The plant may be broken away from its substrate and become free-floating. It then often forms fluffy, yellow green or cream-colored masses on the

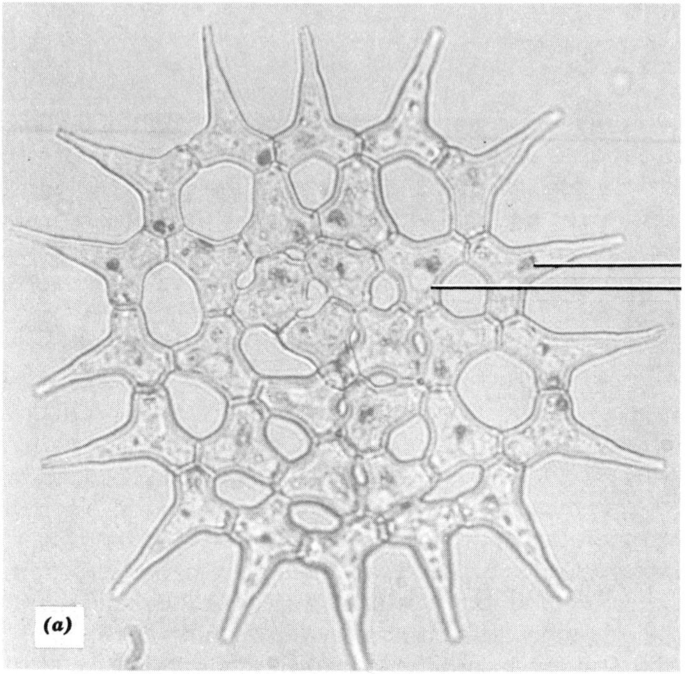

Pediastrum sp.

Figure 17.4
Nonmotile colonial forms. (*a*) *Pediastrum* sp. (750×) and *Hydrodictyon* sp. (156×) produce motile reproductive cells. (*b*) *Coelastrum* sp. (580×) and *Scenedesmus* sp. produce nonmotile reproductive cells.

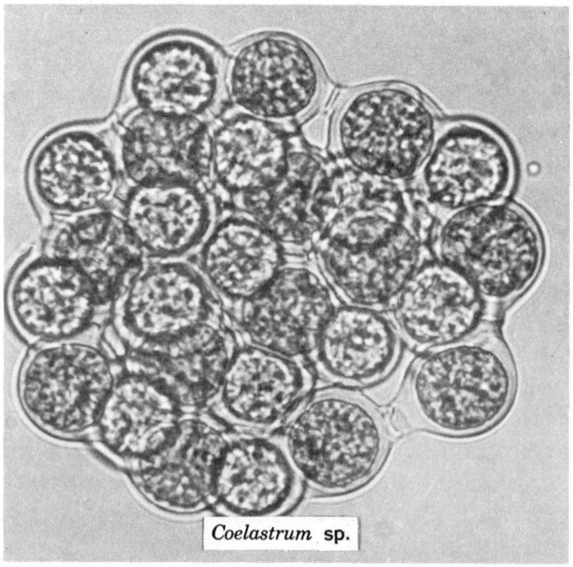

(*b*)

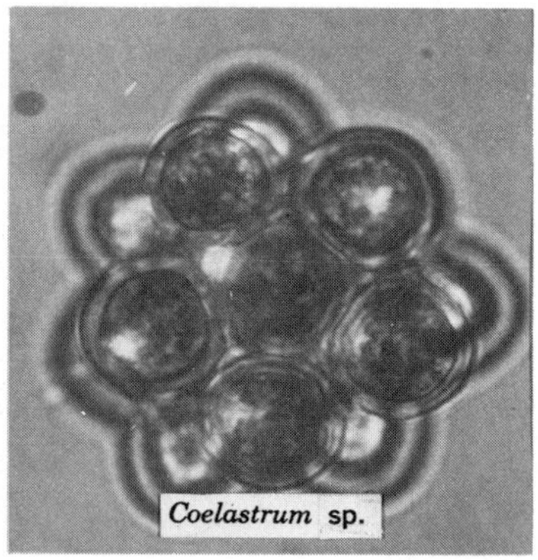

(*b*)

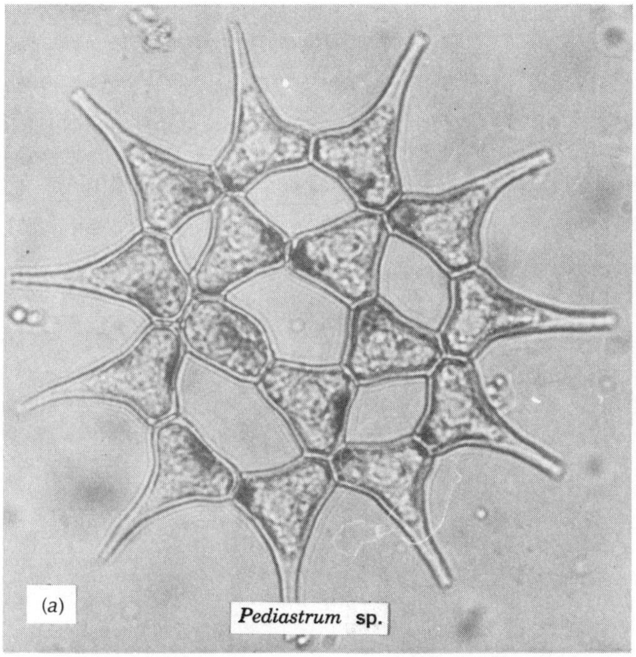

(a) *Pediastrum* sp.

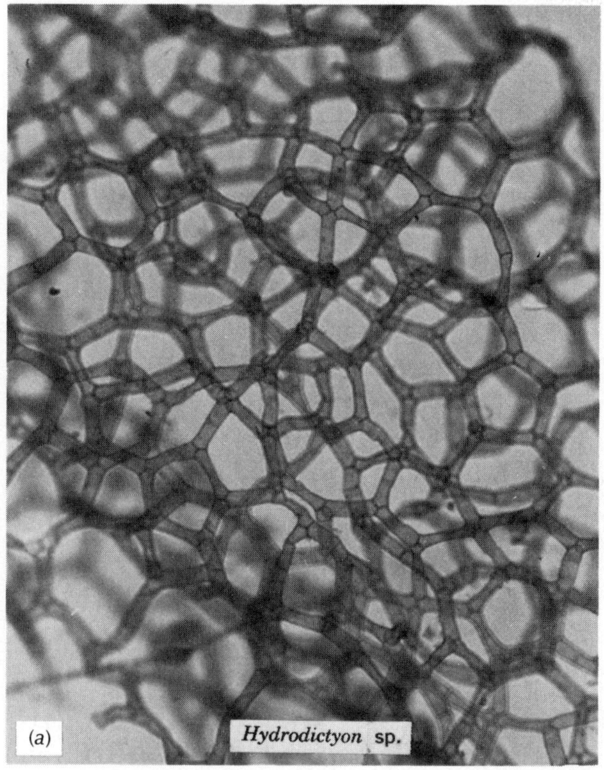

(a) *Hydrodictyon* sp.

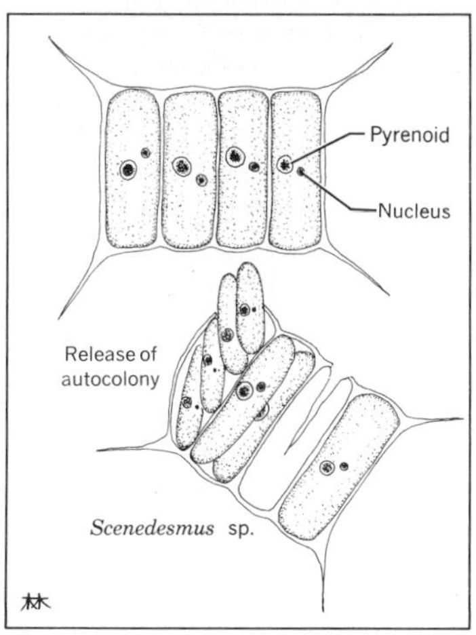

Scenedesmus sp.

(b)

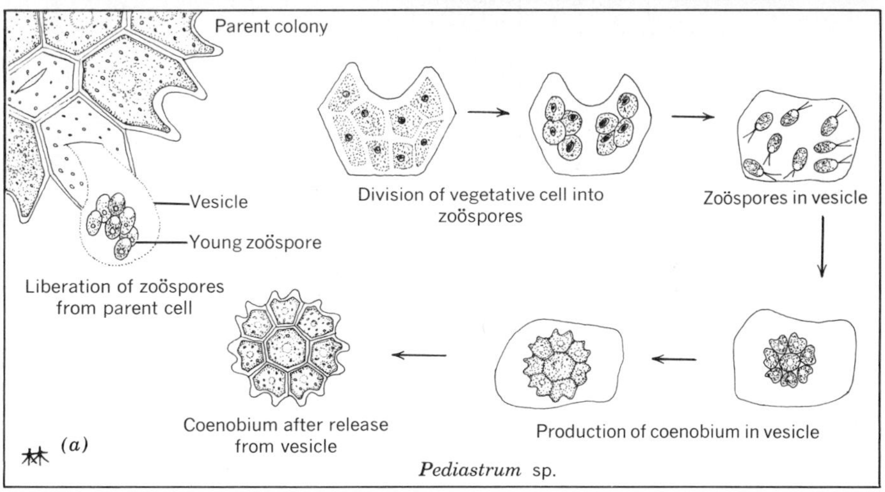

(a) *Pediastrum* sp.

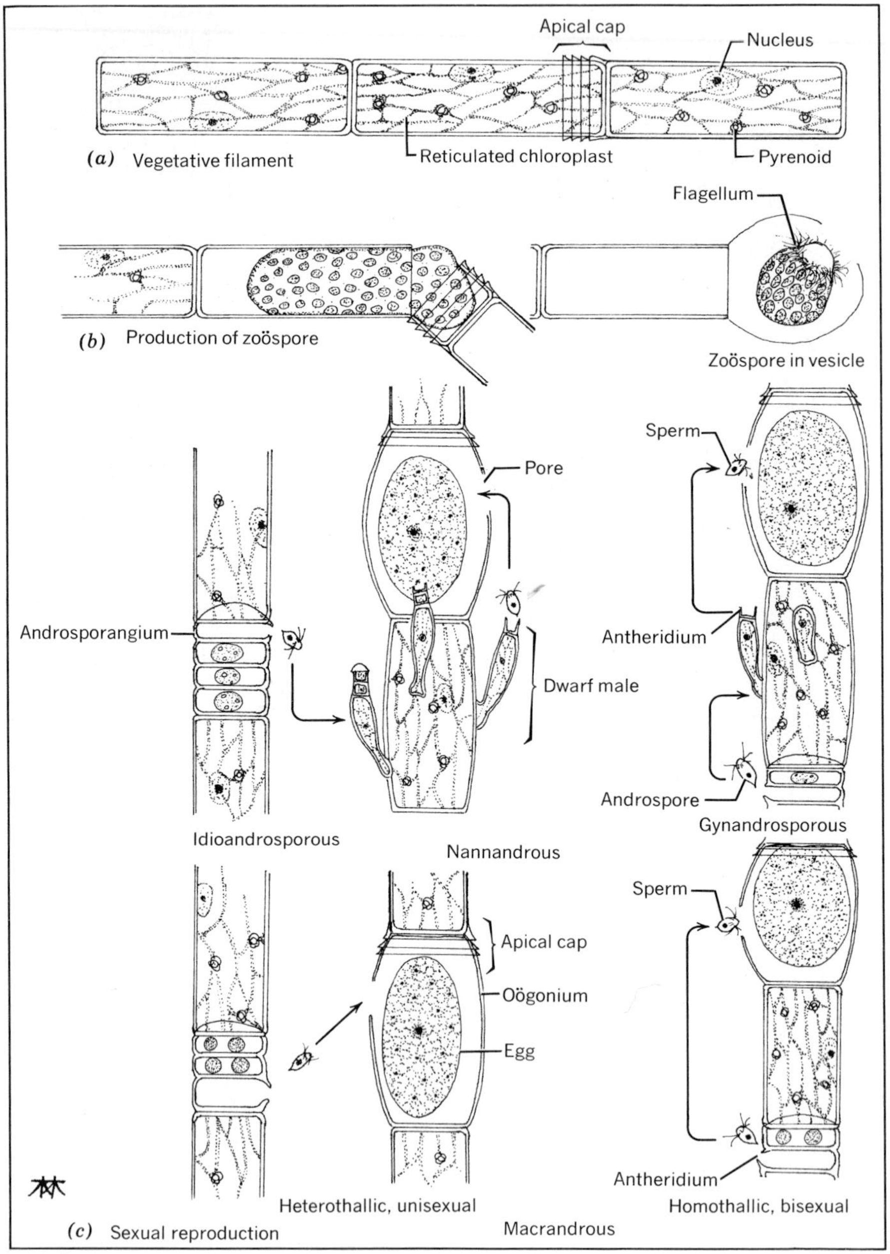
Apical cap
Nucleus
(a) Vegetative filament
Reticulated chloroplast
Pyrenoid
Flagellum
(b) Production of zoöspore
Zoöspore in vesicle
Pore
Sperm
Androsporangium
Antheridium
Dwarf male
Androspore
Idioandrosporous
Nannandrous
Gynandrosporous
Sperm
Apical cap
Oögonium
Egg
Antheridium
Heterothallic, unisexual
Homothallic, bisexual
(c) Sexual reproduction
Macrandrous

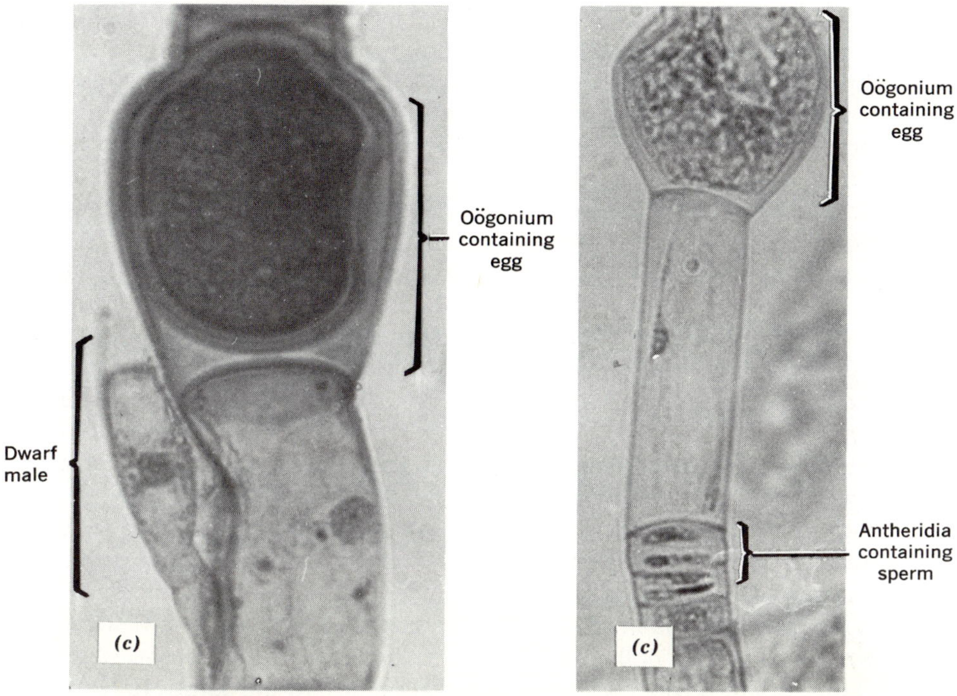

Nannandrous *Macrandrous*

Figure 17.5
***Oedogonium* sp. (*a*) Vegetative plant with apical caps on the cells. (*b*) Asexual reproduction by zoöspores. (*c*) sexual reproduction: Nannandrous (720×), Macrandrous homothallic (720×) and heterothallic.**

surface of the water where decay occurs.

Cladophora spp. (Figure 17.12) is a highly branched green alga that is commonly found attached to piers or along rocky banks of bodies of fresh or salt water, although it is sometimes free-floating. Its vegetative cells are about seven times longer than they are wide and are attached end to end. The branches are usually lateral, but some appear to be almost dichotomous. Branches are produced as lateral outgrowths from the upper end of a cell located near the tip of a filament. The plants are attached by a rhizoidal system that may be perennial although the rest of the plant is usually annual. The chloroplast is a single reticulate sheet that surrounds the outer portion of the cytoplasm. Pyrenoids are found at the intersections of the reticulum. The cells are always multinucleate, karyokinesis being independent of cytokinesis. Various species of *Cladophora* exhibit different types of life histories. Some are quite complicated; others are simple. They will be discussed in detail later in this chapter.

Almost every student who has had even an introduction to the study of biology is at least superficially familiar with the genus *Spirogyra* (Figures 17.13 and 17.14). Although, because of its method of reproduction, it is a very atypical green alga, it is easily found in many fresh-water lakes and ponds, and prepared slides are abundantly available at a reasonable price. The plant is composed of elongate cells that are united end to end forming long, unbranched filaments. The plants often form floating, bright to yellow-green, slimy masses of "pond scum" in the spring (Figure 17.7*b*, p. 254). The filaments produce, except during the time of reproduction, a watery sheath that surrounds their outside wall.

The vegetative cell contains one or more spirally arranged chloroplasts that may or may not almost completely fill the cell. Several pyrenoids are located on the chloroplasts. The cells are uninucleate with the large nucleus suspended in the center of the cell by thick cytoplasmic strands.

Some of the most striking of the green algae (comparable to the delicate beauty of the diatoms) are to be found among the desmids (Figure 17.6), a group that is often included in the same order with *Spirogyra* spp. Desmids are mostly single-celled algae, but some are filamentous. The cells of most of the desmids are composed of two **semicells** that are attached by a region called the **isthmus.** The semicells are mirror images. The nucleus is located in the isthmus. Most desmids possess a chloroplast in each semicell, although some have a single chloroplast that extends throughout the entire cell.

Closterium is a genus that is commonly found in many fresh-water habitats. Although there is no incision in the center of the cell, it is divided into two semicells with a separate chloroplast in each. Each chloroplast contains from one to several pyrenoids. The cell may vary in shape from almost straight to quite curved. The central portion of the cell is often broader than the tapering tips giving the appearance of the horns of a bull. A conspicuous vacuole is present at the tips of each of the semicells. Contained in the tips are granules of gypsum that are kept in constant vibrating motion by Brownian movement.

Species of the genera *Cosmarium* and *Staurastrum* are composed of two semicells that are separated by deep clefts. The semicells are attached by an isthmus however.

The membranous body type of the green algae is probably best illustrated by the flattened, macroscopic alga that is found growing

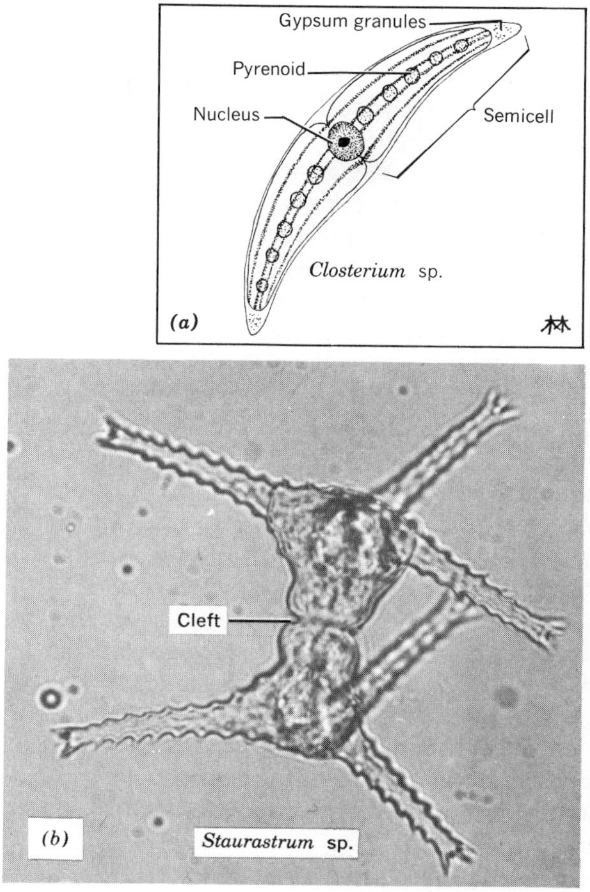

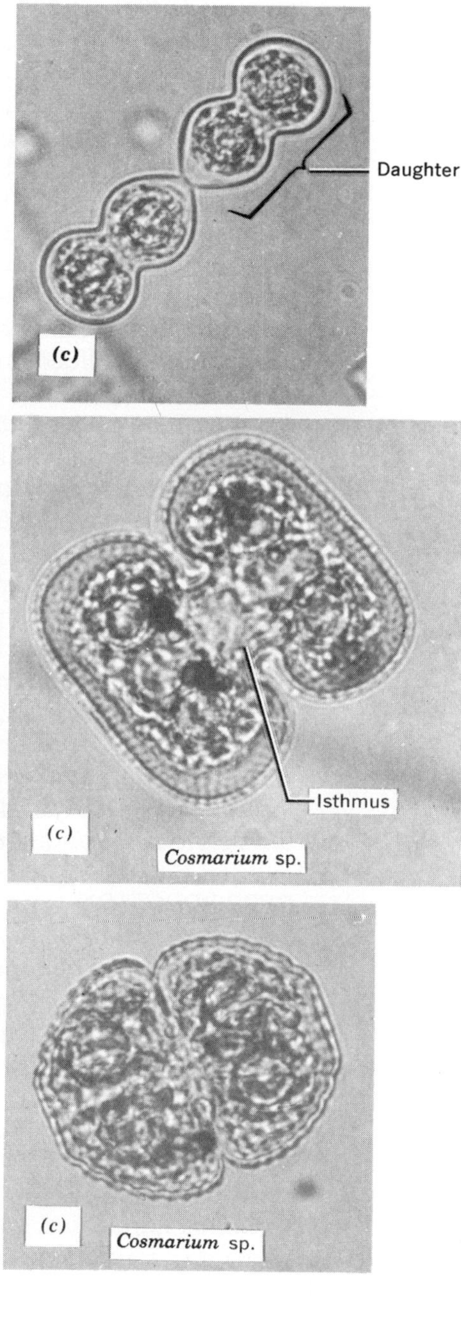

Figure 17.6
Exqmples of desmids. (*a*) *Closterium* sp. (*b*) *Staurastrum* sp. (664×). (*c*) *Cosmarium* sp. vegetative and newly duplicated cells (664×)

on piers and along the edges of marine and estuarian bodies of water. This alga, *Ulva* spp. (Figure 13.1*b*, p. 254) is often eaten directly by human beings and is commonly called "sea lettuce." *Ulva* spp. is a perennial alga that has a widely expanded lobed or undulate blade that may be longer than three feet (Figure 17.7c). The body (or thallus) is attached to the substrate by a multicellular holdfast with rhizoidal outgrowths. The thallus begins its development as a filament, but subsequent lateral duplications produce the expanded blade. A single duplication in the plane parallel to the thallus causes it to be two cells thick. Each cell contains a single cup-shaped chloroplast with a single pyrenoid. All of the cells, except those of the holdfast, are uninucleate. All cells of the thallus are capable of duplication, but the duplications are all perpendicular to the flat plane of the thallus; thus increasing the surface area but not the thickness of the thallus.

The tubular or siphonaceous body type is found among a relatively rare group of marine green algae. They possess body forms from bilateral to radial symmetry. They may be simple or branched. One of the most interesting of these is the genus *Acetabularia* (Figure 17.11), the "mermaid's wine glass." This unique alga is limited to tropical and subtropical waters and, at maturity, is highly calcified. The thallus is composed of an erect, unbranched axis that may be 9 to 10 cm tall. The axis terminates in one or more whorls of raylike chambers that together produce a disc. The entire thallus is one cell, with all of the parts cytoplasmically connected. The thallus is barely mature before the single diploid nucleus duplicates and establishes the coenocytic condition.

REPRODUCTION

The green algae possess many methods of asexual reproduction. Fragmentation of filamentous forms frequently produces juvenile plants and is the only method of asexual reproduction known for genera such as *Spirogyra* and *Zygnema*. Many of the green algae soon die after they are fragmented or detached from their substrate (e.g. *Ulva*).

Zoöspore (Figures 17.4, 17.5, 17.8, and 17.12) formation is quite common among the green algae. The formation of zoöspores is regarded by some phycologists as a temporary reversion to the primitive flagellated condition of the more "primitive" forms. The method of formation of zoöspores shows considerable variation.

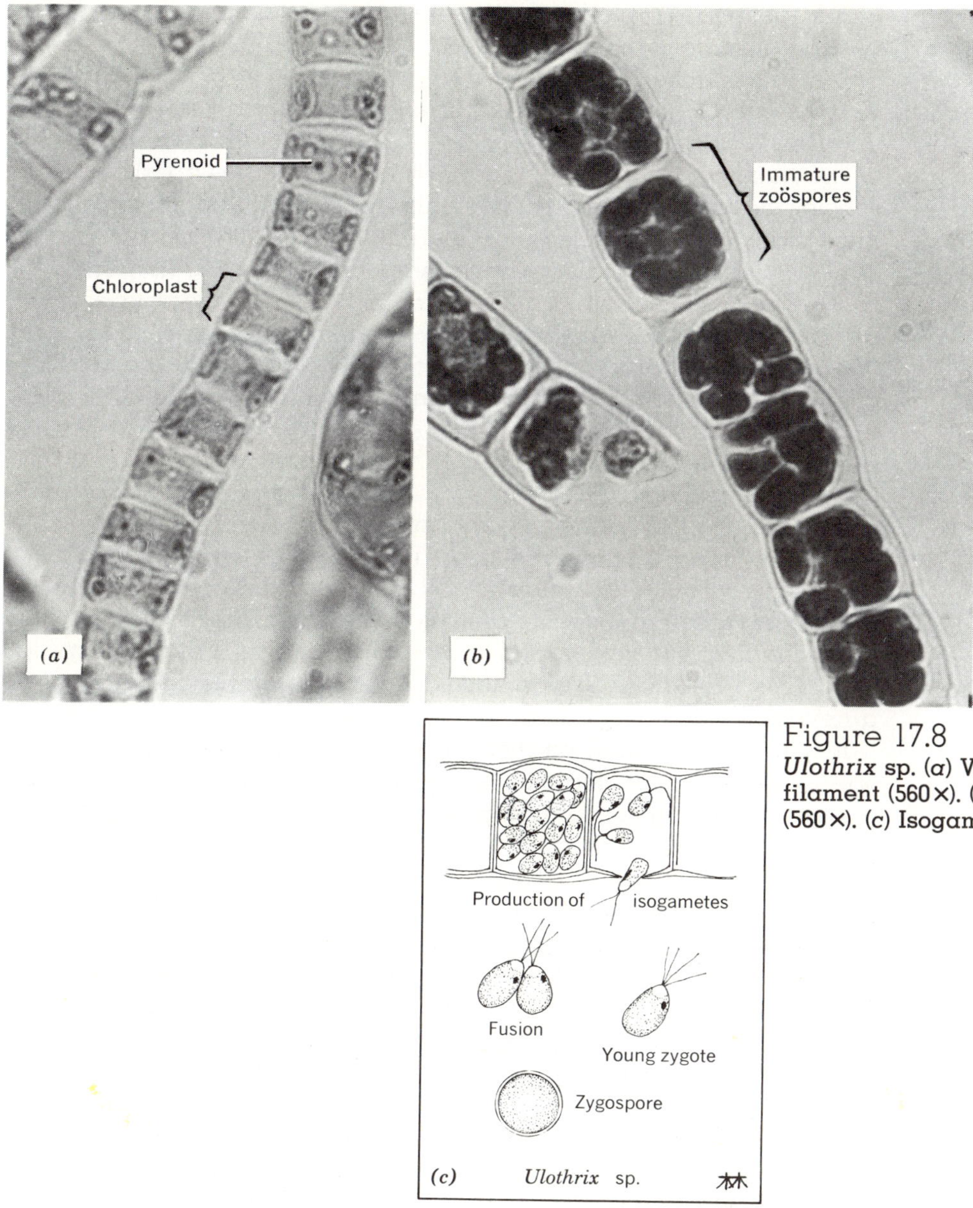

Figure 17.8
Ulothrix **sp. (*a*) Vegetative filament (560×). (*b*) Zoöspores (560×). (c) Isogametes.**

Some species (e.g., *Oedogonium* spp.) produce their zoöspores singly in vegetative cells that show no morphological difference from the other cells (Figure 17.5). Others such as *Cladophora* spp. produce their zoöspores in somewhat enlarged cells called zoösporangia (Figure 17.12).

The number of zoöspores produced in a cell varies from species to species. Several species produce only one per cell, but others produce many. The uninucleate cells produce up to 16 (Figure 17.8) while multinucleate cells produce many more than that (Figure 17.12). The zoöspores are uninucleate, however. Zoöspores are usually quadriflagellate, but *Oedogonium* spp. produces a zoöspore (Figure 17.5) that has a ring of about 120 flagella around one end, and some genera (e.g., *Hydrodictyon* and *Pediastrum*) produce biflagellate ones.

Release of zoöspores is usually through a pore in the cell wall, but some are released only when the cell wall breaks. After release, the zoöspores swim about freely, but their motion is usually in a positive phototaxic direction (towards light). The length of time that they swim varies from 2 to 4 minutes to 2 to 3 days. During this swimming period, the zoöspores do not possess a cell wall. After swimming stops the zoöspore settles onto a substrate, loses its flagella, secretes a wall, and becomes a vegetative cell. Growth of this cell produces a new plant.

In some genera such as *Hydrodictyon* and *Pediastrum* (possessing species with multinucleate cells), the zoöspores are produced by progressive cleavage of the protoplast until finally uninucleate portions of the protoplast are produced (Figure 17.4*a*). Each of these portions develops flagella and begins to swim actively inside of the parent cell. The zoöspores may never be released from the parent cell (as in *Hydrodictyon* spp.) or they may be released but be surrounded by a vesicle formed from the old plasma membrane (as in *Pediastrum* spp.). In either case, the movement of the zoöspores is restricted. After a short swimming period within the parent cell or the vesicle, the zoöspores lose their flagella and assume the same positions in relation to each other as occupied by the vegetative cells of their parent plant. By increase in cell size, but not cell number, the daughter colony then breaks out of its old parent cell or vesicle and produces a new coenobium.

An example of the Chlorophyta that does not produce any motile cells is the unicellular alga *Chlorella* spp. (Figure 17.3). *Chlorella*

Figure 10.1
Blue-green algae growing on rocks in a waterfall.

(a)

(b)

Figure 13.1
(a) *Fucus* sp. growing on rocks in the intertidal zones; exposure to the atmosphere occurs during low tide. Kittery, Maine. (b) *Macrocystis integrifolia* spread on rocks to show stipe, holdfast, blades and air bladders. Some *Ulva* sp. (a green alga) is present. Monterey Bay, California. (c) *Laminaria* sp. spread on rock to show stipe, holdfast, and blade. Monterey Bay, California. (d) *Ectocarpus* sp. growing in the water. Other brown algae are floating. Kittery, Maine.

(c)

(d)

Figure 13.2
(*a*) *Postelsia palmaeformis* in the intertidal zones at Point Lobos, California. (*b*) *P. palmaeformis* photographed to show the thick stipe, the stout holdfast, and the many blades. (*c*) *Fucus furcatus* spread on drift wood to show stipe, holdfast, and blades. Blake Island, Washington. (*d*) *Fucus* sp. Note the inflated receptacles at the tips and the ostioles which open into the conceptacles.

Figure 14.1
Two examples of red algae. (*a*) *Polysiphonia* sp. (*b*) *Porphyra* sp. (From T. E. Weier, C. R. Stocking, and M. G. Barbour. *Botany*, New York: Wiley.)

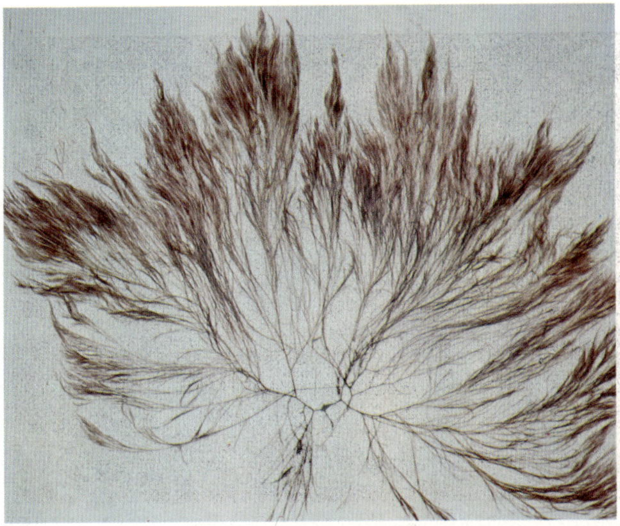

(*a*)

(*b*)

Figure 16.3
Surface film produced from *Euglena* sp. pamelloid colonies in small pond. Red color is caused by hematochrome pigments. (Photograph by Dr. Hugo A. James).

Figure 17.7
(*a*) *Protococcus* sp. on north side of beech tree. (*b*) *Spirogyra* sp. forming mat on surface of small pond. (*c*) *Ulva* sp. (From T. E. Weier, C. Ralph Stocking, and Michael G. Barbour. *Botany: An Introduction to Plant Biology*, New York: Wiley.)

spp., by a series of internal cellular duplications, produces 4 to 8 small cells that are morphologically identical to the parent cell except in size. Because they are identical to the parent cell and have no capacity for motility, the daughter cells are called **autospores.** As they form cell walls and begin to expand, the cell wall of the parent cell breaks and they are released into the water where they grow.

Asexual reproduction in *Scenedesmus* spp. is quite similar to *Chlorella* spp. because each cell of the coenobium produces cells that are like the parent cells. The vegetative cell produces the same number of tiny cells that is contained in the parent colony. These are in the same arrangement as the cells of the parent colony. Such a daughter colony is called an **autocolony** (Figure 17.4), and is released by the breakdown of the parent cell wall. After release, continued growth and differentiation (e.g., of the spines on the end cells) produces a mature colony like the parent colony.

The vegetative cells of the members of the Volvocales produce daughter colonies that contain the same number of cells in the same relationship as the parent colony. The cells of the daughter colonies (or daughter cells in the case of *Chlamydomonas* spp.) become motile before they leave the parent cell.

The vegetative cells of *Chlamydomonas* spp. produce 2 to 8 daughter cells within the parent cell wall by a series of cellular duplications (Figure 17.9). These daughter cells are released by rupture of the parent cell wall and may be released as free-swimming vegetative cells at that time. However, under certain environmental conditions, the daughter cells may secrete a gelatinous matrix and not become immediately flagellated. The gelatinous mass of nonmotile vegetative cells is known as the palmella stage (after the genus *Palmella*) and cellular duplication may continue while the cells are in this stage. Another method of palmella stage formation occurs when motile vegetative cells come together, lose their flagella, and settle onto a substrate. The cells secrete a gelatinous sheath about groups of 2 to 6 vegetative cells. Adjacent groups may partially coalesce forming large gelatinous masses. Continued cellular duplication and production of gelatinous material may produce large masses containing hundreds of cells. The vegetative cells may become actively motile at any time by the production of flagella and escape from the gelatinous matrix. Production of palmella stages is not limited to *Chlamydomonas* spp.; many other algae (e.g. *Euglena* spp.) produce them also. The various palmella stages are difficult to

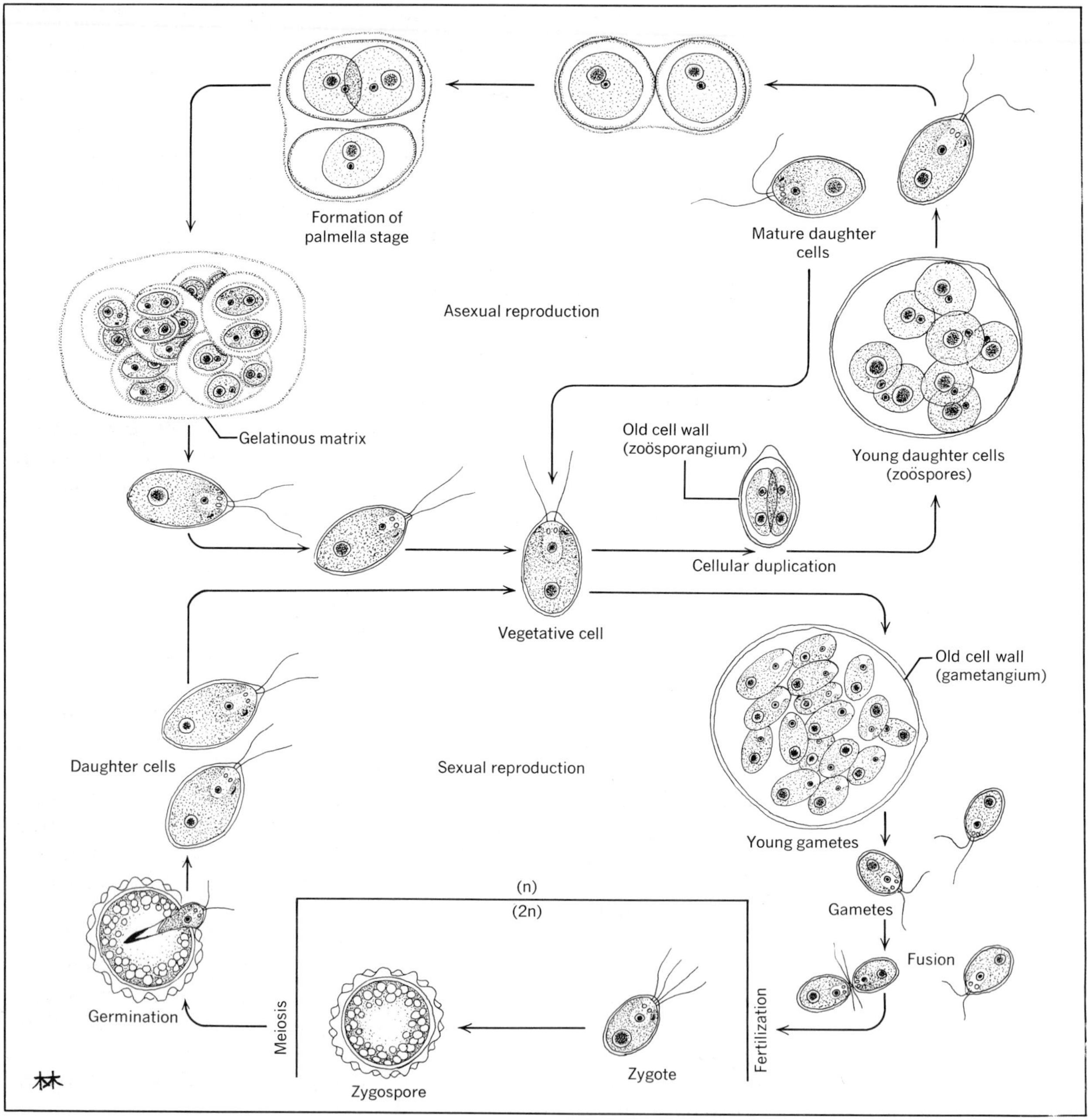
Formation of
palmella stage
Mature daughter
cells
Asexual reproduction
Gelatinous matrix
Old cell wall
(zoösporangium)
Young daughter cells
(zoöspores)
Cellular duplication
Vegetative cell
Old cell wall
(gametangium)
Daughter cells
Sexual reproduction
Young gametes
Gametes
Fusion
(n)
(2n)
Germination
Meiosis
Zygospore
Zygote
Fertilization

Figure 17.9
Life history of *Chlamydomonas* sp., a haplobiontic, haploid, unicellular alga.

separate from each other, and may also be difficult to separate from such genera as *Palmella* and *Gloeocystis*.

Pandorina spp. and *Volvox* spp. are colonial forms that will be used to illustrate another kind of daughter colony formation. Each of the vegetative cells of *Pandorina* spp. simultaneously produces a daughter colony containing the same number of cells produced in the parent colony (Figure 17.2). As the daughter colonies mature inside of the parent cell, they develop flagella and begin to slowly revolve. The parent wall eventually breaks and the daughter colonies are released.

Volvox spp. colonies contain cells that are alike when they are young. As the colony ages, certain of the posterior cells enlarge (Figure 17.10). These specialized cells are called **gonidia** and become slightly depressed into the hollow sphere. Progressive cellular duplications produce a curved plate of cells called the **plakea.** Eventually the plate becomes curved to the point of forming a hollow sphere with a pore, the **phialopore.** Continuous duplications produce a colony of 500 to 60,000 cells. Theoretically the daughter colonies produce the same number of cells possessed by the parent colony. Eventually the entire colony turns itself inside out through the phialopore and each of the cells develops two flagella. The young daughter colony then moves into the hollow sphere of the parent colony. Several daughter colonies may reside inside the hollow sphere at one time. The daughter colonies are finally released by the rupture of the parent colony.

Sexual reproduction among the Chlorophyta includes all three methods: isogamy, heterogamy, and oögamy. While it may be valid to contend that these represent a progression from simple to complex methods of sexuality, they do not correlate with the increase in complexity of the vegetative plant. That is, within the "simple" genus *Chlamydomonas* are to be found all three types, not just isogamy. Homothallic and heterothallic members are also found throughout the Chlorophyta.

Biflagellate gametes are usually produced in vegetative cells of the isogamous and heterogamous species. Indeed, in some unicellular forms (e.g., *Chlamydomonas* spp.), the vegetative cell may also act as a gamete. Cells that have some morphological specialization produce the gametes in the oögamous species. These cells are called gametangia, those that produce sperm are called antheridia, those that produce eggs, oögonia.

Figure 17.10
Life history of *Volvox* sp., a haplobiontic, haploid, colonial alga.

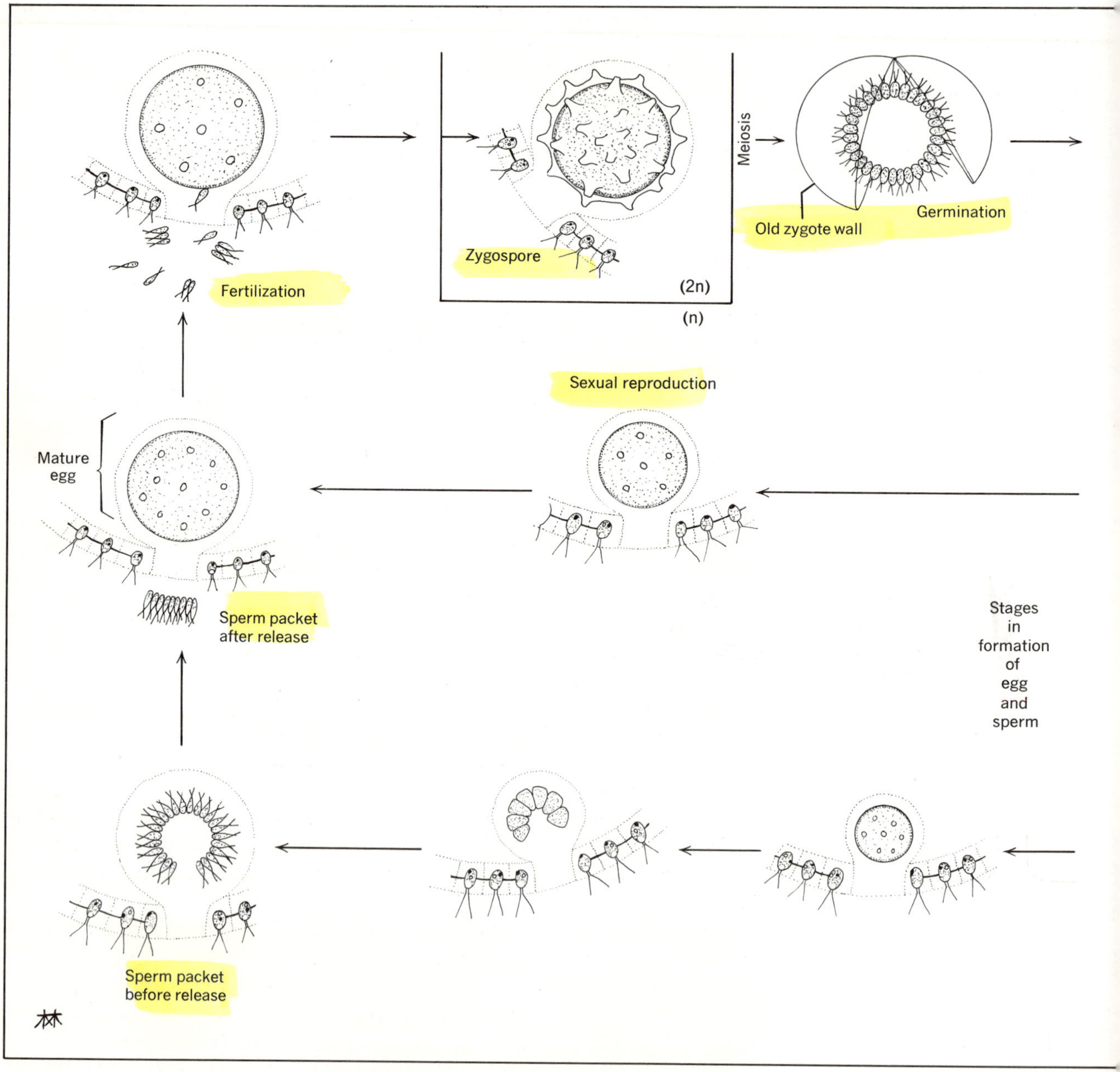

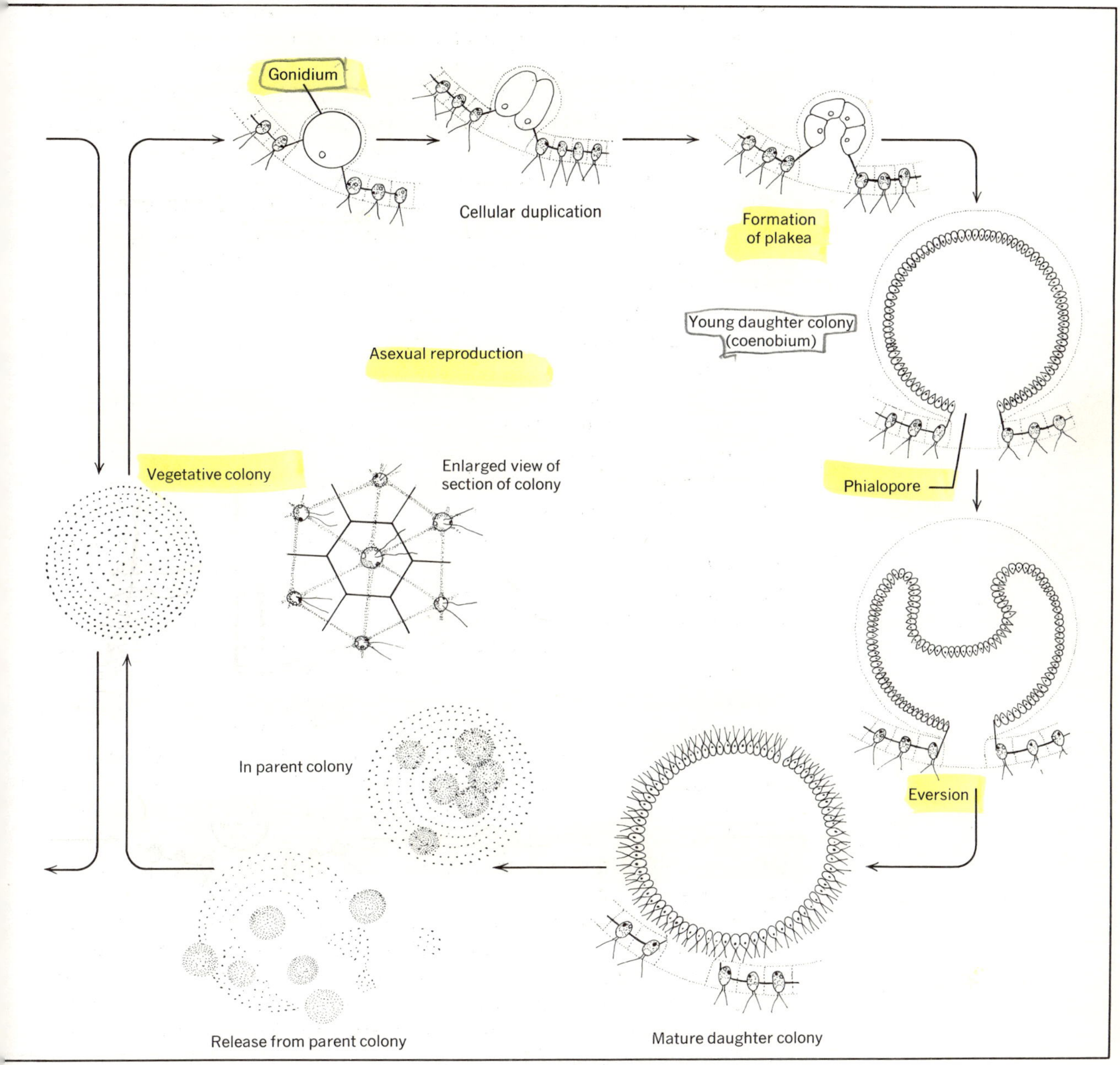
Gonidium
Cellular duplication
Formation
of plakea
Young daughter colony
(coenobium)
Asexual reproduction
Vegetative colony
Enlarged view of
section of colony
Phialopore
In parent colony
Eversion
Release from parent colony
Mature daughter colony

Isogamous and heterogamous (Figure 17.2) species discharge their gametes into the water where fertilization occurs. The zygote then develops away from the parent plant. The oögamous species (Figure 17.5) rarely discharge the female gamete, but retain it inside of the oögonium which the sperm enters through a pore. Fertilization is internal and the zygote may then be discharged, or it may be retained in the oögonium where further development occurs.

Sexual reproduction in the Zygnematales is unique to that group. Amoeboid gametes (aplanogametes) are produced in vegetative cells and unite through a **conjugation tube** as in *Spirogyra* spp. (Figures 17.13 and 17.14) and *Zygnema* spp. In many of the desmids (Figure 17.15), however, the aplanogametes are discharged into the water in close proximity to each other and external fertilization is accomplished.

Although the genus *Chlamydomonas* contains all three methods of sexual reproduction and may have both homothallic and heterothallic members, the most common type observed by beginning students is the union of heterothallic isogametes. When plus and minus strains are mixed together on a slide, one can soon see clumps of the cells beginning to form. Further examination will reveal that the cells are actually uniting in pairs at their anterior, or flagellar, ends. The ends of the cell walls dissolve and the two protoplasts move toward each other out of their parent cell walls. Fusion between the naked protoplasts occurs and the flagella are lost. A new wall then forms around the zygote (Figure 17.9).

Sexual reproduction in *Volvox* spp. is oögamous (Figure 17.10) and may be either homothallic or heterothallic. The sperm is produced by enlarged cells toward the posterior part of the colony. Multiple cellular divisions produce a curved plate of biflagellated sperm that may contain as many as 512 cells. This sperm "packet" is released from the parent colony as a unit and remains together until it is chemotactically attracted to a colony that contains eggs. The packet then breaks up into the individual sperm and fertilization occurs.

Eggs that may function as gonidia unless sperm are available to fertilize them arise from certain cells in the colony. Some colonies produce both eggs and sperm while others produce only one or the other. Fertilization occurs inside of the parent colony and the zygote develops a thick wall. The protoplast produces hematochrome and the zygospore becomes orange-red.

Sexual reproduction in *Pediastrum* spp. and *Hydrodictyon* spp. is by biflagellate isogametes that resemble their zoöspores. The isogametes are released from their cells, however, and fertilization occurs in the water.

Sexual reproduction among the filamentous forms that produce flagellated gametes will be illustrated by a discussion of *Ulothrix* spp., *Oedogonium* spp., and *Cladophora* spp. *Ulothrix* spp. (Figure 17.8) produces from 18 to 64 biflagellate isogametes in its vegetative cells. Each of the vegetative cells except the holdfast cell is capable of producing isogametes, but not all of the cells of a filament do so simultaneously. The gametes are pear-shaped and each contains an eyespot. Most species appear to be heterothallic.

Oedogonium spp. is oögamous but complicated patterns of gamete production have evolved within the various species (Figure 17.5). Some species produce their antheridia on filaments that are the same size as the vegetative filaments. These species are said to be **macrandrous.** Species that produce their antheridia on special dwarf male filaments are **nannandrous.**

Macrandrous species may produce their antheridia and oögonia on the same filament. This condition is described by some phycologists as being bisexual. These species may be either self-sterile (heterothallic) or self-fertile (homothallic). One must culture the specimen to determine whether it is homothallic or heterothallic. Those species that produce their antheridia and oögonia on different filaments are known as unisexual species. They are, of course, heterothallic. The antheridia of the macrandrous species may produce only one sperm per antheridium, but the usual number is two. The sperm are, like the zoöspores, multiflagellate, but they have fewer flagella and are smaller in size. The protoplast of the oögonium becomes a single egg. The oögonial wall develops a small crack through which the sperm will enter.

The nannandrous *Oedogonium* species produce what might be described as a two-part male. In these species, specialized zoöspores (**androspores**) are produced in specialized zoösporangia (**androsporangia**). Although the androsporangia superficially resemble antheridia, only one androspore is produced and its fate is quite different from the fate of the sperm. Upon release, the multiflagellated androspore swims freely until it is attracted to a developing oögonium. It then attaches itself to the oögonium wall or to the wall of the cell supporting the oögonium. It then germinates forming a dwarf male filament. These tiny filaments are usually only one cell long and produce one or two antheridia at their apex. Each of these antheridia produces two multiflagellated sperm cells that are released. They then follow a positive chemotactic response, enter the pore in the oögonial wall and fertilize the egg. Those species that produce both androsporangia and oögonia on the same filament are known as **gynandrosporous;** those that produce the two structures on different filaments are known as **idioandrosporous** species.

Cladophora spp. reproduces sexually by isogamy (Figure 17.12). The gametes are biflagellate and are produced in cells near the ends of branches. The gametes are released into the water where fertilization occurs. All species are probably heterothallic. *Cladophora* spp. possess an alternation of a haploid, gamete-producing generation, with a diploid, zoöspore-producing generation. These two plant forms will be further discussed in the next section of this chapter.

Undoubtedly the alga that is best known among beginning students is *Spirogyra* spp. (Figures 17.13 and 17.14). This common freshwater plant does not produce flagellated gametes, but generates amoeboid gametes that unite through specialized tubes called **conjugation tubes.** Species may be either homothallic or heterothallic. When two compatible filaments come to lie close together, short protuberances are produced from opposing cells. These protuberances elongate until they meet and their end walls dissolve. Each of the cells of a filament is capable of gamete production and a large number of conjugation tubes may be simultaneously produced. The protoplasts of the two connected cells round up forming gametes (aplanogametes). It is the usual case for all of the gametes of one filament to cross the conjugation tubes and fertilize the gametes in the other filament. The motile gametes are then usually referred to as male and the nonmotile ones as female. This condition has been referred to as morphological isogamy but physiological heterogamy. These distinctions become arbitrary, however, when one realizes that in some species the gametes move from the filaments into the conjugation tubes where fertilization occurs. In still others, the gametes of every other cell in a filament moves to the other filament and the cells between the motile gametes remain nonmotile. The opposing cell of the nonmotile gamete produces a motile gamete that moves through the conjugation tube. Therefore, every other cell in each filament contains a zygospore at maturity, and its opposing cell of the compatible filament is empty. These three types of fertilization certainly indicate that the system is a complex rather than a simple heterothallism. The type of conjugation in which two filaments are connected by a series of conjugation tubes is known as **scalariform** (ladderlike) conjugation (Figure 17.14*b*).

Lateral conjugation (Figure 17.14*b*) occurs in homothallic species of *Spirogyra*. In these species, a conjugation tube occurs between every other cell in a single filament. The gametes of every other cell move into the cell with which it is connected by the conjugation tube and fertilization occurs.

LIFE HISTORIES

The members of the Chlorophyta possess life histories that show a wide variety of form and function. The following examples merely demonstrate the variety and are by no means exhaustive of the types found in the Chlorophyta.

Chlamydomonas spp. is a **haplobiontic, haploid** organism that is extremely common (Figure 17.9). The asexual part of its life history has been previously described. Very large populations of the plant result from the production of 2 to 16 haploid zoöspores from each parent cell.

Sexual reproduction occurs when two compatible cells meet and unite. It should be remembered that any vegetative cell has the potential of becoming a gamete (facultative gamete), and that some species of *Chlamydomonas* are heterothallic while others are homothallic. This union may be either isogamous, heterogamous, or oögamous. The observation that the members of the "simple" genus *Chlamydomonas* spp. possess all three types of sexual reproduction demonstrates that there is no correlation between the evolution of "advanced" sexual reproduction as represented by oögamy and "advanced" vegetative structure as evidenced by *Oedogonium* spp.

The zygote is the only diploid part of the life history; it is quadriflagellate and may be motile from several hours to several days. Eventually, however, the flagella are lost; the zygote develops a heavy wall (it may then be called a zygospore), and undergoes a period of dormancy. The zygote germinates with the production of flagellated daughter cells. Meiosis occurs just prior to germination; thus each of the daughter cells is haploid. Most species produce only four daughter cells but 8 to 32 have been observed for some species. Each of the daughter cells becomes a mature vegetative cell and reproduces asexually.

Volvox spp. is also **haplobiontic and haploid** (Figure 17.10). Asexual reproduction by daughter colonies and sexual reproduction by oögamy have already been described. The orange-red zygospore undergoes a prolonged dormancy after release from the colony. Meiosis occurs prior to germination and from one to four zoöspores are produced. The production of only one zoöspore by some species probably indicates the degeneration of the other three nuclei produced by meiosis. Mitotic duplications take the developing colony through a sequence of events that strongly resembles the plakea produced during asexual reproduction. This first colony produced by the zygote usually contains only 128 or 256 cells when it is released from the zygospore wall. The new colony always reproduces only asexually and the resulting colonies from the next few generations contain a progressively larger number of cells. Finally the number of cells that is characteristic for the species is reached and the entire sequence of the life history may begin again.

Examples of **haplobiontic, diploid** green algae are rather difficult to find. *Acetabularia* spp. enlarges to the full size of the mature plant before the zygote nucleus divides mitotically (Figure 17.11). The diploid nuclei migrate throughout the coenocytic plant with some of them moving into the gametangial rays that form the disk at the tip of the plant. Each of the nuclei in the gametangial rays becomes the center of a protoplast that secretes a wall and becomes a **cyst.** The cysts enlarge to several times their original size and become multinucleate by several mitotic duplications. A meiotic division of the nucleus finally occurs and thousands of biflagellate isogametes are produced within each cyst. After their release, the isogametes unite outside the parent plant and initiate a new zygote nucleus. *Acetabularia* spp. is homothallic.

The zygote germinates within a few days by the production of a short tube. One end of the tube grows erect and forms the axis; the other pole becomes the rhizoidal system. The zygote nucleus remains in the latter until the plant is mature.

Cladophora is a genus that possesses a **diplobiontic** life history (Figure 17.12). That is, two forms of this plant are found in the natural environment. In the case of *Cladophora* spp., both kinds of plants are morphologically similar and are thus said to be isomorphic (heteromorphic diplobiontic life cycles are exhibited by *Laminaria* spp., for example). In *Cladophora* spp., the two types of plants are free-living and independent of each other. They vary in their chromo-

Figure 17.11
Life history of *Acetabularia* sp., a haplobiontic, diploid alga.

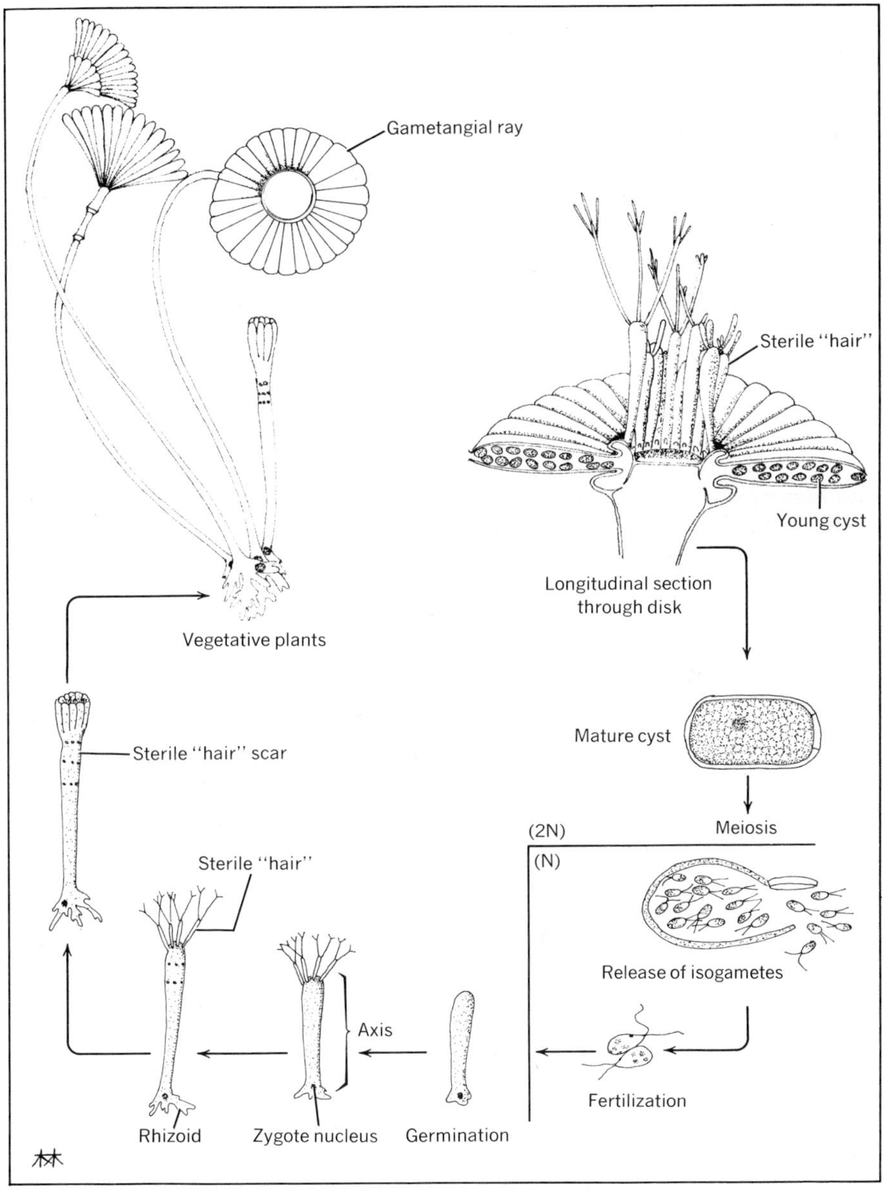

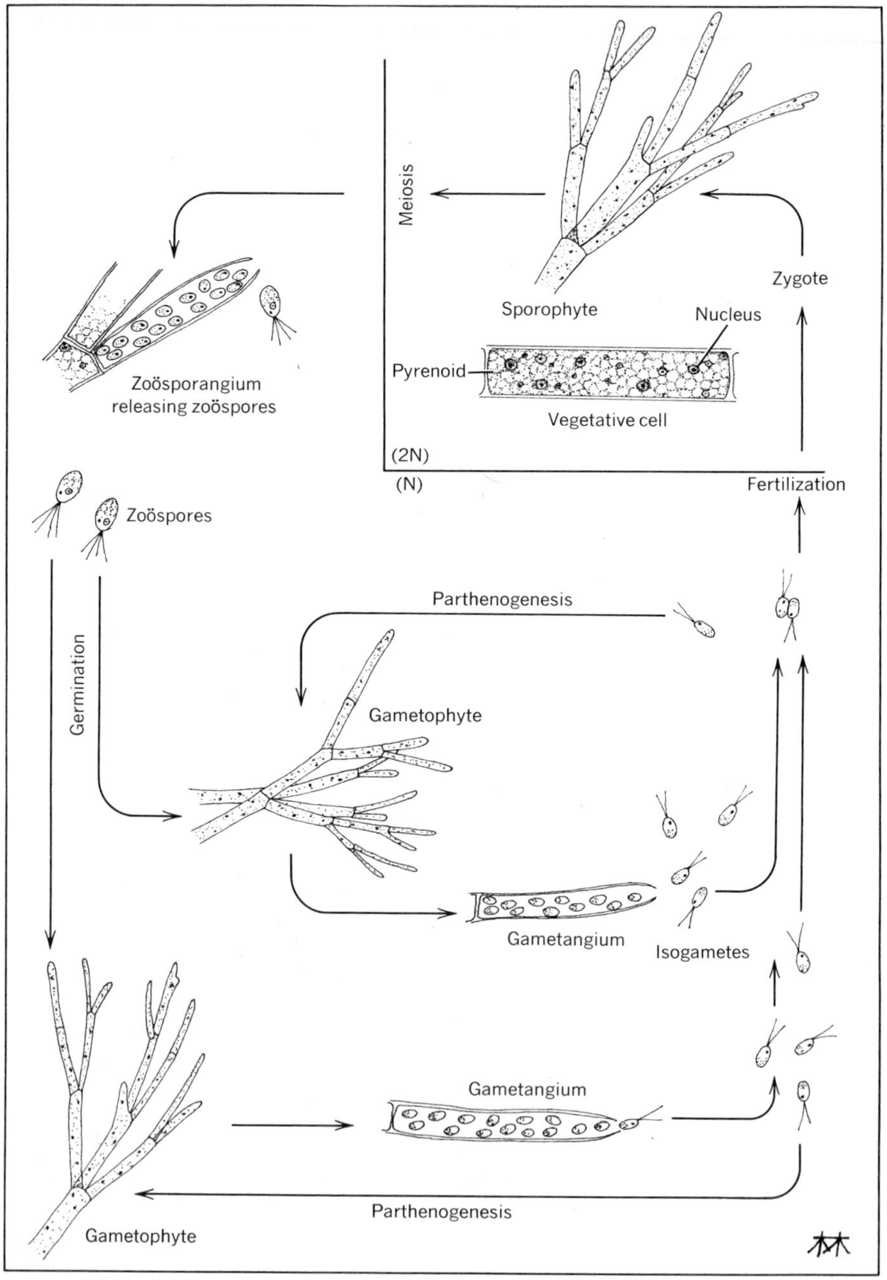

Figure 17.12
Life history of *Cladophora* sp., a diplobiontic alga.

some complement, one being haploid and the other diploid. The haploid and diploid plants represent an alternation of generations, although the alternation is not obligate. The gametes of the haploid plants have been reported to develop parthenogenetically into new haploid plants.

The diploid plant (sporophyte) produces quadriflagellate zoöspores in the zoösporangia that are usually terminally situated. The zoöspores are typically haploid as a result of meiosis that occurs just prior to their formation. After their escape through a pore in the zoösporangium, they soon settle onto the substrate and germinate. The plant produced by zoöspore germination is the haploid gametophyte. The gametophytes are probably unisexual and thus sexual reproduction in the plant is heterothallic. The gametophytes finally produce isogametes in much the same manner as the zoöspores were produced by the sporophytes. The gametangia are also usually terminal but the gametes are biflagellate. Union of the isogametes occurs after their release into the water and the zygotes quickly settle onto the substrate. Germination occurs within one or two days and a new diploid sporophyte is produced.

The common alga *Spirogyra* spp. possesses the **haplobiontic, haploid** type of life history (Figure 17.13). The genus possesses no method of asexual reproduction other than fragmentation, and its sexual activity has been previously described. After **conjugation** has occurred, however, the heavy zygote wall may not be completely produced for several weeks. The wall eventually matures with markings and coloration characteristic of the species and is released by the decay of the filament. Germination of the zygospore does not immediately occur; indeed it may be a year before germination of most zygotes takes place. The zygote nucleus divides by meiosis just prior to germination of the zygospore. Three of the haploid nuclei degenerate. The zygote then produces new cells by mitosis followed by cytokinesis and a new filament is formed.

Cosmarium spp. is a desmid that is commonly found in the phytoplankton (Figures 17.6*c* and 17.15). It is **haplobiontic and haploid,** but will be described here as an example of a unicelled green alga that produces amoeboid gametes. *Cosmarium* spp. has a deep sinus dividing the two semicells with the nucleus lying in the isthmus that connects the two semicells. The isthmus elongates during asexual reproduction and mitosis is followed by cytokinesis. Each of the old semicells regenerates a new one.

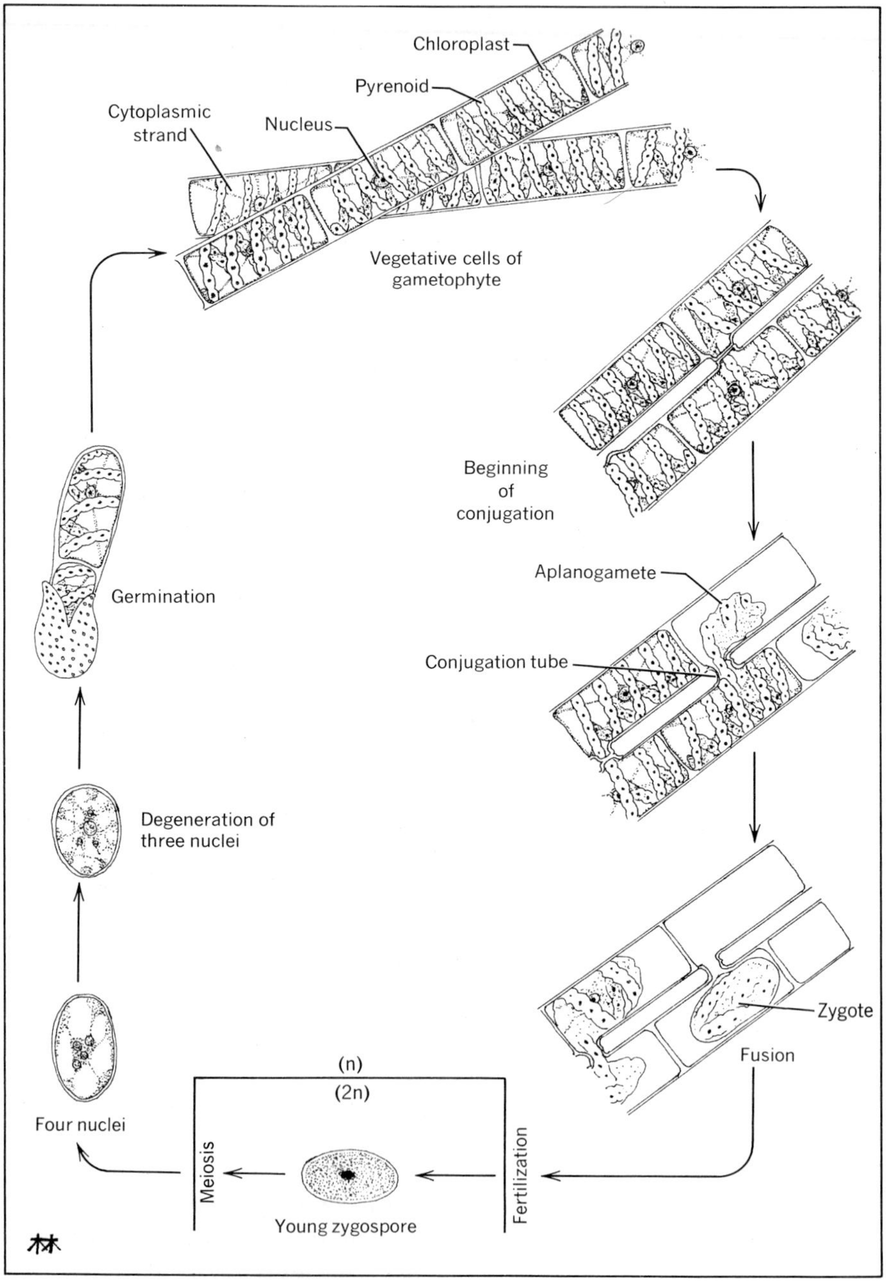

Figure 17.13
Life history of *Spirogyra* sp., a haplobiontic, haploid, nonflagellated alga.

When compatible strains of *Cosmarium* spp. come together, they produce a common pectinaceous sheath around themselves. The two cells in the sheath break open at the isthmus and the protoplasts move together in amoeboid fashion. Fusion of the gametes produces a zygote that develops a heavy, spiny, or warty wall and becomes a zygospore. The zygospore undergoes a dormant period but eventually germinates with the production of one or two vegetative cells.

Meiosis occurs just prior to germination, and some of the nuclei degenerate. If three of the new haploid nuclei degenerate, only one new vegetative cell is produced. It is often the case, however, that only two of the nuclei degenerate and then two new vegetative cells are produced.

This type of sexual reproduction is usually called conjugation although no conjugation tube is produced. The sexual activity has usually been observed to occur only between fully developed individuals, but conjugation between immature ones has been seen. It is also known that the members of a conjugating pair are sometimes sister cells produced from the same parent. Therefore, some species of *Cosmarium* are homothallic. Other species are known to be heterothallic.

HUMAN RELEVANCE

As one might expect after being introduced to the large numbers and the great variation found in the green algae, the members of the group have an important impact on the affairs of man. Although it is logical to assume that photosynthesis by green algae contributes significantly to the oxygen that is present in water supplies and the atmosphere, it is difficult to determine how much oxygen is produced by a particular phylum of algae under natural conditions. Several studies of photosynthetic rates, oxygen production and consumption, carbon dioxide production and consumption, and biomass production have been made under natural conditions. The separation of the component parts of these measurements that can be attributed to green algae is extremely difficult if not impossible. Many laboratory experiments to study photosynthesis in various greens (*Chlorella* spp. and *Cladophora* spp., for example) have yielded data that indicate problems inherent in such studies. Rates of photosynthesis (and thus gas exchange) have been shown to vary greatly with such factors as light intensity, carbon dioxide and oxygen concentration in the water, temperature, nutrients, water move-

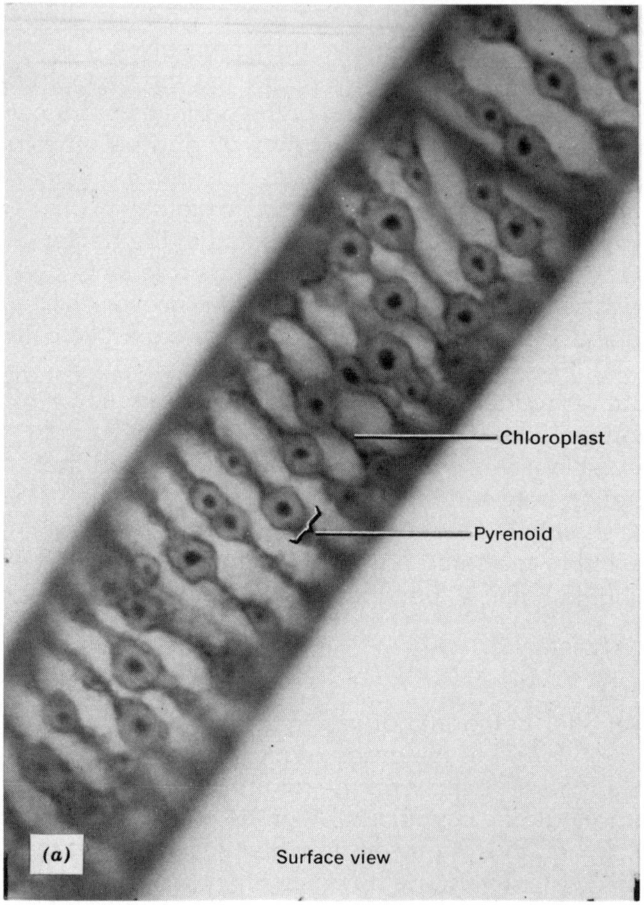

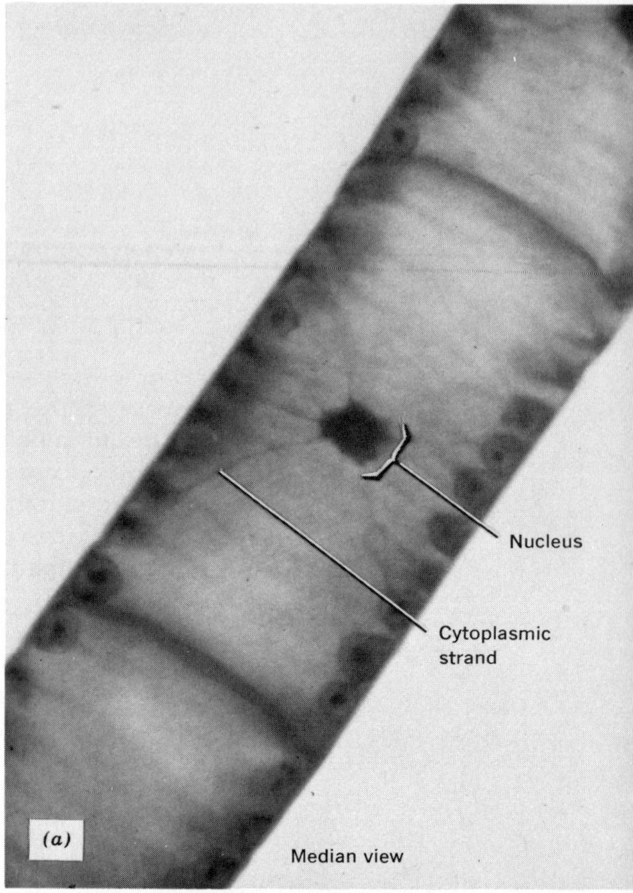

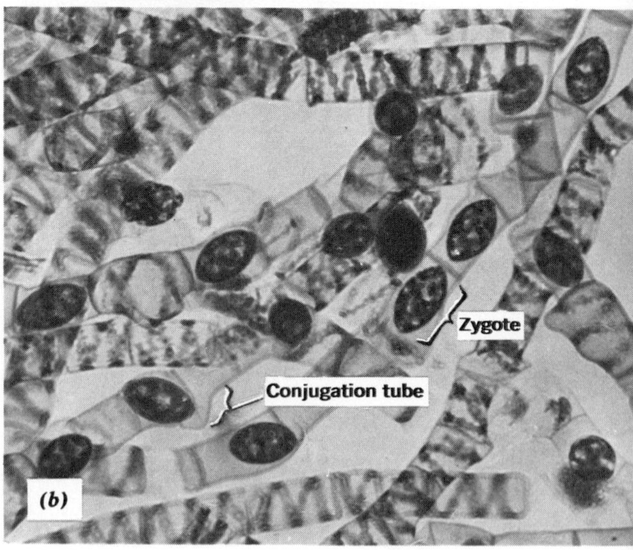

Lateral conjugation

Figure 17.14
Spirogyra sp. (*a*) Vegetative cell. Median (1130×) and surface views (900×). (*b*) Sexual reproduction. Scalariform (250×) and lateral (170×) conjugation.

ment and the condition of the chlorophyll contained by the alga.

Relatively few of the green algae have been used by man as a direct food source. Members of the genera *Ulva* (Figure 17.7), *Enteromorpha*, *Codium*, and *Monostroma* are used as salads or relishes, particularly among Oriental people. These algae have a distinct "fishy" flavor and a slippery gelatinous texture that may make them unpalatable to the uninitiated palate. The Hawaiians used *Codium* spp. and *Ulva* spp. (along with genera of other phyla) in a food that is given the general name of "limu." Limu may be eaten raw or boiled with fish or shrimp and may be used to wrap a dog or a pig for underground roasting. Dried *Spirogyra* spp. and *Oedogonium* spp. are used in India as a rice supplement. They may add some flavoring and a few vitamins and minerals, but very little bulk would be added to the diet.

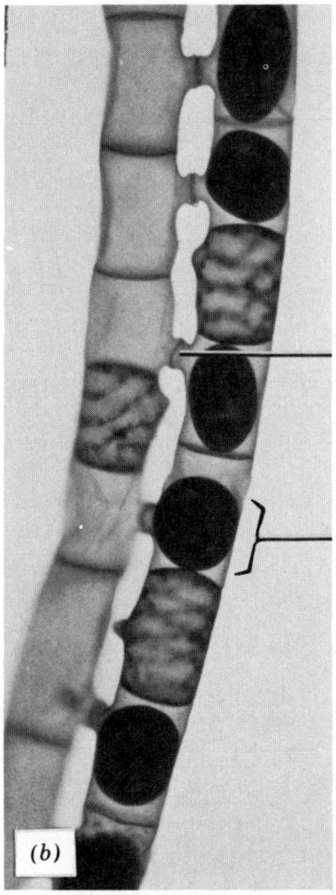

Sclariform conjugation

No discussion of the uses of green algae by man (or indeed of the algae generally) would be complete without a discussion of the genus *Chlorella* (Figure 17.3). This tiny unicell is being investigated as a source of food for both man and other animals. Reports on the digestability of *Chlorella* spp. by man vary, but it is usually agreed that the process is still inefficient. However, if man could establish the microorganisms that would produce the proper enzymes in his intestine, the potential use of *Chlorella* spp. as a food is high. The cells may produce as much as 50% dry weight of protein which contains all of the amino acids essential to man. Some species produce as much as 85% lipids. Various analyses of *Chlorella* spp. show that they usually produce 30 to 40% protein, 27 to 30% carbohydrates, 7 to 15% lipids, and 5 to 19% ash. *Chlorella* spp. also has a high vitamin content. *Chlorella* spp. has been used in making bread, ice cream, cookies, and noodles, which have been eaten experimentally by man. Experimental culture plants have demonstrated the potential productivity of *Chlorella* spp., because the alga has been able to produce 100 lb/acre/day, which is about 17 tons/acre/year. The estimates of the cost of this production vary but the usual cost is about 30 cents per pound.

Chlorella spp. and many other green algae that are found in the phytoplankton contribute to man's food supply in an indirect way. They form the basis for many food chains or food pyramids. Again, it is difficult to establish where a species precisely fits in a food chain.

Figure 17.15
Life history of *Cosmarium* sp., a desmid.

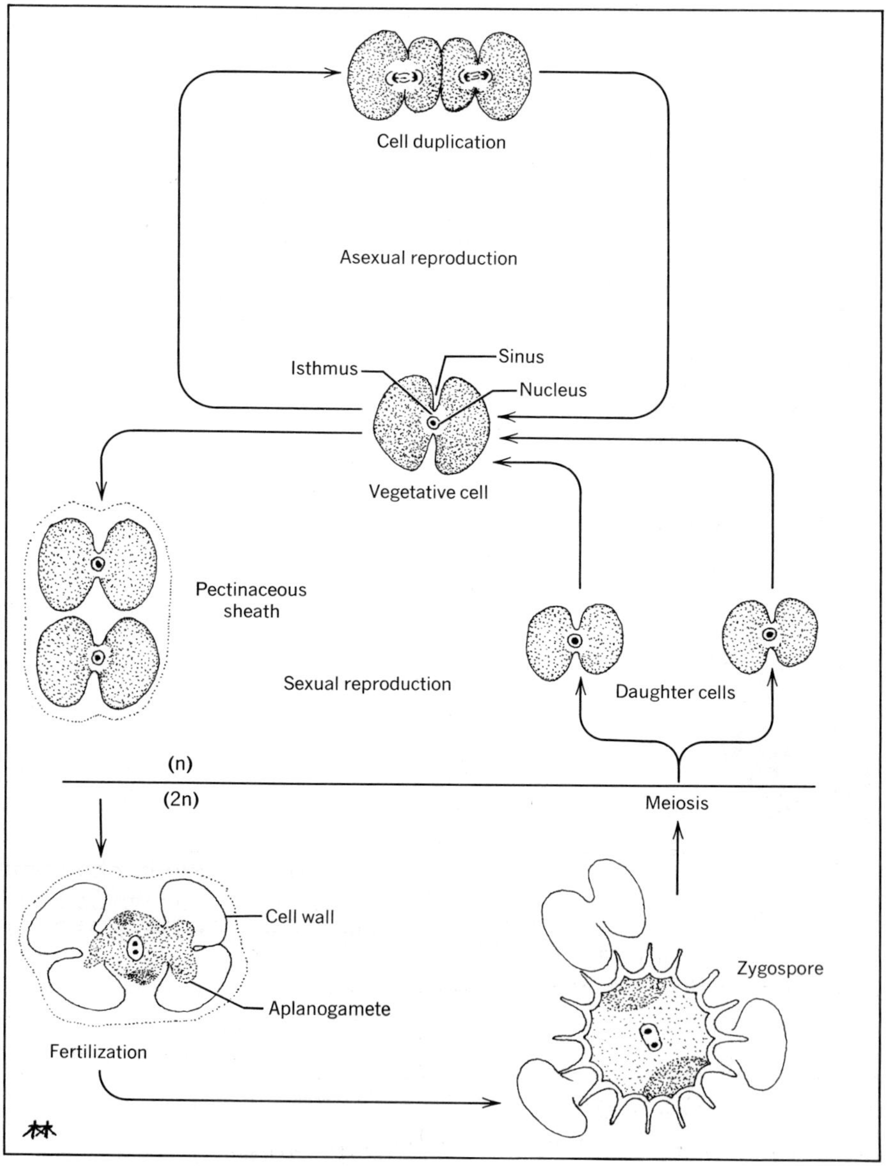

Various experiments to determine the exact relationships have been carried out. In these experiments, particular species of algae (e.g., *Chlamydomonas* spp.) have been fed to particular species of zoöplankton (e.g. *Infusoria* spp.). The resulting activities and populations of the zoöplankton give some indication as to the value of the particular alga species to the zoöplankton species. The problem, of course, of this kind of experimentation is that it is carried out under very artificial conditions. Although a particular zoöplanktonic species may eat a particular phytoplanktonic species under laboratory conditions, under natural environmental conditions the animal may eat a large variety of algae, perhaps to the exclusion of the one fed to it in the laboratory. Thus, this sort of experiment, while establishing what a zoöplankton *can* eat does not necessarily demonstrate what it *does* eat under natural conditions. However, it is important to recognize that algae, as primary producers, do form the first step between inorganic and organic compounds in most ecosystems. The zoöplankton that feed upon them are eaten by large animals that are eaten in turn by still larger animals. This process continues until the animal that man uses eats its final meal.

Man has attempted to introduce algae as a foodstuff for livestock. Most such animals do not find the pure algae palatable. However, it has been shown that when algae has been mixed with grains (e.g., barley) at a concentration of about 10% algae, the mixture produces a valuable feed for sheep, swine, and cattle. The production of the algal components of such feeds costs about $60 to 80/ton.

More important than food production is the use of the green algae by man to clean up his polluted water. The organic wastes produced by man's excessive population are reaching alarming proportions. If organic wastes can be pumped into shallow ponds where bacteria and fungi can break them down into plant nutrients that would be reused by algae in photosynthesis, a cycle could be established in which sewage would be fed into a system that would produce a food usable by man and his animals. Various studies in which the algae are harvested and fed to fish instead of livestock have also been conducted. In the long run, systems that shorten the food chain to algae → insect larvae → fish may prove to be the most beneficial to man. *Chlorella* spp. is often used as the principal alga in such a system. In large, shallow tanks, its rapid photosynthesis produces enough oxygen for use by the bacteria that break down the sewage during aerobic respiration. Various problems must be overcome before a system that efficiently links sewage disposal with food production can be produced. It does appear, however, that the green algae may offer a partial answer to cleaning up the environmental mess in which man has placed himself.

Because of the ability of some of the green algae to purify human waste products, both carbon dioxide and sewage, some phycologists throughout the world have been interested in using them in a closed ecosystem that would support human life on long space flights. Of the many species of algae studied, *Chlorella pyrenoidosa* appears to be the most promising. Experiments that are designed to test the feasibility of using the alga to produce a continuous supply of oxygen and food by using the carbon dioxide and mineral nutrients produced by the astronauts are being pursued. So far, space travelers have had to depend upon systems other than photosynthesis to supply their needs, but long-range plans indicate the need for a photosynthetic system.

An interesting relationship between the productivity of rice paddies and the presence of green algae has been observed. If green algae are not present in the paddy, the rice plants do not tend to do as well as when the algae are present. Also, without the presence of algae, the rice plants are more often attacked by pathogenic fungi. Two possible explanations are readily available. The first is that the algae increase the oxygen content of the water and thus enable the rice plants to grow better. The other is that the algae produce antibiotics that have an antagonistic effect upon the fungi. As with so many other biological systems, the answer may lie with both, either, or neither speculation.

Antibiotics are known to be produced by some green algae. The genus *Chlorella* produces an antibiotic called **chlorellin** that is considered to be a decay product of the alga. It does, however, inhibit the growth of several species of bacteria and other algae.

Green algae have been used as indicators of particular water conditions (**bioassay**). The content of phosphorous and nitrogen in certain water samples has been assayed by using *Carteria* spp. Potential fertility of water has been measured by using *Selenastrum capricornutum* as a test species. *Pandorina morum* and *Eudorina elegans* have been shown to be indicative of small ponds that are polluted with sewage. *Spirogyra varians*, *S. adriata*, and *Mougeotia parvula* are inhabitors of hard-water lakes. Other species of these same genera are indicative of different water conditions. *Spirogyra* *crassa* and *S. decimina*, for example, are indicators of alkaline bog lakes. *S. communis* and *Mougeotia parvula* are indicators of organically polluted waters and *Spirogyra salina* is found only in brackish waters. Thus, as more and more of the specific requirements of specific algae are known, they will become more important as indicators of particular water conditions.

Zoöchlorella is a green alga that has been found in the tissues of marine invertebrates. Members of this endozoic genus appear to form a mutualistic relationship with sea anemones because the anemones position themselves in such a way that the *Zoöchlorella* is advantageously exposed to the light. It would seem that the anemones derive a food advantage from the alga, and that the alga derives a positional and protective advantage from the anemones. Coral that has formed a mutualistic relationship with *Zoöchlorella* is able to extract calcium from the water only during the time that the alga can carry on photosynthesis. The exact nature of these relationships is not as yet understood but other such relationships have been observed with sponges, protozoans, and *Hydra* spp.

Several filamentous algae that form mats on the surface of water have been compressed and the water evaporated from them. These mats form a coarse paper and have been produced from *Oedogonium* spp. and *Spirogyra* spp. as well as the chrysophyte *Tribonema* spp.

The green algae contribute their share of problems to man. Increased contamination by man's wastes of our streams and lakes lead some phycologists to speculate that all but our coldest and swiftest waterways will be clogged with algae within 17 years if man does not do something to prevent it. Prevention lies in denying nutrients to the algae. The most frequently listed limiting nutrients is organic phosphate. Action by environmental groups has begun to decrease the use of phosphate compounds (e.g., laundry detergents that contain phosphates) in the home. Similar action has begun to demand adequate sewage treatment to remove the phosphates before the effluent is allowed to enter the waterways. Even when carbon dioxide has been found to be limiting, the presence of sewage seems to provide the carbon needed to stimulate the growth of *Chlorella* spp. During certain periods of the year, particularly when runoff adds a higher concentration of dissolved compounds to the water, mats of species of *Spirogyra*, *Mougeotia*, and *Zygnema* may clog waterways and filter systems. Other green algae that have been frequently reported to create problems in water supplies are species of *Pediastrum, Scenedesmus, Staurastrum, Cladophora, Pithophora*, and *Chlamydomonas*.

The list of green alga that cause surface bloom, filter-clogging, and other reservoir-related problems is a very long one. Some greens that create a slimy taste in water are *Chaetophora* spp., *Spirogyra* spp., and *Tetraspora* spp. Discolorations in water are caused by species of *Chlorella, Chlamydomonas*, and *Cosmarium*.

Although one usually considers the blue-greens or the dinoflagellates when he thinks about toxic algae, some of the greens are also poisonous to animals. Mice have developed hepatic necrosis when fed *Scenedesmus* spp., and have developed diarrhea and died when fed *Chlorella* spp. *Chlorella* spp. has caused hens to lose weight, develop diarrhea, and suffer reduced egg-laying ability. These and various other symptoms were noted when rats were fed with *Scenedesmus* spp. and *Chlorella* spp. When rats were fed *Chlorella* spp. they developed abnormalities in the pancreas, salivary glands, kidneys, and lens of the eye.

Sensitivity of skin and mucosae have been reported in human beings who have been exposed to airborne species of *Chlorella, Chlorococcum, Scenedesmus*, and *Ankistrodesmus*. Some green algae have been associated with mycoses. *Chlorella* spp. or *Chlorococcum* spp. have been reported to be isolated from lesions resulting from actinomycoses and blastomycoses. The relationship of the algae to the mycoses is not completely understood, but further study is certainly indicated. Also, when species of *Chlorella* and *Scenedesmus* were fed in slowly increasing amounts to volunteers, they developed abdominal distensions, flactulence, bowel problems, nausea, vomiting, abdominal and rectal pain, headache, and weight loss. These observations and investigations certainly indicate that a great deal more study is needed before the green algae can be considered as a reliable food supply for man. Reactions to algae apparently are quite varied, and it is certainly wise to recognize their potential value as an addition to man's larder, but such an addition is not to be considered an unmixed blessing.

the lichens | a symbiotic relationship

18

Although lichens produce bodies that are distinctive and consistently recognizable, they are not really plants. The term lichen is really a designation of a **symbiotic relationship** between an algal component (the **phycobiont**) and a fungal component (the **mycobiont**). Even though the relationship produces a consistent entity, the lichens have been variously classified within the plant kingdom. The mycobiont of the association often produces "fruiting structures" that are the same as the structures produced when the fungus is grown separately from the alga; therefore, the lichens have often been classified with the fungi. This type of classification does not recognize the equal value to the relationship of the alga, but it has become a convenient method of lichen classification. It must be recognized that any classification system for lichens is an attempt to place the various relationships into workable categories. Approximately 15,000 of these distinctive associations have been described.

Lichens are found in an extremely wide range of habitats. They invade bare surfaces of exposed rocks, the surface of frozen soil in the Arctic regions, and the surfaces of living, dead, and decaying wood. Many are epiphytes that hang in great quantities from trees. Some lichens are submerged aquatic plants that are occasionally exposed to the air. Some can grow under extremely xeric conditions while others require a great deal of readily available moisture.

THE "PLANT" BODY

One must recognize that when dealing with lichens he is really dealing with three entities: the fungus, the alga, and the composite form (Figure 18.2). The fungal and algal components may each grow separately and function quite independently of each other. When they establish the association, however, they produce a "plant body" that differs from either of the components and that can live in environmental situations that are unsuitable for either of the separate components.

It is this separate entity that is of interest to beginning students of the plant kingdom. Although the relationship is certainly a symbiotic (two organisms living together in a close, long-lasting relationship) one, its exact nature has been variously interpreted. One interpretation indicates that the fungus is weakly parasitic on the alga. Evidence for this interpretation comes from the fact that the fungus often produces haustoria that enter the algal cell. Also, the fungus can live as a saprobe on any of the algal cells that die. However, the fungal and algal components may be grown separately on artificial media; therefore the relationship is not obligate. Also, experimental evidence has shown that the fungal components of the lichen are not specific to their algal components. The lichen association may be established by a single alga and different fungi or by a single fungus with different algae.

A more widely accepted interpretation of the association is that each of the components receives a benefit from the other (**mutualism**). In this type of relationship the mycobiont provides the alga with protection from desiccation and high light intensities, a ready supply of water and minerals, and anchorage to the substrate. The phycobiont provides the fungus with organic compounds from photosynthesis. The fungi are also notably deficient in vitamins, particularly biotin and thiamine, which are secreted by some phycobionts. Further evidence that the relationship is a mutualistic one comes from the fact that the lichen association survives in an environment that is unsuitable for both components. If conditions shift in favor of either of the components, the association breaks down with the favored component being the survivor. It would seem, then, that the lichen is an association that allows algae and fungi to survive in a mutualistic symbiosis under environmental conditions unsuitable for either. However, when the environment becomes favorable to either, the symbiosis can shift toward parasitism.

Lichen associations produce bodies that are usually divided into three categories: **crustose, foliose,** and **fruticose** (Figure 18.1, p. 302). The crustose lichens are those that adhere very closely to their substrate by means of rhizoidal hyphae. Indeed, they are usually so strongly attached to the substrate that the collector must take some of the substrate in order to get all of the lichen. A cross-section of the thallus (Figure 18.2) reveals an upper hyphal layer (the **upper cortex**) which is heavily gelatinized and forms a protective layer. Under this

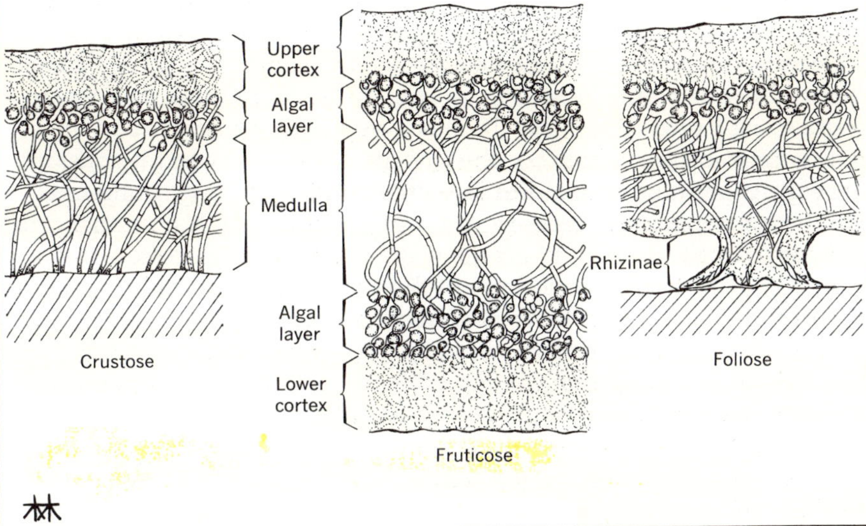

Figure 18.2
Cross sections of the lichen symbiosis. (Redrawn mostly from V. Ahmadjian, *The Lichen Symbiosis*, Blaisdell.)

layer is found the **algal layer** which is composed of loosely interwoven, thin-walled hyphae surrounding clumps of algal cells. The **medulla** is under the algal layer and is a thick layer of loosely packed hyphae that are in direct contact with the substrate. Some of the crustose lichens have upturned portions of their margins but most do not.

The foliose lichens (Figure 18.1) are usually "leafy" in form and are much more loosely attached to their substrates. They have the same three layers as described for the crustose lichens but in most of them the medulla is subtended by a **lower cortex** that is quite similar to the upper cortex (Figure 18.2). Specialized bundles of absorbing hyphae (the **rhizinae**) that rapidly conduct water by capillary action extend from the lower cortex to the substrate. The non-"leafy" foliose lichens have an umbilicate form that is circular in outline and attached to the substrate only by a single central mass of rhizoids.

The largest of the lichen plant bodies are produced by the fruticose lichens (Figure 18.1). These may be long, pendulous strands or erect, hollow stalks. The cortex is really an outer cortex (instead of upper or lower) and the algal layer is thinner and more scattered although it may subtend the complete outer cortex. The center of the thallus is composed of a thick, loosely woven medulla (Figure 18.2). The lichen may grow very loosely attached to the substrate or it may grow attached to the substrate by special rhizinae.

Lichens are among the slowest growing entities among the plants. There are few accurate records of growth rates, but the average appears to be between 0.1 and 10 mm a year. Growth occurs at the tips of lobes or branches, but the thickness remains almost constant. Growth rate varies considerably with variations in environmental conditions, especially temperature and moisture. High humidity and cool temperatures seem to produce the fastest growth rate. Growth rate may vary from lobe to lobe of the lichen and certainly with the age of the lichen body. However, because lichens may live for long periods of time, they may reach very large sizes. For example, some foliose lichens are known to reach 3 feet in diameter and pendulous fruticose lichens may reach 6 feet in length. Some estimates of age of temperate region lichens indicate that they may live from 100 to 200 years. However, some of the arctic species are estimated to be 1000 to 4500 years old.

REPRODUCTION

All lichen reproduction is probably asexual. As the lichen thallus increases in size, the older portions die leaving the younger portions as small, isolated plants. Several lichens produce special vegetative bodies composed of algal cells surrounded by fungal hyphae (Figure 18.3). These bodies (**soredia**) drop or are broken off and are dispersed by air currents. When they reach suitable environmental conditions, they can continue to grow producing a new thallus. Several of the other lichens may reproduce vegetatively by fragmentation.

The mycobionts of lichens are usually Ascomycetes that may produce apothecia (Figures 18.3 and 18.4) or perithecia when they are in the lichen association. Other fungi belonging to the Fungi Imperfecti and Basidiomycetes form several associations. Spores typical of the fungal group are found when the fungus is in the lichen association; most are sterile when not in the association. The spores are usually shot from the lichen, and they have been germinated and grown on artificial media. Therefore, although the fungal spores may reproduce the mycobiont, there is a question as to their value as reproductive mechanisms for lichens. In order to be effective, they would have to land in an area favorable to their growth and in

Figure 18.3
Asexual reproduction by the production of soredia. Fungal apothecium produced in lichen thallus.

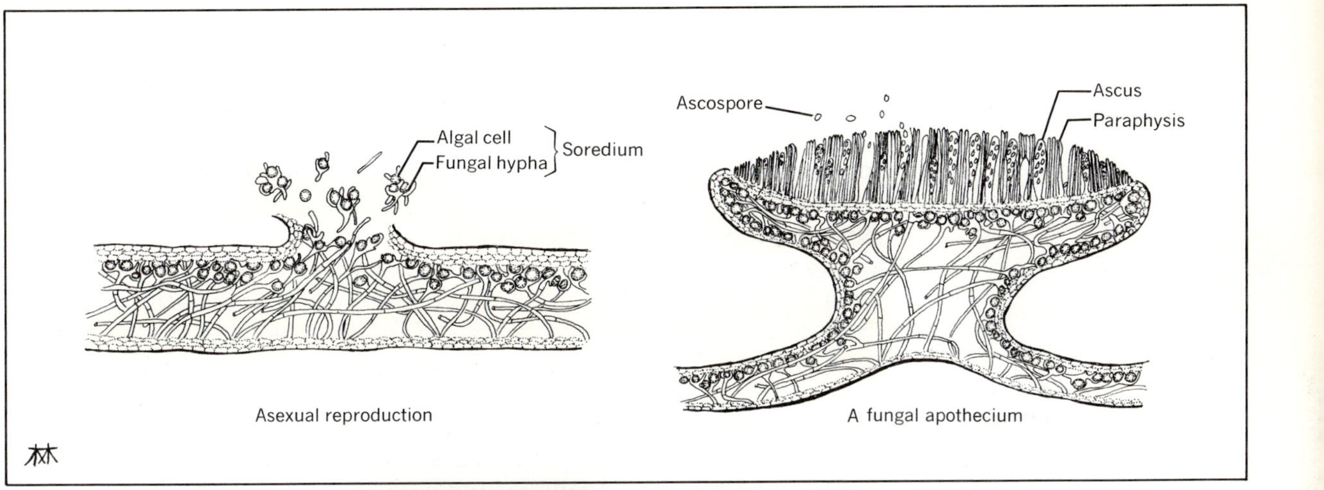

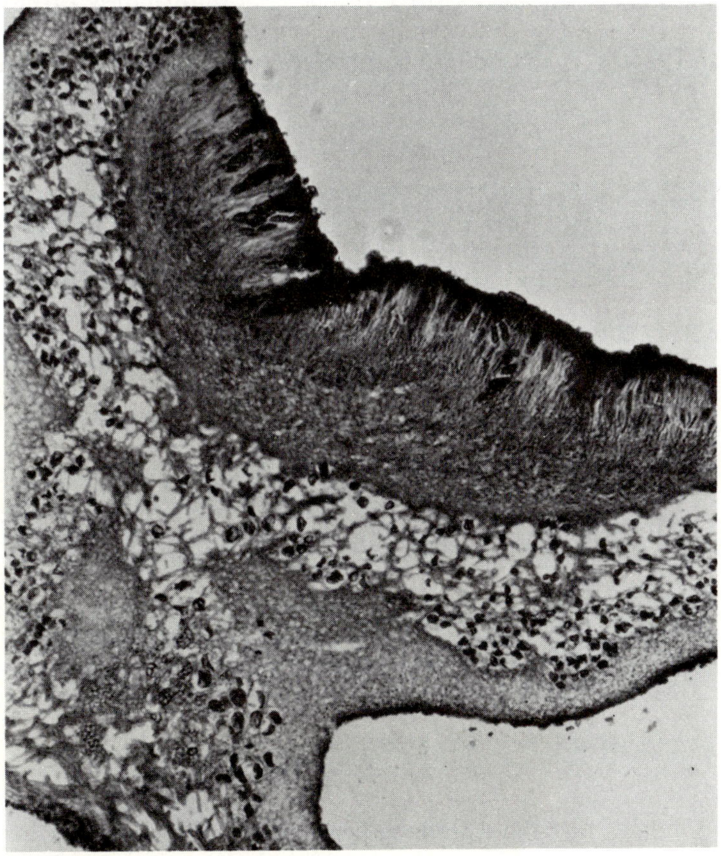

Figure 18.4
Physicia sp. Section through apothecium. Note the hymenium and the various layers of the thallus (160×).

close proximity to an alga that would be compatible for establishment of the lichen association. The chances of this happening with any sort of regularity would seem to be remote. Several hundred mycobionts have been isolated and cultured, but only a few have been induced to produce conidia.

Twenty-six algal species have been isolated from lichen associations. Most of these belong to the green algae with several belonging to the blue-green algae and one belonging to the yellow-green algae. The reproduction of the blue-green phycobionts in the association is by cell duplication, heterocysts, akinetes, and hormogonia. The green phycobionts reproduce by cell duplication and aplanospore production in the association. When some of the greens are isolated into liquid culture, they produce zoöspores.

Because of the low probability of a compatible fungal hypha and algal cell coming into close enough proximity under natural conditions to produce the lichen association, it would seem that lichen reproduction depends upon vegetative mechanisms. This is easily demonstrated for many of the foliose and fruticose types. They are known to produce various vegetative mechanisms. Similar mechanisms have not been noted for several of the crustose forms. Therefore, lichen reproduction is still far from being completely understood.

HUMAN RELEVANCE

Lichens are among the very first plants to invade new habitats such as exposed bare rock. They are able to gain anchorage because of the various acids that they produce, and they can withstand the demanding habitat because of the photosynthates produced by the phycobiont and the water retention abilities of the mycobiont. Their growth is very slow, but they do aid in the breakdown of the rock. They are probably more important as colonizers and soil-formers because they provide habitats for snails, mites and other invertebrates that eat parts of the upper cortex and algal layers. The lichens also provide organic compounds from their bodies when they die. In addition, the lichen bodies provide places for air-borne soil particles to lodge. These processes are slow, but eventually enough soil and organic matter are deposited that mosses and ferns can invade the new area. Such a method of soil formation takes a long time, but in many environments the method is an effective one.

Other animals in addition to invertebrates use lichens directly as a food source. Extensive areas in subarctic conifer forests are covered with reindeer "Mosses" (*Cladonia* spp.; *Cetraria islandica*), which are large fruticose lichens. *Cladonia* spp. is also the dominant lichen on the "floor" of the Scrub Oak and/or Sand Pine Scrubs in Florida. These lichens are often collected, dried, stained green, and used by model railroad buffs as "trees" and "shrubs." Caribou and reindeer use these "plants" as a food source during the winter months. The people of the area even harvest the lichens to use as animal food. The Berber tribes of the Libyan deserts use the crustose lichen *Lecanora esculenta* on which to graze their sheep. This plant is also another candidate for being the manna from heaven for the Israelites (see Chapter 10). The people of the area grind the lichen into a meal from which they make bread. It is reported that one of the rock tripes (*Umbilicaria* sp.) was boiled by Washington's troops at Valley Forge to provide a sort of soup thickener. Various other lichens have been used as foods in other areas of the world from time to time, but usually only on an emergency basis. Although they are either flat-tasting or acid, they have been demonstrated to possess more calories than equal amounts of honey, hominy, and corn flakes.

One of the most important industrial uses to which lichens have been put is the dye industry. Records of their use for this purpose are found in the writings of the ancient Greeks and other early Mediterranean peoples. It was expected that they would lose much of their value with the discovery of the coal-tar dyes in the 19th century. Although they did lose much of their commercial value many people still collect them to use for home-dyeing, hobby occupations. Various colors may be derived from various lichens (sometimes by fermenting them in stale urine!), or from different treatments of the same lichen and from mixing of the lichens. These colors range from browns to blues, yellows, purples, reds, and violets. Lichens are still used in the preparation of litmus paper which turns blue in an alkaline solution and red in an acid one.

Some lichens are still used in the perfume industry. They contain essential oils that are used in soap manufacture and are important as odor fixatives in good perfumes.

Lichens were used during the Middle Ages as medicinals for lung diseases, cranial maladies, and rabies. According to the Doctrine of Signatures, several of the lichens resembled parts of the human body and were used for treatment of maladies of that part of the human body. Thus Old Man's Beard (*Usnea* sp.) was used as a hair restorer, Lungwort (*Sticta* sp.) was used to treat chest diseases, Yellow-wall (*Theloschistes* sp.) was used to treat jaundice, and the Dog Lichen (*Peltigera canina*) was used to treat rabies. Linnaeus took the latter use into consideration when he named that particular lichen *Lichen caninus*. They are still used by some peoples for similar purposes. Antibiotics are produced by some lichens. Extracts from these lichens have inhibited the growth of such bacteria as *Staphylococcus aureus* and *Bacillus subtilis*. The Finnish people extract usnic acid from reindeer "moss" and include it in a salve that is effective in the treatment of external wounds and burns.

Lichens seem to cause man very little harm. Although many of them live upon other plants as epiphytes, most appear to do no harm to the plant. Some fruit trees are apparently slightly parasitized by some species when the lichen grows on young portions of the tree. The older portions of the tree do not seem to be affected although the lichen may serve as a habitat for insects that may be harmful to the tree, and may cause too much shading for the leaves. A rather unique damage is caused by several European crustose lichens when they colonize stained glass church windows and etch their surfaces.

19

Although the fungi and the algae may produce terrestrial members, the bryophytes are considered to be the first plant group to contain members that can produce complicated aerial structures and successfully live on land. This transition from a predominantly aquatic habitat to a predominantly terrestrial one required evolutionary changes that enable all of the cells of the plant to have an adequate water supply although some may not be immersed continuously in water.

Although some members of the bryophytes have developed aerial structures that may extend several inches above the soil, they are still very much limited in size. They produce no roots and, indeed, contain no specialized transport system that would enable water and dissolved minerals to travel from the soil to their uppermost cells, even if they had roots. Therefore, they usually either lie prostrate on the soil or live in a habitat that is quite moist most of the time.

All of the bryophytes also require the presence of water before fertilization can occur. The flagellated sperm must either swim or be splashed by rain to the proximity of the egg. Therefore, although members of this group have been successful in moving to a much dryer habitat than those plants previously discussed, they are still very much dependent upon a water supply that is available directly to all the cells of their body and that is adequate to supply a medium of transportation through which their sperm can swim before sexual reproduction can take place. It should be pointed out that asexual reproduction by fragmentation and specialized structures of the gametophyte (**gemmae**) is the dominant means of propagation and colonization for many species.

The life histories of the bryophytes are very much alike. They usually produce a haploid gametophyte that alternates with a diploid sporophyte. Meiosis occurs in the sporogenous tissue of the sporophyte, and gametophytes are produced from the haploid spores that result. The gametophytes produce gametangia in which gametes are produced. Sexual reproduction is always oögamous; the sperm must swim to the egg. Fertilization re-establishes the diploid condition and the resulting zygote produces a new sporophyte.

The bryophyte life history is very similar to the generalized life history of the vascular plants. The major difference between the two groups is the relative positions of the gametophyte and the sporophyte. The sporophyte produced by the bryophytes is extremely dependent upon the gametophyte; indeed, it is often parasitic upon the gametophyte. Although it may develop chlorophyll and produce some food in some members of the group, it remains mostly achlorophyllous. The opposite condition is true of the vascular plants. In them, the gametophyte becomes progressively reduced as they become more highly evolved. Finally, in the higher vascular plants, the gametophyte is contained within the sporophyte and is completely dependent upon it.

One should not be surprised to learn, however, that there are exceptions to this generalized and usual life history. Under both natural and laboratory conditions, diploid gametophytes and haploid sporophytes have been noted. The former condition results when portions of a sporophyte regenerate producing a gametophyte. This process is called **apospory** and results because of the extremely good powers of regeneration possessed by the bryophytes. The diploid gametophytes produce gametes that are diploid; when they unite a tetraploid sporophyte results. Haploid sporophytes are produced by **apogamy,** the loss of the sexual process.

The bryophytes are usually divided into three main groups. These groups may be considered as classes within the phylum Bryophyta or as separate phyla. In the latter case, the liverworts and hornworts are placed in the phylum Hepatophyta and the mosses occupy the separate phylum Bryophyta. This purely arbitrary treatment will consider the members as three classes within the Bryophyta.

The class Hepaticae includes the liverworts of which there are about 235 genera and 9000 species. Many of the liverworts possess vegetative bodies that are thalloid and that usually lie closely appressed to the substrate. Others, however, possess a plant body that produces a division into **phyllidia** (leaves) and **caulidia** (stems) and that are referred to as the "leafy" liverworts. The latter group is by far the largest, being composed of about 8000 species and 180 genera. The former group is probably better known by the beginning student, however. Members of this group are the ones most often used in general biology and botany classes as the example of the liverworts. *Marchantia* spp. will be discussed as an example of a thalloid form and *Porella* spp. as a leafy form. The hornworts (class Anthocerotae) is a relatively small group containing only about 100 species distributed in five or six genera. Only the genus *Anthoceros* will be discussed. The true mosses are grouped into the class Musci which contains about 675 genera and 14,500 species. Most of the plant body is divided into a "stem" and "leaves," although no vascular tissue is produced. Various body forms are produced within the class, but only *Sphagnum* spp. and *Polytrichum* spp. will be discussed as examples.

THE LIVERWORTS (HEPATICAE)

GAMETOPHYTE GENERATION

Marchantia polymorpha is the species most often used as an example of the thalloid liverwort (Figure 19.1). Its gametophyte is a flat, dichotomously branched body that grows on moist soil (Figure 19.13*a*, p. 302). Growth of the gametophyte continues from the apex by means of an apical cell or cells, depending on the species. The thallus may grow in excess of four inches; often the older portions of the thallus die and decay leaving the newer growing portions of the original plant disconnected from each other in such a way that new plants are produced. The upper portion of the thallus (Figure 19.2*a*) is divided into a number of air chambers each of which is supplied with a pore (Figure 19.2*b*), while the bottom portion is composed of closely packed parenchymatous cells. These cells contain few or no chloroplasts and are usually considered to be storage cells. Most of the photosynthesis is carried on by the **chlorenchyma cells** (parenchyma cells that contain chloroplasts) that line the bottom and sides of the air chambers.

Rhizoids are produced on the ventral surface of the thallus. These structures anchor the plant to the soil and absorb water from it. Scales, which are also produced on the ventral surface, aid the rhizoids in distributing the water by capillary action.

In addition to the vegetative reproduction brought about by the death and decay of older portions of the gametophyte, structures called **gemmae cups** (Figures 19.2*b* and 19.13*a*, p. 302) provide another method of asexual reproduction. These small cups contain a number

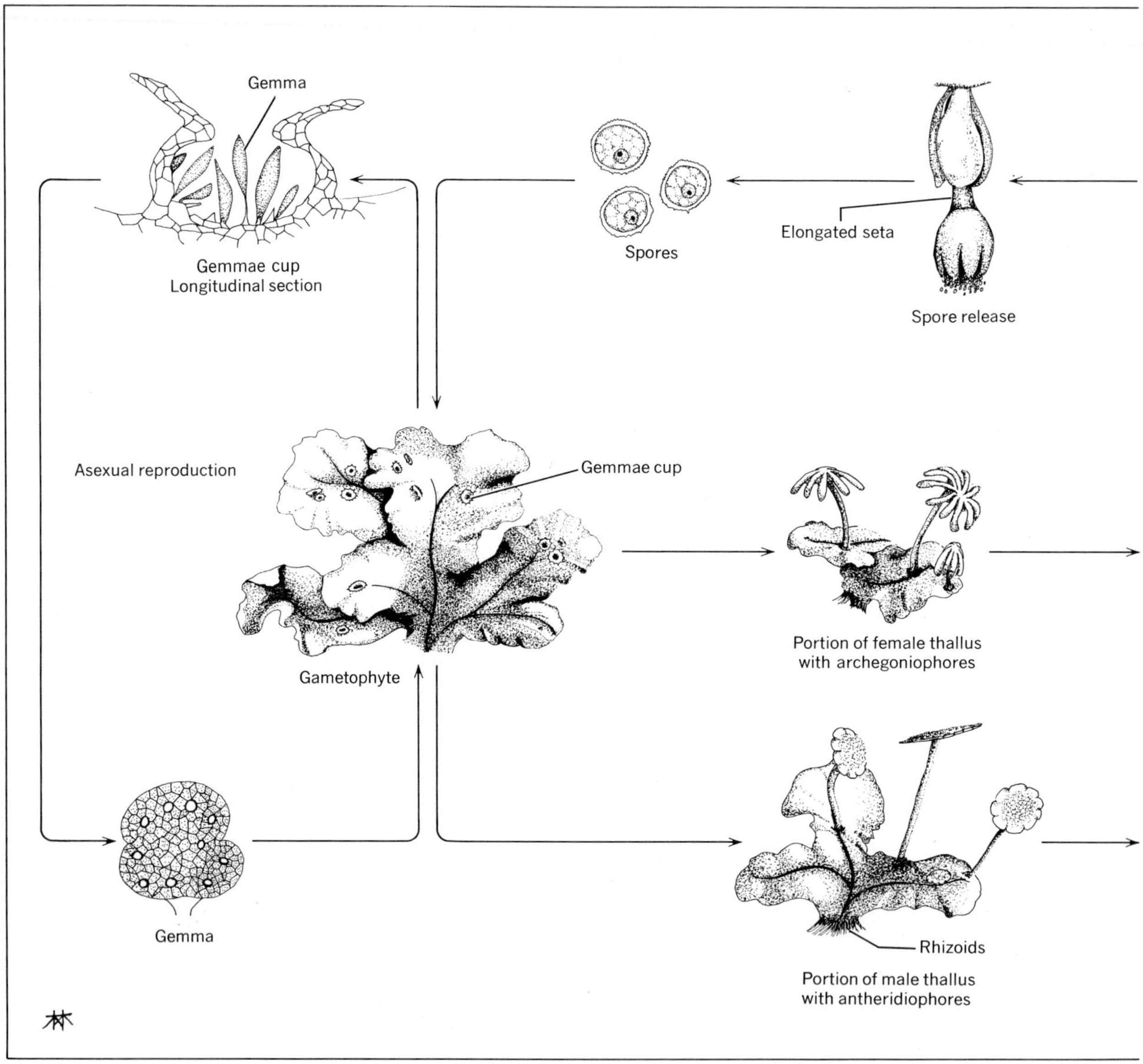
Gemma
Gemmae cup
Longitudinal section
Spores
Elongated seta
Spore release
Asexual reproduction
Gemmae cup
Gametophyte
Portion of female thallus
with archegoniophores
Gemma
Rhizoids
Portion of male thallus
with antheridiophores

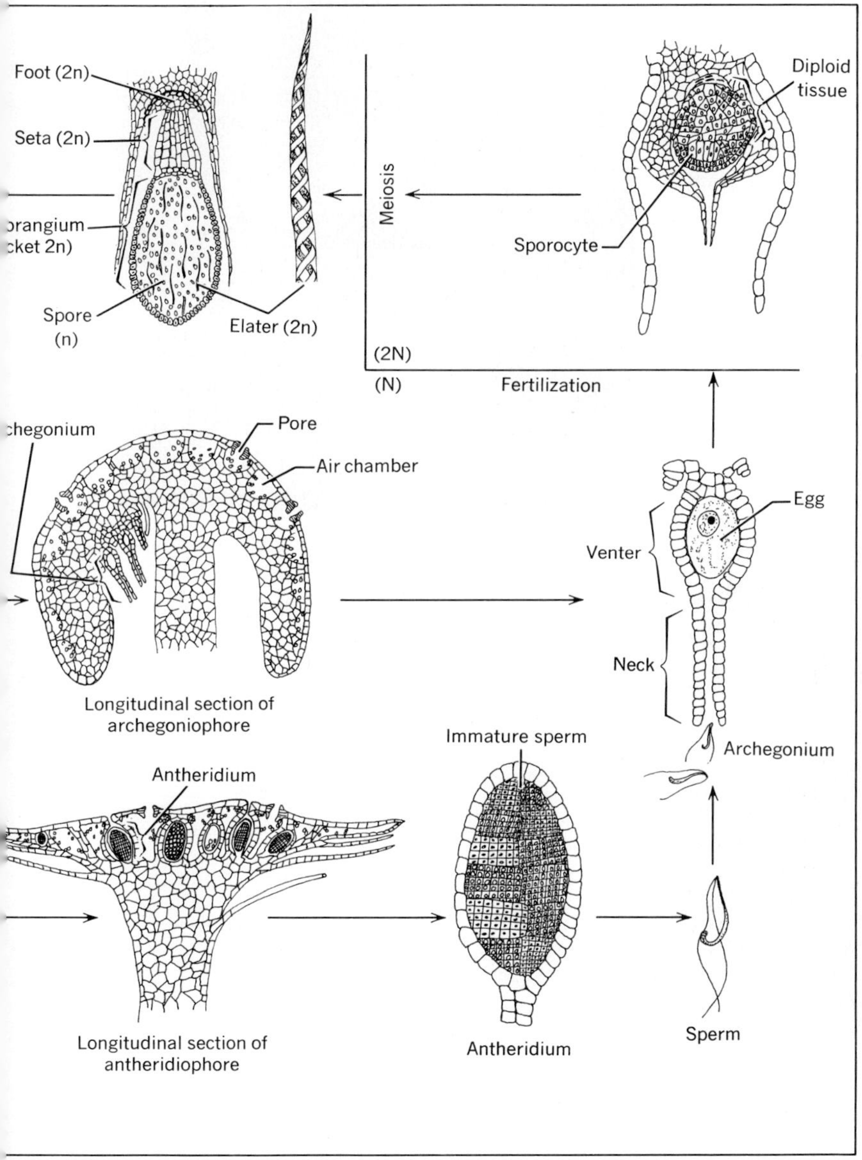

Figure 19.1
***Marchantia polymorpha*. Life history.**

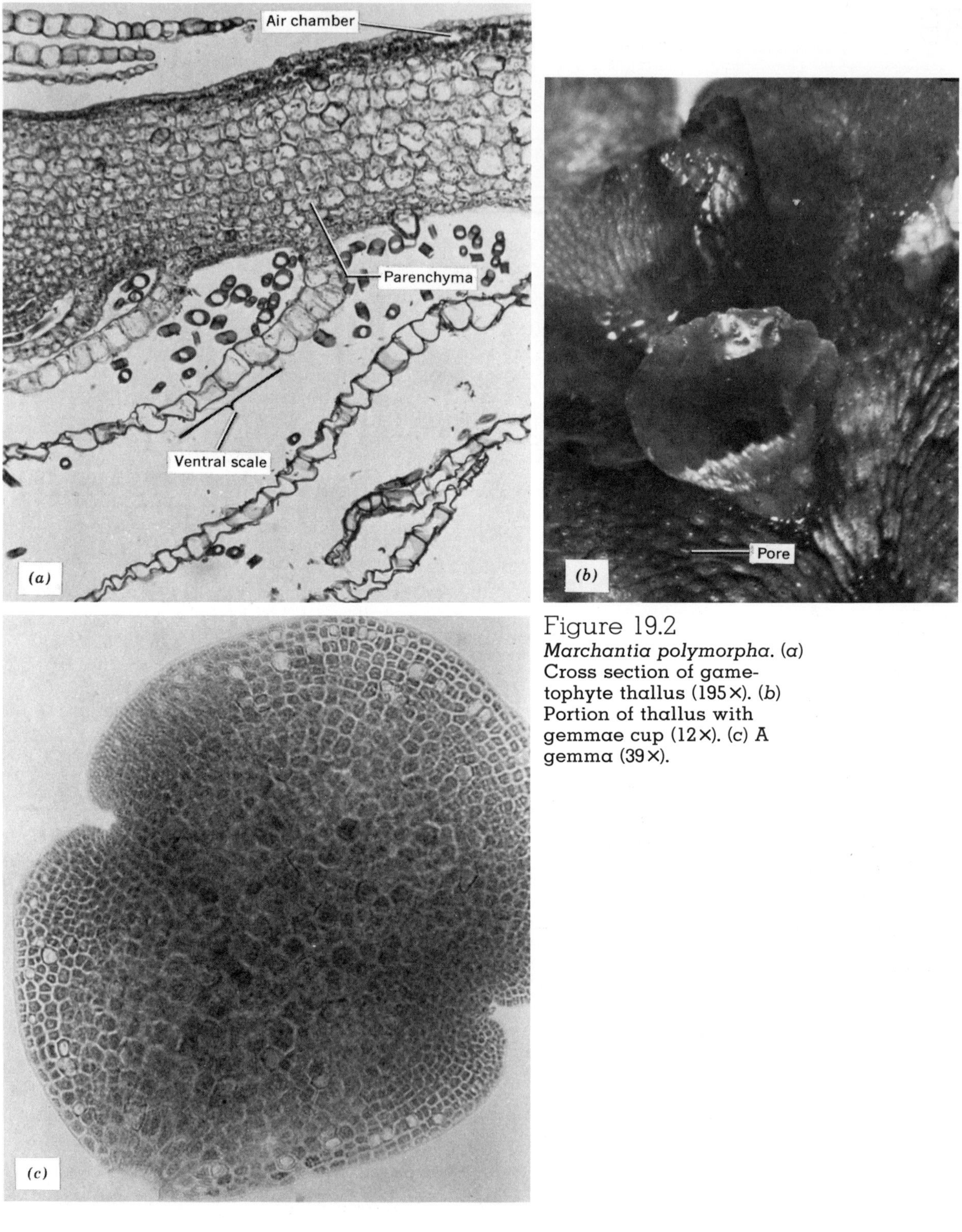

Figure 19.2

Marchantia polymorpha. (*a*) Cross section of gametophyte thallus (195×). (*b*) Portion of thallus with gemmae cup (12×). (c) A gemma (39×).

of multicellular structures, called **gemmae** (Figures 19.1 and 19.2*c*), that are attached by a short stalk to the base of the cup. Each of the gemmae is produced from a single superficial cell from the bottom of the cup. Most of the cells of the gemmae contain chloroplasts but some of the cells contain oil. Therefore, they can either produce their own food or at least have enough stored food to last for a period of time. Specialized cells within the cups produce mucilage that swells rapidly when it absorbs water. This swelling breaks the gemma away from the stalk and allows it to float free. Raindrops landing in a gemmae cup cause a splash that may distribute the gemmae quite a distance away from their parent plant (e.g., 1 meter). Each gemma is capable of germination and thus the gametophyte may be asexually reproduced.

Rather specific environmental conditions stimulate the process of sexual reproduction. Each plant body produces only one type of sex organ. The sex organs are produced on specialized branches, **gametophores** (Figure 19.3), of the gametophyte thallus. The antheridia are produced in an **antheridiophore** (Figure 19.3*a*) which is a flattened, disc-shaped head that is at the tip of an elongate stalk. The archegonia (Figure 19.3*c*) are produced on a stalked **archegoniophore** that has a top that resembles a spoked wheel without a rim.

Antheridia (Figure 19.3*b*) and archegonia (Figure 19.3*d*) of different ages are produced before the stalks of the gametophores elongate. The antheridia are produced in radiating rows just below the surface of the disc with the younger ones toward the outside. The antheridia retain this relationship as the stalk of the antheridiophore elongates. The antheridia mature over a relatively long period of time and some of the biflagellate sperm are released even before the disc gets elevated very far above the thallus.

The archegonia begin their development in radiating rows also, but as the archegoniophore elongates the central tissues grow more than the peripheral ones. This causes an inversion of the tissues and the mature head of the archegoniophores have the archegonia hanging from their ventral surface with the less mature archegonia located toward the center of the disc.

Fertilization of some of the eggs within the archegonia may occur before either gametophore has elongated. Flooding with water is necessary before the sperm can swim to the egg, but orientation of the sperm is undoubtedly accomplished by a chemotactic response

Figure 19.3
Marchantia polymorpha gametophores. (*a*) Habitat view of mature antheridiophores (5×). (*b*) Longitudinal section through antheridiophore (53×). (*c*) Habitat view of young archegoniophores (3×). (*d*) Longitudinal section through archegoniophore (156×).

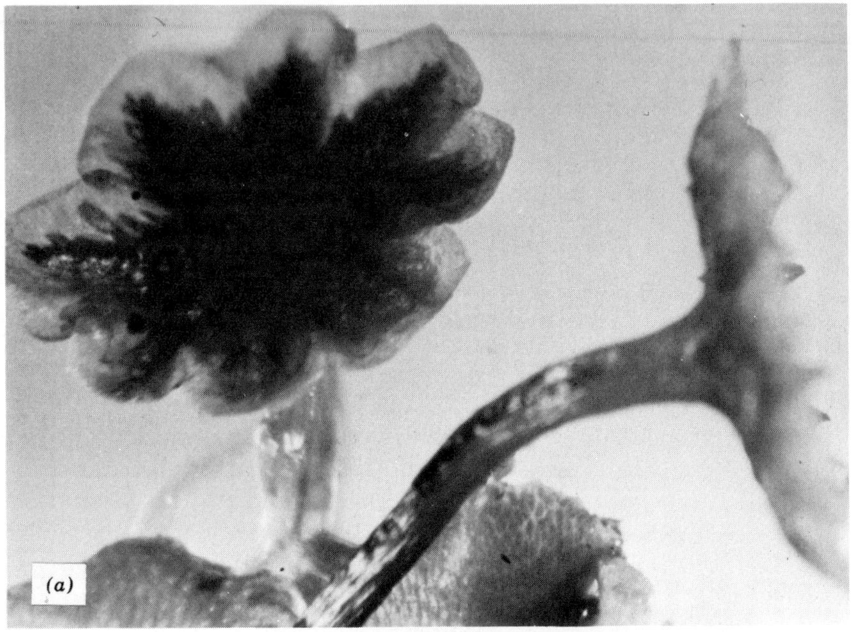

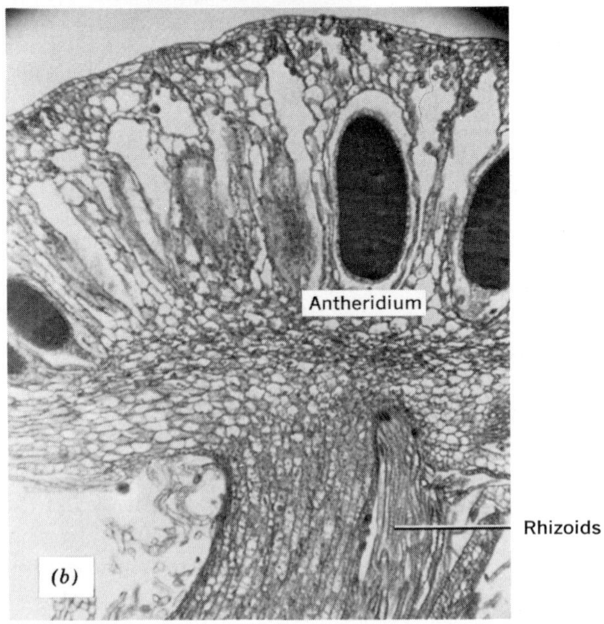

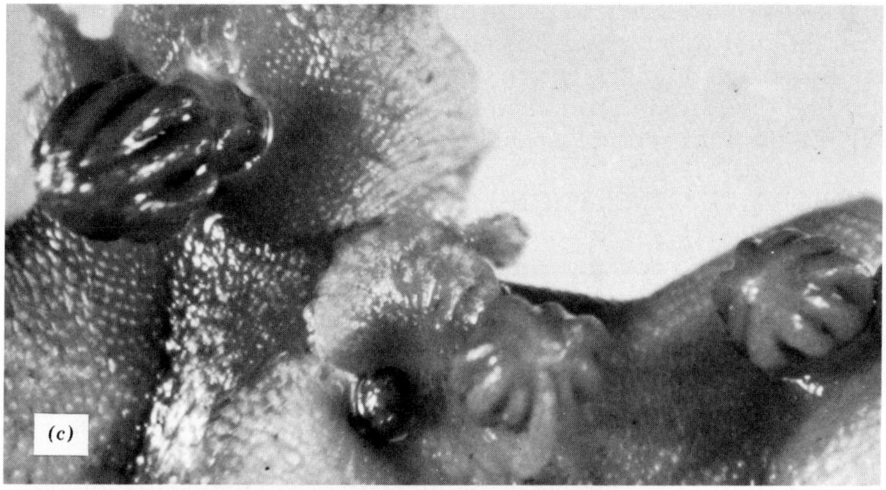

(c)

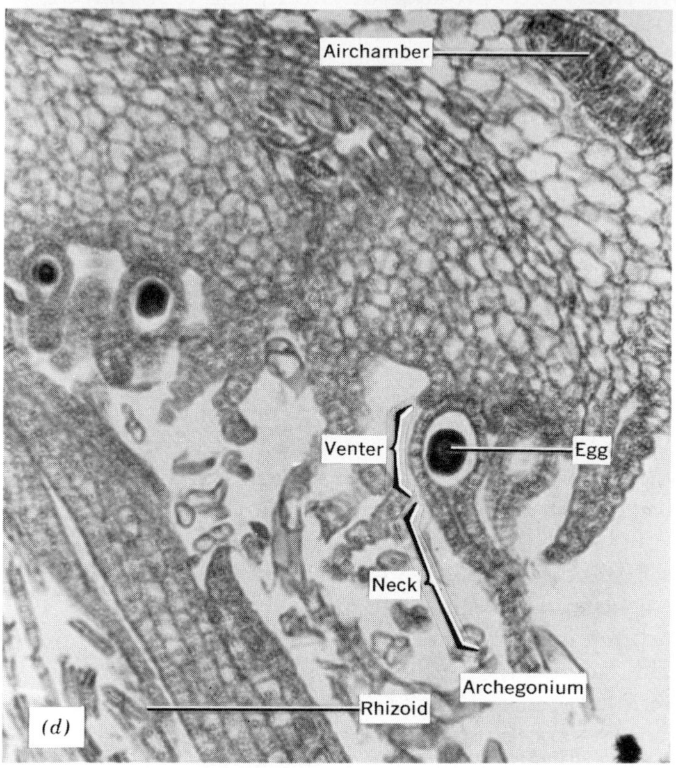

(d)

of the sperm to secretions of the archegonium. Fertilization of eggs after the archegoniophore has elongated is probably accomplished by the swimming of the sperm through surface films of water or by flooding. Also, raindrops hitting antheridiophores have been observed to splash sperm for quite a distance.

SPOROPHYTE GENERATION

Not all of the eggs on an archegoniophore are fertilized and not all of those fertilized develop. However, after karyogamy has occurred between the sperm and the egg nucleus, a mass of diploid tissue is produced in the bulbous base (the **venter**) of the archegonium (Figure 19.1). The diploid cell is, of course, the first cell of the new sporophyte generation.

Cells from the sporophyte grow through the base of the center and form a connection with the storage tissue of the archegoniophore (Figure 19.4). These cells produce an organ (the **foot**) that absorbs food and water and anchors the sporophyte in place. Elongation in the other direction produces a short, intermediate, sterile region called the **stalk** or **seta,** and the distal, cylindrical, fertile portion, the **capsule** or **sporangium.** Although the sporophyte absorbs materials from the gametophyte through the foot, it may not be totally dependent on the gametophyte. Many of its cells contain chloroplasts and it does carry on some photosynthesis.

Periclinal duplications of the cells within the capsule produce an outer layer of cells called the **amphithecium** and an inner mass of cells, the **endothecium.** A sterile jacket of cells is produced from the amphithecium. The endothecium continues to duplicate until a large mass of sporogenous tissue is produced. Further differentiation of the sporogenous tissue produces cells that are spherical and cells that are elongate. The spherical cells are spore mother cells (**sporocytes**) and undergo reduction division (meiosis) producing a tetrad of haploid spores. The elongate cells secrete spiral thickenings, do not undergo any sort of nuclear change (either meiosis or mitosis) and mature into **elaters.** The sharp pointed elaters are extremely **hygroscopic** (sensitive to changes in moisture) and by their expansion and contraction aid in spore dispersal.

During the period of spore maturation, the seta continues to elongate. This elongation pushes the capsule away from the archegoniophore disc. The spores are released as a result of the breaking (**dehiscence**) of the capsule. When the spores reach condi-

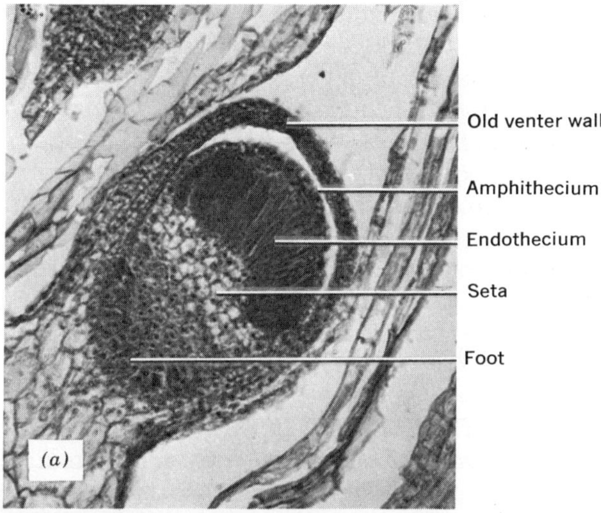

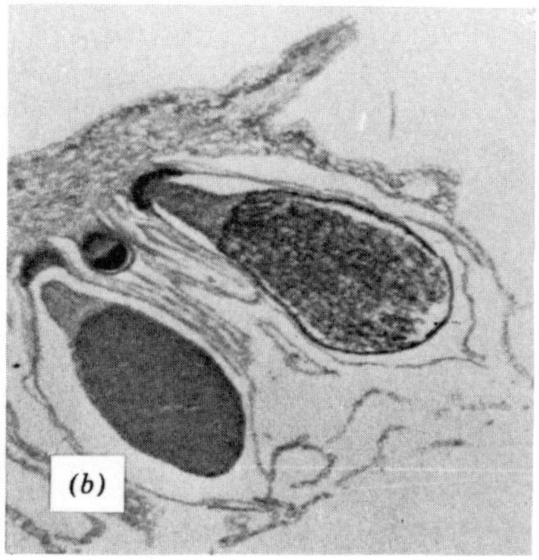

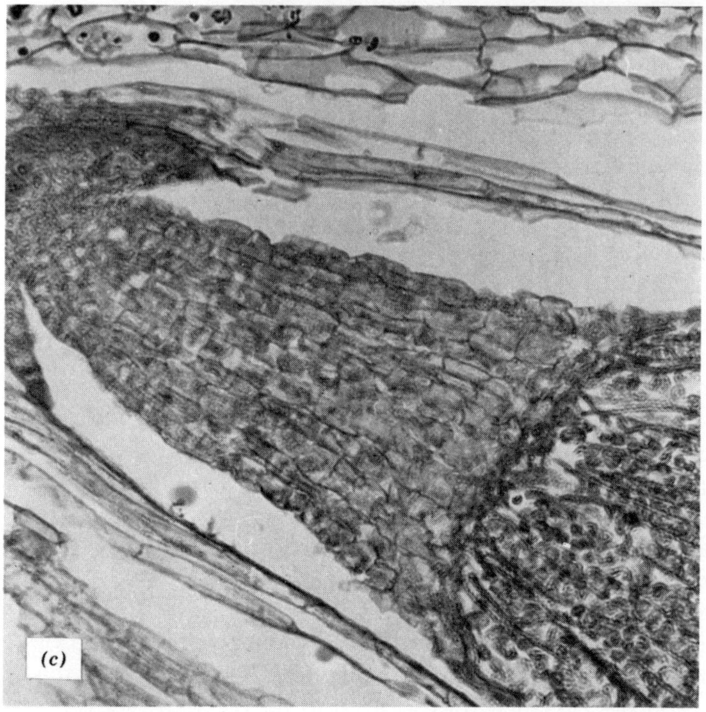

Figure 19.4
Marchantia polymorpha. (*a*) Young sporophyte (139×). (*b*) Portion of archegoniophore with three sporophytes of different ages attached (15×). (*c*) Portions of foot, seta, and capsule of mature sporophyte (150×).

tions suitable for germination, they produce an irregular filament that eventually produces a new gametophyte thallus. Presumably, two spores from each tetrad produce female gametophytes and the other two spores produce male gametophytes.

The genus *Porella* is representative of a huge group of liverworts called the "leafy liverworts." This group of plants is widely distributed in the tropics where they appear to thrive on the high humidity and heavy rainfall. Many more species of leafy liverworts exist than do thallose ones, and on that basis they can be said to be more "typical" of the hepatics. However, because the casual observer often confuses them with mosses, they are likely to go unnoticed by most people.

GAMETOPHYTE GENERATION

The vegetative body, or gametophyte, of *Porella* spp. (Figure 19.5) is definitely divided into a branching axis called a stem that gives rise to flat appendages called leaves. It should be pointed out that the terms "leaves" and "stems" do not indicate a type of complex organ as that which occurs in the vascular plants. Also, no vascular tissues have developed within this group. The plant body results from a single apical cell. The foliose appendages result from segments derived from the apical cell.

Although rhizoids are present on the ventral surface of the plant body, most of the water seems to be absorbed directly into the vegetative cells of the stems and leaves. Because the leaves overlap, and because the plant grows in dense clusters, water is held by capillary action. The plant body is extremely dorsoventrally flattened with one row of leaves emerging from either side of the stem, and one row of smaller leaves emerging from the ventral surface of the stem.

The leafy liverworts thrive in areas of high humidity, but many of them, including *Porella* spp. are capable of withstanding a great deal of drying. Although the older portions of the plant may die, the apices are able to become dormant during the dry conditions but resume growth soon after they become moist again. This provides a method of vegetative reproduction.

Sexual reproduction in *Porella* spp. is closely correlated with climatic changes, at least in the temperate zone where production of antheridia and archegonia occur in the spring. The plants are unisexual, but both sexes produce their gametangia on short, leafy branches. Because the male and female plants grow intertwined in

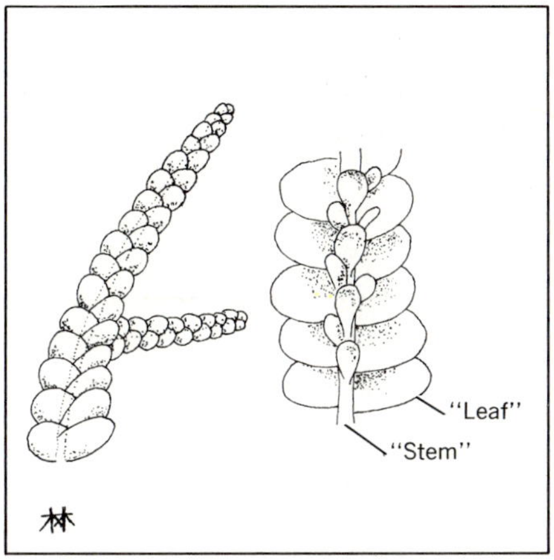

Figure 19.5
Porella spp. Gametophytes

masses, which retain water, the sperm can swim to the archegonia. Many sperm are produced in each antheridium of the antheridial branch; about 8 to 10 archegonia are produced on each archegonial branch. Multiple fertilizations in each archegonial branch may occur, but there is usually development of only one sporophyte.

SPOROPHYTE GENERATION

The zygote develops in the venter in much the same way as occurred in *Marchantia* spp. Eventually it also becomes differentiated into foot, seta and capsule. The cells of the sporophyte become chlorophyllous as they mature, and contribute some nutrition to the sporophyte. However, absorption of nutrients from the gametophyte through the foot does occur. The tissues of the capsule develop a multicellular jacket of sterile cells and an inner mass of sporogenous tissue. The latter tissue produces sporocytes, each of which produces a tetrad of haploid spores after carrying on meiosis. Elaters are also produced from some of the sporogenous cells as in *Marchantia* spp.

After maturation of the spores, the seta elongates pushing the sporophyte away from the gametophyte. Some of the spores may undergo cellular duplications before the capsule dehisces, but they are eventually scattered. Under proper environmental conditions, they germinate producing the new gametophyte.

THE HORNWORTS (ANTHOCEROTAE)

GAMETOPHYTE GENERATION

The gametophytic thallus is somewhat lobed, sometimes dissected and dichotomously branched (Figure 19.6). When growing under

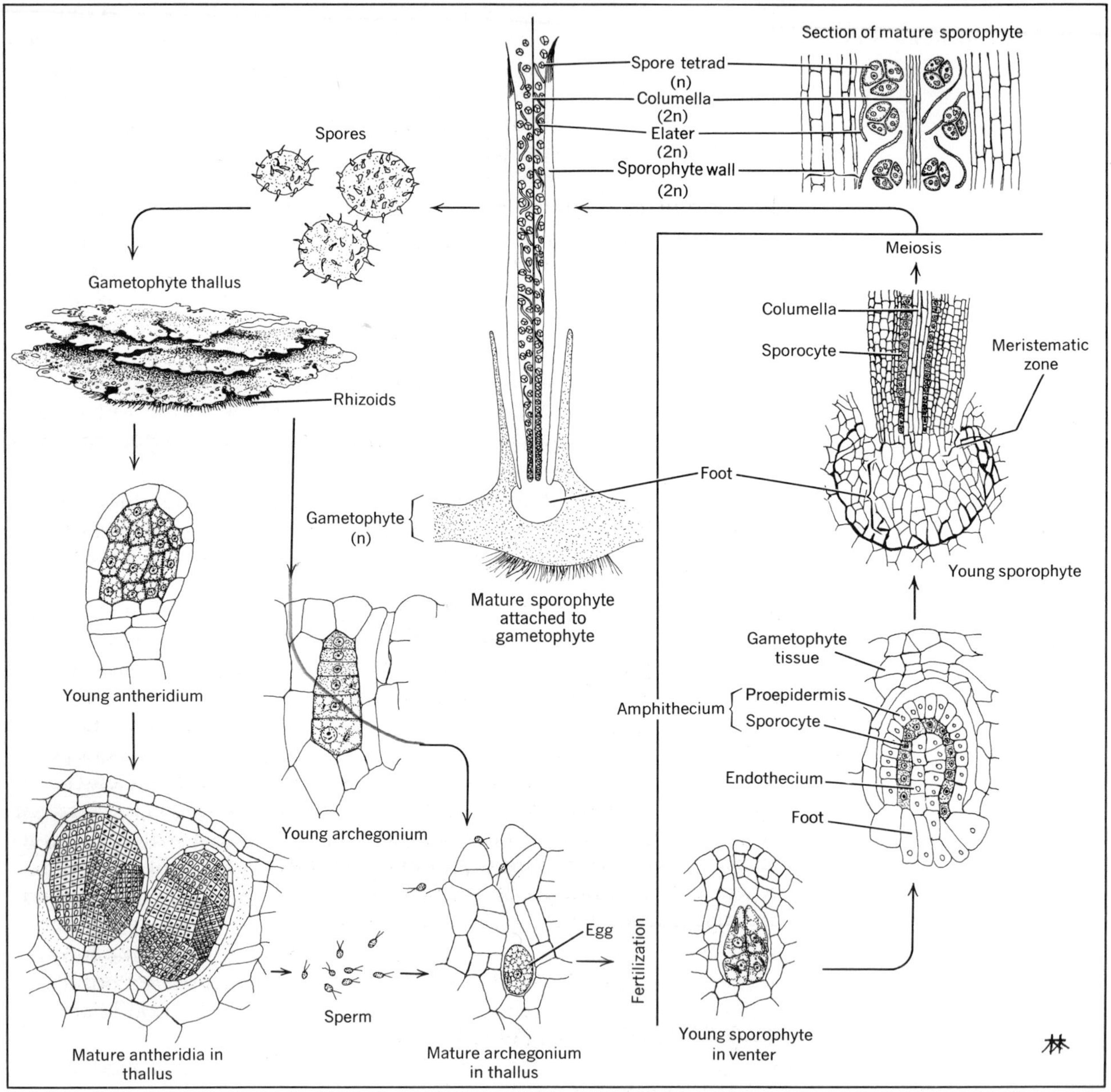
Section of mature sporophyte
Spore tetrad
(n)
Columella
(2n)
Elater
(2n)
Sporophyte wall
(2n)
Spores
Gametophyte thallus
Rhizoids
Meiosis
Columella
Sporocyte
Meristematic zone
Foot
Gametophyte
(n)
Young sporophyte
Mature sporophyte attached to gametophyte
Young antheridium
Young archegonium
Gametophyte tissue
Amphithecium
Proepidermis
Sporocyte
Endothecium
Foot
Egg
Fertilization
Sperm
Mature antheridia in thallus
Mature archegonium in thallus
Young sporophyte in venter

Figure 19.6
Anthoceros sp. Life history.

uncrowded conditions, the thallus may be almost round. It is attached to the soil by ventral rhizoids through which some absorption probably occurs, although absorption also takes place directly into the vegetative cells. The cells of *Anthoceros* spp. contain one large chloroplast that produces a number of pyrenoid segments.

The gametophyte produces internal chambers that contain the blue-green alga, *Nostoc* spp. Because of the alga, the plants have a rather dull, gray-green appearance. The exact relation between *Nostoc* spp. and *Anthoceros* spp. is unknown. A mutualistic relationship in which the algae fixes nitrogen and *Anthoceros* spp. provides water and protection has been suggested. However, other studies have indicated that *Anthoceros* spp. actually does much better when it does not contain the alga. Because the gametophyte is chlorophyllous, it does not depend upon the alga for carbohydrates. It may be that the symbiotic relationship is mutualistic under certain environmental conditions, but parasitic under other conditions.

As in *Marchantia* spp. and *Porella* spp., the older portions of the gametophyte may die leaving the young, actively growing tips effecting vegetative reproduction. Asexual reproduction may also be accomplished by the separation of marginal lobes that become thickened. These lobes are able to withstand extreme environmental conditions and later produce a new gametophyte.

Antheridia (Figure 19.7*a*) and archegonia are produced on the same gametophyte (**bisexual**) in some species of *Anthoceros* or on different gametophytes (**unisexual**) of other species. Both kinds of gametangia arise from superficial cells on the dorsal surface, but are buried in thallus tissue at maturity. Through a rather complicated series of cellular duplications, several antheridia are produced within a single cavity within the gametophyte. When water enters a cavity containing mature antheridia, the antheridial wall breaks down releasing hundreds of tiny biflagellate sperm. The archegonia are also buried within the thallus, but their necks extend to the surface. Fertilization is accomplished by the swimming of the sperm to the mouth of the archegonium and down to the egg in the venter.

SPOROPHYTE GENERATION

Multiple fertilization produces many zygotes on the same thallus. Each of these may develop into a mature sporophyte. The first duplications of the zygote produce a lower, sterile foot that is sunk into

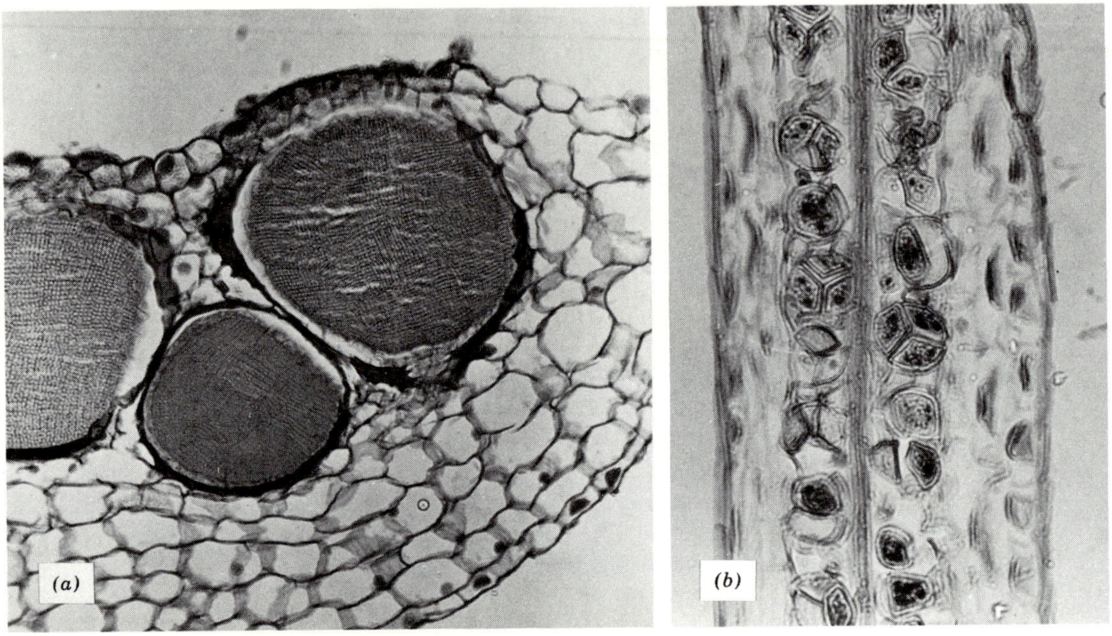

Figure 19.7
Anthoceros sp. (*a*) Antheridia buried in thallus (146×). (*b*) Portion of longitudinal section of mature sporophyte (195×).

the gametophyte tissue and an upper fertile portion that differentiates, by periclinal duplications, into an endothecium and an amphithecium. Until this stage of development, *Anthoceros* sporophytes compare with *Marchantia* and *Porella* sporophytes. Several differences now become evident.

The region that corresponds to the seta of the previously discussed hepatics becomes mitotically active and an **intercalary meristem** is produced. This allows the sporophyte to continue to increase in length by the addition of new cells to its base. This growth may continue for several months producing a slender, elongate sporophyte (Figure 19.6). The endothecium produces a central column of sterile cells, the **columella** (Figure 19.7*b*). The outermost cells of the amphithecium become cutinized and develop stomata that allow for gas exchange in the sporophyte. The innermost cells of the amphithecium produce sporocytes in which meiosis occurs resulting in a tetrad of haploid spores (Figure 19.7*b*). Some of the diploid amphithecial cells produce elaters that differ in structure and cell number from previously discussed elaters. The manner of function of the setal area and the products of the amphithecium and endothecium are quite different from these same tissues produced in *Marchantia* spp. and *Porella* spp.

The cells of the wall of the sporophyte are chlorophyllous and contribute greatly to the nutrition of the sporophyte. Because of the intercalary meristem, the sporophyte continues to grow from its base. Final maturation of the sporogenous tissue occurs at the distal end where the sporophyte dehisces. Because such intercalary growth and differential maturation occurs, mature spores are released over a long period of time. The young spores germinate, each producing a new gametophyte.

THE MOSSES (MUSCI)

The mosses are well-known to the casual observer of nature and are often used as the example of a nonflowering plant in elementary and secondary schools. Their wide range of habitats, their soft texture, and pleasant green appearance make them conspicuous in almost any flora. Some are quite restricted in their habitats, but others are ubiquitous.

Sphagnum spp., the "peat moss," is a genus that grows in swamps, ponds, or other very moist places that are not alkaline (Figure 19.13*b*). Many species grow in waters that are quite acid.

GAMETOPHYTE GENERATION

The first stage of the gametophyte is a green, threadlike structure, the **protonema,** that results when the spore germinates. This structure begins to produce upright branches, the **gametophores,** that are composed of a central axis or "stem" and broad, flattened appendages or "leaves." No really well-developed vascular tissue is present in either structure. The protonema soon dies and the young gametophores are anchored to the soil by **rhizoids** through which they also absorb mineral nutrients and water. Mature gametophores lose their rhizoids but are mechanically supported by their habit of growing in dense, intermingled masses. Absorption also occurs directly into the body of the plant. The stem is differentiated into a central, undifferentiated region surrounded by cortical cells, but evidence suggests that very little conduction occurs through the area. Upward movement of water is aided by the capillary action of the closely packed plants. The denseness of the cluster is aided by the multiple short branches that are produced in the axils of the "leaves" near the tips of the gametophores. Some of the branches droop toward the ground. When the older plant part dies, the upper portion becomes independent. This vegetative reproduction increases the size of the clump or mat.

Sphagnum spp. has a tremendous water-holding capacity (16 to 20 times its dry weight). The cortical cells and tightly packed condition account for some of this capacity, but most of the ability is derived from the huge, barrel-shaped cells in the leaves (Figure 19.8*a*). These particular cells (**hyaline cells**) are the product of an unequal cell division that also produced one vegetative cell. The hyaline cell is dead at maturity, but the smaller vegetative cell remains alive and actively photosynthetic. The hyaline cells retain large quantities of water. The mixture of the two cell types gives the plant a light green, sometimes yellowish, appearance.

Some *Sphagnum* species produce only one type of sex organ on a plant but others produce both sex organs on the same plant. Specialized short lateral branches produce the antheridia in the axils of their leaves. Each antheridium is derived from a superficial cell of the stem and produces many biflagellate sperm. The archegonia are also produced on special, short, lateral branches. There are usually only three archegonia produced on each branch, but they are massive with elongate necks. Fertilization seems to occur in the autumn and winter with the sporophytes maturing during the next May to July.

SPOROPHYTE GENERATION

Although all of the eggs in the archegonial branch may become fertilized, usually only one of the zygotes ever reaches maturity. Transverse duplications of the zygote produce a lower portion that develops into an **haustorial foot** and **seta.** The upper portion produces the **capsule** (Figure 19.8*b*). Although absorption from the gametophyte occurs through the foot, the cells of the sporophyte are photosynthetic and produce most of the carbohydrate necessary for its growth.

An endothecium and amphithecium are produced in the capsule.

Figure 18.1
Examples of lichen bodies. Crustose: Two unidentified lichens on rock (*a*) Note the black apothecia on the gray thallus. Foliose: (*b*) *Parmelia* sp. Note the apothecia. (*c*) *Xanthoria* sp. Bright yellow thallus on base of stone monument. (*d*) Close-up of *Xanthoria*. Note the tiny yellow apothecia. Fruticose: (*e*) *Cladonia alpestris*. Large discrete ball of thallus. (*f*) *C. verticillata*. Note cup-shaped thallus. (*g*) *C. cristatella*. Note bright red apothecia.

(*a*) (*b*) (*c*) (*d*) (*e*) (*f*) (*g*) (*f*)

Figure 19.13
Examples of bryophytes. (*a*) *Marchantia polymorpha*. Female plant with gemmae cups and archegoniophores. (Photograph by D. Mark Cavalcolia.) (*b*) *Sphagnum* sp. Single gametophyte with mature sporophytes. (*c*) *Polytrichum* sp. Many gametophytes with mature sporophytes. (Photos b and c by John W. Cieslewicz.)

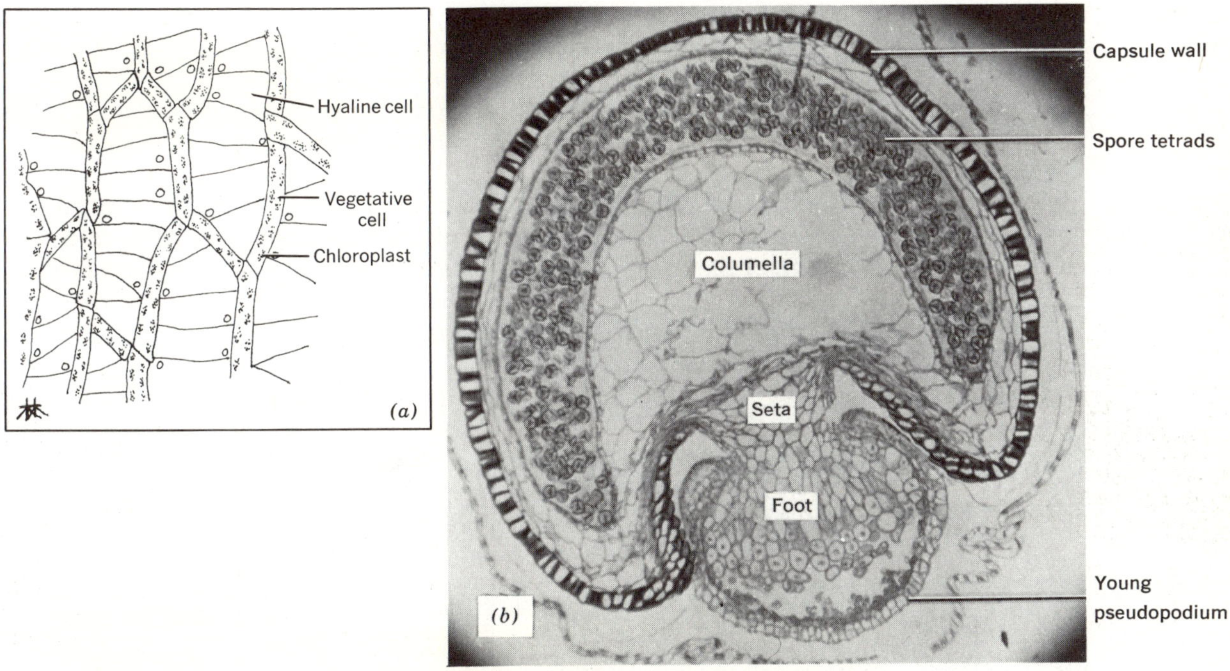

Figure 19.8
***Sphagnum* sp. (*a*) Portion of leaf. (*b*) Longitudinal section through a sporophyte (56×).**

The endothecium produces a massive, dome-shaped columella. The spores are derived from sporocytes that are produced from the innermost layers of the amphithecium. Each sporocyte carries on meiosis; therefore each produces a tetrad of spores, each of which is haploid. No elaters are produced.

Just previous to the maturation of the spores, the **archegonial branch** rapidly elongates, pushing the sporophyte away from the gametophore (Figure 19.13*b*, opposite). This is in contrast to the elongation of the seta noted in the hepatics. The elongate stalk, which is really the archegonial branch, is called a **pseudopodium.** The capsule dehisces along a preformed line (the **annulus**). The sporks are shot out of the capsule through an opening left when the lid (**operculum**) of the capsule is shed. Spores and operculum may be hurled several inches into the air.

Sphagnum spp. spores may germinate immediately when

Figure 19.9
Life history of a moss.

reaching a favorable environment but they remain viable for about six months. A short filament (the protonema) is produced by the germinating spore. This young structure produces rhizoids that anchor it to the substratum. Each protonema produces a single gametophore, but mats of *Sphagnum* spp. are produced by multiple spore germination in a small area.

Polytrichum (Figure 19.9) is a genus of the so-called "true mosses." This genus includes rather large plants that are easily seen, and samples of the moss are readily available in most areas. However, if *Polytrichum* spp. (Figure 19.13*c*, p. 302) is not available for study, almost any moss can be used because the entire group is very much alike in structure.

GAMETOPHYTE GENERATION

The "true mosses" occupy a much broader range of habitats than does *Sphagnum* spp. They may be found submerged in streams or on quite dry land. Most of them, however, grow under mesic conditions rather than the extremes of hydric or xeric conditions.

Like *Sphagnum* spp., the gametophyte starts with the production of a filamentous protonema (Figure 19.10*a*). The protonema of the true mosses is a branching, more persistent structure than that found in *Sphagnum* spp. It eventually gives rise to "leafy," erect gametophores (Figures 19.9 and 19.13, p. 302). The protonemal stage is usually transitory, and dies out soon after producing gametophores. Each protonema can produce many gametophores. The gametophores produce rhizoids from their base. The rhizoids remain present and functional as organs of anchorage and absorption.

The gametophore is divided into a central axis and numerous "leaves." The axis, or "stem," consists of a superficial epidermal layer, a cortex, and a central strand (Figure 19.10*b*). The "stems" of *Polytrichum* spp. may reach 6 to 8 inches tall, but others are usually much shorter. The cells of the "stems" and "leaves" are actively photosynthetic. The "leaves" of *Polytrichum* spp. (Figure 19.10*c*) have a thickened midrib and other internal differentiation, but most moss "leaves" are very simple.

Polytrichum spp. (Figure 19.13*c*, p. 302) grows in clumps or large patches similar to *Sphagnum* spp. Although the "stem" of *Polytrichum* spp. appears to be much more complex than that of *Sphagnum* spp., experimental evidence has demonstrated that there is very little movement of materials through the "stem." Therefore, the cells of the central strand can not be considered as conductive. The moss re-

305

liverworts and mosses—bryophyta

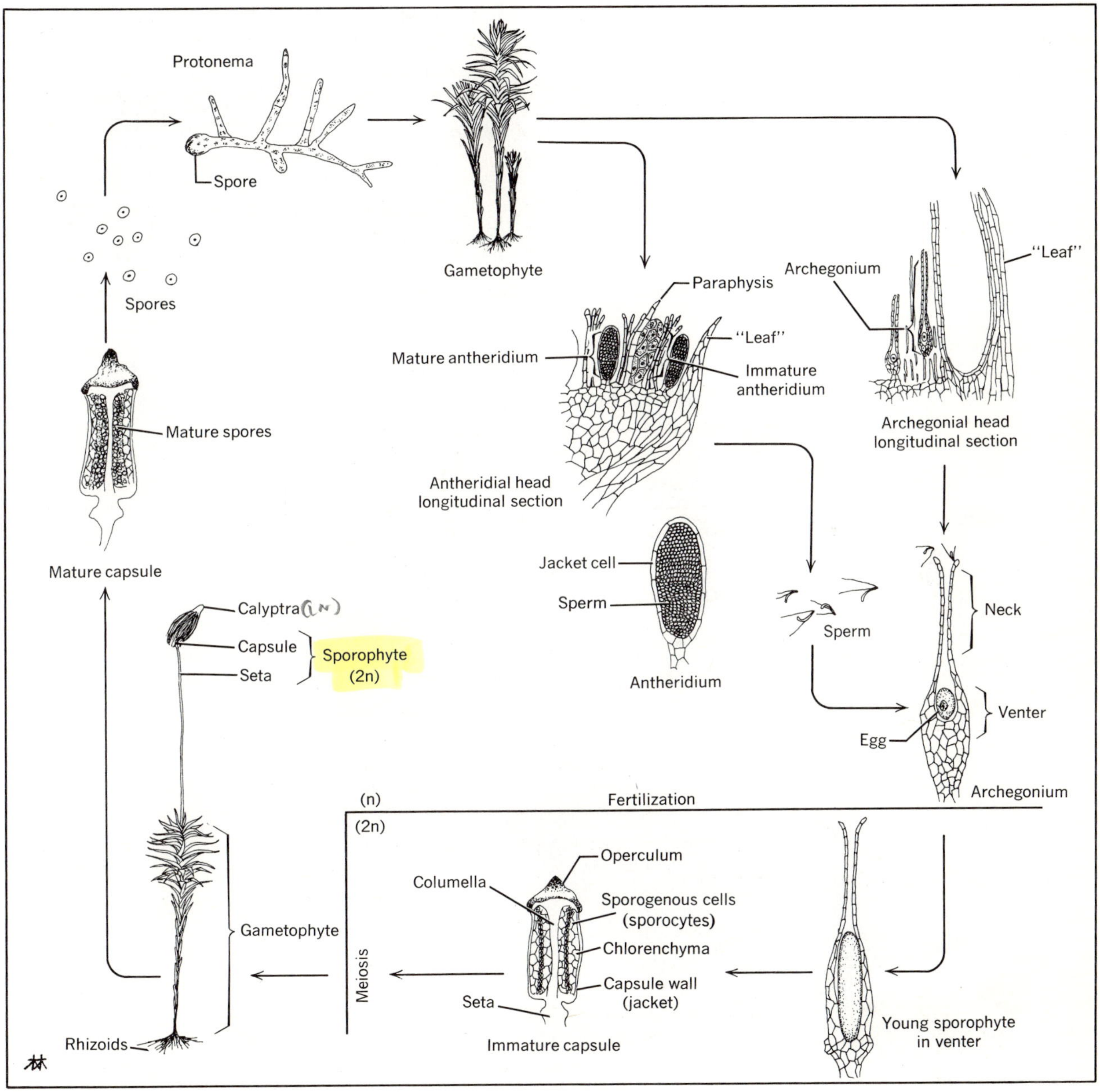

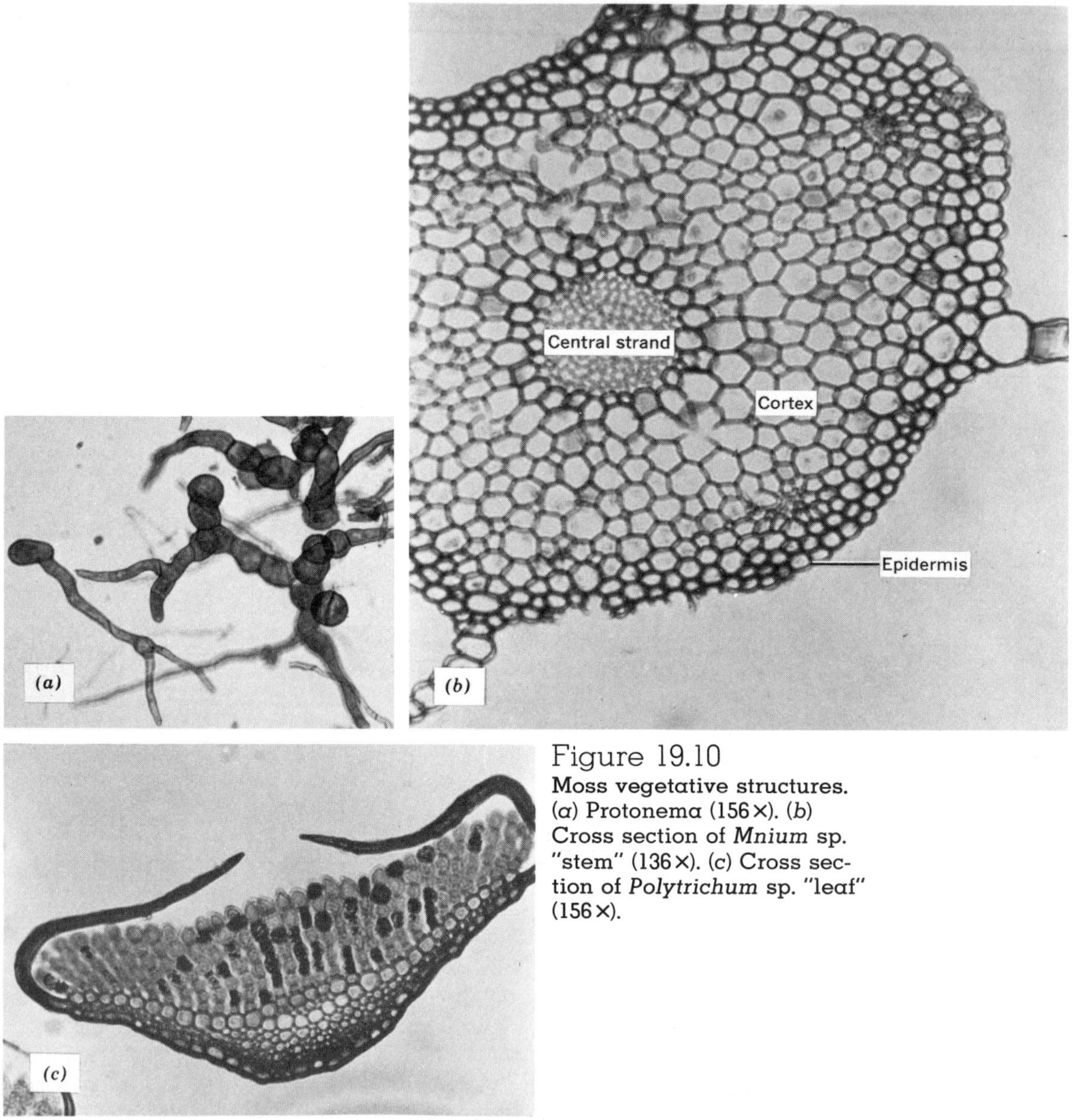

Figure 19.10
Moss vegetative structures. (*a*) Protonema (156×). (*b*) Cross section of *Mnium* sp. "stem" (136×). (*c*) Cross section of *Polytrichum* sp. "leaf" (156×).

tains a great amount of rainwater in its tightly compacted masses; this water moves over the external part of the plant by capillary action.

Many parts of the gametophore may vegetatively reproduce. New gametophytes have been produced from parts of the leaves, stems, rhizoids, and even sterile parts of the sporophyte (**apospory**). The latter gametophytes are, of course, diploid and produce diploid gametes. This can later lead to a tetraploid sporophyte. Some mosses produce gemmae that also reproduce the gametophyte. Many mosses have prostrate vegetative branches in addition to erect gametophores. The prostrate branches may increase the number of plants by death of the older portion, with each of the young tips of the old branch producing a new plant.

Although *Polytrichum* spp. is **unisexual,** many of the other mosses are **bisexual.** *Polytrichum* spp. produces its gametangia at the tips of the main axis, but others produce them on special lateral branches. Sex organs of various ages may be found in the same apex. **Paraphyses** (sterile hairs) are often intermingled among the organs.

Antheridia and archegonia develop from superficial cells at the stem tip. The antheridium (Figures 19.11*a* and *b*) is composed of a sterile jacket of cells enclosing many spermatogenous cells. Active duplication of the spermatogenous cells produces many hundreds of cells that mature into biflagellate sperm. The archegonium (Figures 19.11*c* and *d*) is composed of an elongate neck and a bulbous venter that contains an egg. When the mature antheridium becomes wet, it dehisces at its apex and the sperm ooze into the water where they begin to swim. When they reach the opening into the archegonium, they swim down the neck and fertilization of the egg occurs in the venter.

SPOROPHYTE GENERATION

The basal portion of the zygote grows into the old gametophytic tissue and forms a foot. The upper portion of the zygote produces the seta and capsule. The old archegonial venter may undergo some growth and completely enclose the developing sporophyte. The venter is then known as the **calyptra** and is ruptured as the sporophyte increases in size. The upper part of the calyptra forms a cap that is carried upward on the capsule as the seta elongates. The tissue of the capsule differentiates into endothecium and amphithecium during maturation of the seta. The endothecium produces a

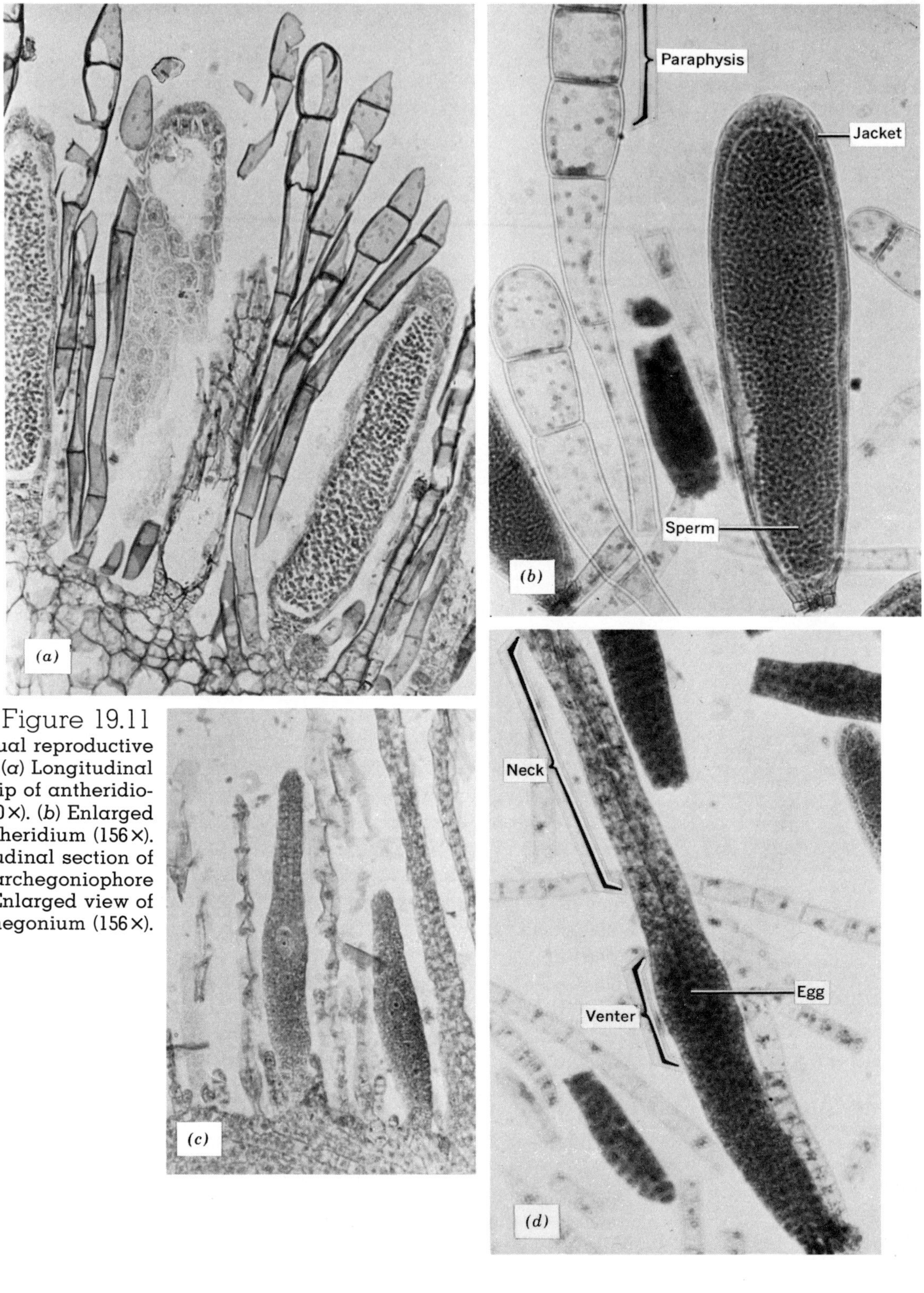

Figure 19.11
Moss sexual reproductive structures. (*a*) Longitudinal section of tip of antheridiophore (130×). (*b*) Enlarged view of antheridium (156×). (*c*) Longitudinal section of tip of archegoniophore (130×). (*d*) Enlarged view of archegonium (156×).

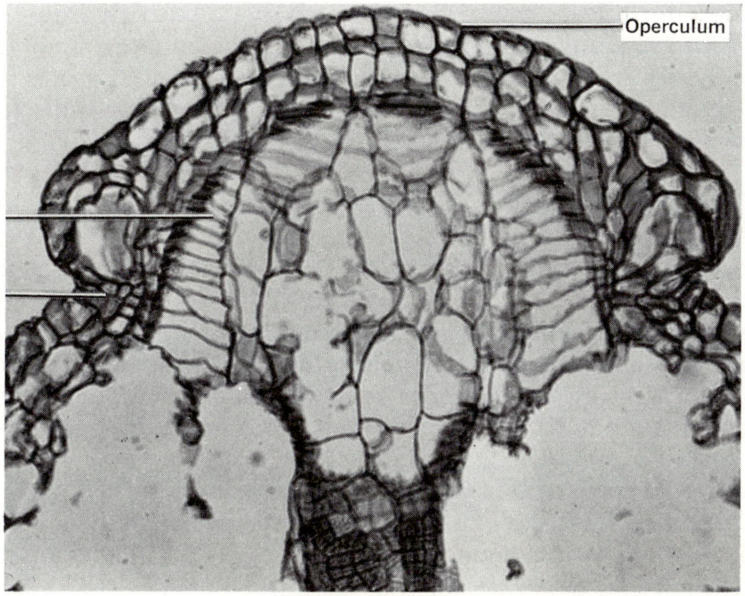

Figure 19.12
Longitudinal section of tip of *Mnium* sp. capsule (143×).

sterile columella and sporogenous cells. The amphithecium produces a sterile jacket. The sporogenous cells undergo meiosis producing a tetrad of haploid cells. No elaters are produced. When the capsule matures, the apical portion thickens forming a lidlike **operculum** (Figure 19.12). The operculum is released by a ring of thin-walled cells, the **annulus,** which circumscribes its base. A ring of toothlike segments are produced just inside the annulus. This ring is called the **peristome.** Each of the segments is extremely hygroscopic and probably aids in spore dispersal as they react to changes in atmospheric moisture. There is a great deal of variation in the numbers and structure of peristomes present in the various genera of mosses. The spores are released by the shedding of the operculum. Although many of these structures (e.g., calyptra) are to be found in the hepatics and *Sphagnum* spp., they were not discussed there. The structures are much more visible and common in the mosses, and more attention will probably be given to the mosses in any beginning consideration.

The sporphyte of many of the true mosses produces chloroplasts from the time it is a few cells in size until the time it matures, sheds its spores, and dies. The photosynthesis carried on by the cells containing the chloroplasts provides much of the carbohydrate necessary for sporophyte development. However, it must be emphasized that the sporophyte receives its mineral nutrition and water from the gametophyte through the foot. It quite likely also receives some dissolved carbohydrates in the same manner.

HUMAN RELEVANCE

The bryophytes are often used as ornamental plants in rock gardens and on stone walls where they aid in the water retention of the soil and help to slow erosion. Many other plants do this, and because of their superficial appearance to mosses (e.g. small, soft and green) are often called mosses. Examples of this type of misidentification abound, but some of the more common mistakes are "moss" rose which is a small flowering plant (*Portulaca* spp.), reindeer "moss" which is a lichen (*Cladonia* spp.), and Spanish "moss" which is another flowering plant (related to pineapple).

The bryophytes are extremely important ecologically because they are often the first "invaders" of bare soil or rock. They aid in the succession of new communities and microhabitats because they hold the water, slow wind and water erosion, aid in the accumulation of soil, and add their own organic materials when they die and decay. These factors allow for the support of "higher" plants such as ferns and lower vascular plants in the area.

Sphagnum spp. is especially important in small lake ecology. Here the plant produces thick mats along the edge of the lake. The mats eventually spread out over the water and may produce a "quaking" bog. For a time the bog will not be very solid and will allow large animals (including the unwary bryologist) to fall through. Eventually enough material will accumulate, allowing other plants to grow. The entire body of water may then evolve to solid soil. Many people of Northern Europe and Ireland harvest this compressed moss and use it as a fuel called peat.

Because of its tremendous water-holding capacity and light texture, peat is also added to soil where it improves the soil texture as well as the soil moisture. It is used to pack nursery stock for long distance shipment. Because of its acidity, *Sphagnum* spp. is somewhat antiseptic and it has been used as packing material for seeping wounds during time of war.

Many of the large, springy mosses have been dried for use as packing materials for crockery.

Because the thallose forms of many of the bryophytes supposedly resembled the lobes of a liver, and because the surface of many of them (e.g., *Marchantia* spp.) are divided into polyzonal areas that appeared to resemble the cross section of an animal's liver, they were thought to have been placed on earth by God to cure "liver" ailments. This belief was only one manifestation of the Doctrine of Signatures which was widely believed by medieval man. The doctrine taught that God had put all of the plants and animals on the earth for man's use, and that man should recognize the use for which they were intended by some structure or form that they possessed. Thus the liverworts were named! No medicinal value to the liver or any other part of the body has been discovered by modern research, however.

2 vascular plant groups

vascular plant groups | introduction

20

The vascular plant groups include those plants with which most people are familiar. Everyone is aware of oak trees, wheat fields, and ferns. There are, however, many vascular plants that are not so common or so obvious, and that may have escaped detection by the casual observer of the plant kingdom. It is helpful, therefore, to approach the vascular plant groups by first understanding the unifying characteristics possessed by the entire group and then considering the differences that separate them into smaller, more closely related groups.

The vascular plants possess three sets of characteristics that separate them from other plant groups. Although there are exceptions, all three sets of characteristics, considered collectively, appear to delimit the group as a fairly well-structured entity.

All vascular plants possess **vascular tissue** in their sporophytes, hence the common name of "vascular" plants (Figure 20.1). The vascular tissue is the transport tissue of the plant, and is composed of two major tissue types, the **xylem** and the **phloem.** The vascular tissue may be variously arranged within a vascular plant, but it is continuous from the root to the top of the tallest tree. In the root and stem, the xylem and phloem form a central rod or column of vascular tissue called the **stele** (Figures 20.1 and 20.2). Other cell types may also be present in the stele, but they are of little concern to us here. Various branches from the stele extend into leaves and other structures, supplying them with vascular tissue. Some vascular plants, notably the monocotyledons (e.g., corn), produce their vascular tissue in discrete strands or bundles composed of xylem and phloem. These bundles are scattered within the stem in an inordinate arrangement.

The part of the vascular system that is produced by the **apical meristem** is the **primary vascular tissue.** This tissue is derived from a growing stem or root tip. These tips are very metabolically active and cell duplication occurs frequently. Such a region of active cell duplication is called a **meristem** and when it occurs at the tip of a root or shoot, it is known specifically as an apical meristem (Figures 20.1 and 20.3). As the cells mature, differentiation produces the distinctive primary xylem and phloem tissues. Other cell types are also produced by the differentiating cells. Some vascular plants also produce a **lateral meristem** called a **vascular cambium.** The vascular cambium is first produced between the primary xylem and the primary phloem, but later may form a complete ring within the stem and root. The cells of vascular cambium duplicate mitotically, the resulting cells of which differentiate into secondary xylem toward the center and into secondary phloem toward the outside of most stems or roots. Some vascular plants produce only primary vascular tissue during their life history, but many others also produce varying amounts of **secondary vascular tissue.**

As has been previously mentioned, the vascular tissues function as a transport system within the plant. There does, however, seem to be a fairly distinct division of labor. The xylem apparently transports water and dissolved mineral nutrients throughout the plant and the phloem mostly transports dissolved foods, although there are some exceptions. No valid comparison can be made between the transport system of plants (the vascular system) and the transport system of animals (the circulatory system). Although there is no completely satisfactory explanation for the movement of materials through the plant's vascular system (especially through the xylem), there is certainly nothing that corresponds to the heart and muscular circulation of animals.

A discussion of the evolution of vascular tissue and its arrangement within the plant is beyond the scope of this text, but it should be repeated that phloem tissue has been found in some of the brown algae (the kelps). Therefore, vascular tissue is found not only in vascular plants but also in some other plant groups. Also, it is believed that vascular tissue probably arose several times during the process of evolution rather than only once.

Second, the life history of all vascular plants is remarkably similar (Figure 20.4). The **sporophyte generation** (usually **diploid**) becomes the part of the cycle that is largest and most obvious, the "dominant" generation. The **gametophyte generation** (usually **haploid**) is much reduced and indeed becomes parasitic on the sporophyte in many

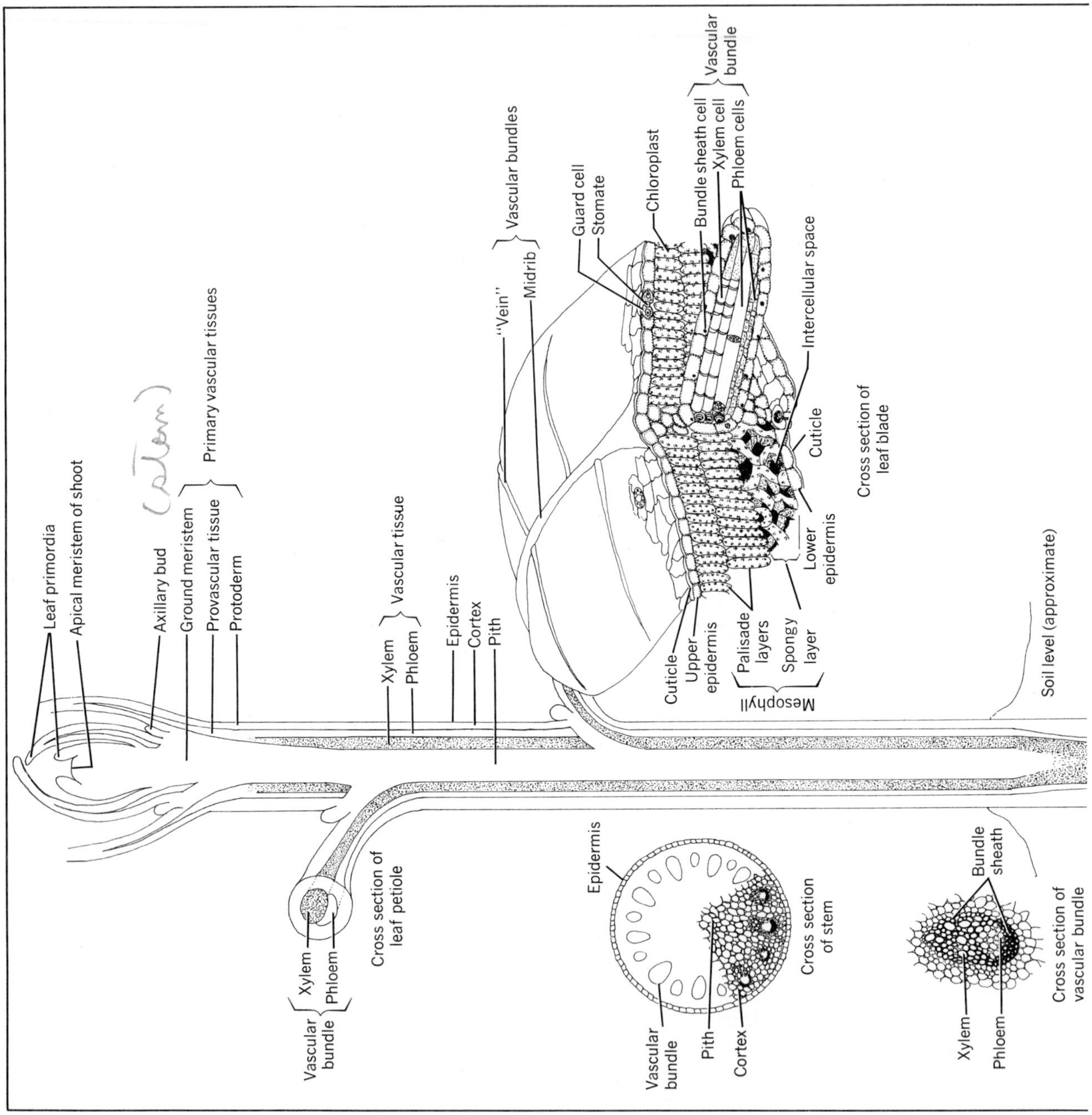
Leaf primordia
Apical meristem of shoot
Axillary bud
Ground meristem
Provascular tissue
Protoderm
Primary vascular tissues
Xylem
Phloem
Vascular tissue
Epidermis
Cortex
Pith
"Vein"
Midrib
Vascular bundles
Guard cell
Stomate
Chloroplast
Bundle sheath cell
Xylem cell
Phloem cells
Vascular bundle
Intercellular space
Cuticle
Upper epidermis
Palisade layers
Spongy layer
Mesophyll
Lower epidermis
Cuticle
Cross section of leaf blade
Soil level (approximate)
Vascular bundle
Xylem
Phloem
Cross section of leaf petiole
Epidermis
Vascular bundle
Pith
Cortex
Cross section of stem
Bundle sheath
Xylem
Phloem
Cross section of vascular bundle

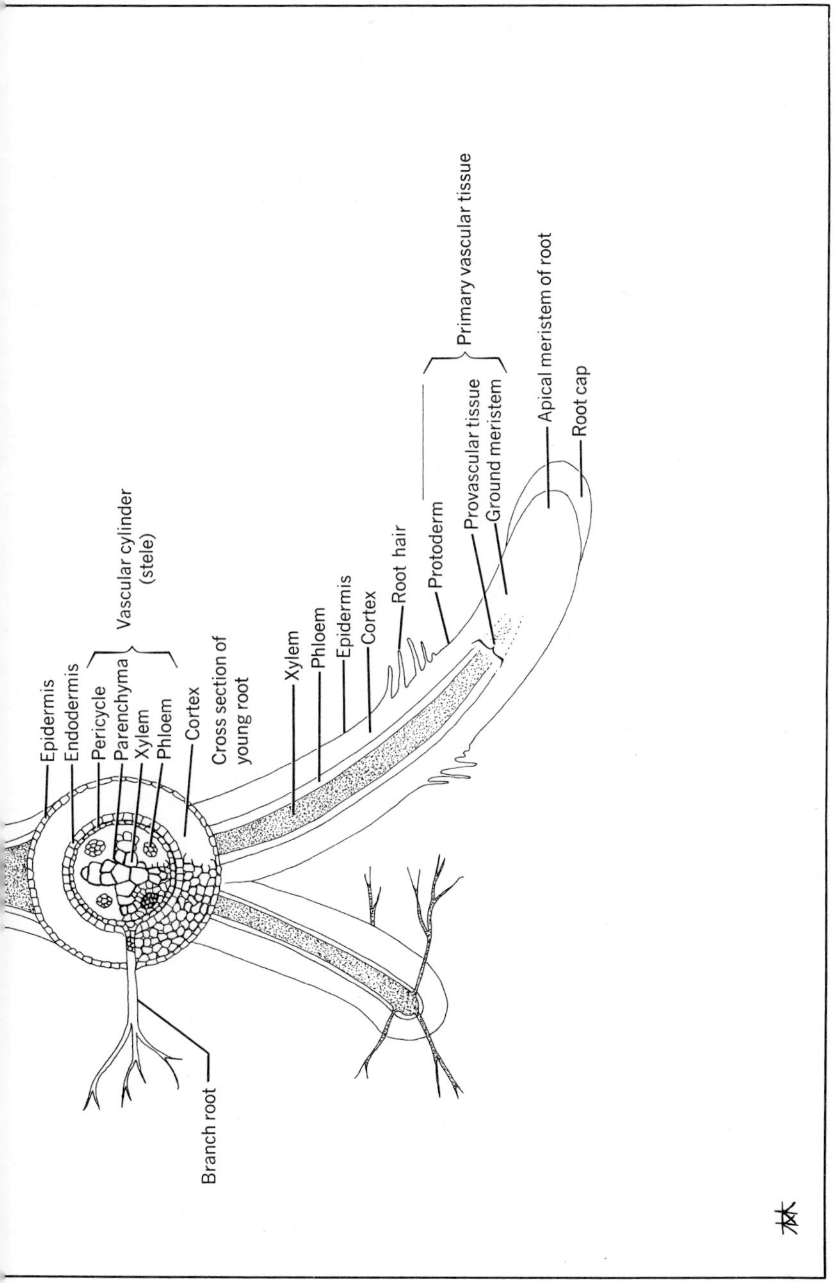

Figure 20.1
A diagrammatic view of the sporophyte of an entire herbaceous dicotyledonous plant. (This diagram will be used as a reference with all discussions of vascular plants.)

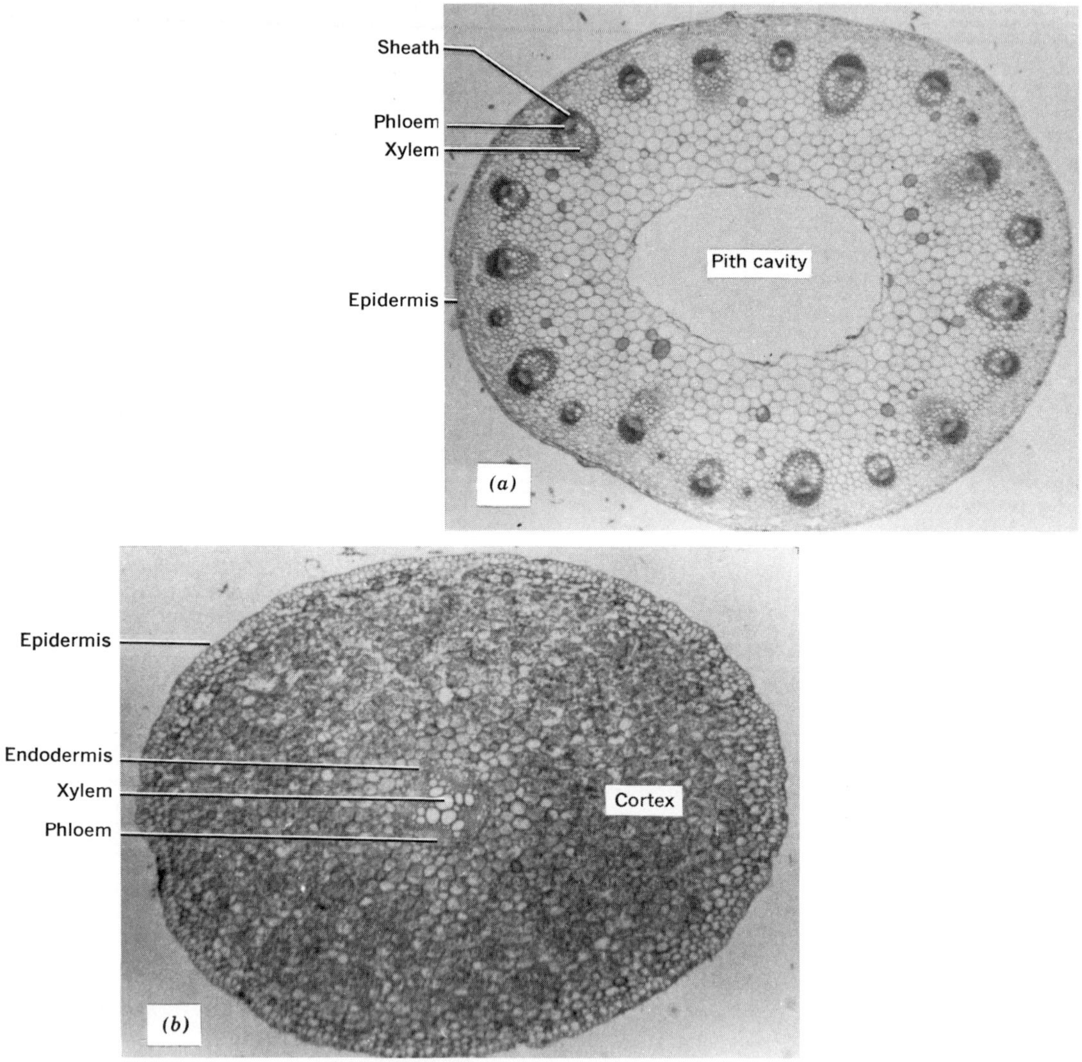

Figure 20.2
Cross section of stem and root of herbaceous dicotyledon (*Ranunculus* sp.) (30 ×). (*a*) Stem. (*b*) Root.

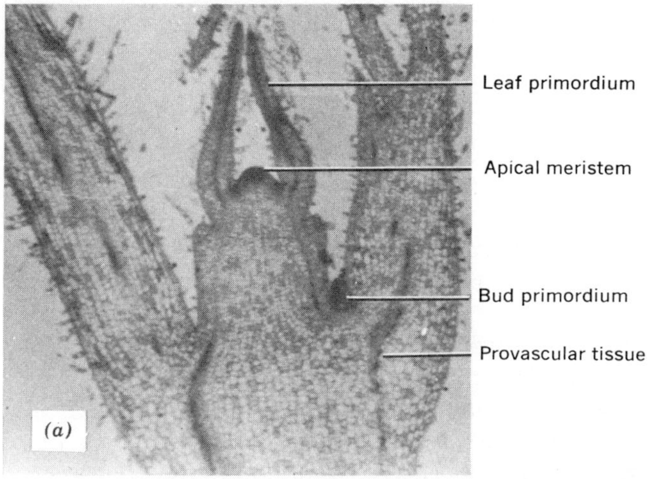

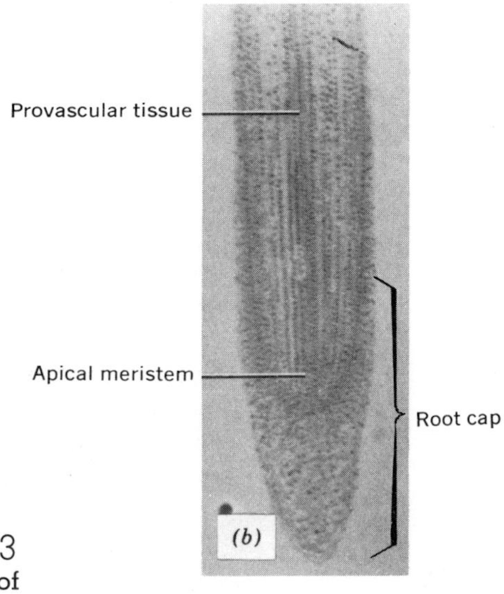

Figure 20.3

Longitudinal sections of stem and root apices. (*a*) Stem (30×). (*b*) Root (20×).

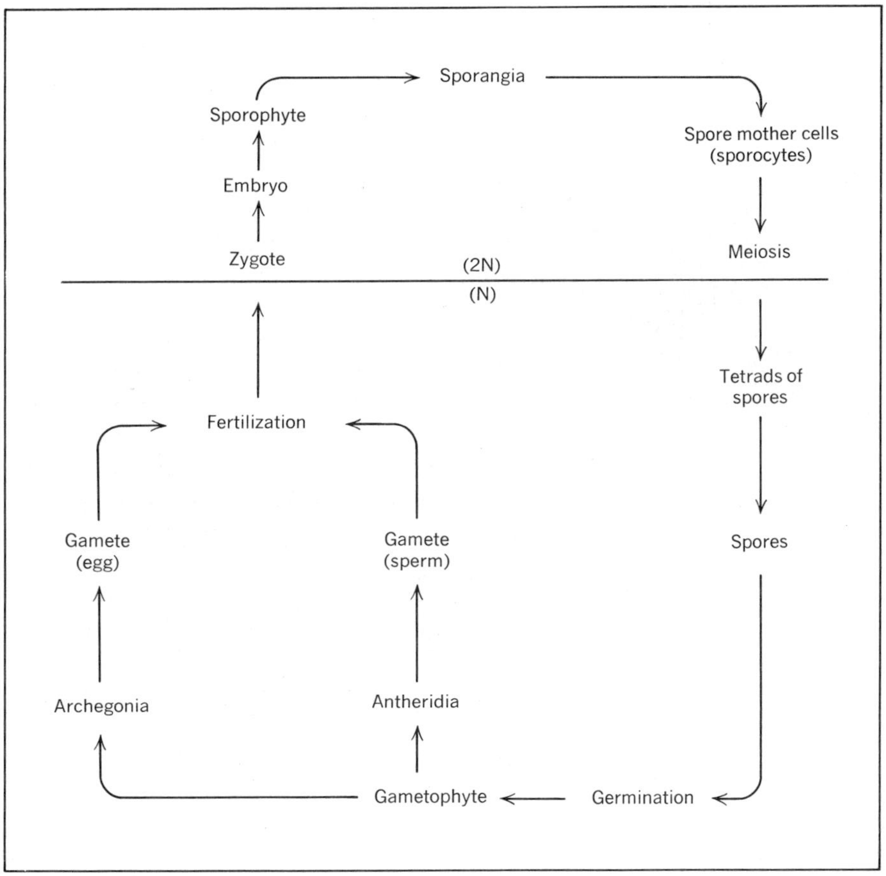

Figure 20.4

A generalized life history for vascular plants.

of the vascular plants. The life history of the vascular plants is thus dramatically different from the life histories of the mosses and liverworts.

The sporophyte produces some sort of **sporangium.** Although sporangial types vary greatly among the vascular plants, almost all of them contain diploid spore mother cells (**sporocytes**) in which meiosis occurs. This nuclear activity, combined with cytokinesis, ultimately produces haploid spores. The germination of the spore results in the **gametophyte** which ultimately produces **gametes.** Two gametes fuse in the process of **fertilization,** and the resulting cell is the **zygote,** the first cell of the new **sporophyte.** The zygote produces an **embryo** that may be protected and nourished for a time by the gametophyte, but it finally becomes completely independent of the gametophyte, producing a new, mature sporophyte.

The parts of the life history have a great deal of morphological variation among the various plant groups. For example, the gametophytes of the "lower" (e.g., ferns) vascular plants produce definite, multicellular structures (gametangia) in which are produced the gametes. On the other hand, the "higher" (e.g., flowering plants) vascular plants have such reduced gametophytes that the gametangia have been completely lost from the life history. Other variations within individual groups will be presented when the group is discussed.

From the above description it would appear that the life history of vascular plants follows a consistent pattern of doubling (producing the sporophyte) and halving (producing the gametophyte) of the chromosome number. Research has demonstrated, however, that even this usual pattern of development has important exceptions. There have been found among the vascular plants (especially among ferns) gametophytes that are diploid and sporophytes that are haploid. Obviously, then, the usual condition of syngamy followed by meiosis does not always occur. Gametophytes may be produced directly from the leaves of sporophytes (therefore they are diploid) without the process of meiosis and spore production. Gametes produced in the gametangia of such gametophytes are diploid, and the zygote resulting from their union is tetraploid. Sporophytes may arise directly from some haploid gametophyte cell or cell group. Such a sporophyte would be haploid and could only produce more haploid sporophytes by vegetative reproduction.

The production of diploid gametophytes by the loss of meiosis and subsequent spore production is known as **apospory;** the production of a haploid sporophyte by the loss of fertilization is known as **apogamy.** It should be noted that if an aposporous gametophyte (e.g., 2n) produces an apogamous sporophyte, the sporophyte will have the usual diploid chromosome number. Aposporous gametophytes and apogamous sporophytes have been produced by laboratory culture techniques, but they have also been known to arise spontaneously in nature. Such variation in chromosome number in the gametophyte and the sporophyte reduces the value of the use of such numbers in the precise definition of the two generations.

The third major characteristic that is used to define vascular plants as a cohesive group is that most of them contain **chlorophyll** and produce their own food by **photosynthesis.** Such a statement is generally true, but several exceptions must be noted. First, the algae and bryophytes also produce their own food by photosynthesis. Therefore, the vascular plants do not have an "exclusive monopoly" on the process. In addition to this exception, most of the vascular plants produce nonphotosynthetic gametophytes that are either parasitic on the sporophyte, saprophytic, or produce mycorrhizal relationships with various fungi. There are also some vascular plants that produce achlorophyllous, therefore nonphotosynthetic, sporophytes. The best known of these are dodder, Indian pipe, and beech drops.

Vascular plants, therefore, have three sets of characteristics that are usually present and that are important when the group is to be defined. These are (1) the possession of a transport (vascular) system composed of xylem and phloem, (2) the alternation of a haploid gametophyte with a diploid sporophyte, and (3) the ability to produce their own food by the process of photosynthesis. None of these characteristics is possessed exclusively by the vascular plants, but all of the vascular plants do possess the transport system.

The vascular plants are often called the "land" plants. Indeed, this designation is warranted if one considers that the possession of the transport system has allowed the vascular plants to greatly increase in size on the land. Such a system is important because water can be transferred to the topmost cells in the tallest tree, and food can be transferred to the end cells of the deepest root. The production of a waxy **cuticle** that covers the epidermal cells of most plant structures aids in water retention by the plant. Some sort of conservation of water would have been necessary before any plant could have grown into an aerial environment. Many vascular plants still retain their dependence on water because they produce flagellated or ciliated sperm that must have at least a film of water through which they swim to the egg, thus accomplishing sexual reproduction. This dependence on water for the completion of their life history is not completely lost until one reaches the **angiosperms** in the evolutionary scale, although some of the **gymnosperms** do not possess motile sperm and therefore do not require the water film through which to swim.

Many plants that are nonvascular (e.g., mosses, liverworts, some algae) are also somewhat successful on land. These, however, occupy specific environmental niches that provide them with a great deal of water. They remain relatively small probably because they lack any sort of transport system through which water and minerals can travel to aerial parts.

Although the vascular plant groups include those plants best known to most people, their classification is also in a state of change. It is not the purpose of this text to present a classification scheme or to indulge in a discussion of the various proposed systems of classification. The student of plant sciences will certainly encounter the various authoritative opinions in his further study. The groupings presented here are simply convenient categories that can be used as units about which can be made various generalizations. It should be easy to transfer the examples to the proper taxonomic categories in any of the various classification schemes the student later encounters.

This group of plants also contains various members that are now extinct but that are well known from the fossil record. It is not the intent of the present work to discuss these fossil members although their importance to the understanding of the evolution of present day plants is fully recognized. Therefore, only brief mention will be made of the various extinct vascular plants when they become necessary to the understanding of the geologic age of a particular group.

21 psilophytes

Although representatives of the psilophytes are to be found as long as 350 million years ago in the Devonian period, only two genera (*Psilotum* and *Tmesipteris*) are still extant in the modern flora. *Psilotum* spp. is widespread in the tropical and subtropical parts of the world and reaches as far north as Florida, Louisiana, and Texas in the continental United States. *Tmesipteris* spp. is much more confined in its distribution (Australia and islands of the South Pacific) and is therefore much less likely to be encountered.

Members of both genera may be found growing in rich soil or among exposed rocks as erect plants. They may grow as hanging epiphytes on the base of trees or on the trunks of tree ferns. In either case their size varies with their location and environment from 8 to 100 cm in the case of *Psilotum* spp. to 5 to 10 cm in *Tmesipteris* spp.

Because *Psilotum nudum* (Figure 21.1, p. 326) is the member of the psilophytes that has the widest range, it is the one to which the following discussion applies. However, a few of the major differences between *Psilotum* spp. and *Tmesipteris* spp. will be pointed out. Members of both genera (Figure 21.2) closely follow the generalized life history for vascular plants that was discussed in Chapter 20.

SPOROPHYTE GENERATION

The sporophyte of *Psilotum nudum* is composed of dichotomously branched aerial stems (Figures 21.2 and 21.3) and underground **rhizomes.** The epidermal cells of the rhizomes produce short rhizoids. Although no roots are present, the plant is anchored by the rhizomes and water and mineral nutrients are absorbed through them. Absorption is probably aided by endophytic fungi that form, with the rhizomes, a mycorrhizal relationship.

Both aerial stems and rhizomes develop from a single apical cell and its daughter cells, all of which remain mitot-

Figure 21.2
***Psilotum nudum*. Life history.**

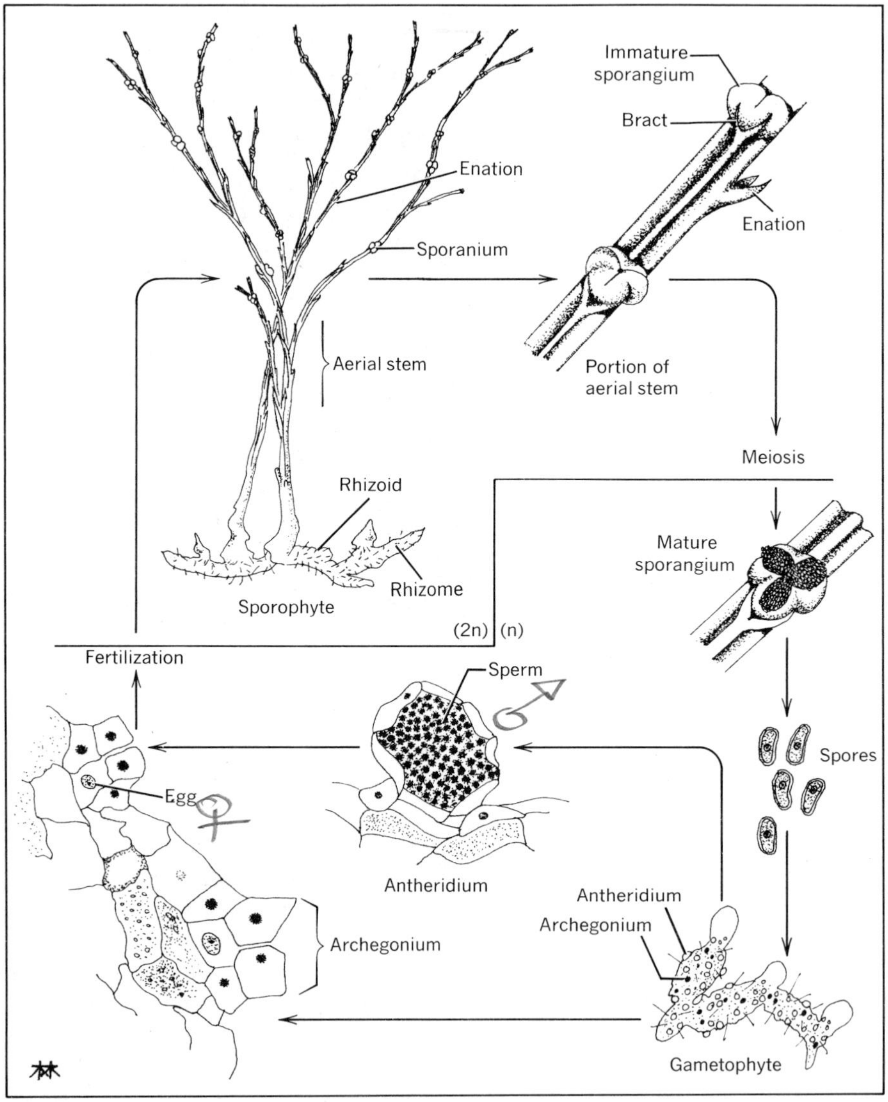

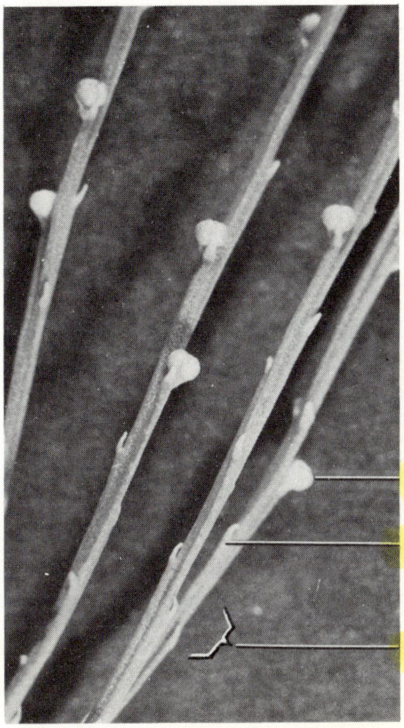

Figure 21.3
Psilotum nudum. Aerial stems. Note dichotomous branching, sporangia, and enations (1×). (Photograph by John W. Cieslewicz)

ically active. The degree of branching of the rhizome is probably related to the mechanical damage inflicted on its apical cell by obstacles (e.g., small stones) encountered under the soil. However, a new aerial branched system may be produced whenever a rhizome tip turns upward. From time to time two apical cells will be produced in the aerial tip and each of them will initiate a new stem, thus producing a new dichotomy. The form of the aerial branches may vary from being circular near the soil to being triangular near the ends of the distal dichotomies.

The apical cell and its derivatives produces three types of primary tissues. These are the outermost single cell layer (the **protoderm**), the innermost mass of **procambium cells,** and the **ground meristem** that is between them. All three primary tissues continue to divide, but eventually they mature in such a way that the protoderm produces the **epidermis,** the procambium produces the central cylinder of vascular tissue (**stele**) and the ground meristem produces the **cortex** (Figure 21.4). The epidermis is composed of cells that have cutinized

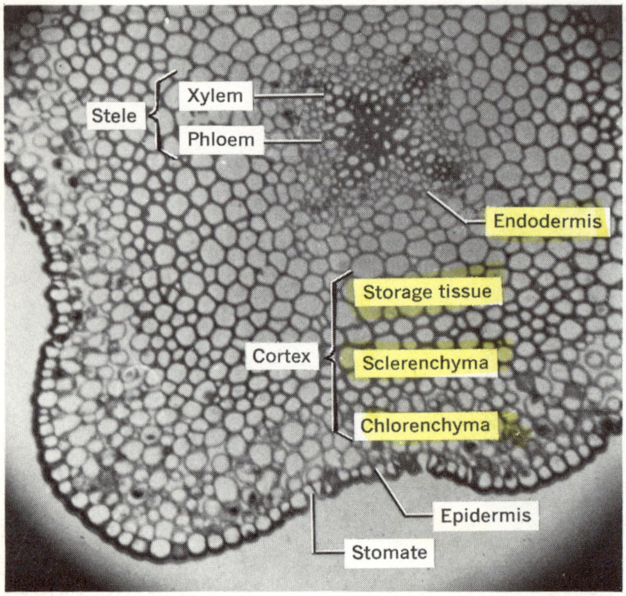

Figure 21.4
***Psilotum nudum*. Cross section of aerial stem. Section taken near the rhizome (55×).**

cell walls and that are covered by a **cuticle. Stomates** are present in the areas between the elongated ridges in the upper part of the aerial stems.

The cortex is just under the epidermis and is separated from the stele by its innermost layer, the **endodermis.** The endodermis is composed of a row of cells that are marked by conspicuous thickenings (the **Casparian strips**) in their walls. Next to the endodermis is a thick layer of parenchymatous storage tissue. The storage layer is surrounded by **sclerenchyma** tissue which is in turn surrounded by several layers of parenchyma cells that contain chloroplasts (**chlorenchyma**). The chlorenchyma tissue is the major photosynthetic tissue of the plant.

The stele is variable at different levels of the aerial stem. In the rhizome the stele has a solid center of primary xylem that may have a variable number of lobes (Figure 21.5). The area between the xylem lobes (or "arms") is filled with primary phloem and some of the primary phloem may surround the ends of the xylem lobes, completing the vascular cylinder. This type of stele, composed of a solid mass of vascular tissue, is called a **protostele** and is considered to be the most primitive arrangement of vascular tissue. The aerial stems may have a large amount of sclerenchyma tissue in the

Figure 21.1
Psilotum nudum (whisk fern) growing in greenhouse.

(*a*)
(*b*)

Figure 23.1
Habitat views of two species of _Equisetum_. (_a_) _E. hiemale_ vegetative plant. (_b_) _E. hiemale_ strobilus (Photographs by John W. Cieslewicz.). (c) _E. sylvaticum_.

(c)

Figure 22.1
(*a*) *Lycopodium clavatum*. Group of plants with young strobili. *L. obscurum*. Single plant with mature strobilus. (*b*) *Selaginella kraussiana*. A creeping form. *S. kraussiana*. Roots visible. *S. emmeliana*. An upright form. (*c*) *Isoetes* sp. Roots visible; one plant axis split.

(*a*) (*a*)

(*b*) (*b*) (*c*) (*b*)

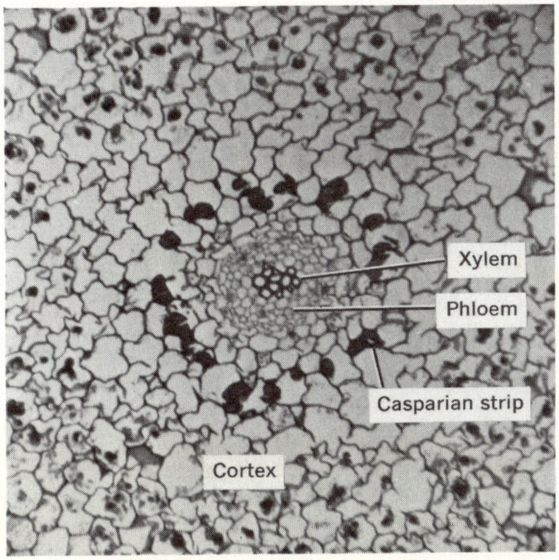

Figure 21.5
***Psilotum nudum*. Cross section of rhizome (55×).**

center of the xylem. If this tissue is interpreted as belonging to the xylem, the stele is still said to be a protostele. However, if it is interpreted to be pith cells that have become sclerenchymatised, the stele is said to be a **siphonostele** (which may be defined as a stele with a central column of pith). This type of variation from rhizome to aerial stem is found in both *Psilotum* spp. and *Tmesipteris* spp. Among the more advanced vascular plants, the protostele (of which there are several variations) is confined to the root system, but the siphonostele (of which there are also several variations) is very common in the stems.

The stem of *Psilotum* spp. possesses many scalelike appendages (**enations**) that are spirally arranged (Figures 21.1 and 21.3). These appendages possess no vascular tissue (Figure 21.6) but in one species a trace of vascular tissue extends from the vascular cylinder to the base of the appendage. *Tmesipteris* spp. differs in possessing a broadened foliar appendage that is supplied with a vascular strand (Figure 21.6) from the vascular cylinder of the stem to the tip of the leaf. Because there is no gap in the cylinder at the point where the trace leaves it, such a leaf is called a **microphyll.** The parenchymatous cells of both enations and microphylls contain chloroplasts, but the enations of *Psilotum* spp. are far too small to be photosynthetically important. The microphylls of *Tmesipteris* spp. are, however, the chief photosynthetic organ for that plant.

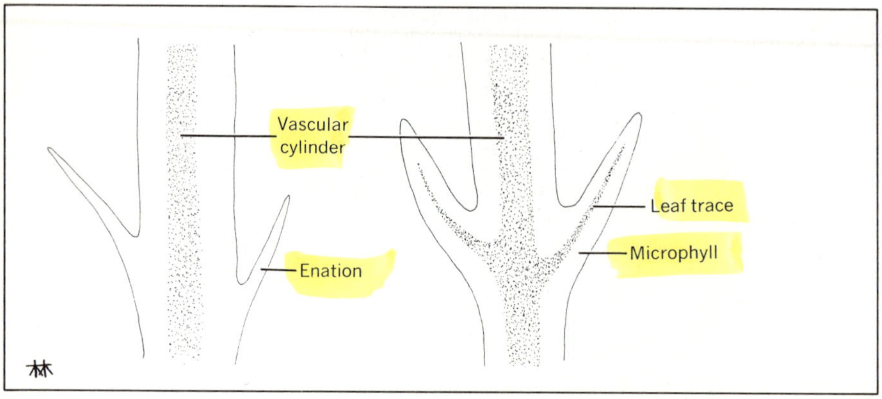

Figure 21.6
Longitudinal section of stem showing relationship between the vascular cylinder and appendage.

The asexual spores are produced in sporangia that are borne at the tips of tiny lateral branches of the stem (Figures 21.2 and 21.3). The sporangium, which has three lobes, is initiated from a group of superficial cells of the lateral branch and is thus **eusporangiate.** Two bracts extend from the main stem and subtend the sporangium and its stem. The spores are produced from spore mother cells (sporocytes) in which meiosis occurs. After cytokinesis, four maturing daughter cells produce haploid spores. Hundreds of spores are produced per sporangium (Figure 21.7*a*). They all are kidney-shaped (Figure 21.7*b*), bilaterally symmetrical structures with netted wall patterns. Because all of the spores are alike, the plant is said to be **homosporous.** The sporangium of *Tmesipteris* spp. is two-lobed, but is produced in the same way as the sporangium of *Psilotum* spp. is produced. It appears to be in the axis of a leaf but this is because its stem tip turns upward during its development. The lobes of the sporangia open at maturity along definite lines of dehiscence.

GAMETOPHYTE GENERATION

Germination of the spores is very slow, and the gametophyte that is produced is rarely more than 2 mm in diameter but may be several millimeters long (Figure 21.2). It may be found growing in rocky crevices, on the trunks of tree ferns, or on old greenhouse pots. It is cylindrical, sparsely branched, achlorophyllous, and often subterranean. The plant is saprophytic and is probably dependent on endophytic fungi for its energy source. There is very little internal differentiation (although some vascular tissue has occasionally been reported) but it is covered with rhizoids.

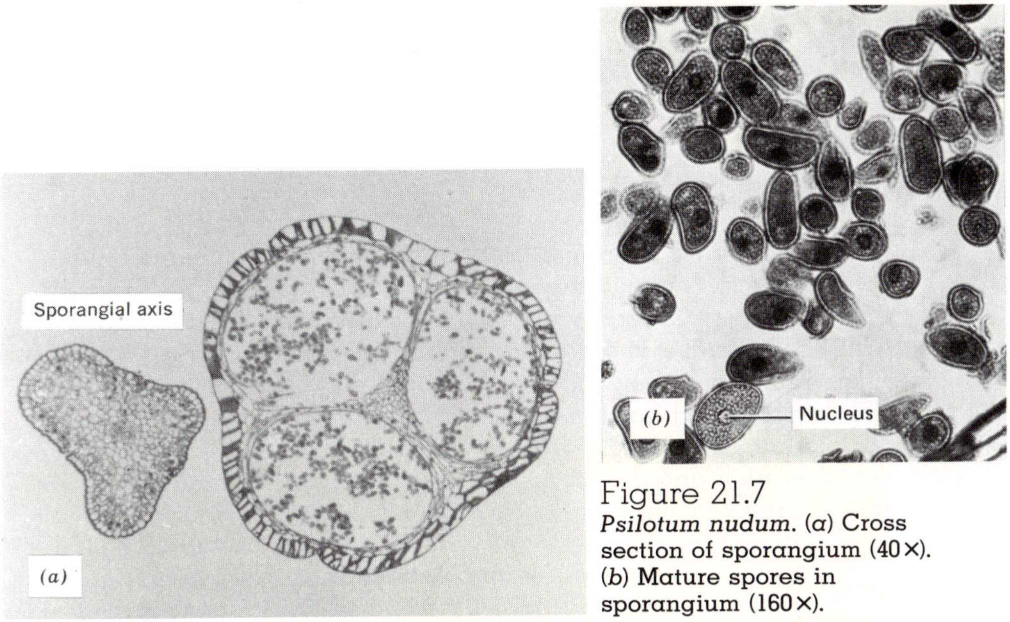

Figure 21.7
Psilotum nudum. **(*a*) Cross section of sporangium (40×). (*b*) Mature spores in sporangium (160×).**

Antheridia and archegonia (Figure 21.2) are both found on the same gametophyte, called a **prothallus,** (making it bisexual). They are randomly scattered over the prothallus, and develop from superficial prothallial cells. The sperm are multiflagellated and must swim from the antheridium to the archegonium where fertilization takes place.

The zygote divides at right angles to the long axis of the archegonium. The upper cell of the two daughter cells produces the new stem. The lower cell produces a foot. The foot absorbs nutrients from the gametophyte, but the new stem contacts the proper fungus as it leaves the gametophyte and the mycorrhizal relationship is formed. Some of the new stem tips become negatively geotropic, thus producing the new aerial portions of the vegetative body. Eventually the sporophyte becomes independent of the gametophyte and breaks away from it, leaving the foot behind.

HUMAN RELEVANCE

The psilophytes have very little or no economic value (except, perhaps, as exotic house plants). The common name applied to *Psilotum* spp., whisk fern, suggests that this plant may have been used as a broom. However, the psilophytes do form an interesting portion of the vegetation wherever they are present.

The real value of this plant group is the evidence that it gives us for understanding the ancestors of modern-day plants. The psilophytes probably represent the most ancient extant vascular plants. Although some botanists believe that *Psilotum* spp. and *Tmesipteris* spp. may represent a reduction series from more complex plants, there is no direct evidence for such an assumption. Therefore, the phylum Psilophyta is usually divided into two orders. The Psilophytales is that order which is set up to include the extinct members and the Psilotales is that order set up to include the extant members (e.g., *Psilotum* spp. and *Tmesipteris* spp.). This type of taxonomic treatment recognizes the primitive nature of the group.

Although xylem cells are only occasionally found in the gametophyte, their presence is sometimes interpreted as being a reduced stele. Such interpretation lends evidence to one of the theories that attempts to explain the alternation of generation in vascular plant life histories. This theory, the **homologous theory of alternation of generations,** proposes the idea that the sporophytes and gametophytes are basically alike and that they arose from ancestral algae in which the two phases were already differentiated, although morphologically similar.

Therefore, although the plants are not important as an economic consideration, they are very important to those who are interested in attempting to understand plants of the earth's ancient past.

ground pine and its relatives

22 lycopodophytes

The lycopodophytes (ground pines, club mosses, quillworts) possess true roots, stems, and leaves (Figure 22.1, p. 326). They are well represented in the modern flora by about 900 species that are grouped into four genera. Three of the genera (*Lycopodium*, *Selaginella*, and *Isoetes*) will be used here as representatives of the group. Fossil lycopods, some of which grew to a height of 100 feet or more (e.g., *Lepidodendron*) are found in the fossil record along with fossil members of the psilophytes. Therefore, they are at least, as a group, as ancient as the psilophytes.

SPOROPHYTE GENERATION

The roots and stems of many lycopod sporophytes are dichotomously branched but others have one main axis from which lateral branches arise (**monopodial branching**). Several possess rhizomes from which roots are produced. This type of root, which arises neither from a primary nor secondary root, but from some other part of the plant, is known as an **adventitious root.**

The leaves of the lycopods vary in their size, shape and arrangement on the stem. They are all, however, microphyllous. That is, they are longitudinally traversed in the center by a single vascular strand that extends from the stele but which leaves no gap at its point of departure (Figure 21.6). Although the cortex of many of the lycopods contains chlorophyllous cells, the leaves are probably the chief photosynthetic organs. The stele of most members of this group is a protostele, although some siphonosteles do occur.

Another characteristic that is unifying for the entire group is the production of sporangia in a definite relationship to particular leaves (**sporophylls**). The sporangia arise from groups of superficial cells (eusporangiate) either in the axis of the leaves or on the surface of the leaf that is toward the axis (**adaxial surface**).

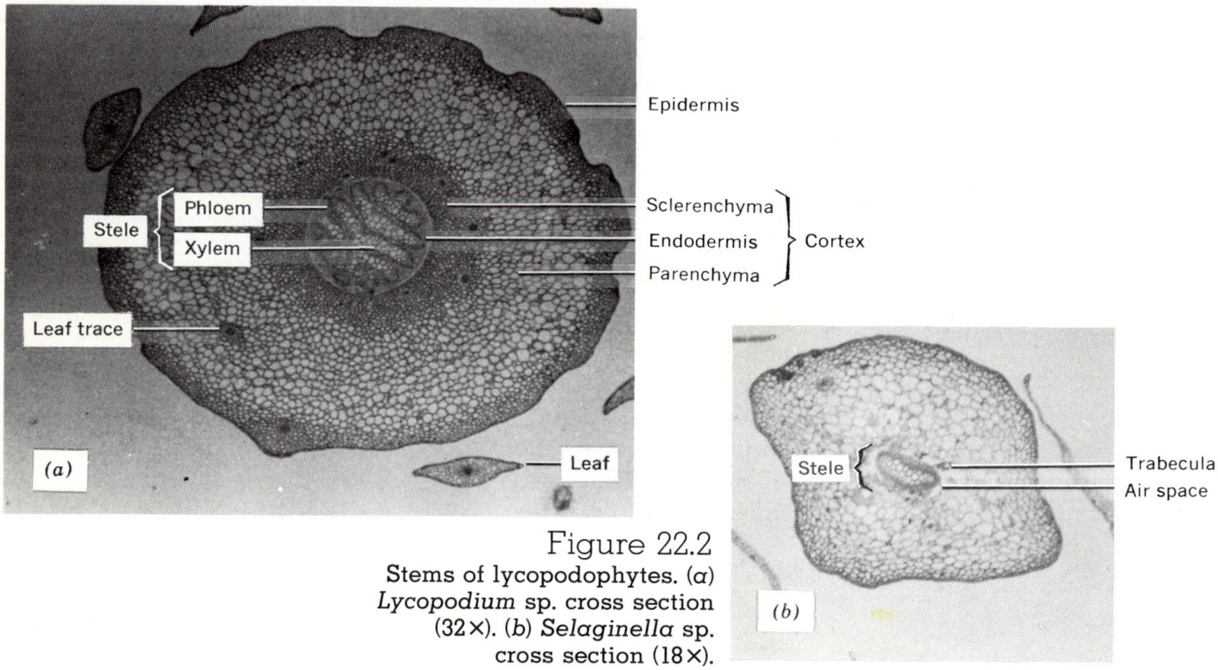

Figure 22.2
Stems of lycopodophytes. (*a*) *Lycopodium* sp. cross section (32×). (*b*) *Selaginella* sp. cross section (18×).

Lycopodium spp. (Figures 22.1*a*, p. 326 and 22.4) possesses a sporophyte that may be erect and shrubby or prostrate and creeping. However, all forms are small. Because of their superficial resemblances to tiny pine trees and mosses, they are often given the common name of "ground pines" and "club mosses" (indeed, some of them were once classified as mosses). They constitute a widely distributed and important part of the tropical and subtropical flora, but several species are to be found in temperate and arctic zones as well. Although most species are terrestrial, many, especially in the tropics, are epiphytic.

Stem and root development (with the exception of adventitious roots) may be traced to a group of apical cells. The development of epidermis, cortex, and stele (Figure 22.2*a*) are much the same as has been previously discussed. The mature epidermis possesses stomates and guard cells. The cortex is quite variable in thickness and, although it is bounded on its inner surface by an endodermis, this layer of cells may be difficult to distinguish in the mature stem. Several rows of cortex cells near the endodermis may become sclerenchymatized. The arrangement of the primary xylem and phloem in the stems is quite varied, both among species and within the same individual. Arms of primary xylem may alternate with patches of primary phloem or the two tissues may alternate with each other in bands. However, by tracing these strands of primary vascular tissue, it can be demonstrated that they are continuous throughout the stem. These different patterns of the tissues, then, are simply variations of a protostele. Small strands of vascular tissue (leaf traces) extend from the stele through the cortex and into the leaves. These strands appear as small patches of vascular tissue in the cortex in a cross section of *Lycopodium* spp. stem. There is no gap in the stele at the point of departure of the leaf trace, therefore the leaf is a microphyll.

Roots (Figure 22.4) are produced adventitiously near the growing tips of rhizomes or aerial stems from a layer of cells (**pericycle**) found between the endodermis and the outer layer of primary phloem. The roots may pass through the cortex for some distance before they break through the epidermis. It is therefore not unusual to find root traces as well as leaf traces in the cortex. The roots usually branch dichotomously.

The sporangia (Figures 22.3*a* and 22.4) are relatively large kidney-shaped structures located near the axis on the adaxial surface of sporophylls. The sporophylls of some species are chlorophyllous and are located in fertile areas that alternate with sterile areas on the stems. Other species produce sporophylls that are small, scalelike, achlorophyllous and that are grouped together in a terminal **strobilus.** All of the spores are morphologically alike, therefore *Lycopodium* spp. is **homosporous.** Some species produce sporophytic vegetative reproductive structures, called **gemmae** or **bulbils,** that consist of a short stem and several leaves. These small structures drop from the tips of aerial stems and produce a new sporophyte.

The genus *Selaginella* (Figures 22.1*b*, p. 326 and 22.5) is composed of more than 700 species and is found most abundantly in the tropics although it is represented in the flora of other areas as well. Some species inhabit damp areas in cold climates while others inhabit a wide range of habitats in the temperate zones. One interesting species, *S. lepidophylla*, lives in the desert areas of the southwestern United States, and has become known as the "resurrection plant." Such a name is applied because when it is in a dried condition it appears as a tight ball of nonliving plant material. However, in the presence of

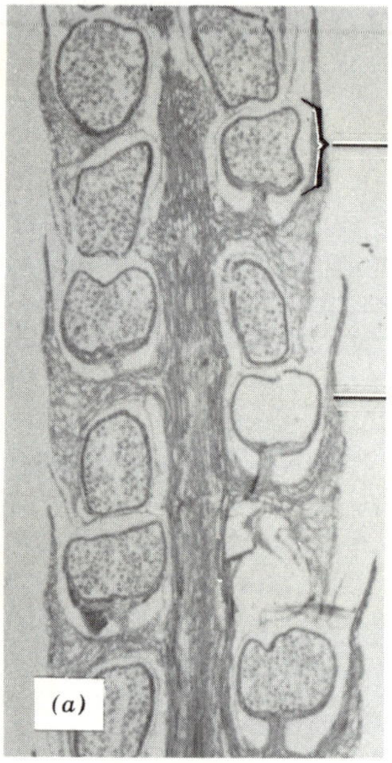

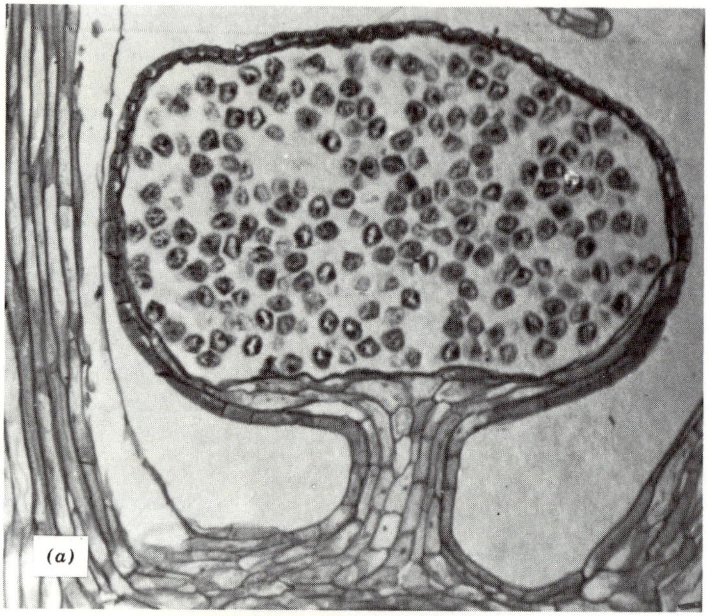

Figure 22.3
Sporangia of lycopodophytes. (*a*) *Lycopodium* sp. longitudinal section of strobilus (40×). Single sporangium with mature spores. (320×). (*b*) *Selaginella* sp. longitudinal section of strobilus (125×). (*c*) *Isoetes* sp. longitudinal section of sporophylls and plant axis (12×).

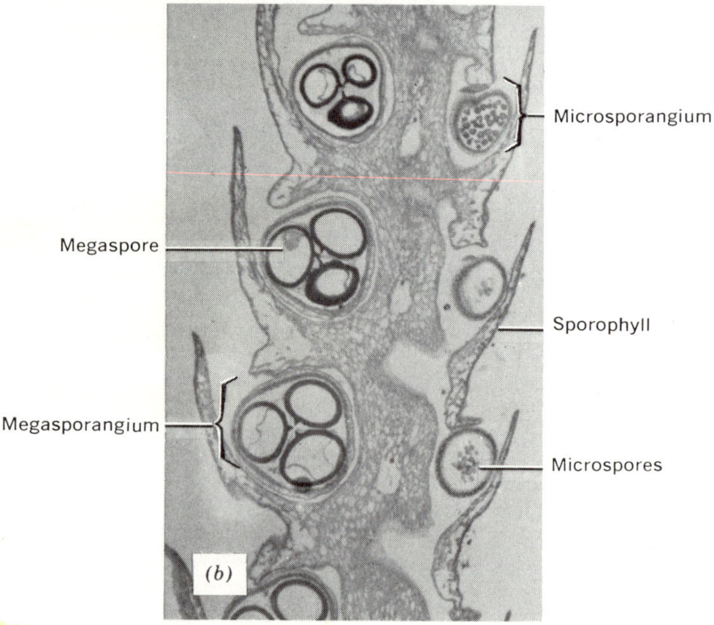

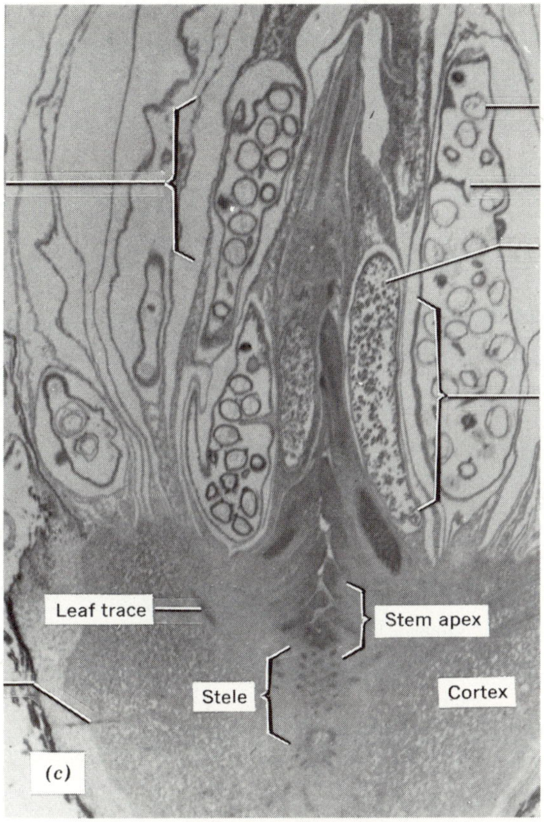

moisture, it will open up, lie flat, and become quite green. The plant is often sold as a novelty, and is sometimes pinned to curtains where the high humidity of the room causes it to open.

The sporophyte varies in its growth form from erect, shrubby tufts to flat, creeping plants that grow along the ground or trail over ledges and other plants (Figure 22.1*b*, p. 326). Various intermediate forms are also found. The stem branches in a dichotomous or monopodial (a system whereby the main axis is dominant over its branch axis) manner. The branching may produce broad, flattened, upright structures that resemble the fronds of ferns.

The leaves of *Selaginella* spp. are microphylls, and may be oval to lanceolate in shape. In most species they are arranged in four rows, and in some species (especially those that are dorsiventrally

Figure 22.4
Lycopodium sp. life history.

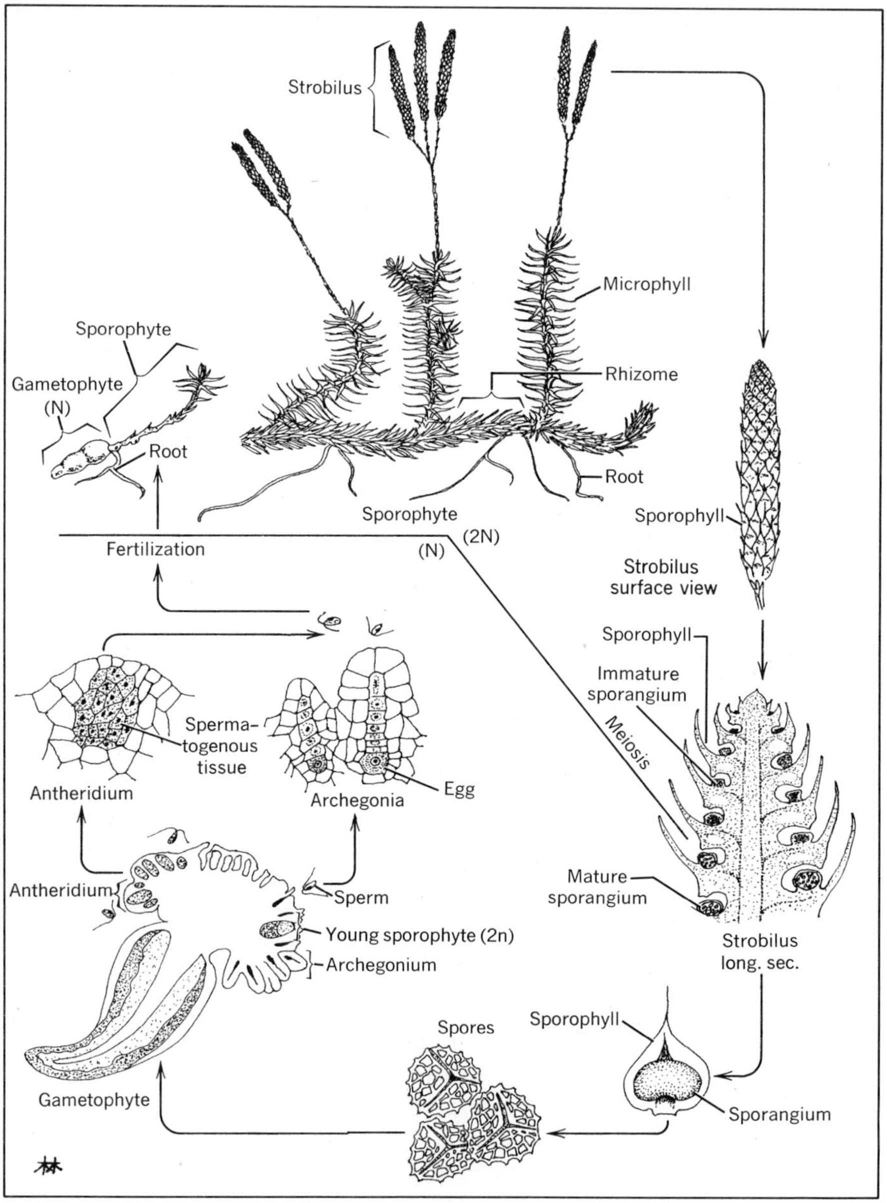

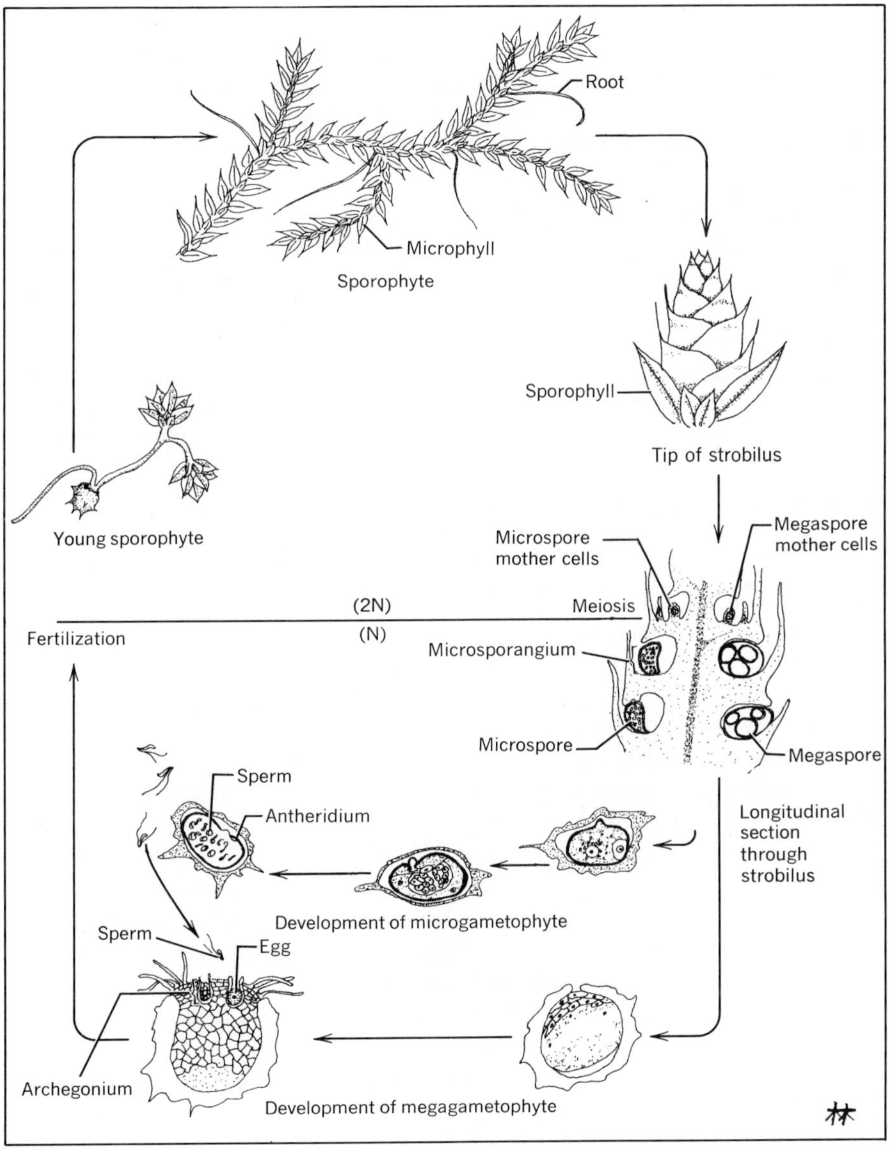

Figure 22.5
***Selaginella* sp. life history.**

flattened) two of the rows are of small leaves and the other two rows are of larger leaves. A small, tonguelike structure (the **ligule**) is found on the adaxial surface of the microphylls. Although the leaves are photosynthetic, the cells of the stem cortex are of at least equal importance in photosynthesis.

The roots are produced at the tips of structures that originate near a branch axis and that extend to the soil (Figures 22.1*b*, p. 326 and 22.5). These structures have been considered to be specialized branches and were given the name **rhizophores.** However, recent studies have indicated that they are really simply adventitious roots that repeatedly branch dichotomously near their tips. The root develops from a single apical cell.

The stem may develop either from a single apical cell or group of cells depending upon the species. At maturity the stem possesses a well defined epidermis, cortex, and stele. These tissues were derived from the apex in the same manner that has been previously described. The vascular arrangement may vary from protostele to siphonostele. Some species may produce a number of concentric bundles of vascular tissue in the stem. These types of stems have been said to be **polystelic,** although interpretations vary. An interesting, and unique, feature in *Selaginella* spp. stems (Figure 22.2*b*) is that the steles are in large air spaces. They are supported in position by radially elongated endodermal cells, called **trabeculae,** which extend from the cortex to the stele across the airspace. The cortex is composed of thin-walled chlorenchyma cells and is extremely important photosynthetically.

The sporangia (Figure 22.3*b*) in all species of *Selaginella* are produced in cones (strobili). The sporophylls that compose the cones may be tightly fitted or they may be loosely compacted and closely resemble the vegetative leaves. However, sporangia are produced on the base of their adaxial surface. The apical meristem of the branch containing the cone may continue its activity and produce a leafy continuation distal to the cone. Because of these factors, the cones are often inconspicuous. The cones of *Selaginella* spp. contain two types of spores (therefore, it is **heterosporous**) that are produced in two types of sporangia. One type of spore is much larger than the other and is known as the **megaspore.** Four is the usual number of megaspores produced in a **megasporangium.** The other type of spore is much smaller, is known as a **microspore,** and hundreds are produced in each **microsporangium.** Because both types of spores are contained on the same plant, the plant is said to be **monoecious.** The arrangement of these two types of sporangia within a cone is considerably varied. Each type of sporangium contains a nutritive layer of cells (the **tapetum**) that is apparently used up during the maturation of the spores. *Selaginella* is the first of the extant "primitive" vascular plants to exhibit heterospory, but the phenomenon becomes much more common as one observes the more "advanced" vascular plants.

Isoetes (Figures 22.1*c*, p. 326 and 22.6) is the final member of the lycopodophytes to be discussed. This genus contains about 60 species and is rather widely distributed in the cooler climates of the world. Most of the species remain immersed in water, but some are terrestrial. Others are immersed part of the time but as their swampy habitats dry up or the water recedes, they become terrestrial (plants with this ability have been called **amphibious**).

The sporophyte is often inconspicuous because its main axis is only a short, stocky, fleshy structure (Figure 22.1*c*, p. 326). The morphological interpretation of the axis is a difficult undertaking that has not been universally agreed upon (Figure 22.3*c*). Essentially, the stele is a central protostele surrounded by an extremely thick cortex. The upper part of the axis produces leaves whose bases are quite close together because of extremely short internodes. The lower part of the axis produces roots. The two portions of the axis seem to be divided by a constriction that occurs just below the center of the cylinder. A unique feature of the axis is that a cambium is produced just outside of the phloem. The morphological interpretation of this cambium tissue is also not understood, but it does produce cells that add to the tissues of the stem. Apparently the cells produced by the cambium become either sieve tubes, tracheids, or parenchyma cells. The parenchymatous cells produced to the outside of the cambium replace the old cortex cells that are sloughed off with the old leaf bases at the beginning of each new growing season.

The leaves of *Isoetes* spp. (Figure 22.1*c*, p. 326) are microphylls although they may be several inches long. They have broad bases that quickly become awl or quill shaped (hence the common name, quillworts). Each leaf contains four air chambers, but the tissue external to the air chambers is chlorenchyma. Although the leaves are the photosynthetic organs of the plant, their bases are often submerged in water or buried in the soil. The leaf bases, therefore, may lack chlorophyll and be photosynthetically nonfunctional. According to

Figure 22.6
Isoetes sp. life history.

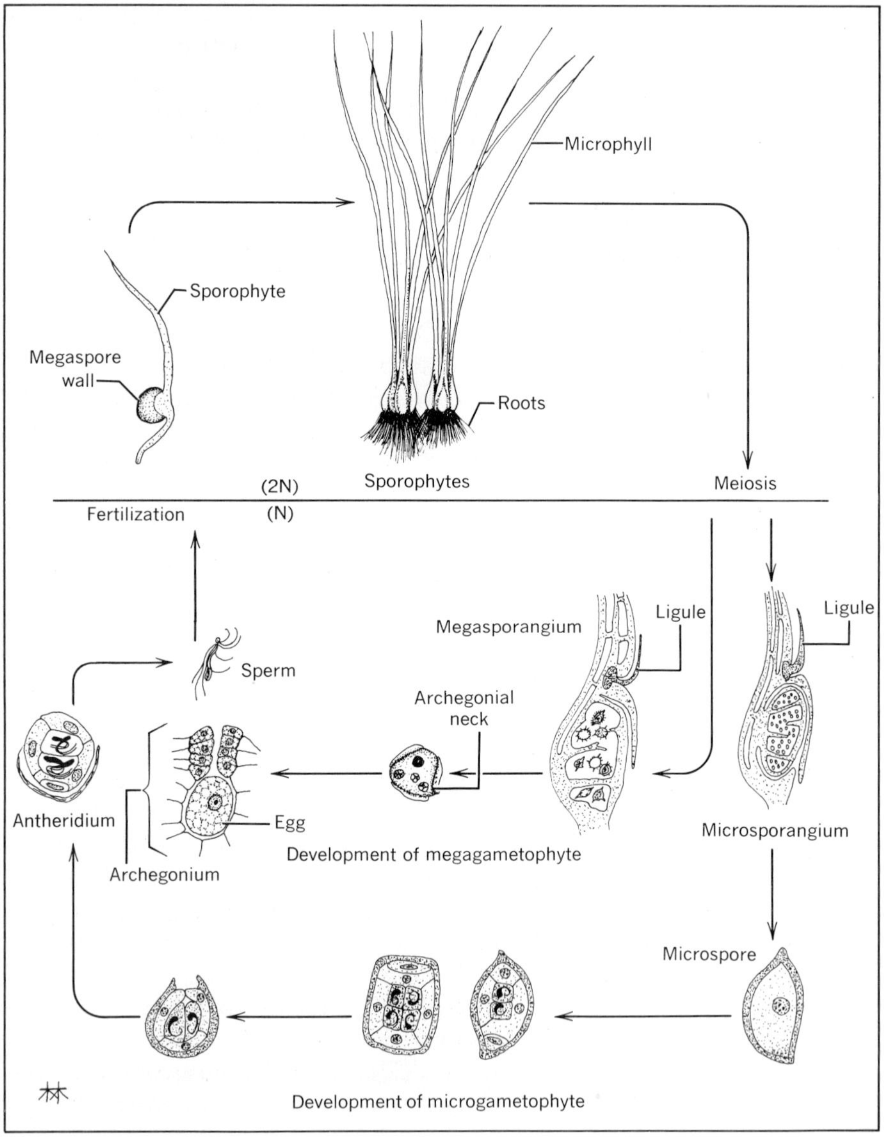

the species, most of the leaves may die at the end of a particular growing season or some leaves may remain attached to the axis.

Although each leaf of *Isoetes* spp. is potentially a sporophyll, the leaves produced earliest in the season are usually sterile. The later leaves produce sporangia on the adaxial surface of their bases (Figure 22.3c). *Isoetes* spp. is heterosporous and monoecious; megasporangia are usually produced previous to microsporangia during the growing season. Leaves produced at the end of the season may be imperfectly developed and produce abortive sporangia. Thousands of microspores are produced in each microsporangium, but the number of megaspores in each megasporangium may number only between 50 to 300.

GAMETOPHYTE GENERATION

The gametophyte of the lycopodophytes is usually inconspicuous and is rarely seen without special effort. The speed of germination of the spores varies greatly among the three genera. *Lycopodium* spp. spores may germinate within a few days or require several years. *Selaginella* spp. spores often begin germination before release from sporangia. *Isoetes* spp. spores germinate only after release, but within a short time thereafter.

Lycopodium spp. gametophytes (Figure 22.4), like those of *Psilotum* grow outside of the spore wall (**exosporic**), and may grow on top of the soil, partially buried, or completely subterranean. The subterranean ones are achlorophyllous, mycorrhizal and saprophytic, but the others are photosynthetic. The gametophytes have an enlarged crown area that produces both antheridia and archegonia (making it bisexual as in *Psilotum* spp.). The large antheridium produces many biflagellated sperm which swim to the egg in the archegonium where fertilization occurs. Several new sporophytes may be produced by one gametophyte, and remain attached to it for quite some time. Eventually, however, the gametophyte disintegrates.

The gametophyte of *Selaginella* spp. (Figure 22.5) results from a new consideration in the vascular plants: heterospory. Because of the new condition, **microgametophytes** (male) are produced from the microspores and **megagametophytes** (female) are produced from the megaspores. Therefore, because the sexes are produced on separate plants, *Selaginella* spp. is unisexual. Also, as contrasted with *Lycopodium* spp., development of both of the gametophytes takes place completely within the spore wall (**endosporic**). They depend for nutrition upon food stored within the spore.

The microgametophyte produces antheridia which in turn produce numerous biflagellated sperm. The megagametophyte produces several archegonia containing eggs. The sperm swim to the egg where fertilization occurs. Fertilization usually occurs after the endosporic gametophytes have been shed from the strobilus although fertilization before shedding has been rarely observed. The young sporophyte soon emerges from the female gametophyte and becomes independent, although the gametophyte may remain attached for a short time.

Isoetes spp. gametophytes (Figure 22.6) are also unisexual and endosporic, but none of their development occurs before the microspores and megaspores are shed. The number of antheridia and archegonia produced is greatly reduced from *Lycopodium* spp. and *Selaginella* spp. The sperms are multiflagellate instead of biflagellate. Only one sporophyte is ordinarily produced from each female gametophyte, and, although it becomes quickly nutritionally independent, it remains attached to the gametophyte for quite some time.

HUMAN RELEVANCE

There are good fossil remains of extinct lycopods and comparisons with extant members of the group aid us in understanding evolutionary trends among the lower vascular plants. The members are, of course, important as autotrophic organisms in that they participate in the various gas exchanges and energy transfers involved in photosynthesis.

The group does not contain any members that are of great economic importance. Various species of *Lycopodium* and *Selaginella* have been used as ground cover and ornamentals in exterior plantings as well as in greenhouse and home display.

Because of its supposed resemblance to pine seedlings, *Lycopodium* spp. have been called "ground pines," and because they sometimes superficially resemble clumps of mosses, they have also been called "club mosses." Certain species of *Lycopodium* are dried, bleached, and dyed to be used by florists in artificial flower arrangements. Such preparations are also used in Christmas and other holiday decorations. Some *Lycopodium* species are also often sold as living aquarium plants and are planted underwater by the aquarium owner. The spores of *L. clavatum* have been used as water repellents and as dusting powder. This plant has also been listed in pharmacopoeias as having medicinal value.

Because of its growth forms the sporophyte of *Selaginella* spp. is often called the small club moss or spike moss. The various species are popular as ground cover and are often used as border plants in greenhouses. Particular species (e.g., *S. lepidophylla*) are sold as novelties as previously mentioned.

The long tapering leaves of *Isoetes* spp. have given it the common name of quillwort, although it is known in Europe as Merlin's Grass. The short, fleshy axis has been used as food by human beings, and it does contain a lot of starches and oils. Domesticated swine have been observed to eat the plant axis, and cows will eat the leaves. The major importance of the genus is probably as a primary producer that is eaten by water birds, muskrats and other aquatic wild animals.

Therefore, although the lycopods are not significantly important as economic plants, they do fill important places in the environment and lend evidence to the understanding of our more advanced plants.

23 horsetails and scouring rushes | arthrophytes

Fossil arthrophytes have been found in rock strata that is about 325 million years old. Later, about 280 million years ago in the Carboniferous Period, the group attained world wide distribution and reached its greatest diversity in form. However, by about 205 million years ago the numbers of forms had been greatly reduced, and in today's modern flora there remains only one genus (e.g., *Equisetum*) containing about 30 species.

A similar situation was noted in the psilophytes and the lycopodophytes. That is, representatives of all three groups are first found about 325 to 350 million years ago, they reach a peak of geographic distribution and diversity in form a few tens of millions of years later, and are finally reduced to a few genera in the modern flora.

Equisetum spp. (Figure 23.1, p. 326) grow in a wide variety of environments. Some can grow in both moist and dry habitats; others are restricted to marshy situations. They are widespread in their distribution, with the temperate-zone species growing to a maximum of four feet in height (although they are usually much shorter) and the tropical species growing taller (to a maximum of about 15 feet with *E. giganteum*).

SPOROPHYTE GENERATION

The major organ of the sporophyte is the stem which is composed of a photosynthetic, erect, aerial portion and an underground, achlorophyllous rhizome (Figures 23.1 and 23.2). Both portions of the stem have well developed nodes and internodes with whorls of scalelike leaves present at the nodes. Branches, in the species in which they occur, are produced alternately with the leaves. Branches are produced in the leaf axils of all other vascular plants that produce branches at the nodes. The stem is ridged as the result of sclerenchymatous tissue that occurs toward the outside of the vascular

Figure 23.2
Equisetum sp. life history.

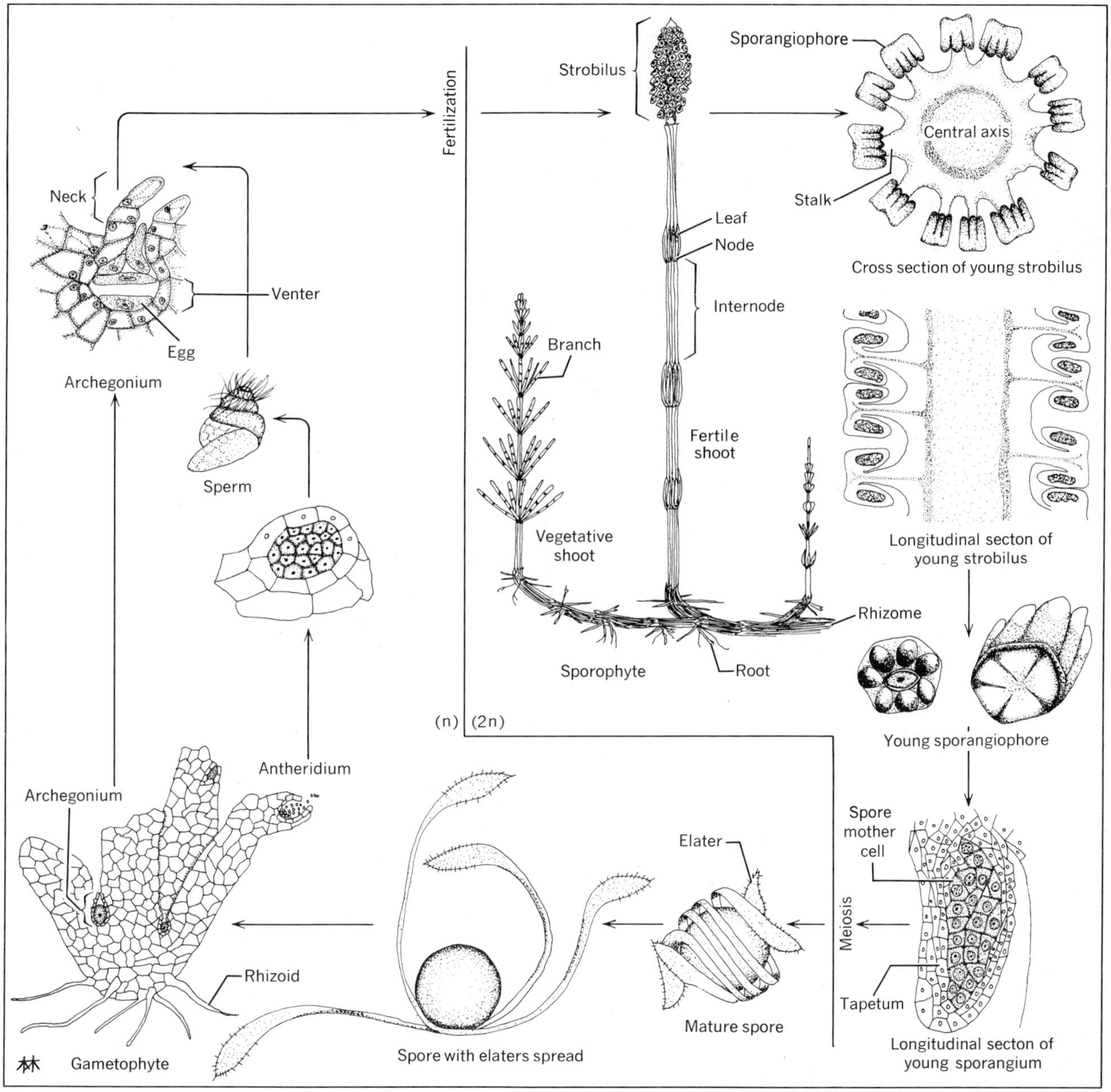

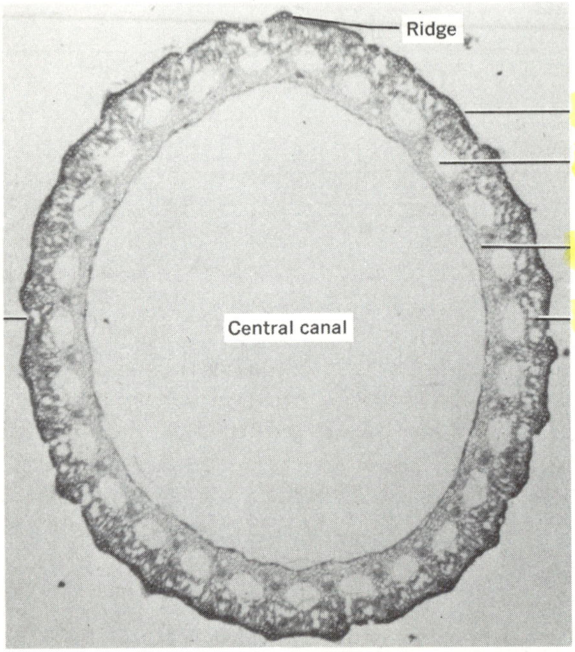

Figure 23.3
***Equisetum* sp. stem. Cross section (30×).**

bundles (Figure 23.3). The walls of the epidermal cells usually contain heavy deposits of siliceous material. Stomata are produced in the epidermal areas between the ridges. The stem is perennial, although the aerial portions may die back each year in harsh environments. New aerial shoots are produced from branch primordia at the nodes on the rhizome.

Roots arise at the nodes of the rhizome and are thus adventitious (Figure 23.2). Branch roots that arise from the pericycle may be produced from the adventitious roots. Roots may also be produced from aerial stems that come to lie on the soil.

Both stems and roots grow from a single apical cell, but the internal structure of the stem is unique (Figure 23.3). At maturity, the stem contains three distinct types of **canals.** The **central canal** occurs in the center of the stem and is surrounded by parenchymatous tissue. External to the central canal is a ring of **carinal canals,** each of which is directly opposite to a ridge on the surface of the stem. The carinal canals mark the position of vascular strands, and also function in conduction. A pericycle occurs just external to the carinal canals, and external to it is produced an endodermis.

The position of endodermal tissue varies with species. In some, endodermal tissue surrounds each vascular bundle; in others, an endodermal layer is produced both internal and external to the row of carinal canals. In still others, only an external endodermis is produced.

The cortex is produced external to the endodermis. It is composed of parenchyma cells, the outer ones of which may contain many chloroplasts. The inner cortex layer contains the third group of canals, the **vallecular canals.** These canals are opposite the depressions between the ridges of the stem.

The vascular bundles in the internodes are discrete, but they unite at the nodes and form a siphonostele. When young, the internodes contain a protostele. **Leaf traces** leave the siphonostele at the nodes and vasculate the scale-leaves. Because no leaf gaps are produced, the leaves are microphylls. They are photosynthetic for a time, but eventually become nonfunctional. By far, most of the photosynthetic activity occurs in the chlorophyllous cells of the outer cortex.

The spores of *Equisetum* spp. are produced in sporangia that are located in strobili at the tips of aerial stems that may be vegetative or strictly fertile. The strobilus is different from the strobili previously discussed because it is composed of a central axis with several shield-shaped, hexagonal sporangiophores (sporangium holder) attached to the axis by a short stalk (Figures 23.2 and 23.4). Five to ten elongate sporangia extend from the undersurface of the sporangiophore toward the axis of the strobilus.

The sporogenous tissue within each sporangium is surrounded by tapetal tissue that provides nourishment for the developing spores (Figure 23.2). The sporogenous tissue undergoes meiotic duplication and hundreds of haploid spores are produced. All species of *Equisetum* are **homosporous;** the spores are released by longitudinal dehiscence of the sporangium. A unique feature of *Equisetum* spores is the presence of four bands of outer wall material, called **elaters,** that are spirally wrapped around each spore. The elaters are attached to the spore wall at a common point. Because they are hygroscopic, the elaters contract and expand with slight changes in humidity: they uncoil with decreasing humidity and coil with increasing humidity. This activity probably aids in their dissemination from the strobilus and their dispersal after they hit the soil. Another unique feature of these spores is that they contain many chloroplasts and are thus quite green when they leave the strobilus.

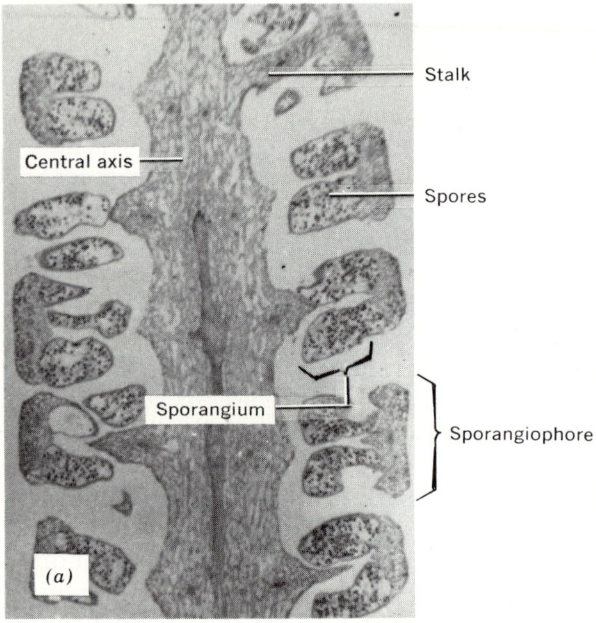

Figure 23.4
Equisetum sp. 1 strobilus (30×). (*a*) Longitudinal section. (*b*) Cross section.

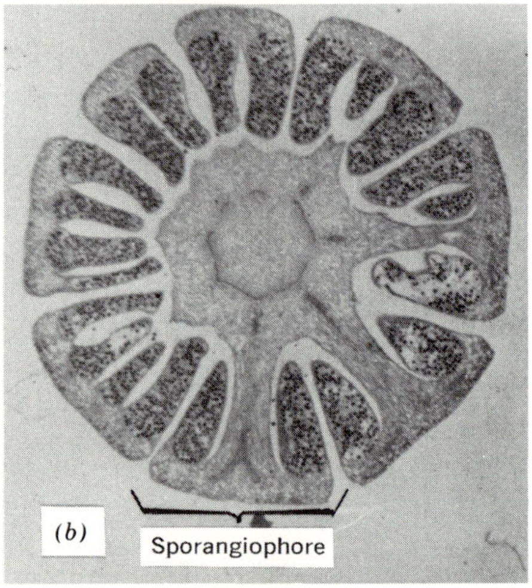

GAMETOPHYTE GENERATION

In a suitable environment, the spores quickly germinate; they are not viable for very long after dissemination. Large clumps of *Equisetum* spp. are probably produced by vegetative (asexual) reproduction. However, the germination of the spores produce an exosporic, cushion-shaped gametophyte that varies greatly in size from that of a pinhead to one that is several millimeters in diameter (Figure 23.2). Rhizoids anchor the plant to the soil.

The chlorophyllous gametophytes produce both antheridia and archegonia on the same plant (bisexual) although the environment may alter the time at which the sex organs appear (Figure 23.2). The age of the gametophyte may also affect the gametangial production (e.g., antheridia may be produced on very young gametophytes). Under laboratory conditions gametophytes from spores of the same species have been observed to produce only antheridia, only archegonia, both, or neither (remain sterile). This sort of varied activity lends confusion to the understanding of gametangial production and should be further investigated.

Large numbers of multiflagellated sperms are produced in each antheridium. They are released by the addition of moisture and swim to the archegonium where they enter the venter and fertilize the egg. Many eggs on the same gametophyte may be fertilized, producing a number of young sporophytes from the same gametophyte. The old gametophyte may persist for some time after the new sporophytes begin to grow. The primary roots grow through the gametophyte tissue into the soil, but do not remain functional for very long. Functional roots are later produced adventitiously from the rhizomes.

HUMAN RELEVANCE

Some botanists believe that members of the genus *Equisetum* may have been present in the Carboniferous in their present form. If this is true, then we have an extant plant group that has undergone very little change during 280 million years of existence. These members, with the fossil remains of those that have become extinct, help us to understand plant evolution.

Equisetum spp. have acquired the common names of "scouring rushes" or "horsetails." These names have undoubtedly been derived from the siliceous materials that are accumulated in their cell walls and that give them a hard, gritty texture. This texture made them valuable to American colonists who used them to scrub their kitchen utensils, hence the name "scouring rushes." The name "horsetails" came from their fancied resemblance to the tail of that noble beast.

Because certain species of *Equisetum* tend to accumulate minerals in their sporophytes, they have been used as indicators during prospecting ventures. The presence of the minerals is determined by either x-ray or chemical analysis. Also, some species are attractive and have been used in quite boggy places as ornamental plants. The plant does, of course, carry on photosynthesis and adds its ecological value in this way to the ecosystem.

Equisetum spp. comes into conflict with man when it grows in and along the banks of his irrigation ditches. Various means to kill the plants and clear the ditches have been found, but the problem is a persistent one.

24 the ferns | pteridiophytes

Unlike the previously discussed groups of vascular plants, the ferns are a widely diverse group that contains over 300 genera and thousands of species. They occupy prominent and important places in the flora of habitats that range from desert to mountain. The greatest diversity in number and morphology is to be found in the tropical rain forests where certain species may be 20 to 60 feet tall. They probably appeared a bit later in geologic time than did the psilophytes, lycopodophytes, and arthrophytes, but good fossil records extend into the Middle Devonian Period (about 285 million years ago).

The ferns may be divided into two classes on the basis of the way in which their sporangia are produced. The first class (i.e., Eusporangiopsida) produces sporangia in the same manner as the psilophytes, lycopodophytes, and arthrophytes; that is, from groups of superficial cells. This class constitutes the fewest and least well-known group of ferns. The other class (i.e., Leptosporangiopsida) produce their sporangia from a single superficial cell and its derivatives. Most of the ferns (about 9500 species) belong to this group, and its members are those that are generally grown for horticultural purposes.

The ferns (Figure 24.1) possess the previously discussed generalized life history of all of the vascular plants. It is within this group, however, that most naturally occurring conditions of apospory and apogamy are found.

Most treatments of the ferns are restricted to members of the leptosporangiate group. Although this is certainly the most important and best-known group, the eusporangiate ferns are also important. Also, within the leptosporangiate group is an often-ignored small group of water ferns. Because the water ferns exhibit the only cases of heterospory among the ferns, they should be at least considered because of their important variation. Therefore, the following treatment will consider examples from each of these three groups.

(a)

(b) (c)

Figure 24.3

Terrestrial fern variations. (a) Tree fern (*Cyathea* sp.) in Golden Gate Park, San Francisco, California. (b) Grape fern (*Botrychium dissectum*). (c) Walking fern (*Camptosorus rhizophyllus*). (d) Staghorn fern (*Platycerium bifurcatum*). (e) Bird's nest fern (*Asplenium nidus*). (f) Boston fern (*Nephrolepis bostoniensis*).

(f)

(d)
(e)

(a) (b) (c)

Figure 24.7
Fern sporangia. (*a*) Interrupted fern (*Osmunda claytonia*). (*b*) Cinnamon fern (*O. cinnamomea*). (*c*) Boston fern (*Nephrolepis bostoniensis*).

Figure 24.1
Fern. Generalized life history.

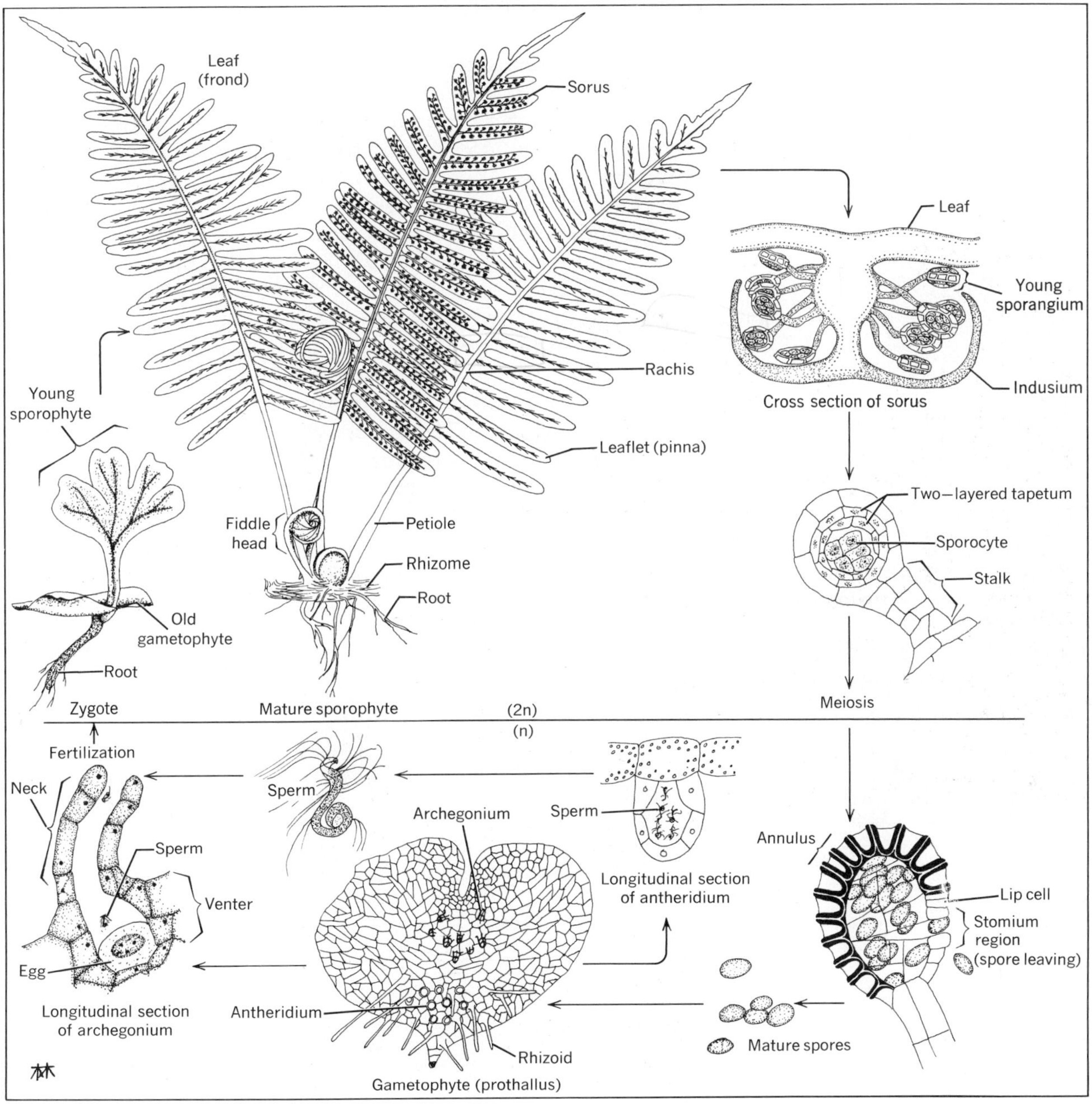

SPOROPHYTE GENERATION

The sporophyte of the various species of ferns varies from the tiny, floating ones of *Salvinia* spp. and *Azolla* spp. (Figure 24.2) to the huge tree ferns (*Cyathea* spp.) that grow to 60 feet tall (Figure 24.3*a*, p. 350). The ferns inhabit a wide range of habitats. They grow on other plants as epiphytes, on the soil, and as floating aquatics. Many of them can grow on almost bare rock and are among the first plants to appear during bare rock succession.

The leaf (**frond**) is the major vegetative organ with most fern sporophytes, and it is in the ferns that the first **macrophylls** among the vascular plants are encountered. A macrophyll (called a **megaphyll** by some botanists) is defined as a leaf whose leaf trace is associated with a **leaf gap** at the point where it branches away from the stele (Figure 24.4*a*). The leaf gap is the interruption, filled with parenchymatous cells, in the stele. In addition, the macrophyll possesses a highly branched vascular system. Some of the genera of ferns have the most complex leaves to be found in the plant kingdom. The leaves of most ferns arise from nodes on rhizomes and are the only portion of the fern usually seen, but other ferns have very large, upright stems.

The stems of the ferns grow from an apical cell and its derivatives in essentially the same manner as described for the psilophytes.

Figure 24.2
Two water ferns, *Marsilea* sp. and *Azolla* sp.

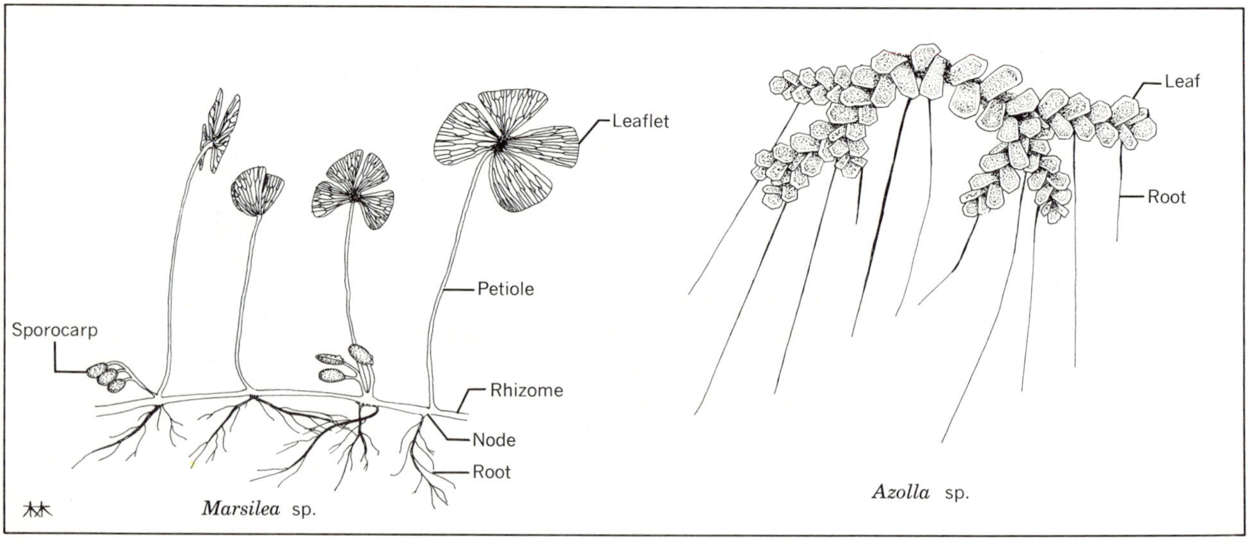

There is no evidence of secondary growth from a vascular cambium in any of the extant or extinct ferns. The protostele is still found among some of the more primitive ferns, but most of them have a siphonostele. The siphonostele may have only an external layer of phloem (e.g., **ecotophloic siphonostele**, Figure 24.4*b*), or it may have both an external layer and an internal layer of phloem (e.g. an **amphiphloic siphonostele**, Figure 24.4*c*). Because the internodes in most fern stems are so short, a cross-section of the stem of many species at any level will show several leaf gaps and leaf traces (Figure 24.4*b*). Based on the amount of dissection caused by the length of the internodes (which allows leaf gaps to overlap), the vascular cylinder may appear to be discontinuous or broken up into discrete strands (Figures 24.4*c* and *d*). It should be emphasized once again, however, that the vascular cylinder is always continuous in some area, and serves as the conducting tissue from root to leaf tip. It can be properly thought of as a tubular network.

The functional roots of the mature sporophytes of the ferns are adventitious. They may arise from the pericycle of a rhizome or an upright portion of the stem (Figure 24.4*c*). If they arise from an upright stem, they often grow through the cortex of the stem back toward the soil. In the case of the tree ferns, this growth through the cortex may cover quite a distance. Root traces are found in transverse sections of stems that have this sort of growth pattern. There is no secondary growth in roots, but lateral roots from the adventitious roots are initiated in the endodermis. The tissues of all of the roots are derived from a large apical cell.

The above generalizations about the sporophyte are true about the vast majority of the ferns. There are, however, some differences among the various fern sporophytes that should be pointed out. These differences occur mostly in the way in which the asexual spore is produced, but other differences will also be noted. The ferns will be divided into three groups. The first of these groups is the **eusporangiate ferns,** the "primitive" ferns. The second group is the **homosporous leptosporangiate ferns,** the most widely dispersed and most generally known group. The third group is the **heterosporous leptosporangiate ferns,** the water ferns.

An example of eusporangiate ferns that is commonly found among temperate zone woodlands is the grape fern (*Botrychium* spp. Figure 24.3*b*, p. 350). Most species of this genus remain less than one foot tall, but some get to be two feet tall under ideal environ-

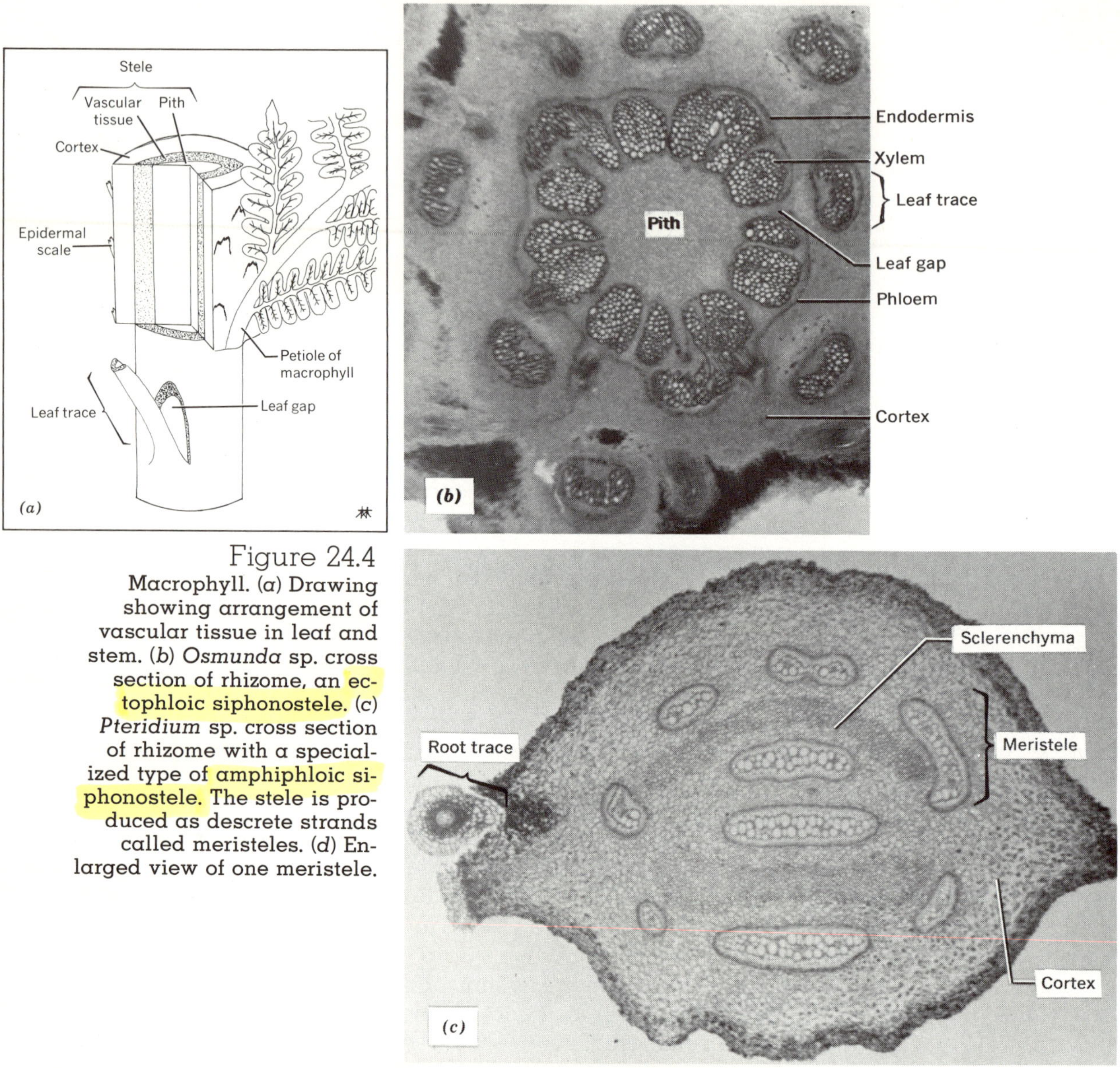

Figure 24.4
Macrophyll. (*a*) Drawing showing arrangement of vascular tissue in leaf and stem. (*b*) *Osmunda* sp. cross section of rhizome, an ectophloic siphonostele. (*c*) *Pteridium* sp. cross section of rhizome with a specialized type of amphiphloic siphonostele. The stele is produced as descrete strands called meristeles. (*d*) Enlarged view of one meristele.

mental conditions. The leaves arise from an underground rhizome, but usually only one leaf is produced during each growing season. The leaves are annual, but the rhizome is perennial. The leaves vary from highly dissected and large to simple and small. The remains of the previous season's leaves surround the tip of the rhizome. A vascular cambium which adds secondary vascular tissue to the stem has been reported for *Botrychium* spp. Because this is in contrast to all other ferns, more critical study is needed.

The spores of *Botrychium* spp. are found in sporangia that are

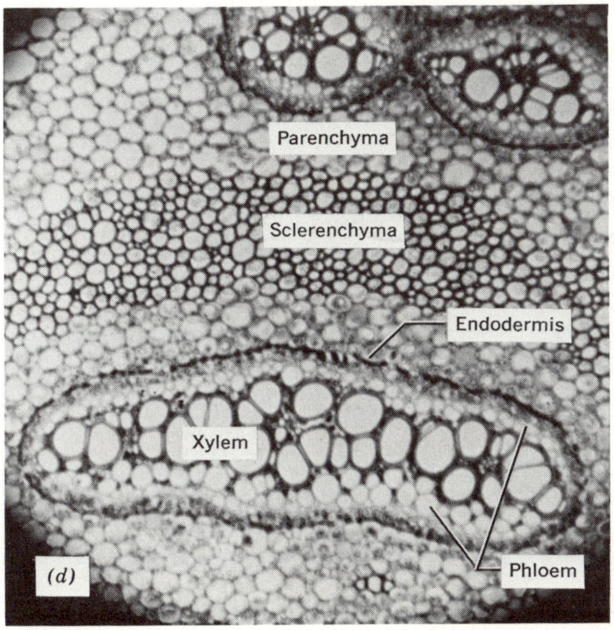

produced on branched or unbranched **fertile spikes.** The spikes arise from the junction of the blade and the petiole of the leaf. The sporangia are arranged close together in rows, and the developing spores derive nutrition from tapetal layers. The haploid spores are released by dehiscence of the sporangium along a predetermined line. Each sporangium may produce as many as 15,000 spores.

The adder's tongue fern (*Ophioglossum* spp.) is a close relative of *Botrychium* spp. that is also found in temperate zones, although in fields and meadows instead of woodlands.

Another group of eusporangiate ferns (the order Marattiales) is exclusively tropical, but they are much more fernlike in their appearance than are *Botrychium* spp. or *Ophioglossum* spp. They have large, fleshy, trunklike stems that may produce huge (20 to 30 feet by 10 feet) fronds. The sporangia are produced on the undersurface of the fronds instead of on sterile spikes. The fronds are coiled like a fiddle-head in this group and unroll in the familiar pattern of most well-known ferns. This arrangement of the leaf in the bud is known as **circinate vernation** (Figures 24.1 and 24.5) and is the most common arrangement among the ferns. *Botrychium* spp. and

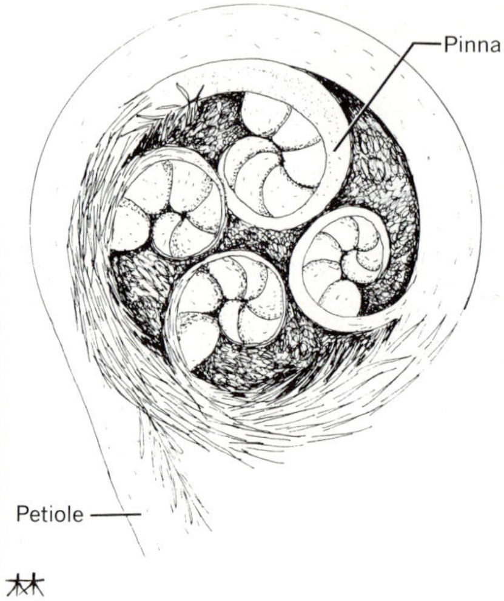

Figure 24.5
A "fiddle-head" illustrating circinate vernation in the pinnae and the entire frond.

Ophioglossum spp. do not, however, exhibit circinate vernation and simply grow straight from the bud.

The most widely distributed and best known ferns are those that are often called the "true" ferns or the homosporous leptosporangiate ferns. There are approximately 9500 species of these ferns that are grouped into about 300 genera. They reach their ultimate variation in the tropics where they occupy environmental niches that include terrestrial and epiphytic. The tree ferns may even produce groves of trees. They are well represented in the temperate zones although the stems are mostly underground rhizomes with the leaves being the only aerial portion of the plant. The division of the homosporous leptosporangiate ferns into taxa is an arduous task that is best left to specialists of fern taxonomy, and they are, by no means, in complete agreement. It will be sufficient for this introductory treatment to simply recognize that most of the homosporous leptosporangiate ferns are included in the order Filicales and the family Polypodiaceae. The examples chosen for discussion are mostly those found in temperate zones although the tree ferns will be mentioned.

The roots, stems, and leaves of this group of ferns are produced in much the same way as was described in the generalized discussion for fern sporophytes. The steles vary from protosteles to highly dissected and specialized siphonosteles. The stems are mostly underground rhizomes from which leaves arise at the nodes. The leaves are the dominant vegetative organ of the plant. All of the functional roots are adventitious, with the exception of the embryonic root.

During their maturation, most fern leaves of this group exhibit the common coiled "fiddle-head" condition (Figures 24.1 and 24.5), but some ferns, such as the walking fern (*Camptosorus rhizophyllus*, Figure 24.3*c*), do not. Although some ferns of the group have **dichotomously branched** leaves (Figure 24.3*d*, p. 350) most are either **simple** (Figure 24.3*e*) or **pinnate** (Figure 24.3*f*). Pinnate forms may be divided many times with their ultimate divisions being known as **pinnules.** The **petiole** attaches the leaf to the stem, and the continuation of the petiole to each of the pinnules (leaflets) of the compound leaf is known as the **rachis.** The venation of the leaf extends from the stele through the petiole and into the pinnules as has been previously described. The venation of the leaves may be quite complex. The more highly divided fern leaves are considered to be primitive, while the simple leaves are considered to be derived or reduced.

The anatomy of a fern leaf is relatively unspecialized. The **mesophyll** may be uniform parenchyma cells, although organization into **palisade** and **spongy** cells does occur (Figure 24.6). An upper and lower epidermis encloses the mesophyll, and in some genera (e.g., *Adiantum*), the epidermal cells contain large numbers of chloroplasts. Stomates are usually produced in the lower epidermis, although in some species the upper epidermis may also have them. The fern leaf is the chief photosynthetic organ of the plant, but in some species (e.g., walking fern), young plants are produced at the tips of the leaves (Figure 24.3*c*) and in others (e.g., *Cystoperis bulbifera*) **bulblets** are produced on the undersides of the leaves.

Although the sporangia of this group are all produced by a single superficial cell (thus are **leptosporangiate**), they vary considerably in their form, groupings, and location on the leaf. Some produce their sporangia on fertile pinnae that differ greatly from the sterile pinnae of the same frond (e.g., Interrupted Fern; *Osmunda claytonia* Figure 24.7*a*, p. 350); others produce sterile and fertile fronds that are separate from each other (e.g., Cinnamon Fern; *Osmunda cinnamomea*, Figure 24.7*b*). Although this **dimorphism** (difference between sterile and fertile portions) is very evident in many members of the group, others produce sporangia on leaves that cannot be distin-

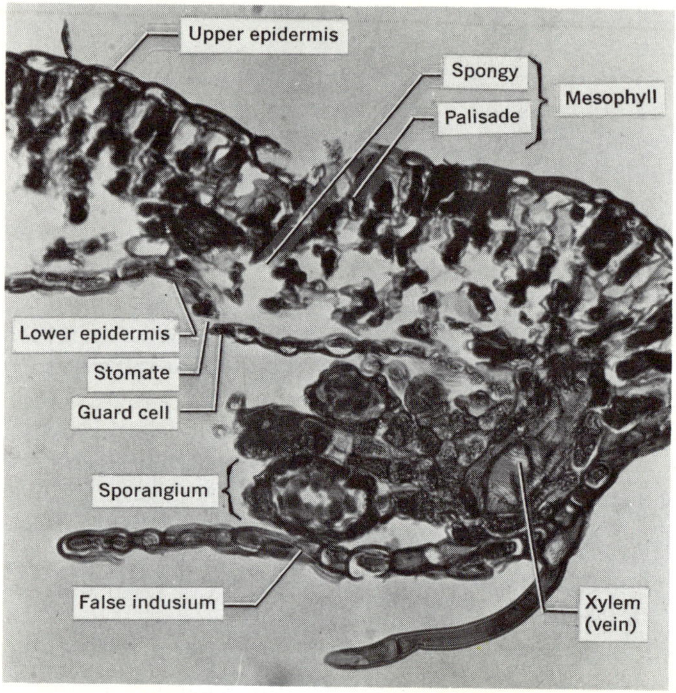

Figure 24.6
***Pteris* sp. frond cross section with marginal sorus.**

guished from purely vegetative ones (e.g., Boston fern, *Nephrolepis bostoniensis*, Figure 24.7*c*). That is, most ferns produce fronds that function both as photosynthetic and reproductive organs. Some of these may exhibit certain differences in the appearance of sterile and fertile portions of the frond.

Most ferns, for example Boston fern, produce their sporangia in groups called sori (Figures 24.1 and 24.6), but some ferns produce their sporangia singly or in marginal zones. The sporangia are commonly located over the end of a vein either at the margin or on the undersurface of the leaf (Figures 24.6 and 24.7*c*, p. 350). The sorus may be naked or covered with a flap of epidermal tissue termed an **indusium** (Figure 24.7*c*). In some ferns, notably *Pteris* spp., the marginal sori are covered by an inrolled portion of the leaf blade called a **false indusium** (Figure 24.6). In some species all of the sporangia within a sorus mature simultaneously; in others, the sori contain sporangia that mature at different times.

The maturing spores within a sporangium are nurtured by one or two tapetal layers. The form of the sporangium (as well as the sorus) varies from genus to genus. However, a very common type of sporangium possesses a partial ring of cells with thickened radial and inner tangential walls (Figure 24.1). This ring of cells is called the **annulus** and aids in the dispersal of the mature spores. When the cells of the annulus lose water through their thin outer tangential wall, they shorten. This shortening process increases the tension on a small group of thin-walled cells, the **lip cells,** located in the **stomium** region between the last annulus cell and the stalk of the sporangium (Figure 24.1). When the tension on the stomium becomes great enough, the lip cells separate and the top of the sporangium is drawn back. Eventually the annulus will cause the top of the sporangium to snap back into place resulting in the spores being "shot" a distance of one or two inches. The process can be easily observed by placing some fresh sporangia under the dissecting microscope and allowing them to dry out. The number of spores in the leptosporangiate ferns is many fewer (48 to 64) than in the eusporangiate ferns. However, the large number of sporangia produced by a single plant allows for the production of a fantastic number of spores.

The third group of ferns, the heterosporous leptosporangiate ferns (Figure 24.2), is composed of aquatic species, although some members may survive on mud when the water level drops. Members of the "water ferns" are distributed throughout the world in tropical and temperate zones. They are not at all fernlike in their superficial appearance and may resemble four-leaf clovers (*Marsilea* spp.) or floating mosses (*Azolla* spp.)

Marsilea spp. (Figure 24.2), produces compound leaves, with four leaflets, that arise from a rhizome growing on or slightly embedded in the mud. The leaves exhibit circinate vernation and may be submerged, floating on the surface, or aerial in the mature plant. The tissues of the vegetative body arise from apical cells as previously described. The stele is an amphiphloic siphonostele. Roots arise adventitiously from the nodes of the rhizome.

Salvinia spp. and *Azolla* spp. (Figure 24.2) are genera that belong to a separate order from *Marsilea* spp. Both are small, floating aquatics, and *Salvinia* spp. is completely rootless. Structures that superficially resemble roots are in reality much reduced leaf segments. *Salvinia* spp. possess leaves in whorls of three. Two of the leaves are floating, but the third is submerged and produces the rootlike segments.

Azolla spp. possesses a very short, branching stem that produces overlapping bilobed leaves. The dorsal lobe is photosynthetic while the submerged ventral lobe is not. The blue-green alga *Anabaena* spp. is usually found in cavities in the dorsal lobe. True roots are produced along the stem. The roots and stems possess protosteles.

The members of both orders of water ferns (*Marsileales* to which *Marsilea* spp. belongs, and Salviniales to which belong both *Salvinia* spp. and *Azolla* spp.) produce their spores in specialized structures called **sporocarps.** Although there are differences that are of evolutionary importance in the ontogeny of these structures, a description of the sporocarp of *Marsilea* spp. will suffice as an example for both orders (Figure 24.8).

The sporocarps are produced on short branches extending from the petiole and they may be produced in clusters or singly depending on the species. Inside the sporocarp, sporangia are grouped in sori that are covered with indusia. Within each sorus, megasporangia are initiated first. Within each megasporangium only one spore (the megaspore) survives although 32 to 64 spores may be produced. The others degenerate and, with the tapetal cells, supply nutritive material to the enlarging megaspore. A large number of starch grains are stored within the megaspore. Microsporangia are initiated after the megasporangia and are much smaller. All of the microspores produced within the microsporangia mature, and although they are much smaller than the megaspores, they also contain starch grains. Thus two types of spores, megaspores and microspores, are produced. These differ considerably in shape, size,

Figure 24.8
Marsilea **sp. development of sporocarp.**

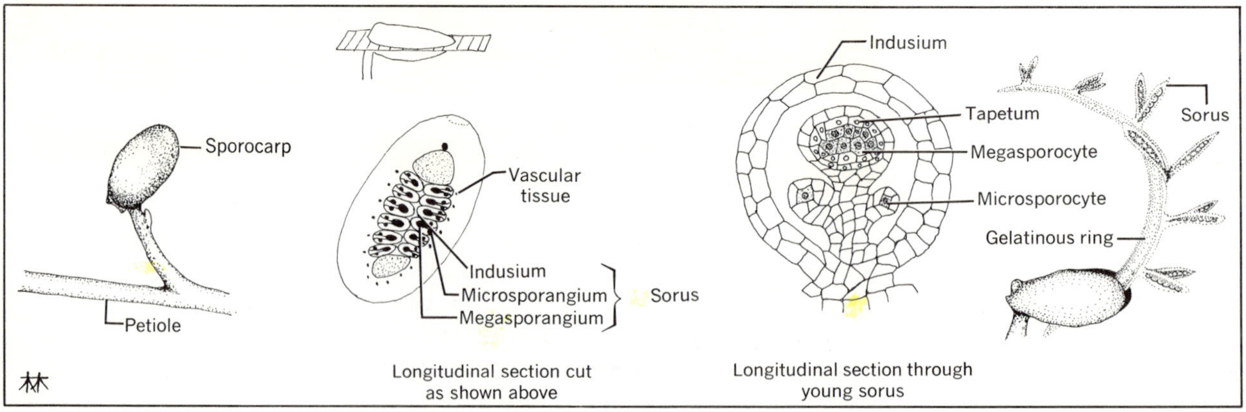

and function and make the water ferns the only group of ferns that contains heterosporous and monoecious members.

After maturation has progressed this far the sporocarp wall becomes very stony. The internal tissues, with the exception of the sori and indusia, gelatinize.

The sporocarp is highly resistant to environmental stresses and may persist buried in the mud for a long while after the parent plant has died. Eventually a portion of the sporocarp wall rots away and allows water to enter. The water causes the gelatinized tissues to expand and creates enough pressure to break open the sporocarp. A gelatinous ring bearing the sori expands into the water (Figure 24.8). Eventually the sporangium wall and the indusium degenerate and release the megaspores and microspores freely into the water.

GAMETOPHYTE GENERATION

With the exception of the heterosporous leptosporangiate ferns, all of the fern gametophytes (**prothallia**) produce antheridia and archegonia on the same plant (Figure 24.1). Environmental factors such as light intensity and light quality apparently influence the speed of germination of the spore and the form of the young gametophyte. Crowding of the gametophytes also seems to stimulate the production of a greater number of antheridia.

The gametophyte of the eusporangiate ferns is non-green and apparently dependent upon an endophytic fungus to maintain a saprophytic existence. It is a fleshy, subterranean, filamentous plant that may be branched and grow to a length of two inches.

The spores of the homosporous leptosporangiate ferns germinate readily when placed under the proper environmental conditions. These gametophytes originate from the spore as filaments of chloroplast-containing cells. Under crowded conditions, the gametophyte may retain its filamentous form. However, when the gametophytes are spaced, they assume the well-known "heart-shaped" form (Figures 24.1 and 24.9). The midrib portion of the gametophyte may be several cells thick, but the outermost part is one cell thick.

Rhizoids in the mature gametophyte are found at the tip of the plant. Antheridia are produced among the rhizoids before archegonia are initiated. Although most of the archegonia are produced near the notch and most of the antheridia are produced near the tip of the prothallium, they are not restricted exclusively to these areas. There is often an intermingling between the two kinds of sex

Figure 24.9
Young fern sporophyte still attached to old gametophyte.

organs in the central part of the prothallium.

Because of the heterosporous condition of the water ferns, as one would expect, unisexual gametophytes are produced. Initiation of the male gametophyte from the microspore and the female gametophyte from the megaspore occurs soon after their release from the sporocarp. As contrasted with the previous two groups where exosporic germination takes place, the water ferns exhibit endosporic germination. The number of antheridia or archegonia on each gametophyte is much reduced among the water ferns.

All of the ferns depend upon the presence of water for fertilization to take place. They all produce flagellated sperm that must swim from the antheridium to the archegonium (Figure 24.1). For a long time, botanists thought that the arrival of a sperm at the neck of an archegonium was by accident because of the large number of sperm produced. It has now been conclusively demonstrated, however, that there is a definite chemotatic response of the sperm to substances produced by the archegonium.

After fertilization, the young sporophyte begins to grow. There is a great deal of variation in the embryological details and in the time necessary to produce an independent sporophyte. The young sporophyte remains attached to the gametophyte until it is photosynthetically independent (Figures 24.1 and 24.9).

HUMAN RELEVANCE

Ferns occupy important positions in the floras found throughout the tropical and temperature zones; indeed, they provide the dominant plant type in some of these areas. They are important, because of photosynthesis, as energy transformers (primary producers) in many ecosystems.

Ferns are beautiful! They provide an enormous amount of variety in anatomical form and growth habit. This variability makes them extremely popular among gardeners and landscape artists. They are used extensively in the plantings of public and private gardens. The graceful tree ferns are used as borders along sidewalks in some areas. Florists use large quantities of fern fronds in floral arrangements. Because many of the ferns can be grown in indirect light, they are often used as house plants or in shady areas of the garden.

Aside from their ornamental value, ferns are of other economic importance. Fibers from old petioles, and pieces of the stem of tree ferns are used for the base on which to grow epiphytes (especially orchids) in plant conservatories. The fibers have been used to stuff mattresses. An essential oil used for perfume oil is extracted from the fronds and rhizomes of some ferns. The male fern (*Dryopteris filix-mas*) has been reported to be an aid in the cure of tapeworm, used as the basis for an ale, and its juices used as a love potion.

Various parts of the fern sporophyte have been eaten by various people throughout the world. Frond tips as well as entire "fiddle-heads" are regularly harvested by some tropical peoples. The **stipules** produced by members of the genus *Marattia* are quite large (about half the size of a fist) and are used by the Central and South American Indians as "corn of the mountains." Stores of grains have been expanded during times of famine in Europe by mixing dried, ground-up fronds of various ferns with the grain. Far from being only an emergency or "primitive" type of food, however, ferns have provided the basis for ecstatic gourmands the world over. "Fiddle-heads" have been imported to some of the best restaurants to be elegantly served as "exotic" food.

Within the past several years, botanists have begun to recognize ferns as valuable research tools. Much work has been done with fern stem apices of the sporophyte in the elucidation of growth and differentiation of plant tissues. Microsurgical methods have been developed that help us to understand the influence of physiology on morphology at the cellular level. Study of the haploid gametophyte has produced important data concerning sexual reproduction, the interactions of light and various internal chemoregulators on seed and spore germination, as well as other physiological, genetic, and evolutionary phenomena. The natural occurrence of the development of a sporophyte from a gametophyte without fertilization (apogamy) and the development of gametophytic tissue directly from sporophyte tissue (apospory) lends evidence to the understanding of the evolution and interrelationships of the sporophyte and gametophyte generations. Two theories have been derived: the **antithetic** or **interpolation theory** and the **homologous theory.** The former theory states that the sporophyte and gametophyte are completely different forms of the same plant. The sporophyte has entered the life history by being interpolated between two successive gametophyte generations by the delay in meiosis. The diploid zygote, instead of undergoing immediate meiosis, undergoes mitosis and produces a diploid plant. The homologous theory states that the sporophyte and the gametophyte are essentially similar. The sporophyte is considered to be simply a modification of the gametophyte. Neither theory is completely satisfactory and experimental evidence is lacking. Botanists vary in their degree of acceptance of either, both, or neither of these theories, and we are faced with the very real possibility that different life history patterns may have arisen in different plants (and animals also) as the result of differing geologic times. The evidence is sketchy, and the conclusions are speculative, but the possibilities are intellectually intriguing.

Extensive interspecific hybridization occurs naturally and can be induced artificially among some ferns, especially *Asplenium* and *Dryopteris*. This hybridization is frequently followed by chromosome doubling. Relatively high chromosome numbers are found among the pteridiophytes. Many ferns have haploid numbers of 100 or more, and one report for *Ophioglossum reticulatum* indicates a diploid number of 1260. These factors lend themselves to speculation about genetic mechanisms, and the ferns provide convenient tools with which to experiment.

25 the naked seeded plants | gymnosperms

The gymnosperms are those plants that have "naked seeds." This means that their seeds are not produced in an enclosure such as the fruit produced by the angiosperms. The seed of a gymnosperm is produced uncovered on some sort of specialized branch. Several of these branches may be grouped together and form a cone (e.g., pine), but no complete enclosure is ever produced.

The gymnosperms are the first plants discussed to possess seeds (Figure 25.1). Because it allows the embryo to survive over long periods of time and extremely unfavorable environmental conditions, and because it is such an important factor in plant dispersal, it is important that the student gain an understanding of the seed. Almost all seeds possess three types of structures, a covering (**seed coat** or **testa**), a young sporophyte (the **embryo**), and a food supply (**female gametophytic tissue, endosperm** and/or **cotyledons**). Indeed, a seed has been poetically described as "a baby plant in a box with its lunch."

The seed coat may be soft or hard, thick or thin, permeable or impermeable to water, dry or fleshy. The possible combinations of these characteristics allows for a great deal of variation. The seed coat does function, however, as a protection to the structures that it contains. The part of the seed that continues to grow upon germination is the embryo. The embryo results from the zygote that is produced when the egg is fertilized. Therefore the zygote is the beginning of the next sporophyte generation.

The food supply is the source of energy for the start and continuation of growth of the young sporophyte until it can become photosynthetic and begin producing its own food. The sporophyte of the gymnosperms grows by digesting the female gametophytic tissue surrounding the sporophyte. The sporophyte of the angiosperms does this also very early in the seed's development, but it has two additional sources of

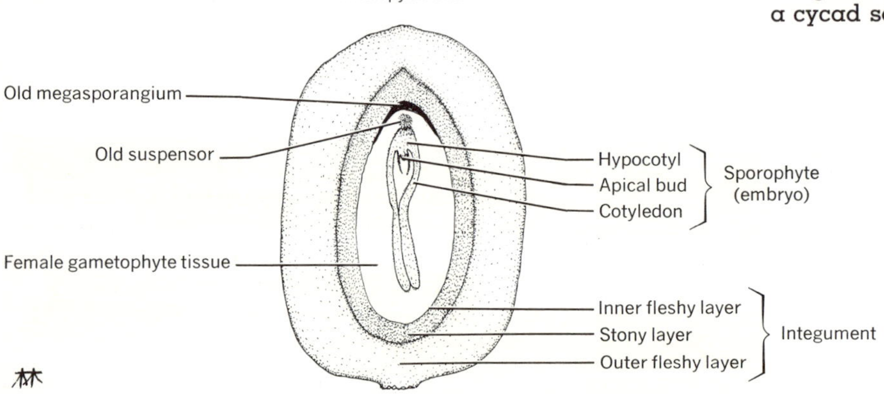

Figure 25.1
Longitudinal section through a cycad seed.

food. These are the **endosperm,** which is produced because of a fertilization in addition to that of the egg, and sometimes the **cotyledons,** which are part of the embryo. Angiosperm seeds will be discussed in more detail at the proper time.

THE CYCADS—CYCADOPHYTES

SPOROPHYTE GENERATION

The cycads are a small group of gymnosperms composed of only 9 or 10 genera and about 100 species. Of these only one, *Zamia* (Figure 25.2*a*, p. 374) is found in the United States and it is confined to Florida. Other members are tropical. Members of the genera *Cycas* (Figures 25.2*b–f*, p. 374) and *Dioon* (Figures 25.2*g* and *h*, p. 374) are often found as ornamental plants and in plant conservatories. Cycads have been found as far back as the Lower Cretaceous Period (125 million years ago), but members of the Cycadophyta are known as far back as the Devonian Period (255 to 275 million years ago).

The sporophyte of the cycads superficially resembles both palms and ferns and, indeed, *Cycas revoluta* (Figure 25.2*e*, p. 374) is often commonly and erroneously called the "Sago Palm." This superficial comparison to palms is understandable because of the large, dissected (pinnatifid) leaves that originate as a crown at the top of a stem. The pinnatifid leaves and the possession of circinate vernation of the leaves and their leaflets explain why some of the cycads are confused with ferns.

The stems of the cycads are usually short and columnar. Most cycads are less than 10 feet tall, but some (*Macrozamia hopei*) may reach almost 60 feet. The apical meristem has the shape of a rounded cone and is used up in the production of permanent structures when an apical cone is produced. A new meristem arising from its base produces another meristem and a new crown of leaves. The leaves are initiated from periclinal duplications in the two outer cell layers of the meristem. Leaves are produced in a closely compacted spiral arrangement at the apex and receive several traces from the stele. These traces form wide girdling traces that are easily observed in cross sections of the stem (Figure 25.3*a*). The traces arise from various places on the stele, and may travel 180° around the stem before they enter the leaf base. The traces leave many gaps in the stele. Therefore the leaves are macrophylls. Under good growing conditions *Cycas* and *Zamia* produce a new crown of leaves each year. As the leaves reach maturity their epidermal layers become thickened, and the leaves become quite leathery and stiff. Leaves may persist for more than one year, but most of the previous year's leaves drop off leaving the stem covered with a thick armor of persistent leaf bases (Figure 25.2*g*, p. 374).

Most of the tissue in the stem is primary. A very sluggish vascular cambium is produced, but it produces very small quantities of secondary xylem and phloem. Therefore, most of the stem is composed of the central pith and a massive cortex (Figure 25.3*a*). The cells of the cortex are important storage organs. **Mucilage canals** are found in both the cortex and the pith.

The roots of the cycads are fleshy (Figure 25.3*b*), but the vascular cambium is more active than in the stem. Therefore, more secondary vascular tissue is produced. In some cycads (i.e. *Cycas*) some of the roots grow above and on the soil. These roots develop nodules which contain bacteria and the blue-green alga *Anabaena*. Such symbiosis is probably a source of nitrogen for the cycad.

All of the cycads produce spores that produce male gametophytes on plants that are different from the ones on which female gametophyte producing spores are borne (Figure 25.2, p. 374). This condition, which has been previously described, is called dioecious (meaning "two houses"). The spores that produce the male gametophyte are known as microspores; those that produce the female gametophyte are megaspores. The male and female spore producing plants cannot be identified during their vegetative phases (Figure 25.2*e*, p. 374).

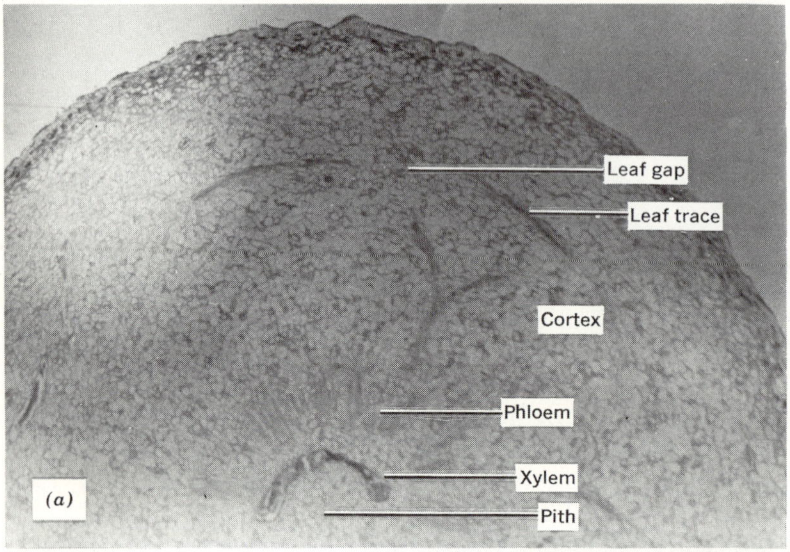

Figure 25.3
Zamia sp. (*a*) Cross section of a mature stem (5×). (*b*) Cross section of a mature root (12×).

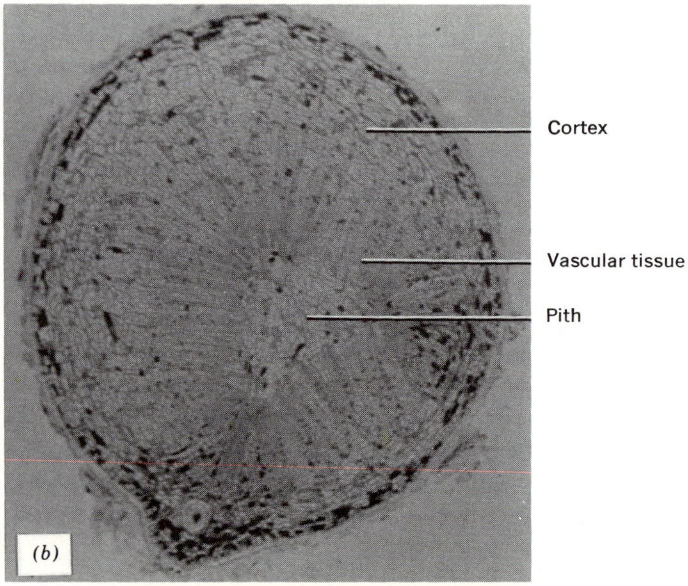

Microspores are produced in eusporangiate **microsporangia** that are located on specialized leaves called **microsporophylls** (Figure 25.4*a*). Many microsporophylls are compactly organized into a "cone" or **microstrobilus** (Figures 25.2*c*, *d*, and *f*, p. 374). The sporangia are produced on the abaxial surface of the microsporophyll. They are clustered together in a manner resembling the sori of ferns. The elongate strobilus (8 to 10 inches) is composed of a central axis to which are attached many spirally arranged microsporophylls (Figure 25.5*a*). The microsporophylls are narrow at their point of attachment to the axis but broaden to an expanded, sterile tip. The number of microspor-

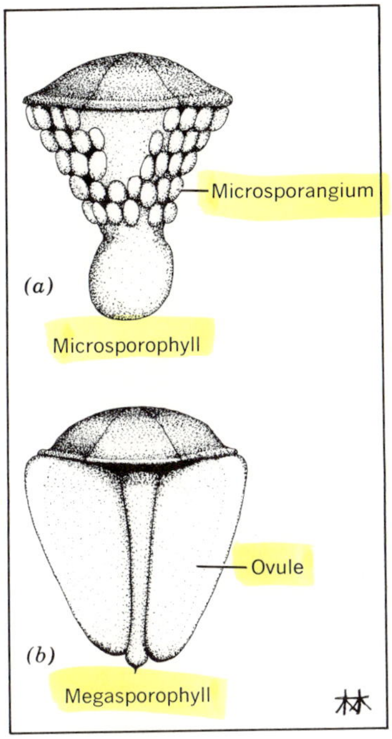

Figure 25.4
Zamia sp. (a) A microsporophyll with microsporangia. (b) A megasporophyll with two ovules.

angia for each microsporophyll ranges from a few in *Zamia* (Figure 25.4*a*) to over a thousand for *Cycas*. As the microsporangium reaches maturity, each of its microsporocytes (Figure 25.5*b*) undergoes meiosis and produces a tetrad of haploid microspores. The outer layer of the sporogenous tissue or the inner layer of the epidermis undergo change to a nutritive layer called the tapetum. The tapetum nourishes the developing microspores. The number of microspores produced in each sporangium varies from 500 in *Zamia* to 30,000 in *Dioon*. Most genera produce only one microstrobilus at a time (i.e., *Dioon*, *Cycas*), but others (i.e., *Zamia*) may produce three or four simultaneously.

Megasporophylls among the cycads vary a great deal in size, shape, and arrangement. For example, in *Zamia* the megasporoophylls are dark-brown peltate structures that are produced from a central axis (Figure 25.4*b*). Each of the megasporophylls produces two orange ovules on its adaxial surface. The strobilus of *Zamia*, therefore, is a tightly packed, easily distinguishable structure. In *Cycas revoluta* however, the megasporophylls are pinnatified, leaf-like structures that produce 6 to 8 ovules on their lateral surface. The sporophyll of *C. circinalis* (Figure 25.6) also usually produces 6 to 8 ovules, but it is not as pinnatifid. One of the largest cones to ever be produced is the female strobilus of *Dioon spinulosum* (Figure 25.2*h*). Most cycads produce only one megastrobilus; a few produce groups of two or three (Figures 25.2*b*, and *h*).

The very young ovule is composed of a massive megasporangium (**nucellus**) surrounded by one fleshy integument and one stony integument (Figure 25.7). The integument is incompletely closed at the end opposite its point of attachment, and a small opening, the **micropyle**, remains. Later, one of the cells buried deeply within the megasporangium enlarges and functions as a megasporocyte. Meiosis of this cell produces a linear tetrad of four haploid megaspores. Of these, only the one farthest from the micropyle becomes functional; the other three degenerate. The functional megaspore enlarges by using the degenerated megaspores and the surrounding megasporangial tissues as nutrients. Thus, at maturity, the megasporangium is much reduced. It is the functional megaspore which produces the female gametophyte.

GAMETOPHYTE GENERATION

The production of the gametophytes in the cycads is similar to the gametophyte production in *Ginkgo biloba* (Figure 25.10). The micro-

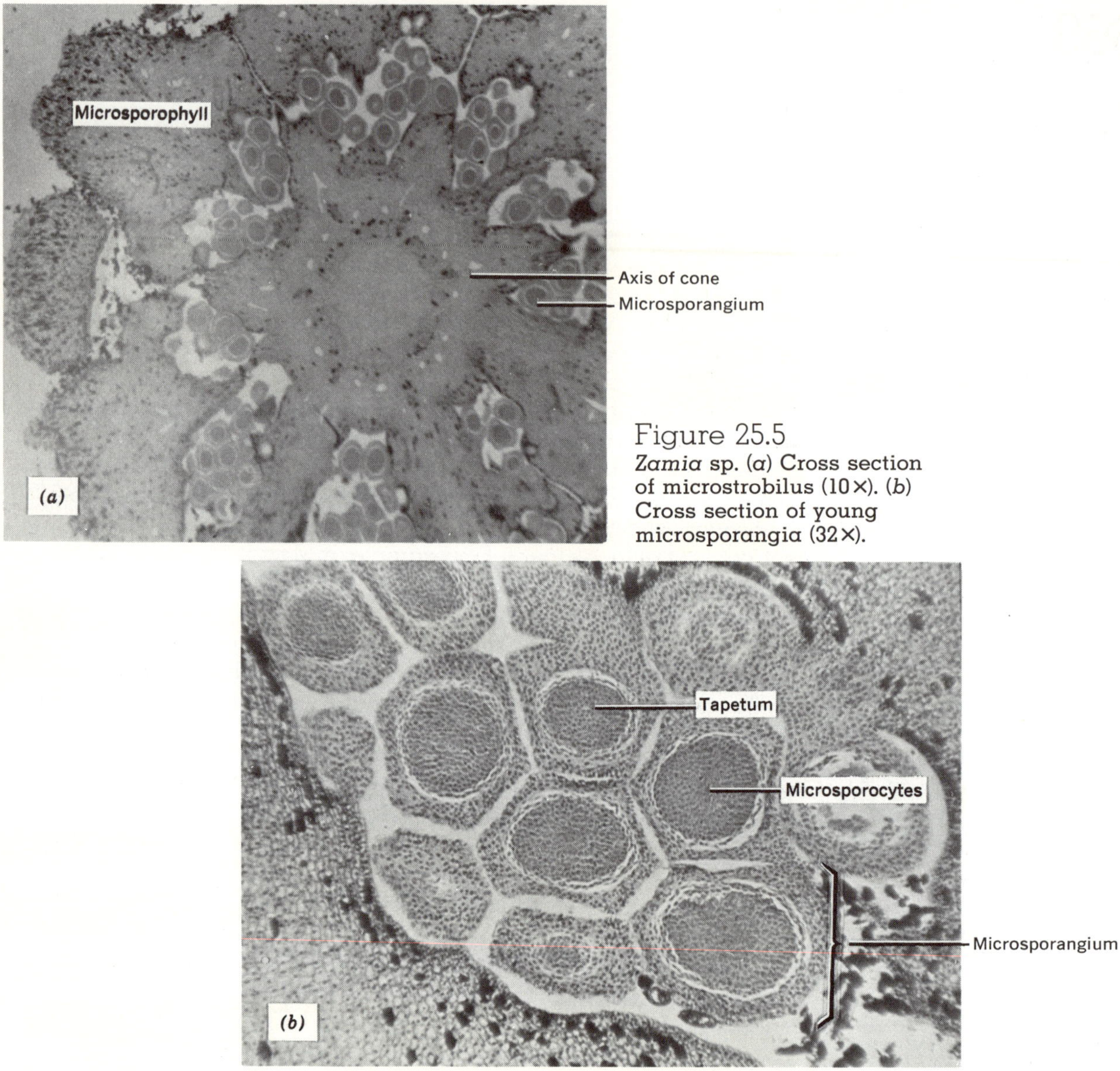

Figure 25.5
Zamia sp. (*a*) Cross section of microstrobilus (10×). (*b*) Cross section of young microsporangia (32×).

spore begins to produce the male gametophyte even before it leaves the microsporangium. Each of the microspores is uninucleate after the tetrad of microspores is produced from the microsporocyte. This nucleus soon carries on mitosis which is followed by cytokinesis. This action produces a basal prothallial cell and a larger cell. Karyokinesis followed by cytokinesis in the latter produces a small **generative cell** and a large **tube cell.** All of this development takes place inside of the wall of the microspore; thus the development is **endosporic.** At this stage the microspores are shed from the microstrob-

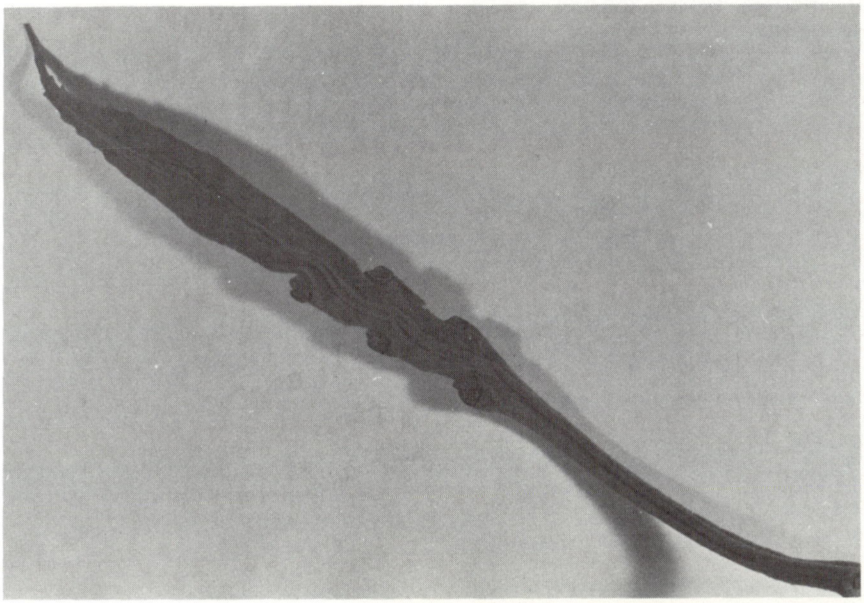

Figure 25.6
***Cycas circinalis*. A megasporophyll with five ovules (0.5×).**

ilus into the wind. The microspore wall containing the three-celled male gametophyte is called **pollen** and the process of transfer to the female strobilus is called **pollination.** The megastrobilus opens slightly during pollination and the pollen grains sift to the micropylar end of the ovules. A droplet of fluid, called the **pollination droplet,** results from the breakdown of a few of the cells of the megasporangium. The pollen grain sticks to the drop and is pulled into the cavity produced by the disintegrated cells as the pollination droplet evaporates. This cavity is called the **pollen chamber** (Figure 25.7*b*). Many pollen grains may adhere to the same pollination droplet, and several may be drawn into the pollen chamber.

An exosporic portion of the male gametophyte is now produced (Figure 25.7*b*). A tube begins to extend from the end of the pollen grain containing the tube cell. This tube invades the tissue of the megasporangium and forms an haustorial relationship with the tissue. These are called **pollen tubes** and they absorb some of the material digested from the megasporangium tissue. The generative cell divides again, producing a **stalk cell** and a **body cell.** The latter cell finally divides once again and produces two multiflagellated **sperm cells.** The flagella are arranged in a spiral around one end of

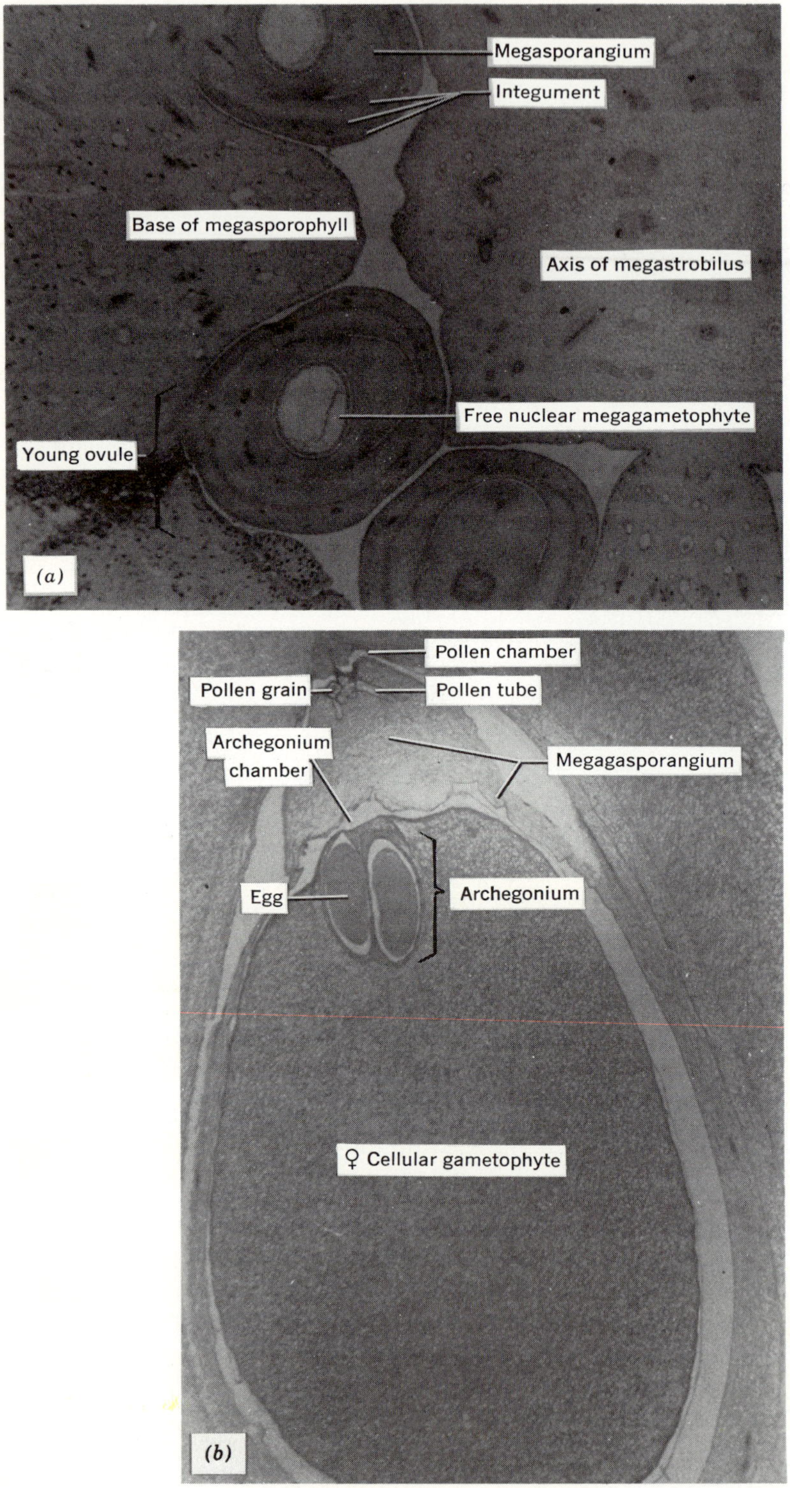

Figure 25.7
Zamia sp. (*a*) Cross section of young megastrobilus (9×). (*b*) Longitudinal section of young ovule (7.5×).

the sperm. The mature sperm are the largest ones known in the plant kingdom (300 microns in diameter). They actively swim down the pollen tube. Several pollen grains may grow from a single pollen chamber.

The female gametophyte is a bit more complicated than the male gametophyte. All of its activity is, however, contained within the megasporangium. The megaspore increases in size and its closely surrounding tissues become vacuolated. The megaspore undergoes a number of **free nuclear duplications** until 1000 free nuclei may be produced (Figure 25.7*a*). The vacuolated tissues surrounding it produce one or more layers of cells that surround the young gametophyte. These layers, the **"endosperm jacket,"** probably function as a tapetum and supply food materials to the gametophyte from the old megasporangial tissue. Eventually, walls that enclose individual nuclei and associated cytoplasm are produced and a cellular gametophyte is formed (Figure 25.7*b*).

A few cells at the micropylar end of the gametophyte now begin to produce archegonia (Figure 25.7*b*) as in *Ginkgo*. The first cell division of these archegonial initials produces a **primary neck cell** and a large inner **central cell.** The central cell enlarges and the cytoplasm becomes vacuolated. The gametophyte cells that surround the archegonium form a jacket that aids in the movement of food from the gametophyte to the archegonium. The central cell nucleus eventually duplicates producing a **ventral canal nucleus** and an **egg nucleus.** The former nucleus soon disintegrates. At maturity, then, the large archegonium contains a huge egg cell (3.0 × 1.6 mm.) with a nucleus that may be 550 microns in diameter. A single female gametophyte may contain from 3–10 functional archegonia. Although *Zamia* spp. usually has only two. From even this elementary discussion of the male and female gametophytes in the cycads, one can easily see that they are greatly reduced from the gametophytes of the mosses or the ferns.

Fertilization occurs in some of the cycads 4 to 6 months after pollination. During the final maturation of the archegonia, the micropylar end of the female gametophyte becomes depressed. This creates a small chamber (the **archegonial chamber**) between the gametopyte and the overlying megasporangial tissue. When the pollen tubes enter this chamber, their ends rupture, and the sperm cells and some liquid are released into it. The sperm actively swim about and exhibit some ameboid movement. Eventually they enter

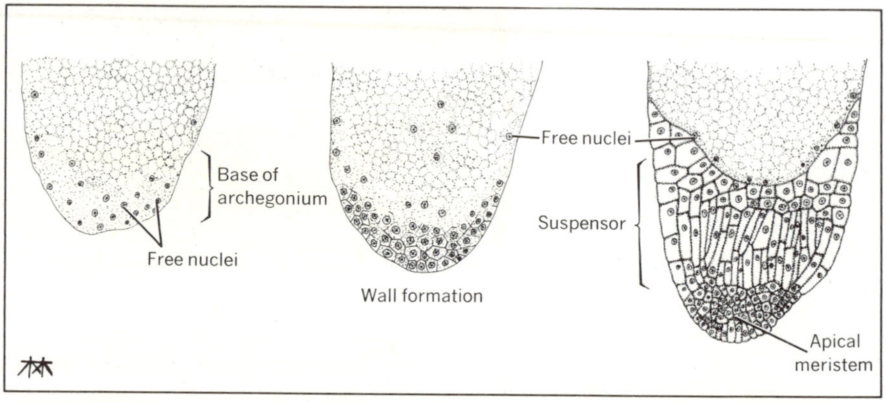

Figure 25.8
***Zamia* sp. Early embryo development inside of the female gametophyte.**

the archegonium and the spiral of flagella is shed. The sperm nucleus then fuses with the egg nucleus, accomplishing fertilization. Many sperm may enter the archegonium, but only one fuses with the egg. The others finally disintegrate.

After fertilization, the zygote undergoes a series of free nuclear duplications producing from 256 to about 1000 nuclei. Wall formation begins at the base of the archegonium and the individual nuclei with some of their associated cytoplasm are enclosed within single cells (Figure 25.8). Some genera retain the upper part of the embryo in the free nuclear state; others become completely cellular. The free nuclear portion finally disintegrates. The cellular portion of the embryo becomes meristematic and produces two separate areas. The region toward the micropyle produces elongating cells that extend through the archegonium into the old female gametophyte tissues. These elongating cells produce an organ known as a suspensor that aids in the digestion and absorption of food materials from the gametophytic tissue (Figure 25.8). The other part of the embryo is pushed down into the nutritive tissue of the gametophyte; it is from this area that the main organs of the embryo arise. The embryo uses the female gametophytic tissue as a food source. The young embryo produces two **cotyledons** which are on either side of the apical bud (**plumule**). The portion of the embryo that is closest to the suspensor produces the embryonic **root** or **radicle.** The axis of the embryo is composed of a **hypocotyl** that extends from the cotyledonary node to the radicle. Although fertilization of more than one egg may occur in a particular gametophyte, only one embryo usually reaches maturity; the others abort (Figure 25.1).

Figure 25.2
Cycads. (*a*) *Zamia florida*. A young vegetative plant. (*b*) *Cycas mexicana*. A female plant with strobili around its base. (*c*) *C. mexicana*. A male plant with a strobilus at its base. (*d*) *C. mexicana*. A single male strobilus with mature microsporangia visible. (*e*) *C. revoluta*. View into crown of vegetative plant. (*f*) *C. circinalis*. Top of tall male plant with mature strobili present. (*g*) *Droon edule*. Young female plant with persistent leaf bases and young strobilus. (*h*) *D. spinulosum*. Mature female plant with huge strobilus.

(a)

(b)

Figure 25.9
Ginkgo biloba. **(a) Three female sporophytes. (b) Branch with bilobed, fan-shaped leaves. (c) Short shoot with megastrobili. (d) Short shoot with microstrobili. (e) Branch with short shoots bearing nearly mature seeds.**

(c)

(

(e)

(a) (b) (c)

Figure 25.18
(*a*) *Podocarpus macrophylli*. Insert: Branches with young female cones. (*b*) Red pine (*Pinus resinosa*). Branch in early spring with new terminal growth, many microstrobili and one megastrobilus. The strobili were initiated during the previous growing season; the megastrobilus is ready for pollination. (c) *Pinus* sp. Branch in early spring with new terminal growth and three megastribili. The two larger strobili were pollinated during the previous spring and fertilization will occur this spring. The smaller strobilus is ready for pollination. (*d*) Yew (*Taxus canadensis*). Female plant with some ovules surrounded by bright red arils. (*e*) Yew. Male plant in early spring with microstrobili. (*f*) Monkey puzzle tree (*Araucaria araucana*). Young tree. Insert: Branch with stiff, flat leaves. (*g*) Coast redwood (*Sequoia sempervirens*). Logs on trucks.

(e)

(f)

(g)

Figure 25.12
Representatives of the conifers. (*a*) Eastern white pine (*Pinus strobus*). Insert: Branches with mature female cones. (*b*) White spruce (*Picea glauca*). Insert: Branches with mature female cones. (*c*) White fir (*Abies concolor*). Insert: Branches with flat, upright leaves. (*d*) European larch (*Larix decidua*). Insert: Branches with mature female cones. (*e*) Bald cypress (*Taxodium distichum*). Insert: Branches with mature female cones. (*f*) Northern white cedar (*Thuja occidentalis*). Insert: Branches with mature female cones. (*g*) Eastern red cedar (*Juniperus virginiana*). A shrub form.

(a) (b) (c) (d)

(*d*)

(*e*)

(*f*)

(*g*)

During this time the entire ovule has been increasing in size and the integuments produce a three-layered seed coat. Thus the mature cycad seed is composed of an embryonic sporophyte that is buried in female gametophyte tissue surrounded by the remains of the old megasporangium. The latter two tissues are used as a food supply. Finally, the massive seed coat surrounds all of the tissues, and the mature seed is ready to be shed (Figure 25.1).

There appears to be no fixed dormancy period in cycad seeds; if the environmental conditions are right, the seeds may germinate immediately after being shed. The stony integument of some cycads, particularly *Zamia* spp. restricts germination. Scarification will increase the germination rate to nearly 90% as well as speed up the germination time. This appears to happen only rarely under natural conditions, thus the natural germination rate of *Zamia* is very low. The seed germinates through its micropyle by first putting forth its radicle which is enclosed in a protective sheath of cells, the **coleorhiza.** The cotyledons continue to absorb food from the surrounding tissues for a few weeks. Eventually the shoot apex with its leaves appear. However, most of the cotyledons remain inside of the seed and absorb the remaining food materials.

HUMAN RELEVANCE

The cycads do not occupy a very important position in the present day flora, but they were extremely important during the Mesozoic Era. There is evidence that during that time the young leaves provided food for the dinosaurs. Because of the length of time that members of the group have been on earth, the present day representatives are regarded as living fossils. They present us, then, with some idea as to how the ancient plants looked and functioned. They also help us to better understand the fossil remains of other ancient plants.

About the only commercial use of the group is as ornamentals. Although *Zamia* is the only member that is native to the United States, *Cycas* will grow quite well when planted outdoors in Florida, Louisiana, and southern California. Other genera are grown in green-houses and plant conservatories. Potted *Cycas* is the plant often seen decorating the stages of conventions and graduations and has been sometimes used as "palm" leaves on Palm Sunday. It has been reported that the natives of Asia use *Cycas* in the preparation of a vegetable gum. The starchy rhizone of *Zamia* spp. contains cycasin, a water soluble toxin. However, after a time-consuming process the cycasin can be removed and the rhizome used as a flour. The pre-Seminole tribes of Florida and the later Seminole tribes used the rhizome in this manner and knew *Zamia* as the white bread plant. The present-day Seminole people still process the rhizome by the ancient method and use it in much the same manner. The seeds of *Zamia* spp. are sometimes eaten after the poisonous alkaloids are removed. However, the most important present day use of the cycads is the light they shed on plants of the past.

THE MAIDENHAIR TREE—GINKGO BILOBA

The maidenhair tree (*Ginkgo biloba*) is another plant that is often called a living fossil. This species is the only modern day representative of a group of plants that fossil remains show extended back 200 million years into the Triassic or even the Permian Periods. It, therefore, is the only living member of the order Ginkgoales, the family Ginkgoaceae, and the genus *Ginkgo*.

The plant was first studied in Japan in 1690 by a European, who proposed its generic name. Although it is now widely cultivated, it probably is no longer present in native stands. The last native home of *Ginkgo* is China; some botanists believe there is evidence of its continued existence since the Permian in southeastern China. Fossil fragments indicate that there may have at one time been about 50 species of the genus. Some of these are known to have existed in other parts of the world including Australia, England, Japan, Alaska and Oregon. Many oriental religious orders cultivated the tree in their monastary and temple gardens; it is because of this practice that we still have the plant today.

SPOROPHYTE GENERATION

The sporophyte of *Ginkgo* may reach 100 feet or more in height (Figure 25.9*a*, p. 374). It has one central axis or trunk from which arise many branches. Young trees are usually cone-shaped, but older ones become more rounded. Two kinds of branches are produced by the tree. The long branches are of indeterminate growth (**long shoots**) and produce many leaves at widely scattered internodes during the first year of their growth. The short branches (**spur shoots**) are of very limited growth, develop from lateral branches on the long shoots, and produce a terminal cluster of leaves during each growing season (Figures 25.9*b–d*, p. 374). Growth of all of the shoots is initiated by an apical meristem.

The leaves of *Ginkgo* are quite distinctive (Figure 25.9*b–d*). They possess a relatively long petiole and an extremely flattened, wedge-shaped blade. The blade is usually notched (Figure 25.9*b*) in the center forming two roughly equal lobes (hence the species epithet, *biloba*). The notch may be very shallow or quite deep; some blades have no notch (Figure 25.9e) at all. The leaves of one tree may possess all of the variations. Their venation is dichotomous, and because they superficially resemble the leaves of the maidenhair fern (*Adiantum* spp.), they are commonly called the maidenhair tree. The leaves are shed each year (**deciduous**) with new ones being produced during the next growing season. They are **megaphylls** because leaf gaps are left in the ectophloic siphonostele when the traces leave the stele and enter the petiole.

In contrast to the cycads the older stems of *Ginkgo* have a small pith and cortex and a great deal of secondary vascular tissue. An extremely active vascular cambium is produced between the primary xylem and the primary phloem by the apical meristem. A second lateral cambium, the **cork cambium,** is produced in the cortex. The latter cambium produces the cork cells which eventually cover the outside of the plant.

The extensive root system is also derived from apical meristems which are covered by a protective root cap. The roots also produce a vascular cambium that adds secondary vascular tissue.

Like the cycads, *Ginkgo* is also **dioecious** (Figure 25.10). Its spore-bearing structures do not resemble the strobili of the cycads, however. The microsporangia are produced in pairs at the tips of microsporophylls (Figures 25.11*a* and *b*). Many microsporophylls are attached to a central axis forming a loose pendulous strobilus. Several (4 to 5) microstrobili are produced among the leaves of spur shoots (Figures 25.9*d*, p. 374 and 25.10). The microsporangia are eusporangiate in origin. They contain microsporocytes (Figure 25.10) that undergo meiosis producing four haploid microspores just as occurred in the cycads. During their development they are nourished by a tapetum (Figure 25.11*b*) that was derived from the outermost layers of the sporogenous tissue. The microspores (pollen grains) are released in the spring.

The megastrobilus is much reduced in *Ginkgo* (Figures 25.9*c*, p. 374 and 25.10). It is composed of only a central axis at the end of which two **ovules** are usually produced (Figures 25.9*c* and 25.10). Each

Figure 25.10
***Ginkgo biloba*. Life history.**

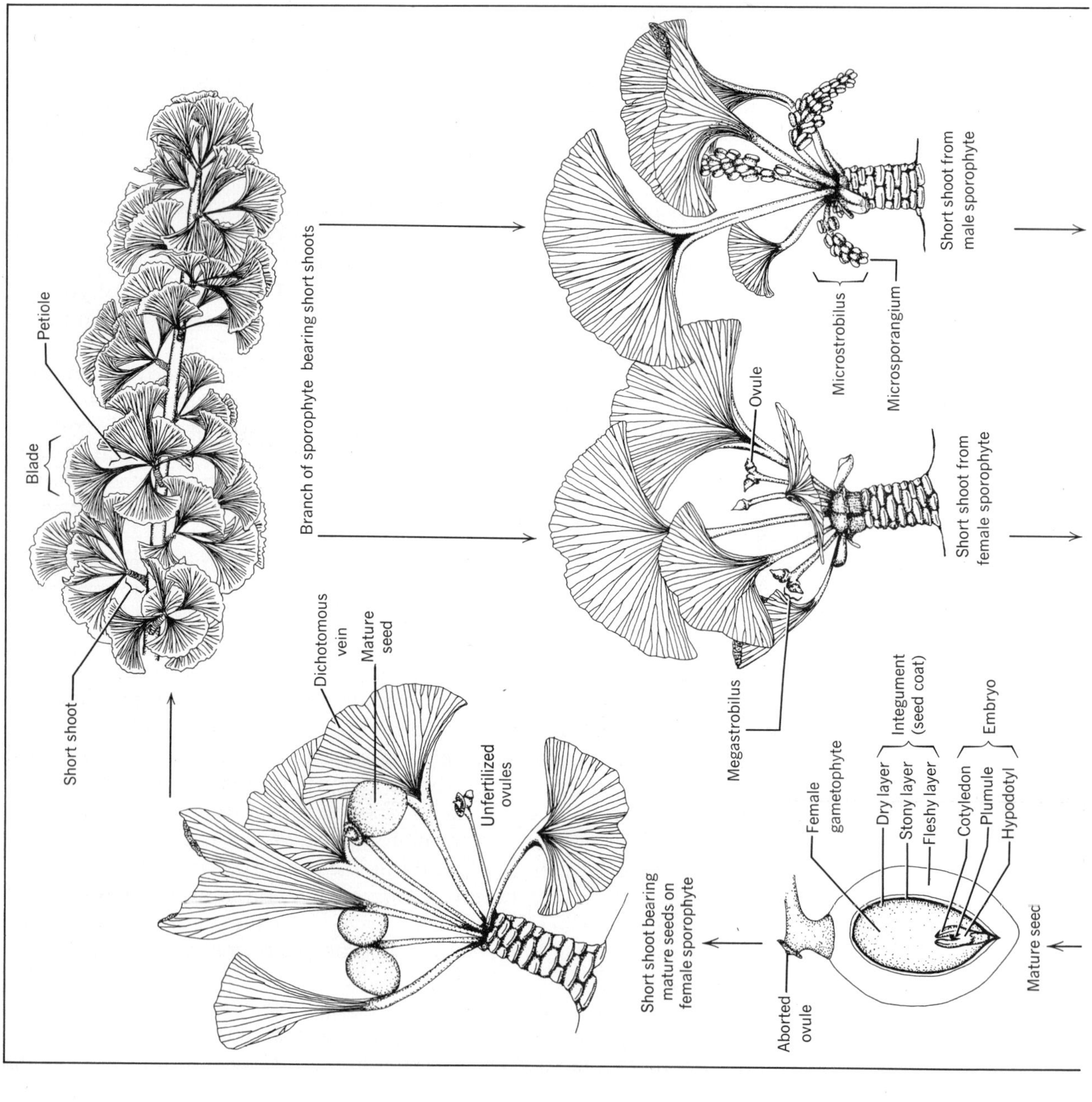

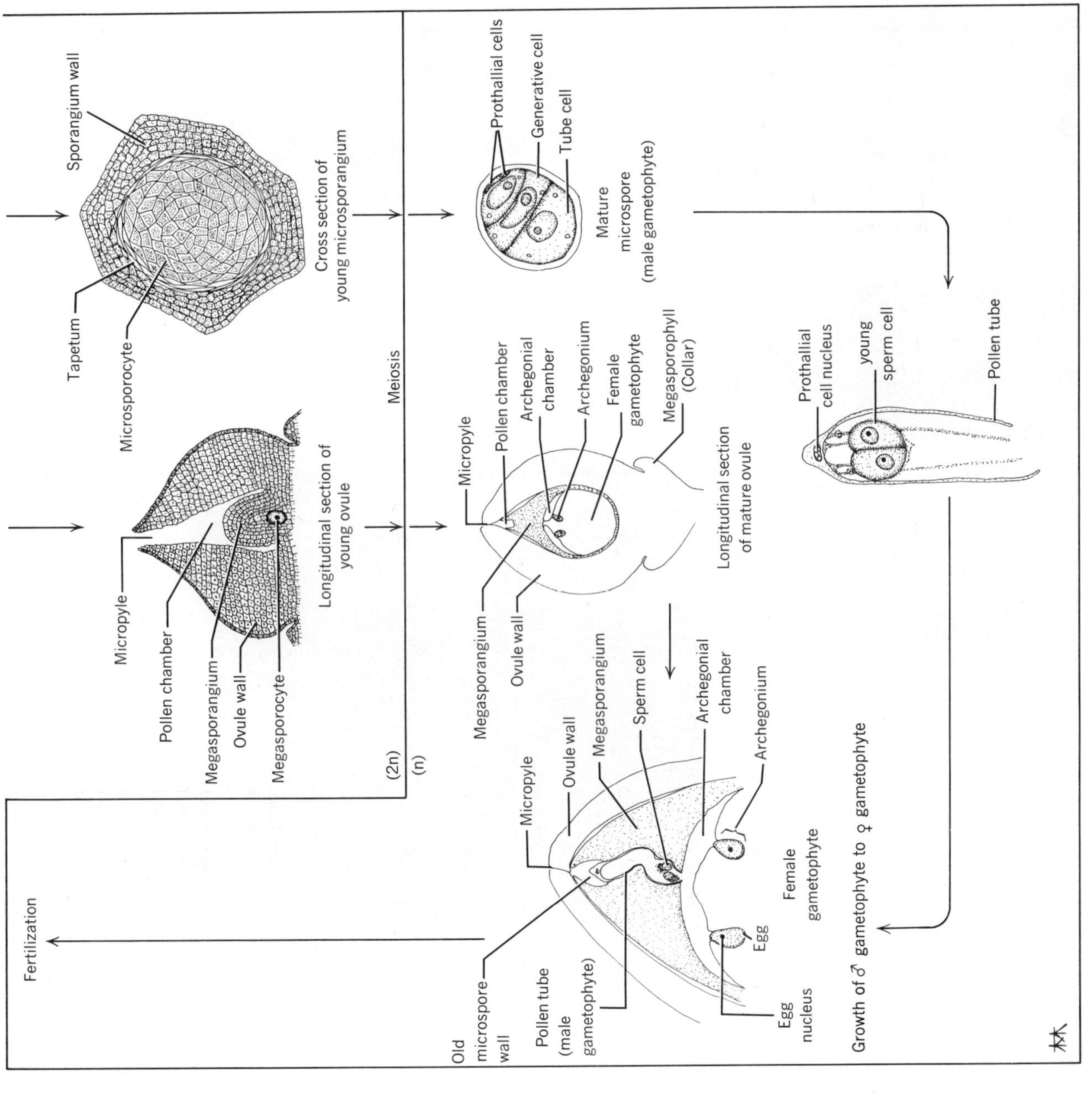
Sporangium wall
Tapetum
Microsporocyte
Cross section of
young microsporangium
Micropyle
Pollen chamber
Megasporangium
Ovule wall
Megasporocyte
Longitudinal section of
young ovule
Meiosis
(2n)
(n)
Prothallial cells
Generative cell
Tube cell
Mature
microspore
(male gametophyte)
Micropyle
Pollen chamber
Archegonial
chamber
Archegonium
Female
gametophyte
Megasporophyll
(Collar)
Longitudinal section
of mature ovule
Megasporangium
Ovule wall
Prothallial
cell nucleus
young
sperm cell
Pollen tube
Micropyle
Ovule wall
Megasporangium
Sperm cell
Archegonial
chamber
Archegonium
Female
gametophyte
Egg
Egg
nucleus
Old
microspore
wall
Pollen tube
(male
gametophyte)
Fertilization
Growth of ♂ gametophyte to ♀ gametophyte

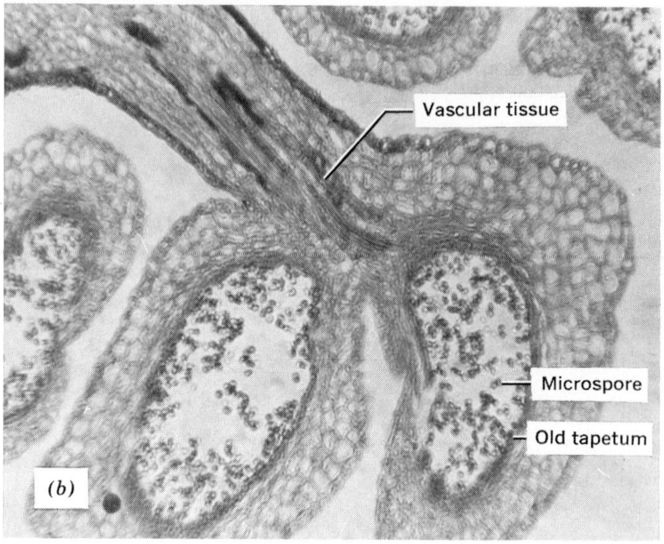

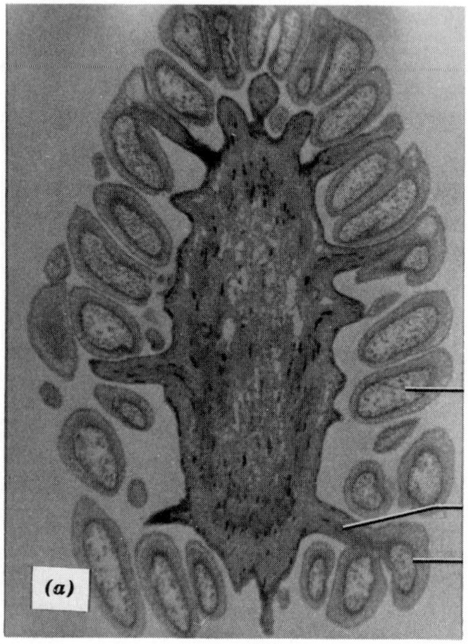

Figure 25.11
Ginkgo biloba. (*a*) Longitudinal section of microstrobilus (10×). (*b*) A single microsporophyll (56×). (*c*) Longitudinal section of megastrobilus before production of archegonia (8×). (*d*) A single ovule with a female gametophyte in the free nuclear stage (56×).

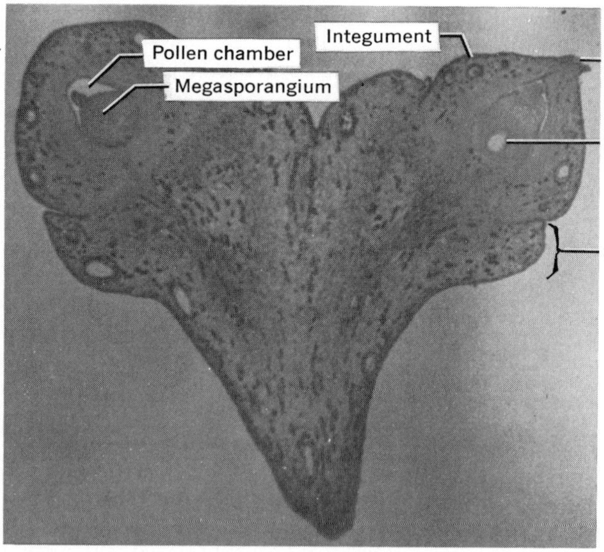

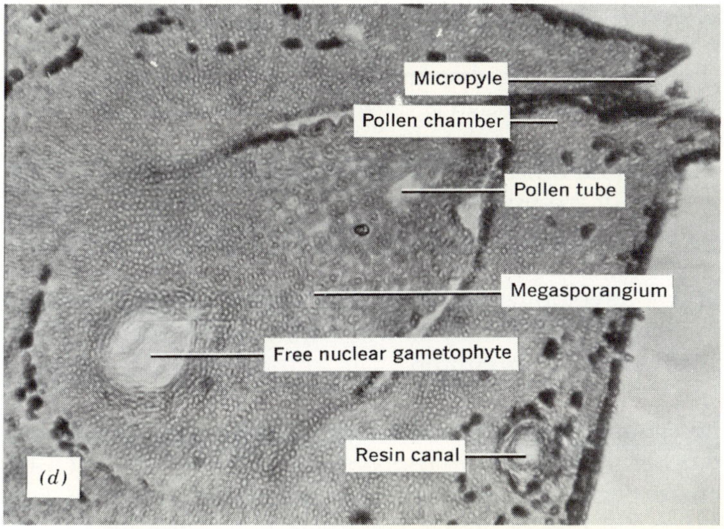

ovule rests on an expanded, flattened base that has a rimlike outgrowth (Figure 25.11*c*). This outgrowth probably is the reduced and vestigial sporophyll because it becomes leafy at times. Several (4 to 8) megastrobili are produced in certain of the spur shoots during the spring (Figure 25.9*c*, p. 374). Although there are certain minor differences, megasporogenesis in *Ginkgo* is like that in the cycads. A megasporocyte (Figure 25.10) carries on meiosis with the production of four megaspores. Of these, three abort; the survivor produces the female gametophyte (Figures 25.10 and 25.11*c* and *d*).

GAMETOPHYTE GENERATION

The gametophytes of *Ginkgo* (Figure 25.10) are very much like the gametophytes of the cycads. The male gametophyte begins its development while still within the microsporangium. The pollen grains are carried by the wind to the micropyle of the ovule where they are sucked into the pollen chamber by the action of a pollination droplet (Figures 25.10 and 25.11*c* and *d*). Although most of its development is endosporic, an exosporic pollen tube grows to the archegonium. Two flagellated sperm are produced from a body cell and fertilize the egg as in the cycads (Figure 25.10).

The functional megaspore produces the female gametophyte through a series of free nuclear duplications (Figure 25.11*d*) and at the expense of the sporangial tissue. The period of free nuclear duplication may last two months with the formation of about 8000 nuclei. Cell wall formation produces a cellular gametophyte. The vegetative cells of the gametophyte develop chlorophyll and become quite green. Some photosynthesis may occur within these cells, but most of the nutrition required by the developing gametophyte is probably derived from the surrounding sporangial tissue. Usually only two archegonia develop at the micropylar end of the gametophyte (Figure 25.10). The development of the archegonia in *Ginkgo* corresponds closely with that in cycads.

During all of this development, the ovule has been enlarging and the pollen tube has been digesting its way through the megasporangium toward the archegonium.

Five months may lapse between **pollination** and **fertilization** in Ginkgo. The two sperms are released into the archegonial chamber and both may enter the neck of an archegonium (Figure 25.10). Only one functions in fertilization, however. Fertilization may occur while the ovule is still on the tree or after it has fallen to the ground.

The embryo of *Ginkgo* develops much like the embryo of the cycads. A period of free nuclear duplications (about 256 nuclei) is followed by cell wall production until a completely cellular embryo is produced. There is no well-defined suspensor produced, but the cells at the base of the archegonium grow into the gametophyte tissue. These same cells organize into a meristem from which the cotyledons and shoot apex are derived. The radicle develops from the cells of the embryo that are toward the neck end of the archegonium. Although the eggs of both archegonia in an ovule may be fertilized, only one usually develops. During this activity of the embryo, the integuments increase in size and produce the seed coat. The inner layer of the seed coat is dry and papery; the middle layer is stony; the outer layer is fleshy (Figure 25.10). The fleshy layer accumulates butyric acid as a result of bacterial action. It may even cause nausea and skin problems in sensitive individuals, and its odor has been described as smelling like warm vomit or rancid butter.

The seed of *Ginkgo* (Figures 25.9e, p. 374 and 25.10) is composed of a three-layered seed coat, a developing embryo, and a food supply derived from megasporangial and gametophyte tissues. There appears to be no obligatory dormancy period, and the seed may germinate as soon as it is placed in favorable environmental conditions. The cotyledons remain within the seed, serving as organs of digestion to the food supply and translocating the food to the young embryo. The first seedling leaves are deeply bilobed and are produced on the apex of the embryo before germination.

HUMAN RELEVANCE

Because it is a "living fossil," *Ginkgo* is very helpful in aiding man to understand the evolutionary history of the plants with which he shares the earth. Because they carry on photosynthesis, they contribute their share of oxygen to the atmosphere and serve as primary producers in the ecosystem.

Ginkgo has long been noted for its beauty and used as an ornamental. Indeed, it is probably because of this that the ancient Chinese preserved it by planting it in their temple gardens. The Japanese once believed that it exuded water during a fire. This belief was suggested by the fact that *Ginkgo* is more resistant to fire than are many other trees. The inheritance that provides *Ginkgo* with fire resistance has apparently provided it with the ability to resist the environmental pressures that would have, through the centuries, made it extinct. Indeed, this same inheritance has enabled it to withstand the atmospheric pollution that has rendered many other trees unfit for use as ornamentals in municipal plantings. Therefore, many cities are now using this species in their city forestry work. Because of the odor of the fleshy layer of the seed coat, the male tree is preferable for such use. If a method of definitely determing the sex of the plant can be derived, the tree will probably be much more widely used.

In spite of the repugnant seed coat, the starchy gametophyte, megasporangial tissue, and the embryo are valued as a food source by many Oriental people. It is highly nutritious and the seeds are often imported to the various "Chinatowns" in the United States for consumption by their populations. The fleshy covering is removed by fermentation and the rest of the seed is then boiled or roasted. If eaten raw they are mildly poisonous and may cause a skin rash.

THE CONIFERS—CONIFEROPHYTES

The conifers are the best known and most carefully studied gymnosperms. The most common of the conifers are pines (*Pinus*, Figure 25.12*a*, p. 374), spruce (*Picea*, Figure 25.12*b*), fir (*Abies*, Figure 25.12*c*), Douglas fir (*Pseudotsuga*), larch or tamarack (*Larix*, Figure 25.12*d*), hemlock (*Tsuga*), bald cypress (*Taxodium*, Figure 25.12e, p. 374), coast redwood (*Sequoia sempervirens* Figure 25.22), giant redwood or "Big Tree" (*Sequoia* or *Sequoiadendron*), arbor vitae (*Thuja*, Figure 25.12*f*), and juniper (*Juniperus*, Figure 25.12*g*). This listing includes only a small representation of the 40 to 50 genera and the 500 to 550 species.

The evolutionary history of the conifers extends back at least to the Permian Period (180 to 205 million years ago), and the group may be even older than the cycads and *Ginkgo*. Evolutionary pressures have treated the groups quite differently, however. Whereas the cycads have been left with about 10 genera and 100 species and *Ginkgo* with only one species to represent them, the conifers have changed in such a way that they are an extremely diversified group that is almost ubiquitous in its distribution. Most of the conifers grow best in temperate zones; some reach their peak development in areas where winter ice and snow may bend and break their branches. Huge forests of conifers are to be found in North America, Europe, and Asia. Australia and New Zealand have extensive stands of certain conifers. Very few conifers are strictly tropical in distribution, and most of these occur at high altitudes.

The best known genus of conifers in the Northern Hemisphere is undoubtedly *Pinus* (Figure 25.12*a*, p. 374). Members of the genus will be used to discuss and illustrate the various phases of the life history (Figure 25.13) of the conifers. However, reference to other conifers will be made from time to time. The complete life histories of many of the conifers have not been worked out. Generalizations about so diverse a group as the conifers, taken from a single genus, must be viewed with suspicion. Therefore, the student should regard the following discussion as valid for *Pinus* and simply illustrative of what probably happens in other genera.

SPOROPHYTE GENERATION

The sporophytes of conifers are usually trees; some of them get to be huge. The largest of the conifers belong to the genus *Sequoia*. Some members of *Sequoia sempervirens* (Figure 25.22) are known to be over 300 feet tall and have diameters greater than 15 feet. Members of this species are known to be 1000 years old. *S. gigantea* does not get as tall as *S. sempervirens* but its diameter may reach 35 to 40 feet. Some of these specimens are known to be 3500 to 4600 years old. Douglas fir specimens are often 200 feet tall and 8 to 10 feet in diameter. Trees of this size are usually about 400 years old. Other economically important conifers range somewhat smaller in size, but very few of them can be considered as shrubs. One of the most common of the latter are some of the junipers (Figure 25.12*g*, p. 374). There are no herbaceous conifers.

Most of the conifers have tall, straight trunks with branches extending from them. This particular type of branching (**excurrent**) causes the form of the tree to be pyramidal. Some species produce their branches in a diffuse manner, and a main axis is not apparent. Most species of pine possess the excurrent branching form; junipers often have the diffuse form. Other tree forms between these two extremes are found in some genera.

Most of the genera produce only branches of unlimited growth and with scattered leaves in an alternate or spiral arrangement. In other genera (e.g., *Pinus*, *Larix*), both **long shoots** and lateral **short shoots** are produced (Figure 25.13). In *Pinus*, scale leaves are produced on the long shoots. The short shoots are produced in the axils of the scale leaves and produce the only photosynthetic leaves which are the needlelike leaves. The needle leaves are produced in groups containing from one to twenty according to the species. *Larix* species produce photosynthetic leaves on both the long and the short shoots.

Most conifers are evergreen. That means that they do not shed their leaves after each growing season as do deciduous trees. Some leaves are maintained on the tree from 2 to 14 years. The leaves eventually are abscissed however, and fall. Because **leaf abscission** is a continuous process, and because the abscission does not occur simultaneously on a particular tree, the loss of the leaves in evergreen species is not very spectacular. Most of the leaves fall individually, but in pines the entire short shoot is abscissed. In larch the leaves fall but the short shoot remains and produces new leaves during the next growing season. Larch and bald cypress (*Taxodium*) are the only deciduous conifers native to North America (Figures 25.12*d* and *e*, p. 374).

Although the leaves of the conifers are simple, they are **macrophylls.** In addition to the needlelike leaves and the nonphotosynthetic scale leaves already mentioned, many species produce scale leaves that are photosynthetic (e.g., junipers). Some conifers (e.g., *Podocarpus*, Figure 25.18*a*, p. 374) from the Southern Hemisphere produce broad, flat leaves that may have an open, complex dichot-

Figure 25.13
Pinus sp. Life history.

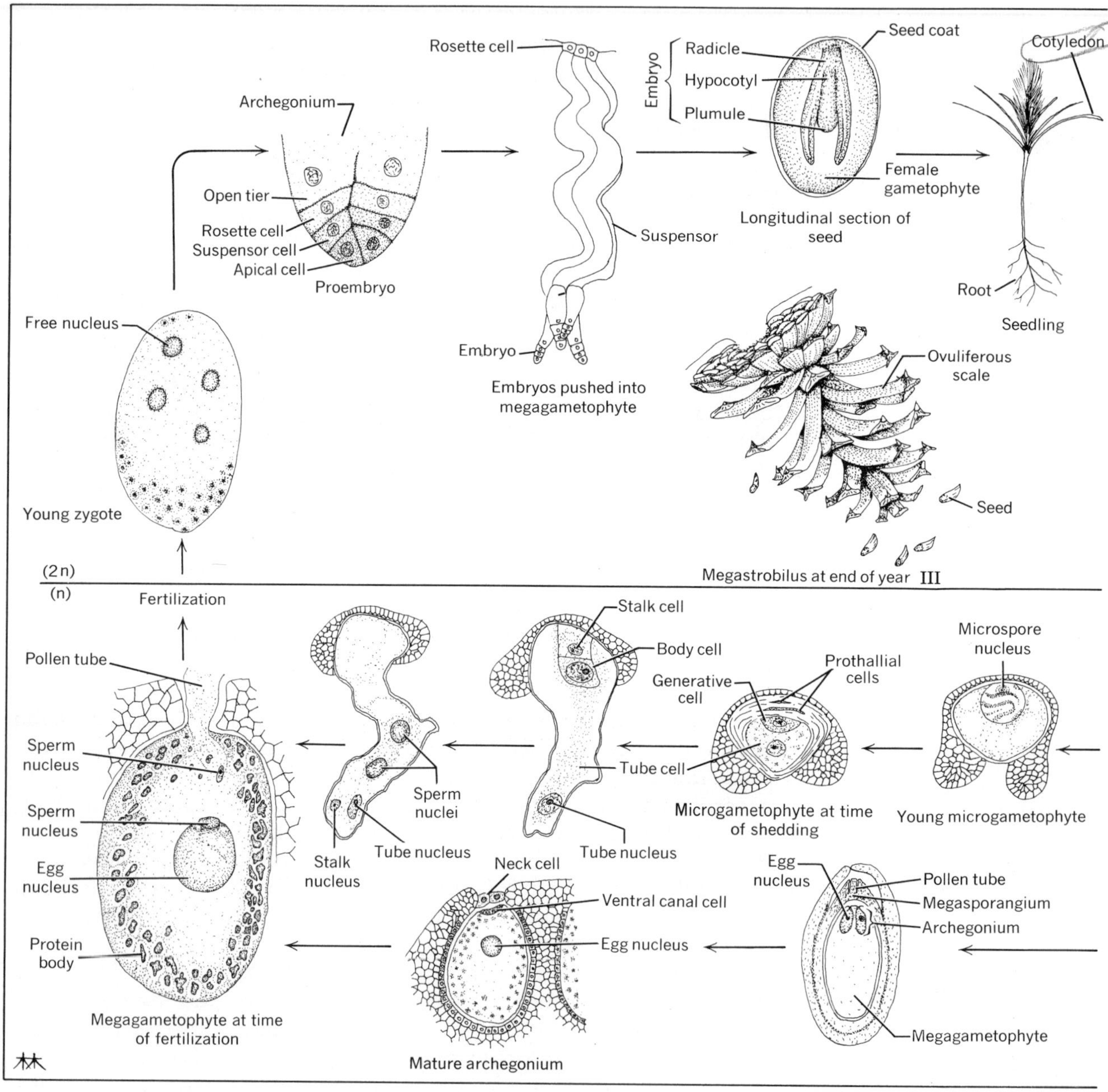

387

the naked seeded plants gymnosperms

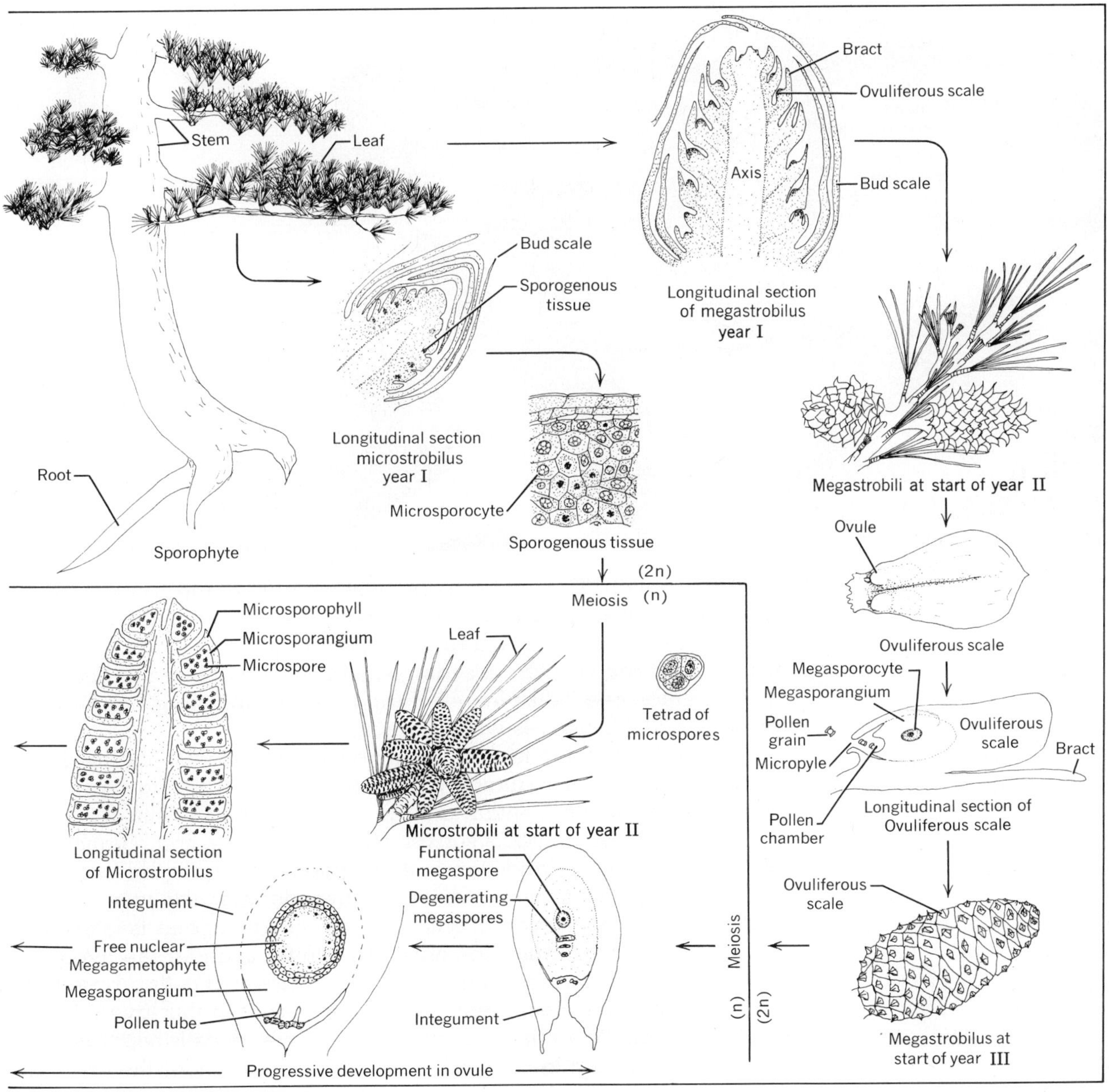

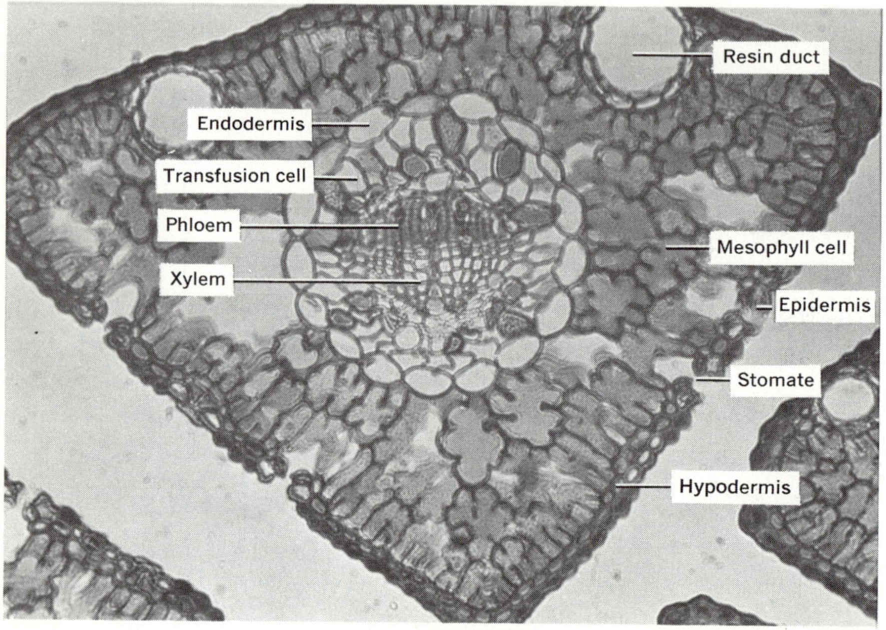

Figure 25.14
Pinus strobus. **Cross section of single leaf (1.56×).**

omous venation. The leaves of the vast majority of the conifers are unveined. Many species of conifers may produce one type of leaf during their juvenile period and another during adulthood. Some of the junipers are very good examples of this type of leaf variation; they produce needlelike leaves during their young stages and appressed scalelike leaves when they mature.

The needlelike pine leaf is a common example to use for the discussion of conifer leaf anatomy (Figure 25.14). The outside layer of cells have walls that contain a large amount of **cutin.** This epidermal layer also contains sunken stomata, and the combination of water resistant cutin and stomatal arrangement allows the leaf to withstand extremely dry conditions. A **hypodermis,** also with thick-walled cells, is composed of one or more cell layers and is located just under the epidermis. The **mesophyll layer** is under the hypodermis and is composed of many tightly compacted cells whose thin walls are infolded. **Resin ducts** are often present in this layer. A wellformed **endodermis,** consisting of one layer of cells separates the mesophyll from the central tissues of the leaf. Under the endodermis is found a large amount of **transfusion tissue** in the center of which are found one or two **vascular bundles,** depending on the

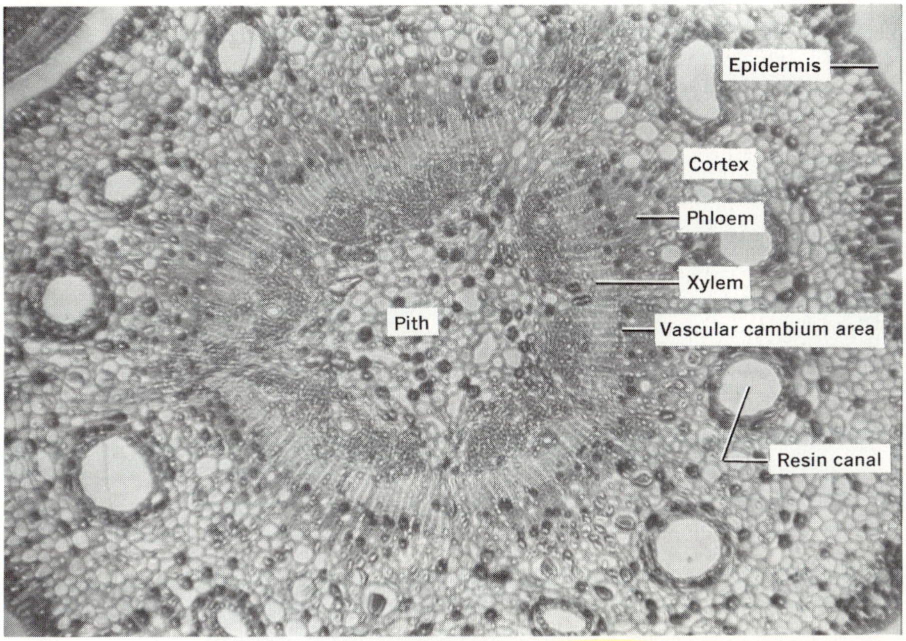

Figure 25.15
***Pinus* sp. Cross section of one year old stem (56×).**

species. The transfusion tissue is thought by many botanists to be modified vascular tissue that aids in the movement of materials from the vascular bundles to the mesophyll. This tissue appears to be unique to the conifers.

The stem grows from a multicellular apical meristem that is active during the spring and summer. The primary vascular tissues are produced in the manner discussed in the introduction to the vascular plant groups (Figure 20.1). The young (Figure 25.15) stem is a siphonostele that contains a rather large **pith.** A thick cortex layer surrounds the vascular tissue, and it in turn is enclosed by the epidermis. Secondary vascular tissues are produced by the action of the vascular cambium which is differentiated between the primary xylem and phloem early in the development of the young seedling. Secondary xylem is produced toward the pith. Because the conifers produce only one kind of element (**tracheids**) in their xylem, the secondary xylem is quite homogeneous (Figure 25.16). The diameter of the tracheids varies, however, depending on whether they were produced during times of abundant water supply or during times when water was scarce. Since water is usually abundant during the spring of the year, the tracheids produced during that time have a

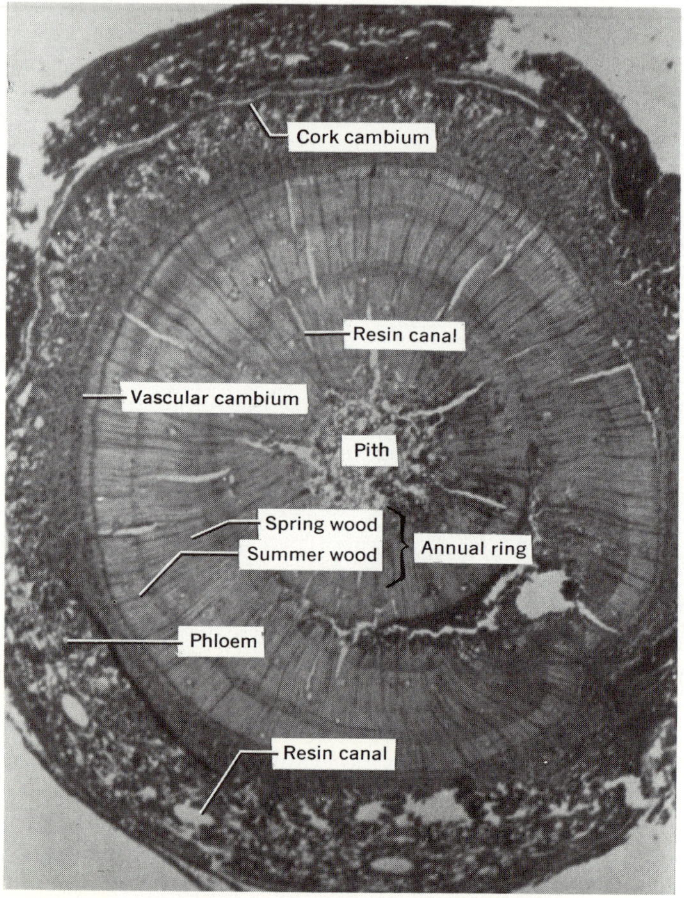

Figure 25.16
Pinus sp. Cross section of three year old stem (20×).

wider diameter; there is a decrease in diameter during the summer months when water usually becomes less available. The difference between the diameters of the previous season's **"summer wood"** and the present season's **"spring wood"** produces a growth ring that is often quite visible to the unaided eye. One complete year's growth is represented by one layer of spring wood and one layer of summer wood. These two layers are collectively known as an **annual ring.** This absolute relationship may be interrupted by a number of factors. For example, defoliation of the tree by insect or fungal infestation may produce a false picture in the rings. Trees that were standing in the battle areas during the American Revolution have been shown to have reduced growth in certain rings. This reduced growth corresponds to the times during which battles were raging; the growth reduction is attributed to the defoliation caused by the battle. Also, in areas where water availability does not vary from spring to summer, the growth rings may be quite indistinct or almost nonexistent. Other factors add variation to the interpretation of growth rings. However, when one takes these variables into consideration, the analysis of the rings of a tree may give a fairly good picture of its history. Resin canals are also found in the secondary xylem.

The vascular cambium also produces secondary phloem toward the epidermis (Figure 25.16). The phloem is also rather homogeneous although it too contains resin canals. The production of the secondary phloem results in the crushing of the cortex cells. However, another meristematic layer, the **cork cambium,** arises in the outer layers of the cortex. This cambium produces cork cells toward the epidermis. The walls of the cork cells are heavily impregnated with a fatty material called **suberin.** The suberin is impermeable to water and gases. As a result of the corky layer, the inner tissues of the stem are protected from excessive mechanical injury and water loss. The tissues external to the corky layers are deprived of water and dissolved food materials, however, and soon die. The external forces of erosion wear away the epidermis and cortex layers. The cells from the cork cambium thus replace the epidermis and cortex cells as the outermost layers of the stem. New cork cambia are produced from it. In this fashion, new cork cells are produced from the inside of the stem as the old ones die and wear away from the outside of the stem. Eventually all of the cortex is replaced with corky tissue. Then new cork cambia are produced from old secondary phloem cells. Some trees produce very smooth corky layers because their cork cambia are large and continuous. Others produce very rough corky layers because their cork cambia are produced in relatively small, irregular patches. When one observes the outermost layer of the woody tree stem, he is looking at corky tissue, but he often calls the tissue the **bark.** The bark really includes all of those tissues that extend from the vascular cambium outward.

As a result of the meristematic activity of the vascular cambium, the woody stem increases in diameter in two directions. These are by the production of the xylem tissue toward the inside and phloem toward the outside. The cork cambium contributes to diameter in-

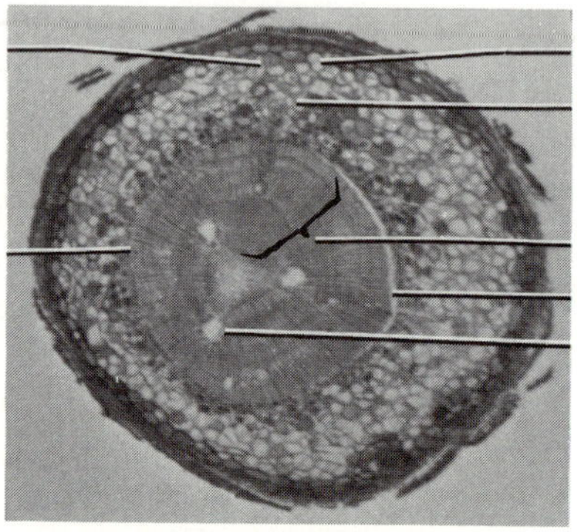

Figure 25.17
Pinus sp. Cross section of four year old root (18×).

crease with the production of the corky cells toward the outside of the stem. The last two types of increase (production of secondary phloem and corky tissue) are about equaled in the mature stem by the death and sloughing off of external tissues. Thus, while the xylem (or woody tissues) continue to increase in diameter, the thickness of the layer of phloem and corky tissue (the bark) remains about constant.

The conifers have a well-developed **taproot system.** The stele is a protostele that is surrounded by cortex and epidermis when the plant is young. A vascular cambium develops and produces secondary xylem and phloem and the root becomes quite woody (Figure 25.17). Cork cambia produce corky tissues as in the stem.

All of the pines are **monoecious** (Figure 25.18*b*, p. 374), as are the majority of the conifers. Two common dioecious members are juniper (Figure 25.12*g*, p. 374) and yew (Figure 25.18*d* and *e*, p. 374). The male cones (**microstrobili**) are produced at the base of terminal buds. They start this growth in the spring or early summer and develop the sporogenous tissue by that winter (Figures 25.13 and 25.19). The production of the mature pollen grain does not occur until the following spring, however. Therefore, the production of a mature microstrobilus takes place over the period of one year. The microstrobilus (Figure 25.20) is simple and closely resembles that of the cycads and *Ginkgo*. It is composed of a central axis that produces a series of **microsporophylls.** The **microsporangia** are produced on the abaxial surface of the microsporophyll in a **eusporangiate** manner. **Microsporocytes** are produced from the sporogenous tissue in the early spring, meiosis follows, and each of the microsporocytes produces four haploid **microspores** (Figure 25.13). The microspores develop extensions, called wings, to their outside wall layer. The mature microspores (**pollen grains**) are released by the longitudinal dehiscence of the microsporangia (**pollen sacs**). The release of the pollen grains occurs during the spring, usually during March and April, depending on the species of pine and the geographic location. The developing microsporocytes and microspores are nourished by a large **tapetum** on the inside of the pollen sac and starch which was stored in the microsporocyte. At the time of release (Figure 25.20*c*), the pollen grain (the **young male gametophyte**) is composed of two **prothallial cells** which are flattened and dead, a **generative cell,** and a **tube cell.**

The female cone (**megastrobilus**) is much more complicated in structure than is the male cone. Comparisons of extant and extinct species of pines and related conifers reveal that the ovule is not borne directly on the **bracts** that are attached to the cone's axis, but do develop on an **ovuliferous scale** that is on the bract (Figures 25.13 and 25.21*a*). The ovuliferous scale has probably been produced as the result of reduction of a fertile shoot. Therefore the megastrobilus is a compound structure and it is inaccurate to consider that the ovules are produced on megasporophylls. A complete discussion of this characteristic is outside the scope of this text. Although almost all conifers produce a definite female cone, some have a much reduced strobilus. The megastrobilus of yew (Figure 25.18*d*, p. 374), for example, is reduced to one ovule.

Two ovules are borne on the adaxial surface of the ovuliferous scales (Figure 25.13). The ovules have a very thick **integument** that surrounds a small megasporangium (**nucellus**). The integument is not closely appressed to the megasporangium at the micropyle, and a **pollen chamber** is produced between the tips of the integument and the end of the megasporangium (Figure 25.21*b*). The further development is much like that in *Ginkgo* and the cycads. One cell deep within the megasporangium differentiates into a **megasporocyte** that carries on meiosis producing a linear tetrad of three or four megaspores. The megaspore farthest away from the micropylar end remains viable; the other three disintegrate (Figures 25.13 and 25.19).

GAMETOPHYTE GENERATION

Much of the endosporic development of the male gametophyte occurs in the microspore before it is released from the microspor-

Year	Season	Microstrobilus; Male Gametophyte	Megastrobilus; Female Gametophyte, Zygote
I	Spring Summer Fall Winter	Initiation of strobili. Sporogenous tissues produced.	Initiation of strobili enclosed within bud scales.
II	Spring	Microstrobilus recognizable. Meiosis-production and shedding of microspores (pollen).	Development of ovule. Megastrobilus recognizable. Megasporocyte produced. Pollination.
	Summer	Germination of pollen grain (male gametophyte). Duplication of generative cell forming stalk and body cell.	Meiosis—megasporogenesis. Production of free nuclear female gametophyte (about 32 nuclei produced).
	Fall		
	Winter		
III	Spring	Stalk and body cells move to lower end of pollen tube.	Production of additional free nuclei (about 2000). Production of cell walls
	Summer	Two sperm nuclei produced from body cell nucleus. Rapid growth of pollen tube through megasporangium.	in female gametophyte. Archegonia production. Fertilization. Free nuclear duplication of zygote. Production of mature embryo.
	Fall		Maturation of seed. Shedding of seed.
	Winter		

angium. The microspores of pine (Figures 25.13 and 25.20c) produce two prothallial cells, a generative cell, and a tube cell before leaving the sporangium, but the pollen grains of junipers, yews and some others are shed while uninucleate. The prothallial cells are considered to be all that remains of a much reduced antheridium. In the latter cases, no prothallial cells are produced, but nuclear duplication after the pollen grain reaches the pollen chamber in the ovule produces a generative cell and a tube cell.

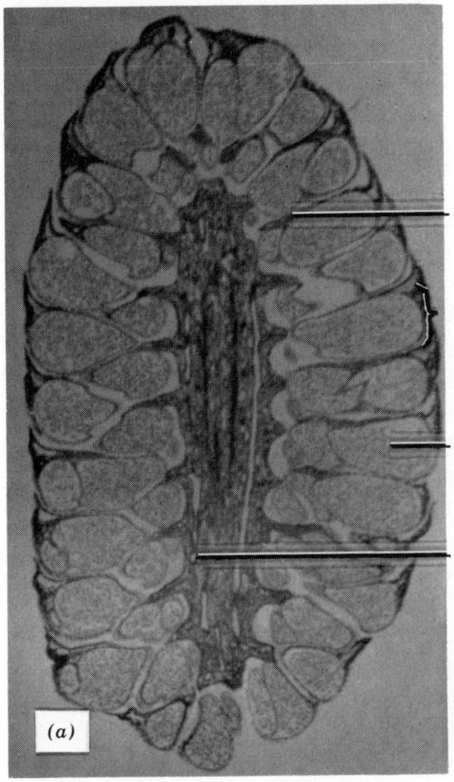

Figure 25.19
Pinus sp. An outline of the significant events that occur during the first, second, and third years of the development within the microstrobilus, male gametophyte, megastrobilus, female gametophyte, and zygote. The times listed for each event are approximate and may vary with geographic location and environmental conditions.

Figure 25.20
Pinus sp. Microstrobilus. (*a*) Longitudinal section (9×). (*b*) Cross section (15×). (*c*) Mature microspores (550×).

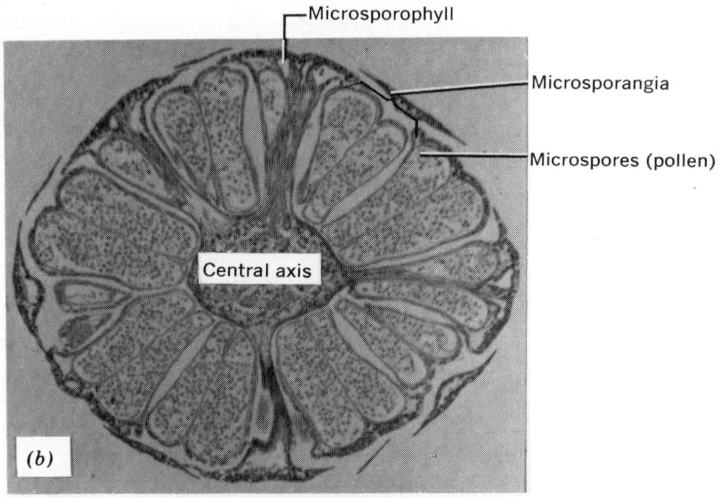

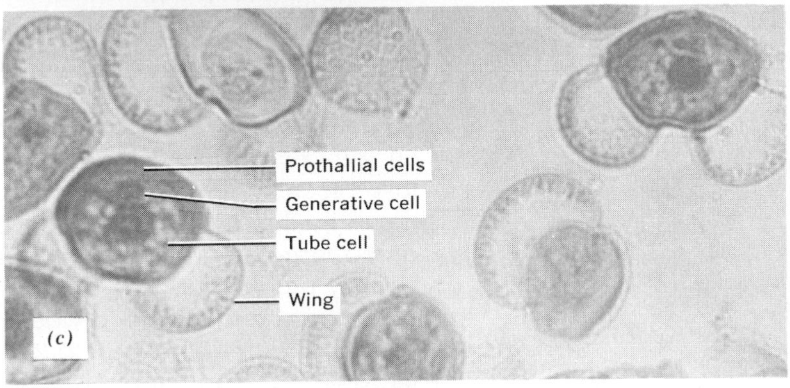

Pollen grains are released in fantastic numbers, often producing what is called a "sulfur shower" in which the air is turned a golden yellow. Most of the pollen grains die; only a few reach the ovules (Figures 25.13 and 25.19). At this time, the ovuliferous scales of the still upright female cones separate slightly (Figure 25.18*c*, p. 374). This separation allows the pollen grains to sift between the scales and come into contact with the micropylar end of the young ovule (Figure 25.21*b*). After pollination, the megastrobilus seals the spaces between the scales (Figure 25.18*c*). The parts of the integument that extend over the pollen chamber produce a sticky pollination droplet to which the pollen grains adhere. As the pollination droplet evaporates, the pollen grains are drawn into the pollen chamber and contact the megasporangium (Figure 25.21*b*). After this contact is made, the exosporic portion of the male gametophyte is produced. This portion is the pollen tube and it begins to digest its way through the megasporangium tissue, using the digested tissue as a food source. During this growth, the generative cell usually divides forming a **stalk cell** and a **body cell** (Figures 25.13 and 25.19).

The female gametophyte has not completed its development at the time of pollination (Figures 25.13 and 25.19). Indeed, in many pines the time interval between pollination and fertilization is about one year. At about the time that the pollen grain begins to grow into the megasporangium tissue, the functional megaspore enlarges and undergoes a series of free nuclear duplications. About 32 free nuclei are produced before the growing season is over and the male and female gametophytes become dormant during overwintering. Resumption of metabolic activity during the following spring stimulates the production of sperm nuclei and archegonia.

Further development of the female gametophyte begins with more free nuclear duplications until about 2000 free nuclei may be produced (Figures 25.13 and 25.21*c*). Wall formations then start and the gametophyte finally becomes a completely cellular body. Archegonia are produced about two weeks before fertilization is to take place (Figures 25.13 and 25.21*d*). These organs are differentiated by the division of a surface cell into a small outer **neck cell** and a large inner **central cell** (Figure 25.21*c*). More neck cells are usually produced by the duplication of the first one. The central cell enlarges and ultimately divides producing a **ventral canal cell** and an **egg** (Figures 25.13 and 25.21*f*). The ventral canal cell finally disintegrates. The number of archegonia produced per ovule may vary from one to six.

The resumption of spring growth stimulates the stalk and body cells of the male gametophyte to move into the pollen tube (Figure 25.13). About one week before fertilization, the nucleus of the body cell duplicates producing two **sperm nuclei.** The pine sperm (Figure 25.13) are much reduced from the sperm of the cycads and *Ginkgo*. They are only nuclei and not complete cells, they are unequal in size, and they are not flagellated. The pollen tube continues to grow through the megasporangial tissue and finally forces itself between the cells of the archegonia neck. There is ruptures and releases the two sperm nuclei, the tube nucleus, and the stalk cell into the egg's cytoplasm. The larger of the two sperm nuclei unites with the nucleus of the egg. The other nuclei disintegrate. Although several pollen grains may be attached to the pollination droplet and pulled into the chamber, only a few continue to develop and reach the archegonium. Apparently only one finally ruptures the archegonium.

At this point, the student is well-advised to review the times at which events occur in the pine life history (Figure 25.19). The microstrobilus of a species is produced during the spring or early summer of a particular year. By winter of that year, the microsporangia containing sporogenous tissue is produced. The megastrobilus is initiated during the same spring and proceeds in development until the axis, lateral bracts, and ovuliferous scale primordia are produced by late autumn of that growing season.

During the next spring, the pollen grains complete their development and are shed. Pollination occurs. During April or May of this spring, the ovules have been initiated in the megastrobili and the pollen grains reach the micropylar end of the ovule. At this time the ovule does not contain a mature female gametophyte. Both gametophytes continue their development throughout this growing season.

Both gametophytes mature during the next spring. About the middle of June, fertilization occurs. Thus the time interval from initiation of strobili to fertilization is about three years in most pines. Other conifers differ from this time interval, but most of them require far longer to complete their sexual cycle than do the angiosperms.

After fertilization, the zygote undergoes a short period of free nuclear duplication, producing only four free nuclei (Figures 25.13 and 25.21g). Successive duplications are accompanied by wall forma-

Figure 25.21
Pinus sp. Megastrobilus. (*a*) Longitudinal section through strobilus at beginning of second year (10×). (*b*) Ovule at beginning of second year (150×). (*c*) Ovule with free-nuclear gametophyte at beginning of third year (11×). (*d*) Ovule with two mature archegonia (7×). (*e*) Tip of mature archegonia (136×). (*f*) Mature archegonium (50×). (*g*) Zygote in free nuclear condition (50×). (*h*) Longitudinal section of embryo with two cotyledons showing (8×).

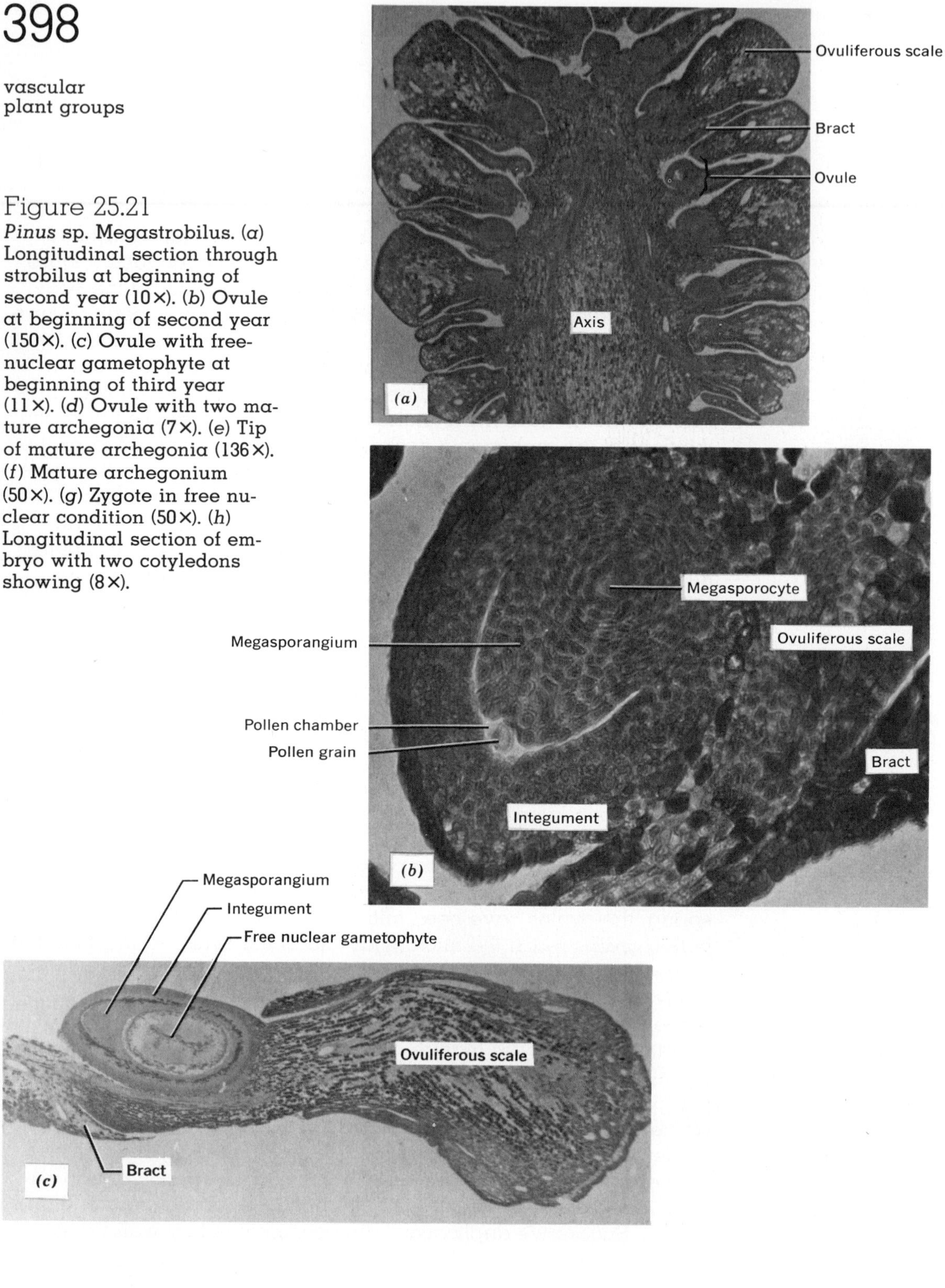

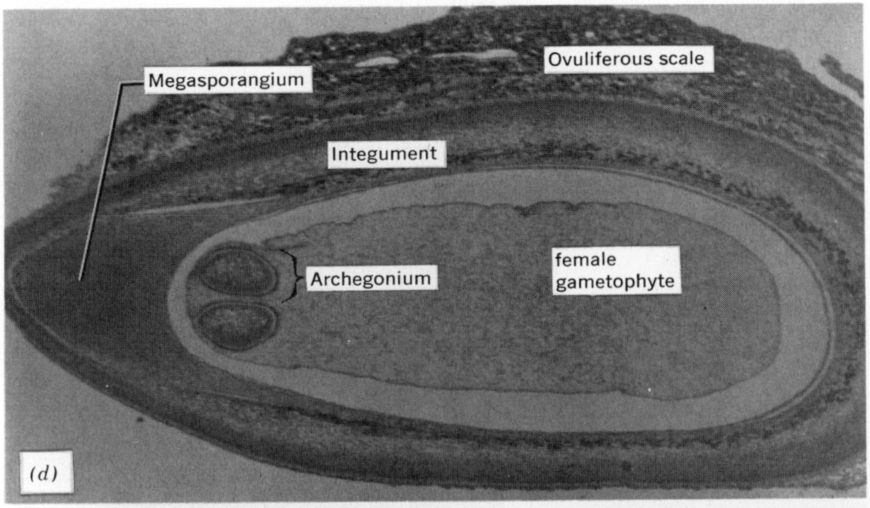

Ovuliferous scale
Megasporangium
Integument
female gametophyte
Archegonium
(d)

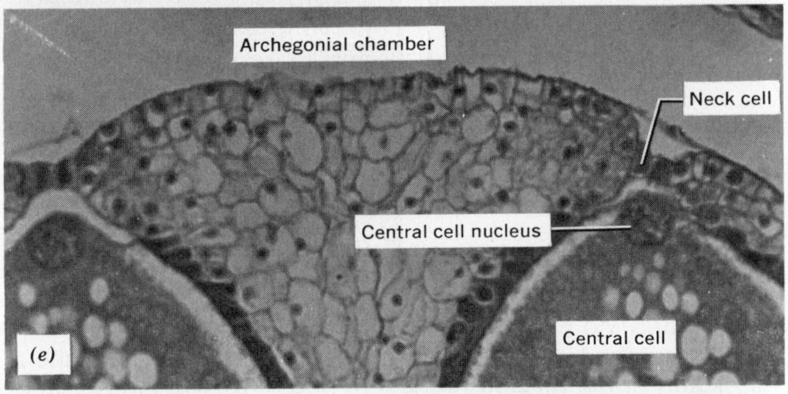

Archegonial chamber
Neck cell
Central cell nucleus
Central cell
(e)

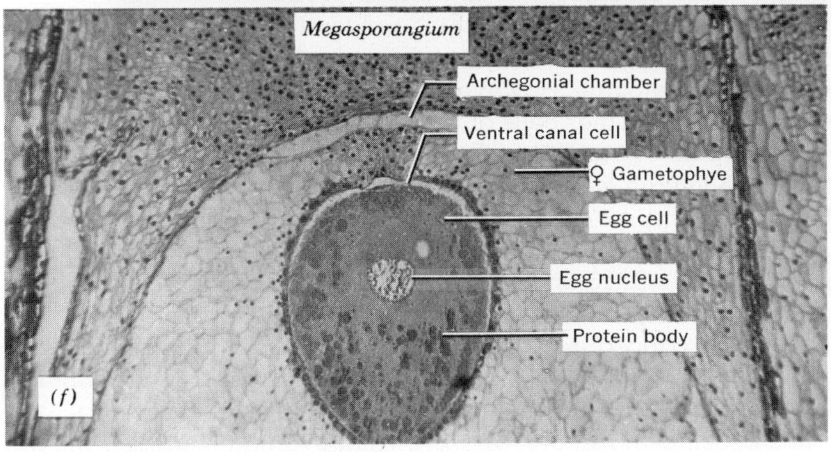

Megasporangium
Archegonial chamber
Ventral canal cell
♀ Gametophye
Egg cell
Egg nucleus
Protein body
(f)

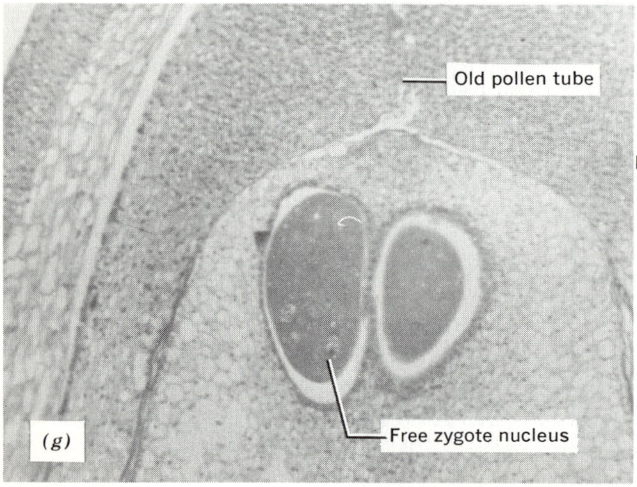

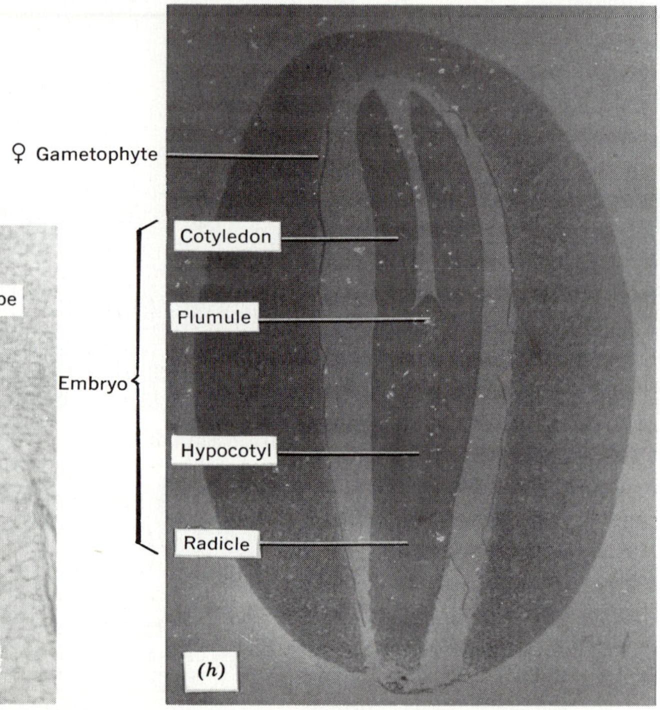

tions. When the number of cells reaches about 16, the structure is called a **proembryo.** Each of the four cells farthest from the micropylar end of the proembryo is capable of producing an embryo (Figure 25.13). Another way in which polyembryony may be accomplished is by the fertilization of the egg of more than one archegonium in the ovule. However, only one embryo usually matures in each ovule. The other cells of the proembryo develop into a suspensor that pushes embryo cells deeply into the female gametophyte. The female gametophyte continues to grow and store food.

The mature embryo is embedded in the female gametophyte and consists of eight (usually) seed leaves (cotyledons) that surround the shoot tip (Figures 25.13 and 25.21*h*). A large embryonic root (radicle) is attached to the shoot tip by the hypocotyl. The hard protective seed coat is produced from the integuments and in many species the seed coat becomes winged. The seed matures about 12 months after fertilization. During the time of seed maturation, the megastrobilus becomes hard, dry, and brown. It eventually turns from an upright to a pendulous position and the seeds are shed when the ovuliferous scales open.

Pine seeds may remain dormant for many years, and some remain in their cones for several years before they are shed. Some pine cones require exposure to fire (e.g., jack pine) before they will open and shed their seeds. The seed germinates by the growth of the radicle through the micropyle and the elongation of the hypocotyl. The cotyledons remain inside of the seed coat and continue to absorb the food through their tips. The seed coat may remain on the tips of the cotyledons for quite some time after germination, but eventually it is shed and the cotyledons begin to carry on photosynthesis (Figure 25.13).

HUMAN RELEVANCE

The conifers are extremely economically important to man. Almost half of the forests of Asia are of the coniferous type. Of these, huge areas are largely unexplored and unexploited. For example, the forbidding environment of Siberia has slowed the use of its huge tracts of coniferous timber. Most reports indicate, however, that although hundreds of millions of acres are occupied by the conifers, the wood is mostly of average timber quality and much of the area is occupied by swampy wasteland. Species of conifers that are found in the area are pine, fir, larch, and spruce. Coniferous forest stands are to be found in Afghanistan on middle and upper mountainous regions, but China has been largely deforested because of her long history of overpopulation denuding her forested areas. India still possesses large coniferous forests in her mountainous northern regions, and the higher plateaus of the Himalayas contain pine, fir, and larch. The forests of the higher elevations of Indo-China have stands of various pines and *Podocarpus* spp. (Figure 25.18*a*, p. 374). Japan possesses large stands of spruce, larch, pine and fir in her northern areas. Therefore, although much of Asia is unexplored and unexploited, denuded because of overpopulation, and naturally sparsely forested with conifers, the area must still be considered as containing huge amounts of timber because of its immensity. Some estimates place the amount of good coniferous timber at approximately 1½ to 2 billion acres.

Europe contains about 10% of the world's forest lands; about three-quarters of this is in coniferous forests. At one time, the land areas of Europe were much more heavily forested. For example, the British

Isles contained extensive forests; the ancient literature of Greece praises her huge stands of magnificent virgin forests. These areas are greatly reduced now because of the demands of large populations using (and misusing) the forests over long periods of time. Large stands of conifers (spruce, pines, larch) are still to be found in northeastern Europe, but the Scandinavian countries contain the best coniferous forests. There are many angiospermous species to be found in these areas also, but Scotch pine and Norway spruce are the species that are the most valuable. Although the European countries were largely deforested by population demands, the Europeans have been very active in reforestation research. Thus, timber production is continually increasing. Much of the knowledge of what is occurring in forestry (as well as everything else) in some parts of Europe is restricted because of the political problems presented by the Iron Curtain.

Good stands of conifers are found at high elevations in Central and South America and the Caribbean Islands. The most important members here are various species of pines, but significant stands of Douglas fir, other firs, and bald cypress do exist. One of the most important coniferous forests in South America is found in southern Brazil. A very unusual conifer, the Parana pine or "monkey-puzzle" tree (*Araucaria araucana*), constitutes the dominant species in the forest (Figure 25.18*f*, p. 374). This species is a very important source of lumber in South America.

The gymnosperms do not make a very significant contribution to the forest lands of Africa. A few species of juniper and *Podocarpus* spp. are to be found in East Africa, but most of the forests are angiospermous. Australia has only a few conifers and these are confined chiefly to coastal areas. Some of the Pacific Islands do contain large coniferous forests composed mainly of pines and *Podocarpus* spp.

The forested areas of the United States and Canada are predominantly coniferous although angiospermous forests do exist. The United States possesses six distinctive forest belts; four of these are coniferous. The Canadian forests are almost exclusively coniferous. The forests of these two countries supply great quantities of timber to the world's lumber market, but the supplies are exhaustible. Therefore, reforestation projects have been undertaken to replace much of the virgin timber that has been removed.

The United States and Canada have a wide variety of conifer species that are distributed in four coniferous forest belts. The Great

Plains divides the North American forests into eastern and western sections. The northern coniferous forest belt is continuous from east to west in its northern Canadian portions and extends northward into Alaska. In the United States, this belt extends from New England to the Great Lakes and Minnesota. Angiospermous species are found in this belt, of course, but conifers are the dominant plants. The economically important coniferous members include four pine species, larch or tamarack, three spruce species, hemlock, fir, and arbor-vitae.

The southeastern coniferous belt extends from Virginia to Florida and westward through southern Georgia, Alabama, Mississippi, Louisiana to Texas and occupies the low, sandy areas of these states. Many angiospermous species are to be found here, also. The economically important conifers include four species of pine, bald cypress, and red cedar.

The coniferous forests of the western United States are also divided into two main belts. The Rocky Mountain forests are fairly well confined to mountainous areas that extend from Mexico to Canada but are scattered over a number of minor ranges in addition to the Rocky Mountains. The belt extends from the western extreme of the Great Plains to the arid Great Basin of Nevada and adjoining states. The valuable conifers to be found in this forest area include four pines, western larch, two spruces, Douglas fir, and two other firs.

The second of the western coniferous forest ranges is the Pacific coniferous belt which extends from southern California to Alaska and from the Sierra mountain ranges on the east to the Pacific Ocean on the west. Large stands of its trees occupy areas of western California, Oregon, Washington, British Columbia, the Yukon, and Alaska. The most economically important tree in this range is the Douglas fir, but two redwoods, three pines, one spruce, one hemlock, eight firs, incense cedar, arbor-vitae, and some species that are commonly called "cedars" but that do not belong to the cedar genus *Cedrus* are also important components.

The student should be reminded at this point that the major characteristic that makes the trees so valuable to man is their wood. This may seem to be a statement of the obvious but it should be realized that the activity of the vascular cambium produces the secondary xylem tissue which is the wood. Wood is, of course, used in many ways and different species produce wood that is valuable for various purposes. In a book such as this, it is impossible to discuss all of the types of wood produced or all of the purposes to which it is put. It is important, however, for the student to be informed of some of the more important species and uses and even some of the exotic ones. Perhaps this exposure will aid him in understanding why he has a very real, vested interest in preserving our forests even if he lives in an urban area. Therefore, although this account barely indicates the value of trees, some specific examples will be discussed. In many discussions of trees and timber, the gymnospermous trees are often referred to as the softwoods and the angiospermous trees as the hardwoods. Distinctions such as these have been avoided in this discussion because many of the gymnosperms produce wood that is harder than that of some angiosperms; therefore the distinctions are almost meaningless.

The pines (genus *Pinus*) have probably contributed more to man's welfare than any other single group of trees. The eastern white pine (*P. strobus*, Figure 25.12*a*, p. 374) was found by the North American settlers throughout the northern coniferous forest belt. This tree occupied hundreds of thousands of acres and, although other species of tree were also found, the white pine often occurred in almost pure stands. In spite of their numbers, cutting of the trees was so extensive that they were reduced to secondary importance by the beginning of the 20th century. However, white pine is now producing good stands of secondary growth over much of its original range.

The white pine was highly prized for use as masts on the sailing ships of the Royal Navy during colonial days, and extensive stands of prime timber trees were reserved for such use. Penalties for the improper cutting and use of these trees by colonists were often severe. This versatile wood has many uses. It is easy to work (being a "soft" pine) and contains little resin. In addition to ships' masts, it was used for structural lumber, boxes, crates, matchsticks, and almost every other conceivable purpose. Although this species is still used when it is available, most of the white pine timber that is currently used is cut from stands of western white pine (*P. monticola*). The western white pine is found in extensive stands in the Pacific coniferous belt and to some extent in the Rocky Mountain belt. This species has wood that is slightly stronger than its eastern relative and is shipped in huge quantities from British Columbia, Washington, Idaho, and Montana to be used as structural timber. Both of these pines are subject to white pine blister rust (*Cronartium ribicola*), an extremely destructive disease fungus. The red or Norway pine (*P. resinosa*, Figure 25.18*b*, p. 374) is found in much of the northern coniferous belt, but the thickest stands are in the Great Lake states. This pine is heavier (denser) than white pine and is used extensively for general construction, crates, pulp, and other common uses. The tree is also being used as a common reforestation species in some white pine ranges because it does not require the same high level of soil fertility that is necessary for white pine.

The jack pine (*P. banksiana*) is not very important as a timber tree because of its stunted growth and persistent low branches. These low branches are necessary, however, for the nesting of an endangered species of bird, the Kirtland's warbler. Apparently the two species, tree and bird, have evolved together in such a way that the bird is completely dependent upon the tree. Also, the tree will grow on very sandy, poor soils, and is important as an invader of such areas. The trees will persist until finally the soil is built up and can support the more demanding species. When this happens, the jack pine is soon shaded out. The species is found throughout the northern coniferous belt, and extends farther north than any other species. It is used in various local areas as a rough lumber tree, but is becoming more and more important in the pulp industry.

Pitch pine (*P. rigida*) is found along the eastern part of the northern coniferous belt. This species is very resinous, and was used as a source of tar and turpentine at one time. Because it produces a hot fire, it was used in wood-buring steam engines. The wood is quite durable and was used as water wheel timber for grist mills. It is now used mainly for rough construction, posts, crating, and pulp.

The southern yellow pine is really the collective name of four species of pine that are found in the southern coniferous forest. These species are longleaf pine (*P. palustris*), slash pine (*P. elliotii*), shortleaf pine (*P. echinata*), and loblolly pine (*P. taeda*). These pines are not usually distinguished from each other on the market, but are used for construction timbers, ties, interior lumber, pulp, furniture, crates, excelsior, and reforestation. This group of trees is also used for the turpentine and resins that it produces. The longleaf pine is particularly important in the production of these materials. The crude product is found in the resin canals of the plant, and is known as **pitch.** Because pitch was used to treat the lumber used in building wooden ships and to caulk the seams, the pitch and its derivatives came to be known as **naval stores.** The pitch is obtained from the tree by tapping or chipping away the outer bark and collecting the exudate. The stumps, tops, limbs, and culled trees are used in a distillation process that yields their naval stores. The raw pitch is steam distilled. During this process the spirits or oil of turpentine is driven off and the resin remains as a heavy residue. The turpentine is used as a solvent and thinner in paints and varnishes and the resins are used in the manufacture of soap, varnishes, waxes, lubricants, and many other products. A fossil resin from an extinct pine is known as **amber.** This resin is presently used as jewelry because of its ability to accept a high shine. In ancient times, however, it was sometimes also used as a medium of exchange.

The ponderosa or western yellow pine (*P. ponderosa*) is an extremely important species found in the Rocky Mountain and Pacific forest belts. Because it may get to be 190 to 200 feet tall and eight feet in diameter, and because its wood is light although hard and strong, it is widely used as a lumber tree. Some specimens have been found that are 500 years old. The pinyon pine (*P. edulis*) is found in many areas of the Rocky Mountain belt and was important because of the pinyon "nuts" (really seeds) eaten by the Indians of the area. Scotch (*P. sylvestris*) and red pine (*P. resinosa*) have become very important in recent years as plantation trees grown for use as Christmas trees. Other species of pines are used for this purpose, along with some spruces and firs, but these two pine species contribute most of the Christmas trees.

The largest pine species is the sugar pine (*P. lambertiana*). Specimens of this tree have been known to reach more than 200 feet in height and 14 feet in diameter. It is confined to certain areas in the Pacific coast belt, and is harvested for the same uses as the western white pine with which it shares its range.

Red spruce (*Picea rubens*) and white spruce (*P. glauca*, Figure 25.12*b*) are the two most important spruces of the northern coniferous forest. They are important as timber trees as well as being used for sounding boards for musical instruments, oars, and Christmas trees. Because they add no flavor to food, they are often used to make the crates for shipping food. White spruce is often used extensively for reforestation. Although many species of trees (including some angiosperms) are used for pulpwood, these two spruces are highly favored for such use.

The most important of the spruces is the Engelmann spruce (*P. engelmannii*) which is found in the high ranges of the Rocky Mountain belt. Although light in weight, the wood is strong and is valuable for aircraft production, doors, sashes, and interior lumber. The blue spruce (*P. pungens*), which occupies a relatively small range in the same forest belt produces new growth that has a definite blueish appearance. This color, and the beautiful symmetry of the tree, makes it valuable as an ornamental.

The largest of the spruces is the Sitka spruce (*P. sitchensis*) which is found in a narrow strip of the Pacific coastal belt. This tree may get to be 300 feet tall and possess a diameter of 16 feet. It is important in the airline industry, for crates, and as pulpwood.

Hemlock (*Tsuga canadensis*) is not important as a timber tree, but has been used in the past as a source of tannin. Its western relative (*T. heterophylla*) in the Pacific belt is important for both timber and pulpwood. Both of these species are finding more and more use as ornamentals.

Balsam fir (*Abies balsamea*) is found extensively in the northern coniferous forest. It is important as a pulpwood tree, but some lumber for interior use is made from it. It is widely used as a Christmas tree because it will hold its needles well. Oils that are used in perfumes and soaps are distilled from its bark and needles. Canada balsam is an exudate that forms blisters on the bark. This material is refined, and because of its optical properties is used to mount specimens onto slides for microscopic examination. The white or Colorado fir (*A. concolor* Figure 25.12*c*, p. 374) is found in the Rocky Mountain belt and is often marketed with some of the Pacific firs. The latter include the noble fir (*A. nobilis*) which produces some of the best fir timber that is useful for boats, boxes, and interior woods such as flooring and trim. In this same range is to be found the largest of the firs, the red fir (*A. magnifica*). It may get to be over 200 feet tall with diameters of 10 feet and age of 700 years. It is used as an ornamental and for rough lumber.

Arbor vitae or northern white cedar (*Thuja occidentalis*, Figure 25.12*f*, p. 374) is a relatively small tree that is valuable as an ornamental. Several horticultural varieties have been developed and these are widely planted. Because its wood will not shrink or warp, it has been used for things such as boats, posts, shingles and cisterns where water contact is inevitable. The western red cedar (*T. plicata*) is found in the Pacific belt and is much larger than the eastern species. Its wood is much more important and is used widely for ornamental wooden shingles.

A conifer that invades the swampy areas in the southern coniferous forest is the bald cypress (*Taxodium distichum*, Figure 25.12e, p. 374). The wood of this tree is extremely resistant to decay and is used for water tanks, ships, ties and coffins. The extensions of the root system that form the cypress "knees" have been much collected, polished, and used in ornamental carvings and lamps.

The eastern red cedar (*Juniperus virginiana*, Figure 25.12*g*, p. 374) is found in the southern coniferous belt and throughout much of eastern United States. It is important in colonizing open areas of land, and will grow on rather poor, sandy, dry soil. The wood is durable and is locally used for fence posts. Because of its fragrance and beauty of grain, the wood is also used for cedar chests, closet lining and paneling, ornamental boxes, wood carvings, and pencils. Several horticultural forms that are important as ornamentals have been developed. The incense cedar (*Libocedrus decurrens*) is found in the Pacific belt. Although it is not a member of the same genus as eastern red cedar, it is put to many of the same uses. One of the unique uses to which the fleshy cones of the junipers (mainly *Juniperus communis*) is put is in the flavoring of gin. The mash used to prepare the alcoholic portion of the drink is composed of a corn-malt-rye mixture, molasses, or almost any carbohydrate source. The aromatic oils from the cones are added after the liquor has been distilled. Other flavors may be added as well, but the juniper "berries" are the basic ingredient. Because the flavoring is added soon after distillation, no unnecessary time is wasted in aging the drink before it may be presented for consumption.

Douglas fir (*Pseudotsuga taxifolia*) is not a true fir (*Abies* spp.) but is probably the major lumber tree in North America today. It reaches magnificent proportions in the Pacific coastal belt. The wood is strong, durable, workable, and attractively grained. It is in great demand for heavy construction, plywood, and interior trim. Prime logs are often used in the production of veneers for furniture and paneling. Because it grows rapidly, it can replace itself in cut over stands rather quickly. Proper reforestation and protection from fire make this species one of our best hopes for future wood products.

Finally we come to what many botanists consider to be the most impressive and magnificent genus of trees (Figures 25.18*g*, p. 374 and 25.22). The genus *Sequoia* was named after an important Indian chief who invented a phonetic alphabet and taught his people to read and write. A more fitting tribute cannot be imagined. There are only two extant species of *Sequoia*. These are the coast redwood

(*S. sempervirens*, Figures 25.18*g*, p. 374 and 25.22) which is found in a narrow coastal belt in northern California and southern Oregon, and the giant redwood or "Big Tree" (*S. gigantea*) found only in a few groves in the Sierra Nevada mountains in California. The latter tree is sometimes removed from the genus *Sequoia* and placed in the genus *Sequoiadendron*. The coastal redwood is much the more economically important of the two species (Figure 25.18*g*, p. 374), although it is the smaller one. The distribution of this tree seems to be dependent on the fog that rolls onto its range from the Pacific Ocean. Although the wood is brittle, it is light, strong, and resistant to decay. It finds important uses in large construction timbers, shingles, doors, coffins, and furniture. New growth from old stumps are often marketable with 50 years after the cutting of the parent tree.

The giant redwood is not a good lumber tree. However, its spectacular size and old age make it an important tourist attraction. The members of this genus are receiving more and more protection from conservationist groups. The "big tree" could soon become extinct without such protection; such a loss would be unforgivable. One needs to spend only thirty minutes in a grove (Figure 25.22*a*) of either of these trees, and he can surely understand the necessity of their continued existence.

This long, but very incomplete, discussion of the relevance of the conifers to man is an attempt to view that relevance from the standpoint of first observing the plant as a living organism rather than from the standpoint of first viewing a product and then trying to relate the product to the plant. It is hoped that this approach, without a detailed account of industrial processing, will stimulate an awareness within the student that these are living organisms with which we must learn to interact, conserve, and properly use or we may soon be without them. Perhaps with a better appreciation of the organism, we can gain a greater understanding of the environment and the necessity of its preservation.

EPHEDRA, WELWITSCHIA, GNETUM—GNETOPHYTES

This group of plants contains three genera that have long been considered to be related although extreme variations in structure have also been recognized. Some botanists place them together in the order Gnetales but divide them into three separate families. Others would separate them at the order level, placing one genus in each

Figure 25.22
Sequoia sempervirens. (*a*) An open grove of mature trees. Note the man standing at the base of one. (*b*) A single tree that has been cut about half way. (Photos courtesy of the Chicago Field Museum of Natural History.)

(b)

order. Because this treatment is not particularly concerned with taxonomy, they will be treated here as a distinctive group, the members of which are probably not very closely related to each other.

The genus *Ephedra* (Figure 25.23) contains about 40 species, and will be the one considered here in detail because it is more readily available for laboratory study than the other two genera. *Welwitschia* (Figure 25.25*b*) contains but a single species (*W. mirabilis*) while *Gnetum* (Figure 25.26) contains about 30 or more species.

Some of the characteristics found among the gnetophytes are also found among the flowering plants. Because of these similar characteristics, some botanists have proposed that the gnetophytes represent a link in the ancestral line to the flowering plants. Close study of the evidence indicates that such an evolutionary pattern probably did not occur, but that the three genera represent the end products of quite diverse and specialized evolutionary lines. Also, this group is quite a bit younger than the flowering plants. The fossil record indicates that it goes back only as far as the Tertiary Period (60 to 63 million years ago) whereas flowering plants are known to have extended back to the Cretaceous and Jurassic Periods (125 to 150 million years ago).

Figure 25.23
Ephedra trifurca. A mature sporophyte. (Photo courtesy of the Chicago Field Museum of Natural History.)

The gnetophytes are considered to be gymnosperms because they all produce naked ovules. In addition, *Ephedra* possesses archegonia and embryogeny much like the gymnosperms. Characteristics that are not gymnospermous are the vessels that are produced in their xylem and the compound male and female strobili.

Most species of *Ephedra* inhabit dry, desert areas and are thus xerophytic. They are found in both the Eastern and Western Hemispheres; in North America six species are found in the deserts of California, Nevada, Arizona, New Mexico, Utah, and Mexico. *Gnetum* (Figure 25.26) species require much more moisture, and they are found in the tropical areas of Asia, Africa, and South America. *Welwitschia mirabilis* is found only in the deserts of the coastal regions of Southwest Africa.

SPOROPHYTE GENERATION

The sporophyte of *Ephedra* (Figure 25.23) is usually a small (2 to 6 feet tall), much-branched shrub although some species are trailing plants and others are climbers. One South American species becomes quite treelike.

The leaves of *Ephedra* are macrophylls but are much reduced and are either opposite or in whorls of three (Figure 25.24*a*). They are produced at each node of the stem and often are united producing a whorled sheath. Because of the segmented appearance that this gives to the stem, the plant often superficially resembles either *Psilotum* or *Equisetum*. The generic name means "horsetail," which would express the resemblance to *Equisetum*. Although the leaves may be green for a time, their photosynthetic contribution is minimal. The younger stems contain chlorophyll and carry on most of the photosynthesis.

The stems grow from an apical meristem. The young shoots have a large pith composed of parenchyma cells and that is surrounded by vascular bundles. Outside of these is a cortex composed of photosynthetic and sclerenchymatous cells. The cells of the epidermis have very thick walls. The stomata are sunken on the slopes of the ribs. These characteristics conserve the water supply of the plant and allow it to live under xeric conditions.

In all three genera a vascular cambium produces secondary xylem that contains vessels. These structures are not produced in the other gymnosperms and in only a few of the lower vascular plants. Most of the xylem of the flowering plants contains vessels, and this

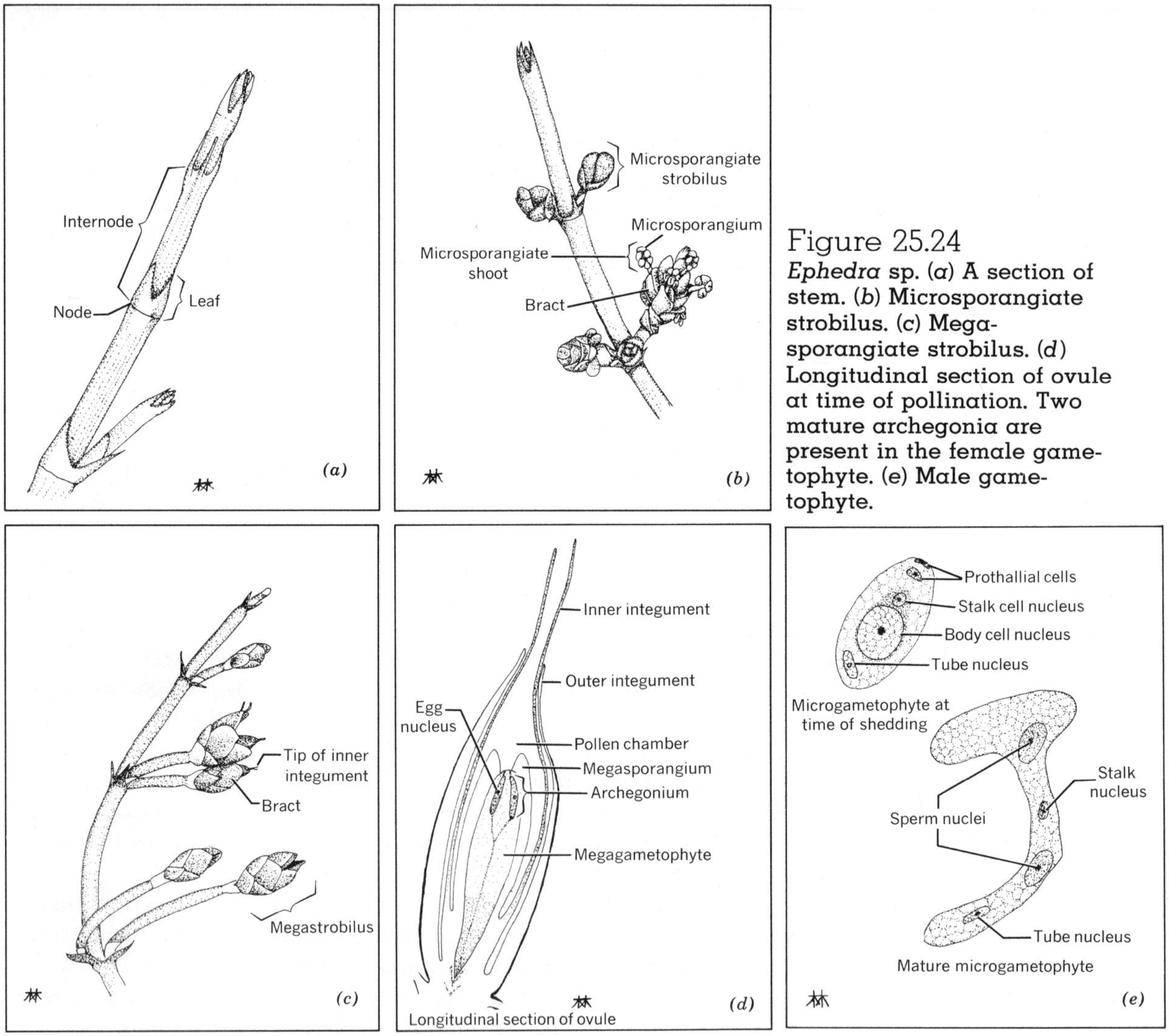

Figure 25.24
Ephedra sp. (*a*) A section of stem. (*b*) Microsporangiate strobilus. (*c*) Megasporangiate strobilus. (*d*) Longitudinal section of ovule at time of pollination. Two mature archegonia are present in the female gametophyte. (e) Male gametophyte.

fact was used as partial evidence for the presumed ancestry of the gnetales for the flowering plants. However, the vessels of the gnetales are produced in a much different way than those of the flowering plants. The stem is attached to the soil by a large taproot and many adventitious roots that arise from rhizomes. New plants may also arise from buds produced on the rhizomes, thus asexually producing a clump of plants. The most unusual sporophyte produced by any plant is probably that of *Welwitshia mirabilis* (Figure 25.25) discovered in 1860 by Dr. Frederic Welwitsch. He sent specimens to Dr. Joseph Hooker who named the plant after him. The sporophyte rarely extends more than a few inches above the soil, and is often almost covered. The stem in top view is an elliptical disc that may be 4 to 5 feet across its widest diameter (Figure 25.25*b*). The rest of the plant axis is composed of an extremely long taproot (Figure 25.25*a*) that extends to the water table far below the surface of the soil. Without the taproot, the plant could not survive because the rainfall in its habitat is usually between ⅓ and 1 inch per year. The stem tip soon stops growing except along the edges of the disc, and the disc becomes depressed. Two huge leaves are produced from the disc. These leaves persist from their production to the time of the death of the plant (estimated to be about 100 years old). A **meristem** at the base of the leaves continues to be active and the leaf continues to increase in length. The leaf possesses parallel venation and is thick and leathery. As it continues to increase in length it splits into long ribbons that flap about in the wind. The cells exposed by the split quickly become **suberized** and the plant is not harmed.

Most of the species of *Gnetum* are vines that climb or trail over other vegetation and may reach the tops of the tallest trees. Some species are trees and a few are small shrubs. The members of this genus produce leaves that are net veined and that make the plant look like a dicotyledon. The leaves vary greatly in size and shape.

Most species of *Ephedra* are **dioecious** (Figures 25.24*b* and *c*). Both the microsporangiate (male cone) and megasporangiate (female cone) strobili are compound. That is, the strobilus axis produces pairs of bracts that are under the microsporangiate or megasporangiate shoots. A microsporangiate shoot develops in the axis of each of the bracts in the male cone. One to eight microsporangia develop at the tip of the shoot. When the pollen is mature, the axis of the shoot elongates and carries the microsporangia out away from

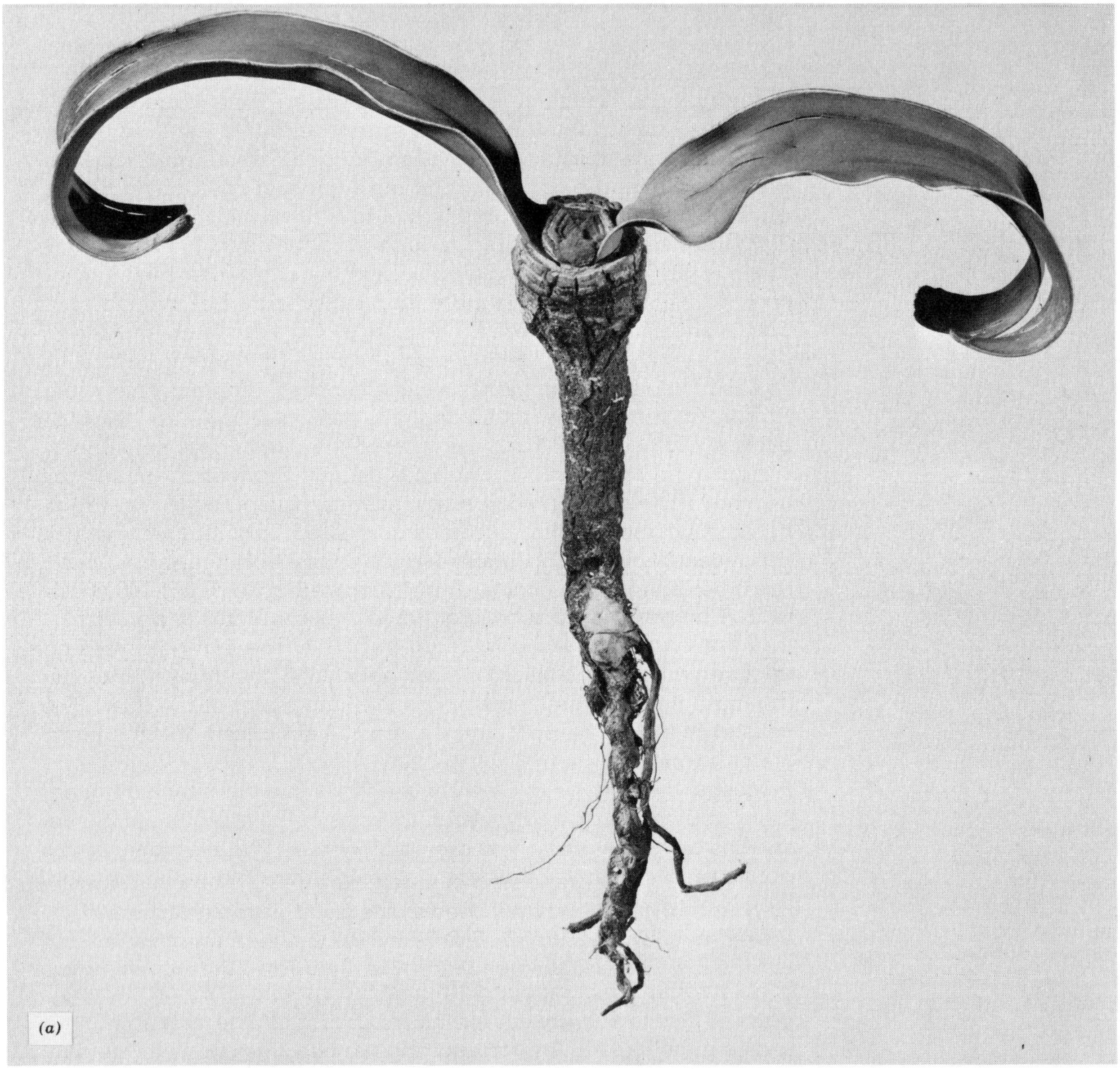

(a)

Figure 25.25
Welwitschia mirabilis. (a) Young sporophyte. (b) Habitat view of display prepared by Chicago Field Museum of Natural History. Female plant is in left foreground and male plant is in right foreground. (Photos courtesy of Chicago Field Museum of Natural History.)

the bracts of the cone. The cone is actually a bit more complicated than described here, but the further explanation would involve excessive morphological detail.

The female cone is also quite complicated. However, without all the detail, it too is composed of a central axis with bracts subtending megasporangiate (ovulate) shoots. The shoot consists of a short-stalked ovule; usually only one or two such shoots are produced in each cone. The integument of the ovule extends beyond the bracts of the cone at the time of pollination (Figures 25.24*c* and *d*).

The microsporangium is surrounded by a two-cell layered wall. The internal sporogenous tissue is separated from the wall by a tapetal layer. Each of the microsporocytes separate from each other, undergo meiosis, and produce the usual tetrad of spores. These spores mature into pollen grains that are released by dehiscense of

Figure 25.26
Gnetum gnemon. Mature sporophytes. From a display prepared by the Chicago Field Museum of Natural History. The branch at the left has microsporangiate branches; the one to the right has megasporangiate branches with maturing seeds. (Photo courtesy of Chicago Field Museum of Natural History.)

the sporangial wall. The ovule consists of a megasporangium (nucellus) surrounded by the integument. A single megasporocyte is differentiated deep within the nucellar tissue, and it finally produces a linear tetrad of megaspores following meiosis. As usual, only one of the megaspores remains functional; the others abort.

GAMETOPHYTE GENERATION

The male gametophyte (Figure 25.24e) has developed inside of the pollen grain (endosporic development) before the pollen grain is shed. At this time it consists of two prothallial cells, a stalk cell, a body cell, and a tube nucleus. Variations on this pattern have been reported for other species of *Ephedra*. The pollen is distributed by the wind, comes into contact with the extended integument and either falls or is drawn into a pollen chamber at its base.

The megaspore (the first cell of the female gametophyte) undergoes a series of free nuclear duplications (producing 256 to 1000 nuclei) and these are finally incorporated into cells by the production of cell walls. One to three archegonia are then produced at the micropylar end of the gametophyte (Figure 25.24*d*). The archegonium of *Ephedra* produces a many-celled neck, a ventral canal cell nucleus and an egg nucleus. Neither *Welwitschia* nor *Gnetum* produce archegonia. The female gametophytic cells outside of the base of the archegonium become filled with stored metabolites. The egg can now be fertilized.

Fertilization usually occurs within 10 hours after pollination, an unusually short time interval for a gymnosperm. The exospore portion of the male gametophyte begins its development within the pollen chamber. The pollen tube begins its growth toward the egg. The body nucleus duplicates producing two sperm nuclei that are nonflagellated. When the pollen tube reaches the egg nucleus, it ruptures discharging the sperm nuclei. One of the sperm nuclei unites with the egg nucleus; the other eventually disintegrates.

The zygote undergoes free nuclear duplication until eight nuclei are formed. Each of these nuclei becomes surrounded by a wall that also encloses some cytoplasm. Each of these new cells may develop into an embryo; they are called **proembryos.** As many as six of the proembryo cells may continue to develop. The proembryo cell produces a tube into which moves the nucleus. Nuclear duplication, followed by cytokinesis occurs. The embryo continues to develop from the lowermost of these two cells; the uppermost one elongates producing a **suspensor** which pushes the developing embryo into the nutritive tissue of the female gametophyte. The embryo then differentiates into two cotyledons and the shoot apex. Although several embryos in a single archegonium may partially develop, only one is present in the mature ovule. The integument develops thick walls around its cells, and the mature seed is shed. About five months have lapsed since the female cone was first recognizable. Under favorable conditions, the seeds immediately germinate.

The life history of *Ephedra* has been used as an example of this group because more is known about it than is known about *Welwitschia* and *Gnetum*. Indeed, all three genera need additional study. However, it should be pointed out that all species of *Welwitschia* and *Gnetum* that have been studied are **dioecious.** It is interesting to note that *Welwitschia* produces its strobili in the axils of its two leaves at the edge of the disc (Figure 25.25). *Welwitschia* is unique among all vascular plants in that tubes extend upward from the female gametophyte toward the descending pollen tube. The tips of the two tubes rupture at their point of contact. One of the female nuclei migrates into the pollen tube, and fertilization occurs there; the fused nucleus then returns to the female gametophyte.

HUMAN RELEVANCE

The gnetophytes are relatively unimportant to man. They do fill particular niches in certain habitats, especially xerophytic ones. They have entered into the folklore of certain peoples. For example, *Welwitschia mirabilis* may have given rise to some of the superstitions about man-eating plants. The long, much split and frayed leaves waving across the desert during a strong wind might appear to an uninformed person that the plant was "seeking" food. If that same person had become entangled in the leaves, he would have what he would consider further evidence for his belief.

Various species of *Ephedra* are important as range plants and have been grazed in some desert areas of the southwestern United States. Some of the low, brushy species are also important soil-binders of the shifting, sandy soil. The American and Mexican Indians boiled the roots and stems to produce medicines for the treatment of genitourinary diseases (hence the scientific name *Ephedra antisyphilitica*). They were also used for a cooling beverage (Mormon tea), and their seeds were used to produce a bitter type of bread. The medicinal properties of *Ephedra sinica*, an Asiatic species, were known by the Chinese as early as 2737 B.C. It is now known that the medicinal properties are attributable to the alkaloid **ephedrin** which is contained in the stem. Ephedrin is used to relieve nasal and bronchial congestion caused by asthma and hay fever. The alkaloid is now mostly chemically produced.

Certain species of *Gnetum* are cultivated as a food plant in some Asiatic countries. For example, *G. gnemon* is raised in Java for its leaves which are cooked in coconut milk. Rope is also made from the plant fibers.

the flowering plants | angiosperms

26

The angiosperms are those plants that produce their ovules inside of a specialized "container" called the **carpel** (Figure 26.1). As the ovule matures into a **seed,** the carpel matures into a **fruit.** This characteristic separates the angiosperms from the previous "naked seed" group, the gymnosperms. The carpel is part of a group of structures called the flower (Figure 26.2). The flower has been variously defined by different botanists, and its evolutionary history is widely disputed. However, for the purposes of the present discussion, the **flower** may be considered as a specialized group of structures that insure the sexual reproduction of the plant. Only two of these structures (**pistil; stamen**) are directly involved with reproduction, but other sterile accessory structures (**sepals; petals**) are often indirectly involved. Not all flowers possess all four kinds of structures, but the discussion of the specific variations will be found in the section dealing with sexual reproduction.

Factors other than the flower hold the angiosperms together in a cohesive, well-defined group. For example, the pollen grain germinates at a distance from the ovule (Figure 26.2). A relatively long pollen tube carries the various male nuclei to the ovule. This situation contrasts with the condition in the gymnosperms where the pollen grain germinates inside of a pollen chamber and in direct contact with the ovule.

The female gametophyte (**embryo sac**) of the angiosperms is usually produced from a single **megaspore** that survives from the meiotic divisions of the **megasporocyte** (Figure 26.3). Although there are variations in the method by which the embryo sac is produced and in the number of nuclei in the sac at maturity, the mature embryo sac contains at least one **egg nucleus** and a nucleus that will combine with a second sperm nucleus producing a food supply, the **endosperm.** Other nuclei are also present, but they abort during the

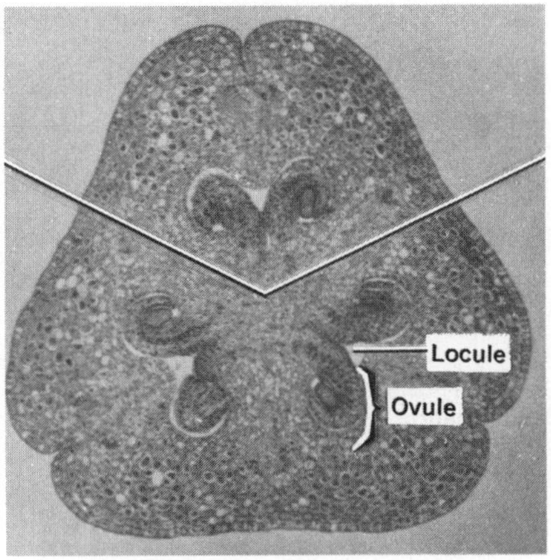

Figure 26.1
Cross section of lily ovary (25×). (Photo by John W. Cieslewicz)

development of the embryo sac. It is important to note that, unlike the female gametophytes of the gymnosperms, the angiosperm does not produce any archegonia.

Finally, the characteristic of **double fertilization** occurs in most of the angiosperms (Figures 26.2 and 26.17). This simply means that in addition to the fertilization of the egg by one sperm, a second fertilization occurs between another sperm and one or more nuclei produced in the embryo sac. This second fertilization produces a food supply (endosperm) that is often triploid (although higher ploidy numbers are sometimes found) and that is used by the developing embryo. Such a food supply is produced only by the angiosperms.

The angiosperms are a geologically young (Middle Jurassic Period; 130 to 140 million years ago) but highly successful group of organisms. Approximately 300,000 species and 12,000 genera are distributed over all the earth's surface in almost every conceivable habitat. They are found in extremely hot and dry conditions as well as submerged in fresh or salt water, and they occupy all of the habitats between the two extremes. Many are epiphytes; some are green but take water and minerals from a host (e.g., mistletoe), while others are nonchlorophyllous parasites (e.g., Indian pipe, Figure 26.5*a*, p. 446). Their sporophytes possess a wide variety of body form. Some are tiny plants that float on top of the water. Others are herbaceous or bushy. Still others produce woody vines or trees. They vary in the length of the life span of the sporophyte. Some live only one

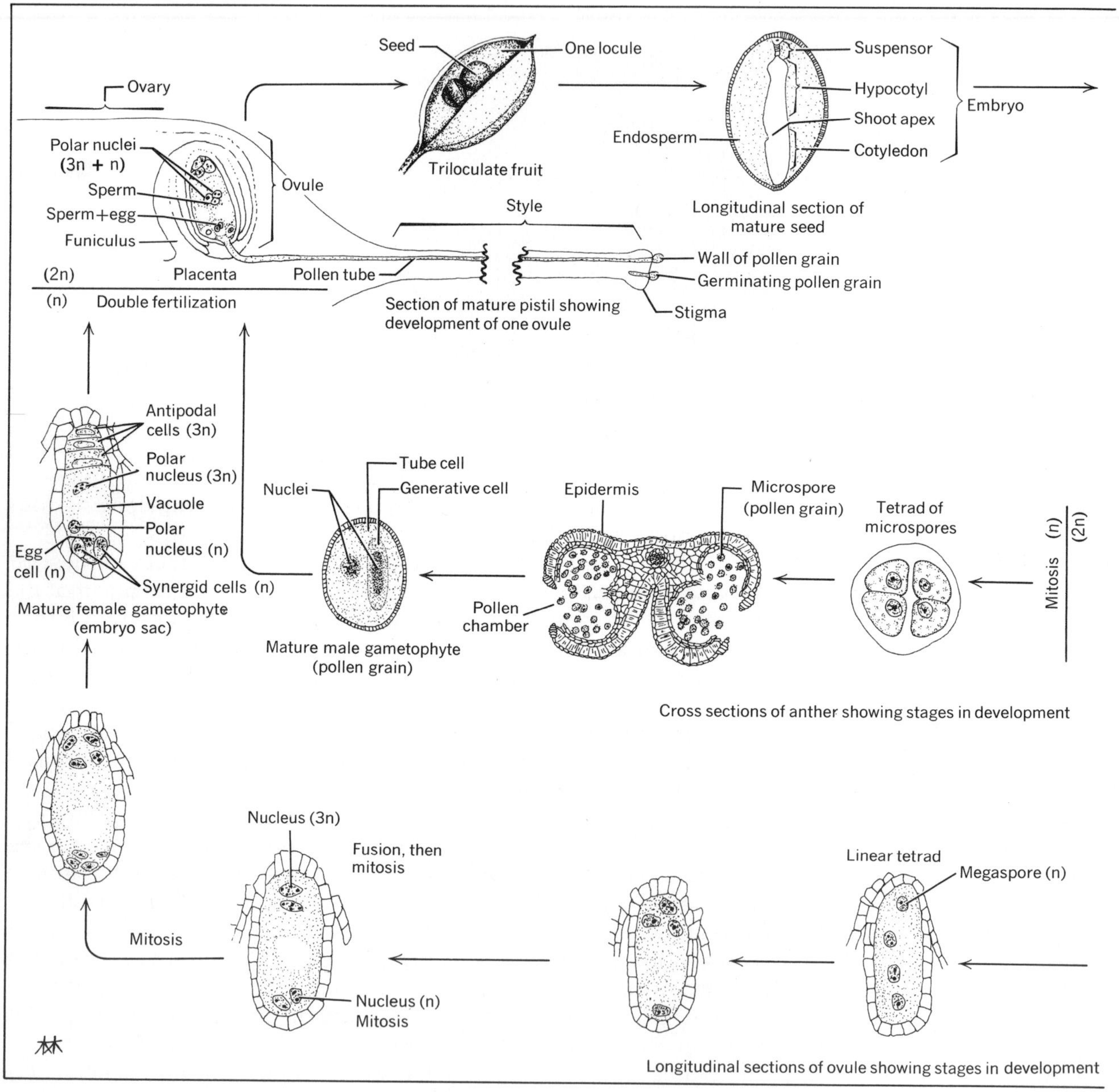
Seed
One locule
Triloculate fruit
Suspensor
Hypocotyl
Shoot apex
Cotyledon
Embryo
Endosperm
Longitudinal section of mature seed
Ovary
Polar nuclei (3n + n)
Sperm
Sperm+egg
Funiculus
Ovule
(2n)
Placenta
(n)
Double fertilization
Style
Pollen tube
Wall of pollen grain
Germinating pollen grain
Stigma
Section of mature pistil showing development of one ovule
Antipodal cells (3n)
Polar nucleus (3n)
Vacuole
Polar nucleus (n)
Egg cell (n)
Synergid cells (n)
Mature female gametophyte (embryo sac)
Tube cell
Generative cell
Nuclei
Mature male gametophyte (pollen grain)
Epidermis
Microspore (pollen grain)
Pollen chamber
Tetrad of microspores
Mitosis
(n)
(2n)
Cross sections of anther showing stages in development
Nucleus (3n)
Fusion, then mitosis
Mitosis
Nucleus (n) Mitosis
Linear tetrad
Megaspore (n)
Longitudinal sections of ovule showing stages in development

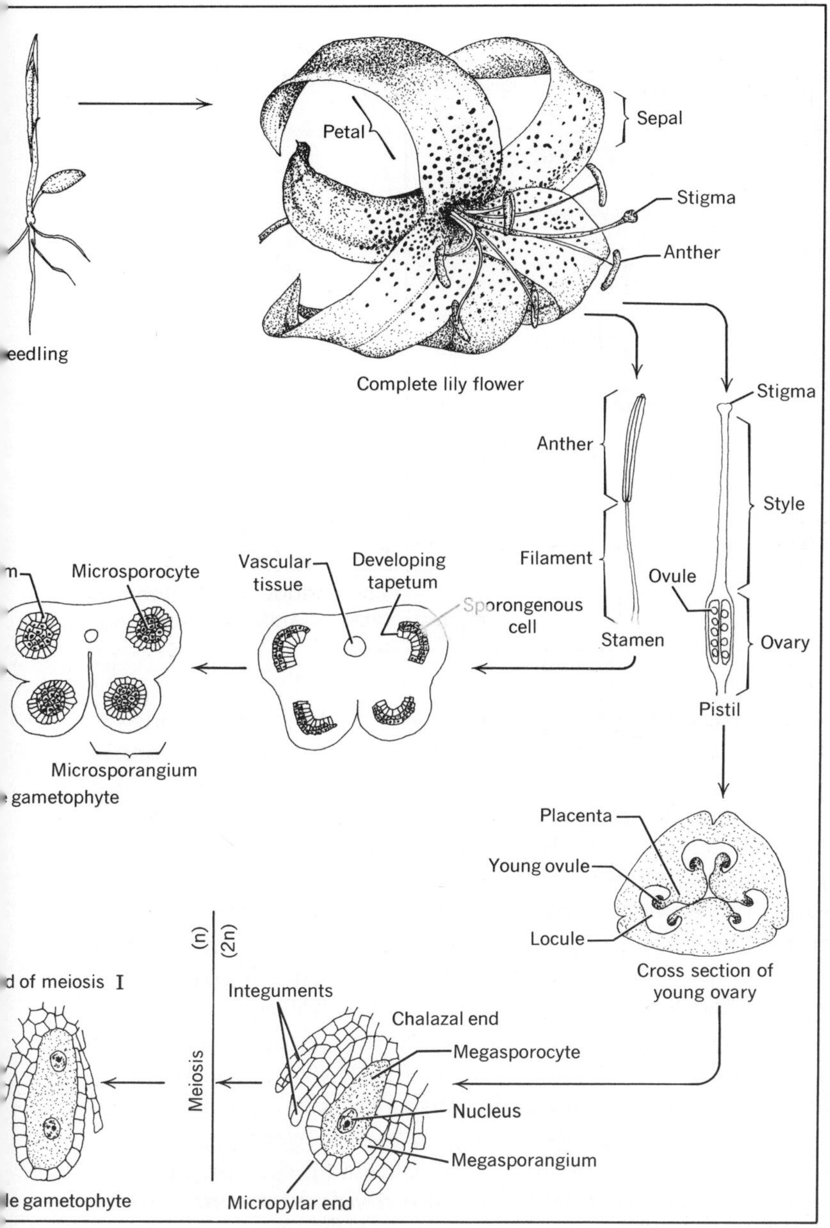

Figure 26.2
Angiosperm life history.
***Lilium* type of embryo sac development.**

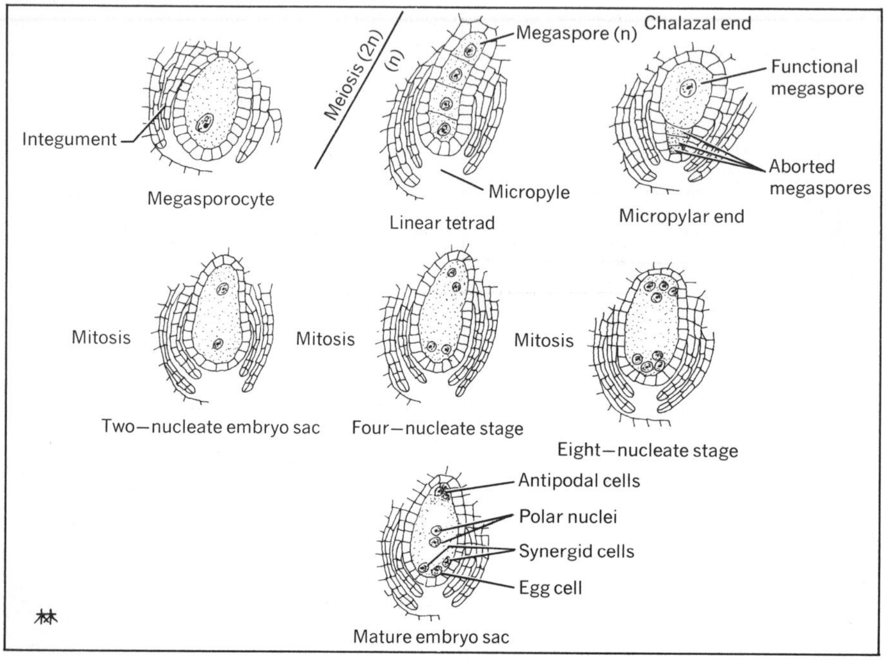

Figure 26.3
The "usual" or *Polygonum* type of embryo sac development.

growing season (**annuals**); others live only two growing seasons (**biennials**), while still others live longer than two seasons (**perennials**).

A group of organisms that possess as much variety as do the angiosperms present great problems in the understanding of their phylogeny. Indeed, it is not at all clear how they originated on the earth. Theories have been advanced that the angiosperms originated from mesozoic seed plants, from gymnosperms, from ancestors common to the Gnetales, and from seed ferns. Even the phylogenetic relationships among the extant plant groups are not very well understood. There are many areas of the world that possess large populations of plants that have never been critically examined by botanists. If these populations are not destroyed by war and agriculture before they have received proper study, man may get more of the answers to the fascinating questions of relationship and origin.

Traditionally the angiosperms have been divided into two major groups, the **monocotyledons** and the **dicotyledons.** These distinctions were originally based on the idea that members of the former group produce only one "seed" or embryonic leaf (cotyledon) and that

Figure 26.4
Seeds and seedlings of three different angiospermous plants. Garden bean: A dicotyledon with food stored in cotyledons. Corn: A monocotyledon with food stored in endosperm. Castor bean: A dicotyledon with food stored in endosperm; the cotyledons become photosynthetic.

members of the latter group produce two cotyledons (Figure 26.4). Additional correlations of structures related to these distinctions were that the monocotyledons produce their flower parts in threes (Figures 26.2 and 26.5*b*, p. 446) or multiples of three and that the dicotyledons produce theirs in fours and fives or multiples thereof. The monocotyledons produce leaves in which the major vascular tissue is produced in bundles (veins) that run parallel to each other (Figure 26.5*c*, p. 446) and the dicotyledons produce leaves in which the bundles radiate from one origin (palmately veined, Figures 26.5*d*, p. 446 and

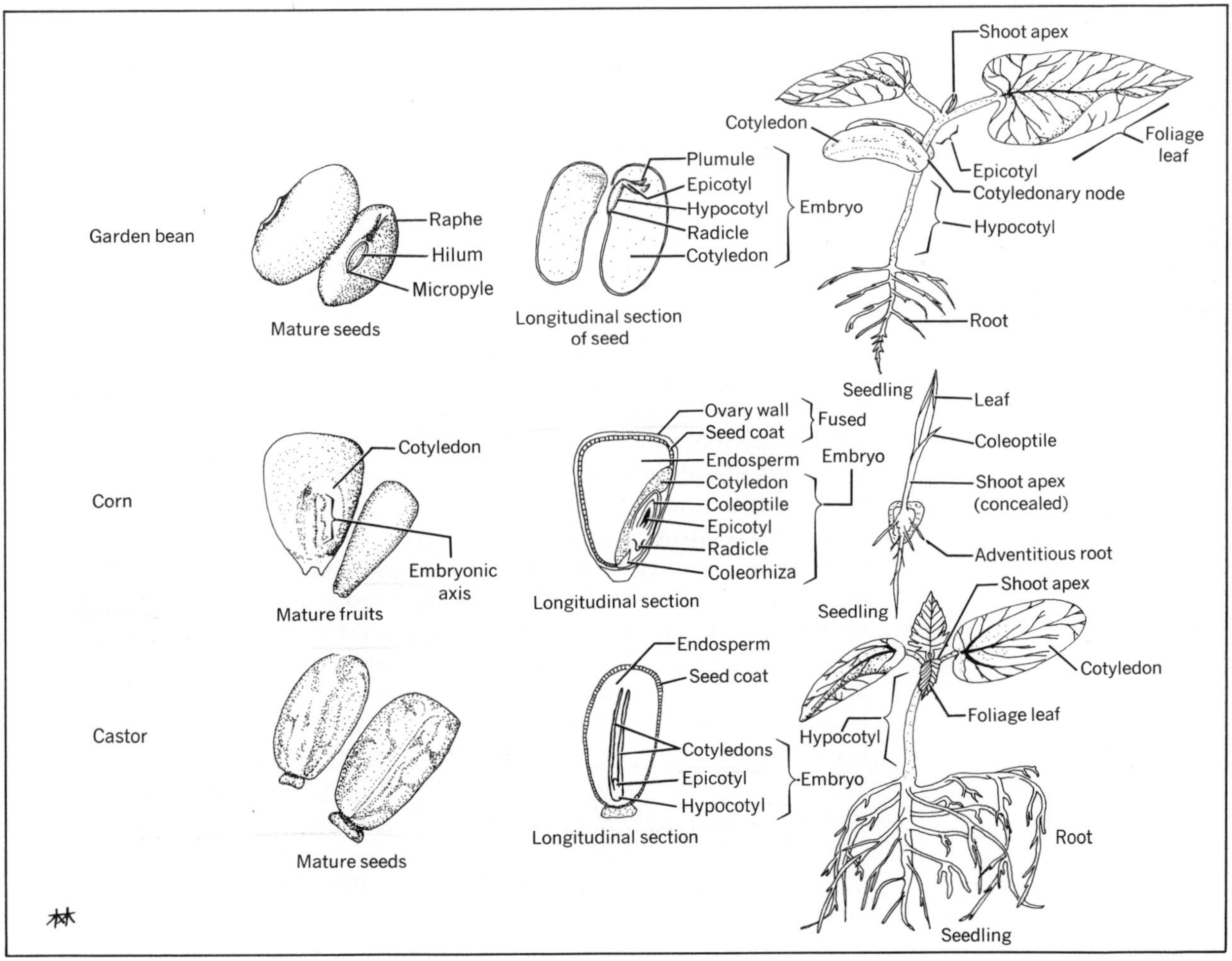

26.21*b*) or branch from a central major bundle (pinnately veined, Figure 26.5e, p. 446). Connecting bundles often produce a netted or reticulate pattern. The vascular anatomy of the stems was considered to be quite different. The monocotyledons possess stems in which the vascular bundles are scattered; the bundles appear in a cylinder in the dicotyledons (Figure 26.6). Although these traditional, classical distinctions may be used to separate most monocotyledons and dicotyledons, many exceptions to them have been found and there is no completely reliable set of morphological characteristics by which to separate the two groups. Neither is there any reason to separate them on the basis of life histories. The outstanding East Indian botanist Maheshwari has clearly stated his view when he says "there are no essential differences between the monocotyledons and dicotyledons as regard the development and organization of the male and female gametophytes and the endosperm, and the process of fertilization is the same in both subgroups." Therefore, the following discussion will be quite general without any distinctions being made about the variations which occur in the various groups of angiosperms. The reader is referred to any one of the many good taxonomic or floristic treatments of the various groups.

SPOROPHYTE GENERATION

As with all of the plant groups since the mosses, the sporophyte is the dominant phase of the angiosperm life history. As has been previously mentioned, the sporophyte may be found in an extremely wide variety of forms and habitats. The sporophyte is almost always composed of three major vegetative regions or structures. These are the **root,** the **stem,** and the **leaf** (Figure 20.1). These three structures have often been defined as plant "organs," but the term does not seem to be appropriate. An organ is usually defined as a group of tissues that work together performing a particular task. In animals the organs seem to be well-defined entities, but in plants the distinction of one structure from another, either structurally or functionally, is much more difficult. For example, the vascular (conducting) bundles are composed of a number of tissue types (e.g., phloem, xylem, parenchyma, sclerenchyma) that function collectively as a means of transporting materials from one part of the plant to another (Figure 26.7). Therefore, the vascular bundle could, by definition, be considered an organ. However, the vascular bundle extends through root, stems, and leaves (as well as flowers when they are

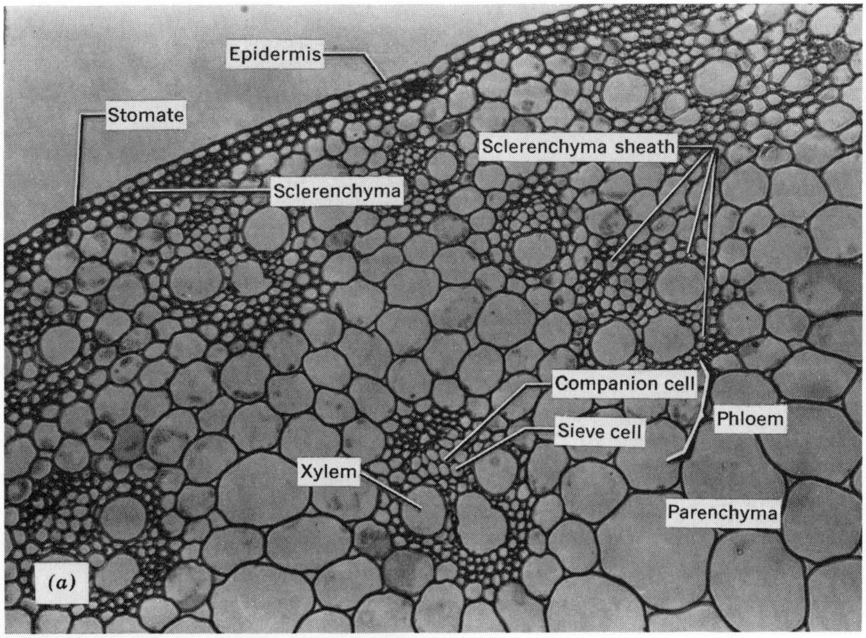

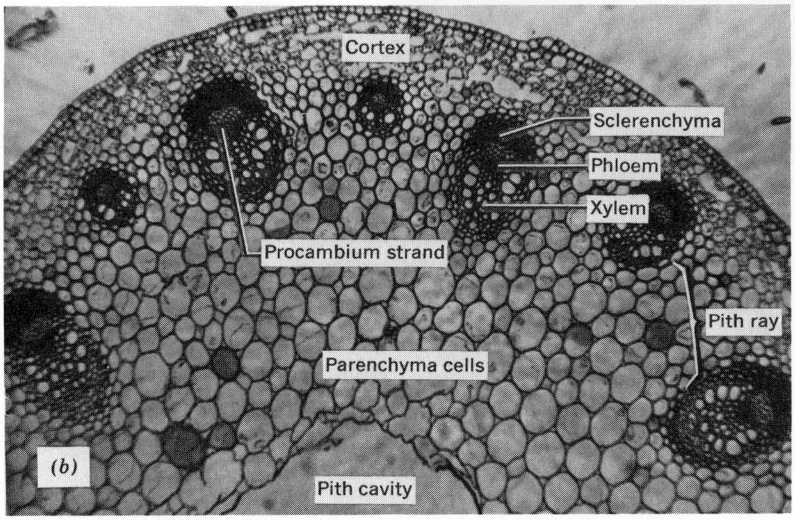

Figure 26.6
Cross section of herbaceous stems. (*a*) Monocotyledon (*Zea mays*) (94×). (*b*) Dicotyledon (*Ranunculus* sp.) (47×).

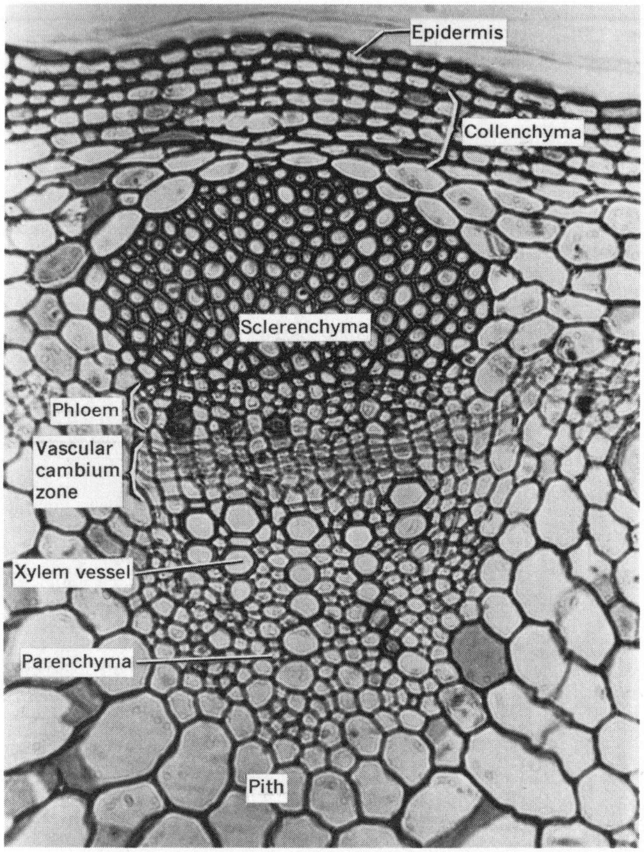

Figure 26.7
Helianthus sp. Cross section of a vascular bundle near the edge of a young woody dicotyledon stem (156×).

present). These structures are the ones that are traditionally defined as organs. Other problems of definition arise when one considers that plants grow in length from the tips of stems and roots. Because of this, a particular region may not remain consistent in either structure or function for very long. The term organ was defined in relationship to animals, but does not fit plants with the same degree of clarity; therefore the structural and functional units of the plant will be referred to as regions rather than organs. The first structure to emerge from most germinating seeds is the embryonic root, the **radicle** (Figure 26.4). The radicle may continue to grow and produce a large, single root (the **taproot**) from which secondary roots may grow (Figure 26.20*a*, p. 446). The radicle may produce a branching root system in which there is no main root. The radicle may also cease growth soon after germination. In this case, most of the plant's root system is produced from parts of the plant that do not originate from the radicle, for example from stems. Such structures that arise from what could be considered unusual or unexpected places on a plant are said to be **adventitious.** Many grasses (as well as other plants) produce most of their root system adventitiously (Figure 26.5*c*, p. 446). One could perhaps argue that since most grasses produce their roots in this way, the location is neither unusual nor should it be unexpected. Therefore, the root system of grasses should not be defined as adventitious. This reasoning points again to the difficulty of definitions in general and the difficulty of the definition of "organ" in particular. A plant organ can certainly not be defined in terms of its origin.

All of the roots grow in length from an apical meristem which is covered by a protective layer of cells called the root cap (Figure 20.1). The outermost cells of the root cap are continuously sloughed off as the root grows through the soil, but new cells are continuously added to the innermost surface by the apical meristem. The other cells produced by the meristem undergo a series of stages in maturation. These stages have been traditionally divided into root zones or regions. That is, the production of new cells occurs in the meristematic zone. As these cells increase in size, they reach a stage in maturation called the **zone of elongation.** As these same cells continue to mature, they begin to differentiate into the various functional types such as xylem, phloem, epidermis, and cambium. This stage of maturation is called the **zone of differentiation.** Finally extensions of the epidermal cells extend into the soil. This stage of maturation is known as the **root hair zone** and it is in this area that most of the absorption of water and minerals by the plant occurs. All of the stages of maturation are present on a young root tip at any point in time, but it should be emphasized that they represent different ages of the individual cells rather than cells that remain structurally and functionally consistent. Cells that are intermediate in their development may be found in the transitional areas between any two zones.

The root posterior to the root hair zone does not function as an absorptive structure. Its outermost cells become cutinized and thus impermeable to water. The internal tissues of this root area are quite functional, however. The differentiation that starts in the zone of differentiation continues until definite tissue areas are recognizable. In the young root, these tissues are the epidermis, the cortex, and the stele (Figure 20.2*b*). The epidermis is composed of a single layer of cells, some of which develop an elongation of their outer wall forming a root hair. The functional life of a root hair is usually only a few days or weeks although some (e.g., redbud) may persist for years. New ones are constantly being produced by the young cells in the zone while the hairs of the older cells continually die. Although absorption does occur through root hairs that increase the surface area of the root, much absorption occurs directly through the epidermal cells that do not produce root hairs. Indeed some aquatic plants and some gymnosperms do not produce root hairs at all and must rely completely on the absorption that occurs through their epidermal cells.

The cortex is a relatively thick layer of parenchyma cells that is found under the epidermis (Figure 26.8). The cells accumulate excess photosynthates, and in many plants large quantities of these foods are retained. Large intercellular spaces are found between the parenchyma cells, and water and minerals diffuse through these spaces (and through the cells) toward the center of the root. The innermost surface of the cortex is a single cell layer called the **endodermis.** In many plants the endodermis cells are surrounded by a suberized wall thickening called the **Casparian strip.** This layer is impermeable to water and thus appears to force materials to go through the endodermal cells instead of around them. Because this forces the water and dissolved materials to pass through the selectively permeable plasma membrane of the endodermal cells, many botanists have insisted that the mechanism provides for a selectivity of materials that reach the center of the root.

The portion of the root that lies within the endodermis is called the vascular cylinder or **stele** (Figure 26.8). The tissues of this structure are the ones that conduct the water, minerals, and previously stored food from the root upward to the stems and leaves and excess photosynthates downward from the photosynthetic organs to the root. The outermost layer (or layers) of cells of the stele is called the **pericycle.** The pericycle is a meristematic tissue that produces branch roots as well as cells that become conductive tissues on its inner surface. The pericycle contributes to part of the vascular cambium in those roots that have secondary growth and also to a cork cambium. The conducting (vascular) tissues are internal to the pericycle. This region is usually composed of a solid core of xylem with "arms" that extend outward toward the pericycle. Patches of phloem are produced in the spaces between the "arms." Most roots do not

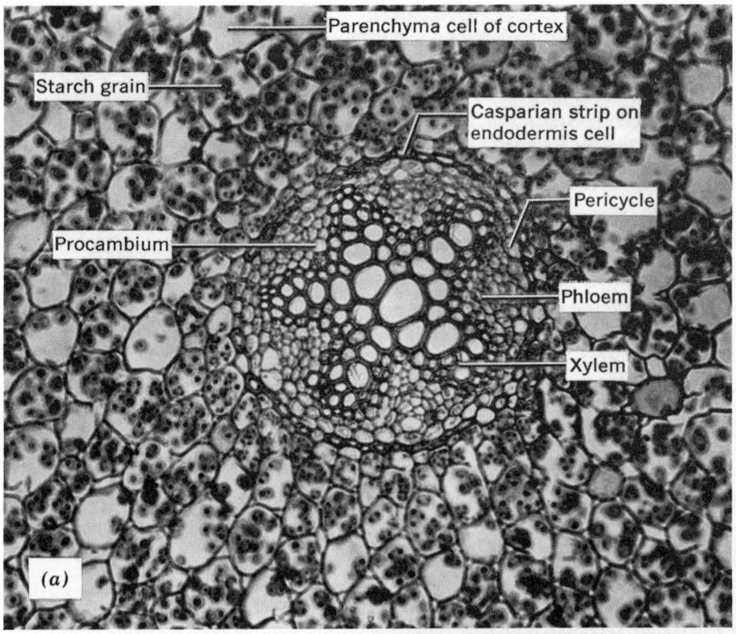

Figure 26.8
(*a*) Cross section of the stele and part of the cortex of a young *Ranunculus* sp. root. (170×). (*b*) Cross section of a three-year-old woody root (15×). (Photo by John W. Cieslewicz).

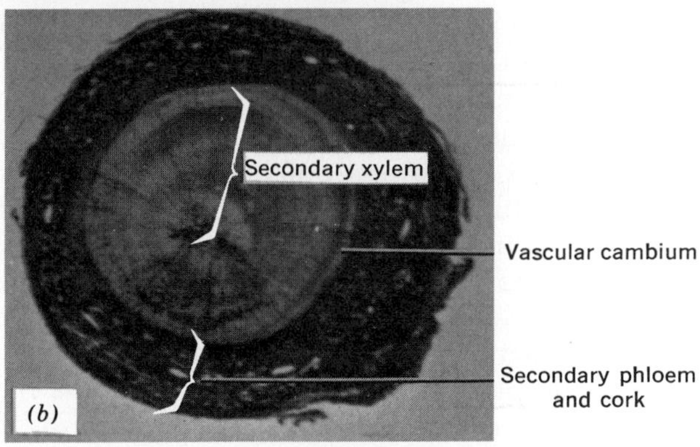

possess pith in their center, but some (especially in some monocotyledons) do. A layer of meristematic cells, the **procambium,** lies between the xylem and the phloem.

As the roots continue to grow, the procambium cells and the pericycle cells at the ends of the xylem "arms" produce a continuous layer of vascular cambium in those roots that will have secondary growth. This vascular cambium functions in the production of xylem cells towards its innermost surface and phloem cells towards its outermost surface. Eventually a solid ring of secondary phloem surrounds a solid core of secondary xylem with the vascular cambium continuing to function between them (Figure 26.8*b*). The epidermis and cortex are finally ruptured and sloughed off. A cork cambium is produced first from some of the old pericycle cells (which is also finally sloughed off) and then later from older secondary phloem. The cork cambium produces cork cells that are impermeable to water, and thus this older root ceases to function as an absorptive organ. Monocotyledonous roots do not produce secondary growth, and such growth in dicotyledonous roots varies from none to large quantities.

The stem is the most complex vegetative structure in the angiospermous plant. Because of the many types of stems that exist, and the extreme kinds of variation that they possess, only representative types of stems will be discussed here. These representatives will include a monocotyledonous stem and a dicotyledonous woody stem. More detailed discussions of stems can be found in general botany, plant morphology, or plant anatomy texts.

Although there is a great deal of variation among stems, they all provide very important functions that enable the plant to survive. They support and elevate the leaves and flowers of the plant. This action places the leaves in a position of exposure to the sunlight and allows for an increased photosynthetic rate. Because the flowers are elevated they are placed in a position more readily available to wind and insect pollination. Also, when the fruits and seeds are produced, they may more easily be caught by the wind or carried by birds and other animals insuring wider distribution. Second, the stems provide a means of transport, through their conducting tissues, between the various plant structures. This allows for the movements of water, minerals, dissolved foods, and hormones from one part of the plant to another. Third, stems serve as a place for the production of new tissues and structures (Figure 20.1). These include new growth of the internal tissues and the production of leaves, flowers, and buds from the apical meristem at the stem tip as well as increased diametrical growth from the vascular cambium. In addition to these three activities, many stems also store food materials.

Except in a few genera (e.g., *Yucca, Agave*) all monocotyledons have **herbaceous stems.** That is, they do not produce any secondary growth and no wood (secondary xylem) is produced. Many of them do not possess the external features that are most commonly considered as present in stems, but others possess the usual node, internode structures with leaves being produced at the nodes and branches and flowers being produced in the leaf axils (Figure 20.1). The grasses are the best known herbaceous angiosperms, and of them, corn has been much studied. Although there are many perennial monocotyledons, corn is annual.

The external layer of a corn stem is composed of an epidermis under which is found a thick layer of heavy walled **sclerenchyma tissue** (Figure 26.6*a*). This tissue gives the stem strength and is often referred to as the **rind.** The rest of the stem is composed of thin-walled **parenchyma cells** (pith). The conducting tissue is scattered in discrete bundles (vascular bundles) throughout the parenchyma in corn stems. However, there is a concentration of the bundles around the edges of the stem with fewer being found toward the center. In some monocotyledons, the vascular bundles are found only in one or two rows around the outside of the stem. The vascular bundle contains xylem and phloem but no vascular cambium and is surrounded by a bundle sheath composed of sclerenchyma cells.

Although many dicotyledons do not produce any woody secondary growth and are thus herbaceous, almost all of the woody plants (with the major exception of the gymnosperms) are dicotyledons. As is true of all stems, increase in length occurs from the stem tip by means of the meristematic apical meristem. This meristem produces three primary tissues, the **protoderm,** the **procambium,** and the **ground meristem** (Figure 20.1). Leaf and bud primordia are also produced from this area. The protoderm develops into the **epidermis,** the procambium which is meristematic produces the **primary xylem** and **primary phloem,** and the ground meristem produces the **cortex** and the **pith.** The procambium usually first appears as strands that are arranged in a definite circle in the ground meristem. The procambium strand produces a small amount of primary phloem toward its outer surface and a small amount of primary xylem toward its inner surface. Thus, in the young dicotyledonous stem (whether herbaceous or woody) the primary vascular tissue is arranged in a definite circular pattern in the parenchymatous ground meristem tissue (Figure 26.6*b*). The parenchymatous tissue between the vascular cylinder and the epidermis is the cortex; the parenchymatous tissue inside of the cylinder is the pith. The tissue between the bundles that connects the cortex and the pith is known as a **pith ray.** In the herbaceous dicotyledons, the development does not progress much beyond that as so far described. There is sometimes a small amount of secondary xylem and phloem produced, but most of the diametrical increase comes simply from the increase in cell size of the primary cells.

In woody plants, however, several additional events occur. The procambium in the bundles continues to produce new xylem and phloem and stimulates the parenchyma cells at each of its ends to become meristematic (Figure 26.7). This process continues until a complete circle of meristematic cells, connecting the procambium cells in each vascular bundle, is produced (Figure 26.9*a*). After the circle of meristematic cells has been completed and begins to produce new xylem and phloem, it is called a **vascular cambium.** All tissues produced from this cambium are **secondary,** and the secondary xylem is what constitutes wood.

The vascular cambium cells frequently duplicate tangentially (Figure 26.9*b*). If the outer daughter cell differentiates, it produces a secondary phloem cell; if the inner daughter cell differentiates it produces a secondary xylem cell. In either case, the other daughter cell remains as a vascular cambium cell. Occasionally the cambium cell will duplicate radially which adds more cells to the circumference of the vascular cambium. This allows the vascular cambium to expand around the ever increasing quantity of secondary xylem.

As more and more secondary xylem is produced, the pith becomes more and more reduced in comparison (Figure 26.10). The xylem elements (mostly **vessels**) that are produced in the spring (**spring or early wood**) are larger in diameter than those produced in midsummer or fall (**summer or late wood**). The differences in size of the two types of xylem produce a discernable ring between the early wood of a particular growing season and the late wood of the previous season (Figures 26.10 and 26.11). This is known as an **annual ring** and is composed of one layer of early wood and one layer of late wood. Since one annual ring is produced for each growing season in trees of temperate zones where temperature fluctuations determine length of growing season, the age of such a tree can be predicted with a high degree of accuracy. In geographical locations where the growing season is determined by rainfall, the tree-ring information must be analyzed in terms of wet and dry season. The annual ring should not be confused with the **sapwood** and the **heartwood.** The sapwood refers to that part of the xylem that continues to transport water and minerals. The heartwood is

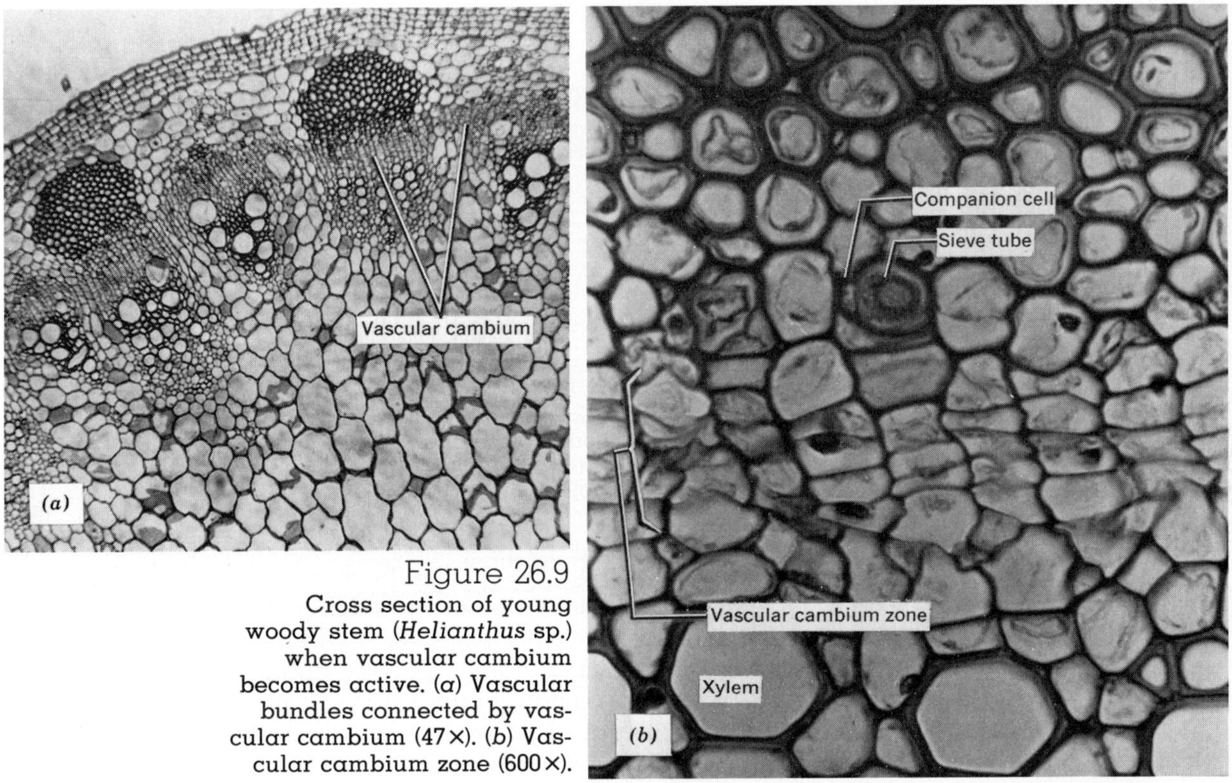

Figure 26.9
Cross section of young woody stem (*Helianthus* sp.) when vascular cambium becomes active. (*a*) Vascular bundles connected by vascular cambium (47×). (*b*) Vascular cambium zone (600×).

that xylem which is no longer functional as a transport system as a result of becoming filled with insoluble waste materials such as **lignins** and **tannins.** As the plant ages, the amount of heartwood increased greatly in relationship to the amount of sapwood.

As secondary phloem is produced toward the outside of the vascular cambium (Figure 26.11), the epidermis and cortex layers are split and eventually sloughed off. Before this happens, certain cells of the cortex become meristematic and produce a **cork cambium** (Figure 26.11) that produces cork cells toward their outside surface. Other patches of cork cambia are produced by old phloem tissue. As in the gymnosperms, the corky tissue blocks the lateral movement of water and minerals from the vascular tissue to the cells external to the corky tissue causing them to die. Eventually all of the cells and tissues external to the corky tissues die and are sloughed off. New cork cambia are constantly produced from old phloem; thus the

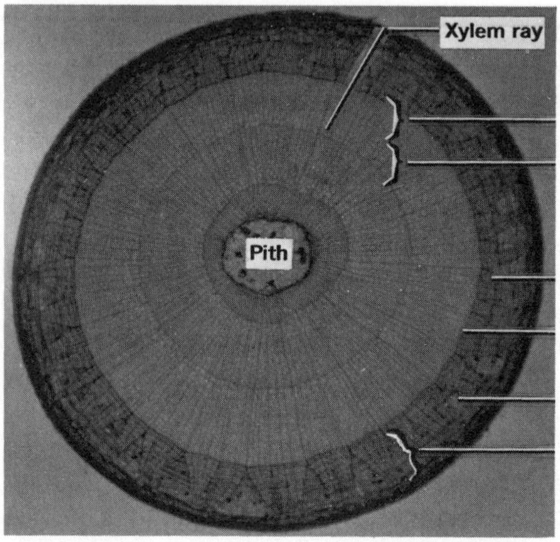

Figure 26.10
Cross section of a three-year basswood (*Tilia americana*) stem (11×). (Photo by John W. Cieslewicz).

process is a continuous one. Therefore, although new phloem cells are continually produced, the relative thickness of the phloem layer remains the same. The total amount of xylem continues to increase, however, because there is no mechanism whereby it is removed.

It should be emphasized that the phloem and xylem tissues are complex. That is, they contain many kinds of cells in addition to the phloem and xylem cells. However, since those topics are best discussed in plant anatomy or comparative plant morphology, their inclusion here seems unnecessary.

The structure in which most photosynthesis occurs is the leaf, although many angiosperms produce stems in which much photosynthesis occurs. Indeed, the photosynthesis of most cacti (Figure 26.12*a*, p. 446) occurs almost exclusively in the stem. As might be expected, the leaf organization and form of leaves are extremely varied. They range in size from tiny and inconspicuous as in the cacti to huge as in the palms and bananas (Figure 26.12*b*). They are variously arranged on the stem, and each of them possesses an axillary bud. Because their vascular tissue leaves a gap in the stele of the stem when it enters the leaf, they are **macrophylls.**

Leaves are often divided into two main parts, the **petiole** and the **blade** (Figure 20.1). The petiole is the stalk which attaches the blade to the stem and through which the vascular tissue travels from the stem to the blade. Many leaves do not possess a petiole and are thus **sessile** on the stem. The blade may be divided into many

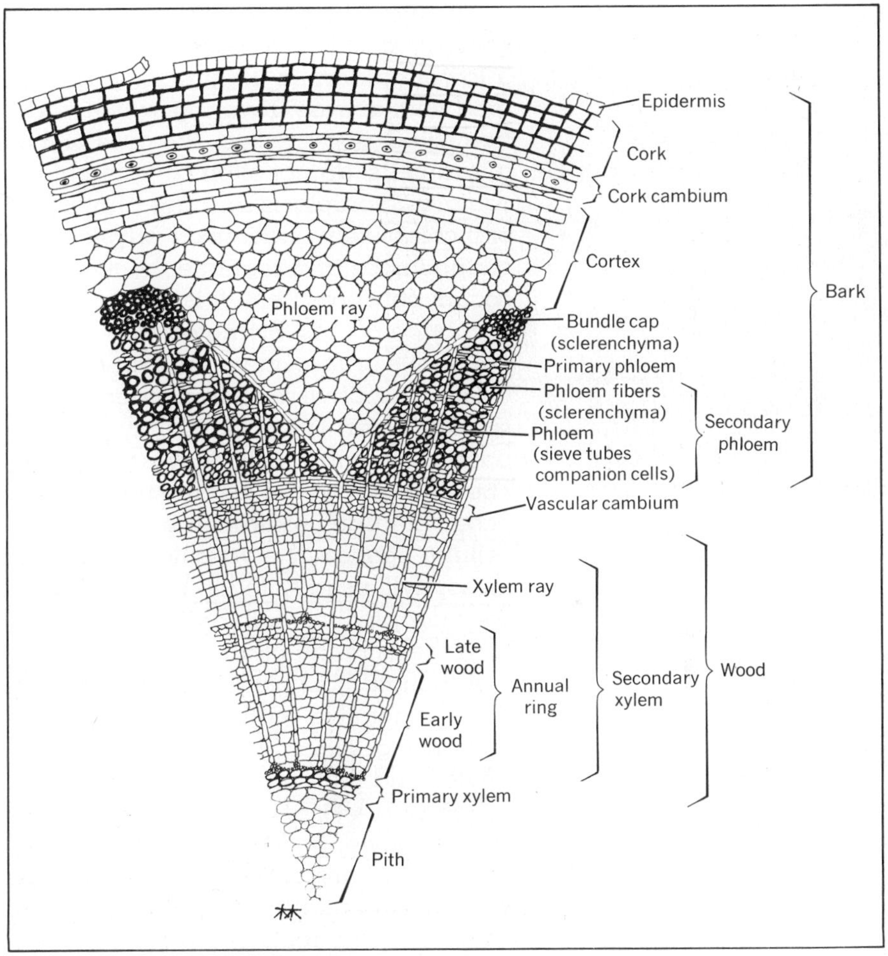

Figure 26.11
A diagrammatic section of a two-year woody stem showing the different cell and tissue types.

smaller parts, called **leaflets** (e.g., walnut, Figure 26.12*c*, p. 446), or it may not be divided (e.g., oak, Figure 26.5e, p. 446) at all. The edge of the leaf may be **entire** (e.g., corn, Figure 26.5*c*), **lobed** (e.g., oak, Figure 26.5e), or **toothed** (e.g., American beech, Figure 26.12*d*). The vascular tissue may branch in various ways after it enters the leaf, but the branching is usually **parallel** (corn and most other monocotyledons, Figure 26.5*c*), **palmate** (castor bean, Figure 26.5*d*), or **pinnate** (oak, Figure 26.5e). Most angiosperms lose all of their leaves at the end of the growing season, but many others (e.g., southern magnolia) are evergreen.

Although the tissues vary in arrangement and relative amounts present, most leaves possess an **upper epidermis,** a **mesophyll** and a **lower epidermis** (Figure 20.1). The epidermal layers often are covered with a waxy covering which varies considerably in thickness from species to species. The cells of the epidermis, except for the **guard cells,** do not often contain chloroplasts. The guard cells do contain chloroplasts and are photosynthetic. Through a long series of incompletely understood physiological reactions, the guard cells change sizes and regulate the size of the stomatal opening that occurs between them. It is through this opening that most of the gas exchange occurs. Stomates may occur in only the upper or lower epidermis or they may occur in both.

The mesophyll is often divided into a **palisade layer** and a **spongy layer.** The palisade layer is usually a single layer of tightly packed columnar cells that is produced under the upper epidermis. The spongy layer is usually under the palisade layer and is composed of large, isodiametric cells that are separated by large intercellular spaces. The vascular bundles are found running through the mesophyll, usually in the spongy layer. The cells of both layers of the mesophyll contain chloroplasts and carry on most of the photosynthesis in the plant. These cells are often the site of food storage.

Although most plants produce their roots below the soil and their other vegetative parts above the soil, there are many exceptions to this arrangement. Some plants such as many orchids grow as epiphytes and their roots never reach the soil. Others such as corn (Figure 26.5*c*, p. 446) produce adventitious roots which grow into the soil from the lower nodes of their stem. These roots always have an aerial portion and are called **prop roots.** Some trees and shrubs such as banyan and mangrove produce prop roots that may be several feet long. The prop roots of the banyan may produce a small dense forest around the parent stem and the mangrove invades the swampy waters with its roots (Figure 26.12*e*, p. 446).

Stems do not always grow above the soil. Many angiosperms (e.g., iris, canna) produce underground stems (**rhizomes**) from which are produced upright flowering shoots. Adventitious roots are also produced from the rhizomes. Some plants such as the Irish or white potato (Figure 26.12*f*, p. 446) produce slender rhizomes that enlarge at their tips and store large amounts of food in a **tuber.** Other plants (e.g., gladiolus) produce thickened underground stems called **corms** and still others (e.g., tulip, onion, Figure 26.13*a*, p. 446) produce **bulbs** which are slender underground stems that are surrounded by thick, food storing leaves. Some plants such as the strawberry (Figure 26.13*b*, p. 446) produce horizontal stems (**stolons**) that run along the top of the soil and produce upright, leafy shoots and adventitious roots that grow into the soil.

Leaves are consistently produced from stems, whether they be above or below the soil. Most leaves are produced above the soil and are thus exposed to sunlight enabling them to carry on photosynthesis. Some plants (e.g., onion), however, produce at least some of their leaves below the soil where they store food. Some of the angiosperms (e.g., Indian pipe, Figure 26.5*a*, p. 446) are parasites on the root systems of other vascular plants and produce no chlorophyll at all in their leaves or stems.

The above examples are merely representations of the extremes of variation that exist among the vegetative structures of the angiospermous plant. These variations pale to almost insignificance when compared to the variability of the portion of the angiosperm that gives the group its name and that appears to be destined by the forces of evolution to insure the survival of the flowering plants. This structure is, of course, the **flower;** the discussion of this extremely important sporophytic sex structure will occupy the rest of the section on the sporophyte.

Although flowers exhibit an infinite variety of color and form, remarkably they are all functionally alike (Figures 26.2 and 26.5*b*, p. 446). Two structures that are produced by the flower function sexually. These are the stamen and the pistil (Figures 26.2 and 26.5*b*). The stamen is usually composed of a stalk known as the **filament** which supports the **sporangia** that make up the **anther.** It is in the anther that the **pollen grains** are produced and thus the stamen is known as the **male** part of the flower. The pistil is the female structure and is usually composed of a flat, sticky tip called the **stigma,** an elongate "neck" called the **style,** and a bulbous base called the **ovary.** The **ovules** are enclosed by the ovary. If a flower contains both of these sex structures, it is called a **perfect flower** (Figures 26.2 and 26.5*b*, p. 446); if it contains only one or the other of the sex structures, it is said to be **imperfect.** A flower that contains only male structures is a **staminate flower** (Figure 26.13*d*, p. 446); one that contains only the female structures is a **pistillate flower** (Figure 26.13*c*, p. 446). Two other sets of structures are also often found in flowers. These are the petals and sepals. They are sterile, but they are often brightly colored

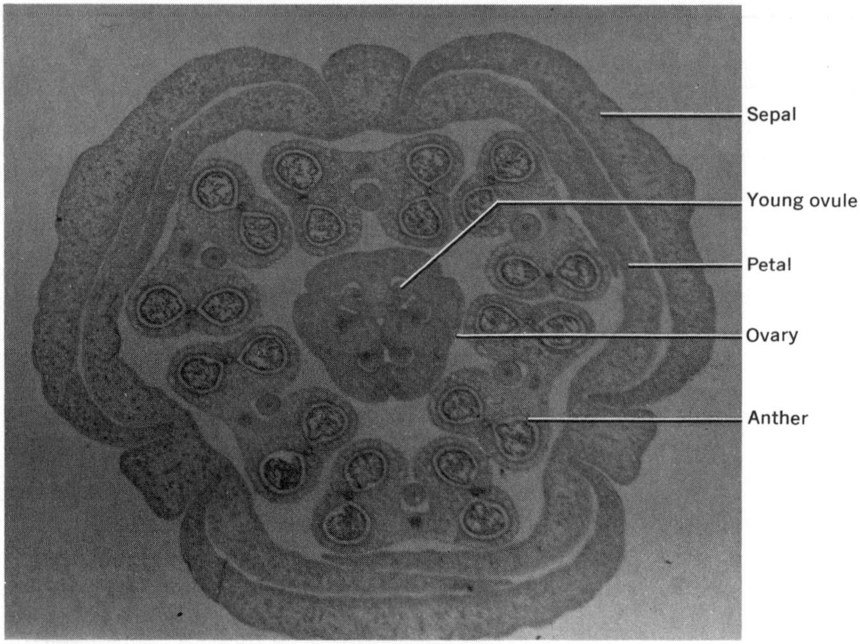

Figure 26.14
Lilium sp. Cross section of flower bud showing all four whorls of floral structures (14×). (Photo by John W. Cieslewicz).

and easily noted. The sepals form the outermost ring of floral structures and the petals form the inner rings until the stamens are reached. The sepals may be of the same shape and color as the petals (Figure 26.5*b*, p. 446), or they may be considerably different (Figure 26.13e). If a flower has all four flower parts (pistils, anthers, petals, sepals), it is a **complete flower** (Figure 26.14); if one or more is missing, it is an **incomplete flower.** The base of the flower, to which the other structures are attached is the **receptacle,** and the flower stalk is the **peduncle.**

The interpretations of the flower, particularly from an evolutionary viewpoint, are varied. However, as used here, the flower is limited to the angiosperms and considered to be distinct because the ovules are enclosed by the ovary. Further, the entire flower is considered to be a reduced shoot that is determinate in its growth and that produces spore-bearing structures (the anthers and pistils).

GAMETOPHYTE GENERATION

The gametophytes are greatly reduced in the angiosperms (Figure 26.2). The male gametophyte is reduced to a three-celled structure, and although the female gametophyte may contain as many as six-

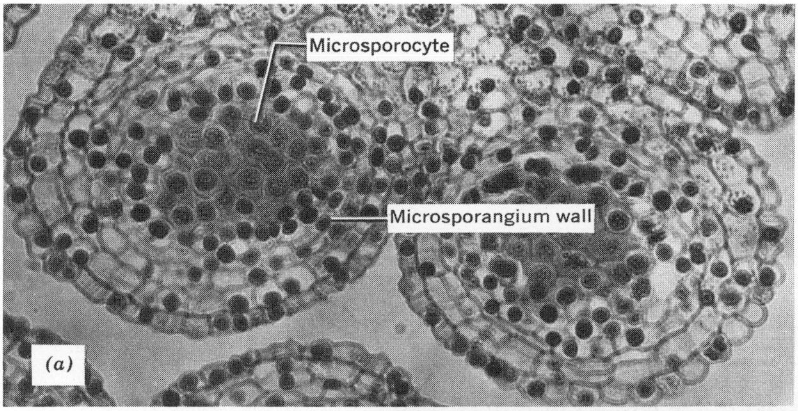

Figure 26.15
Lilium sp. Cross section of anther. (_a_) Two microsporangia with closely packed microsporocytes (135×). (_b_) One microsporangium with microsporocytes undergoing meiosis (170×).

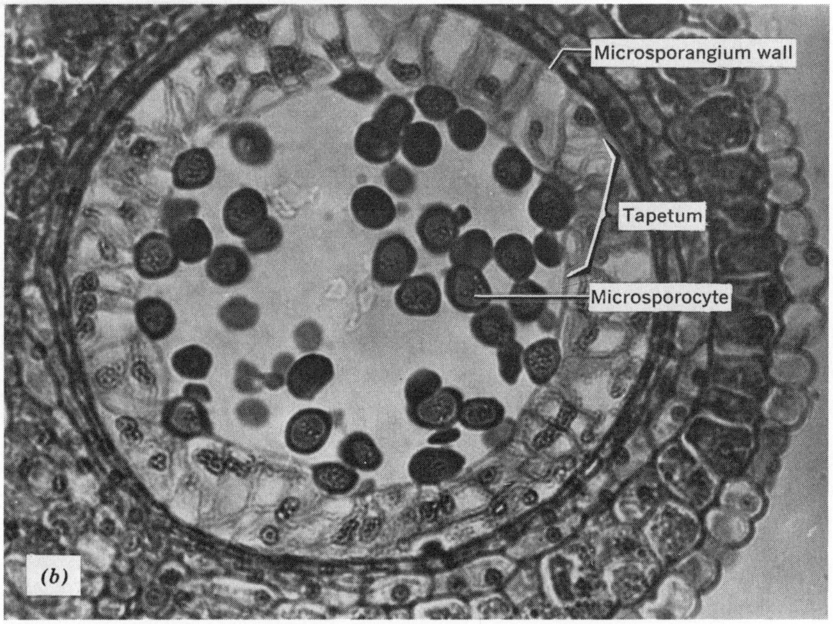

teen cells, eight-celled gametophytes are the most common.

The male gametophyte is produced in microsporangia that are in the anther (Figures 26.2, 26.14, and 26.15). Each anther usually contains four microsporangia that produce a heavy outer wall and an inner tapetum layer. Groups of cells in the center of the anthers are microspore mother cells (microsporocytes) each of which divides by meiosis and produces a tetrad of haploid microspores (pollen grains) which usually soon separates into four individual pollen grains. In the mature anther, the walls of adjacent microsporangia usually break down and produce two large **pollen chambers.**

The pollen grain is the **male gametophyte** (Figure 26.2) as in the gymnosperms, but consists of only three cells when mature instead of the several cells that may be present in gymnospermous pollen grains. The endosporic development of the pollen grain takes place before it is shed from the anther. The first mitosis, followed by cytokinesis, of the haploid nucleus produces two cells. One of these is a vegetative cell called the **tube cell** and the other is a **generative cell.** The latter cell duplicates once more and produces two **sperm cells.** The production of the sperm cells may take place in the pollen grain before it is shed, but it may also occur in the pollen tube. The walls of the anther eventually split, and the pollen grains are released. They are then transferred by wind, water, birds, insects or other means to the sticky, receptive stigma of the pistil (Figure 26.2). This transfer, however, accomplished, is called **pollination.**

The female gametophyte is more complicated than the male in the angiosperms, but not as complicated as the female gametophyte of the gymnosperms. However, as in the gymnosperms, it is produced in an ovule (Figure 26.2). The ovule arises inside of the ovary by developing from certain superficial cells (the **placenta**) on the inside wall of the ovary (Figures 26.14 and 26.16*a*). The cells eventually differentiate into a stalk (**funiculus**) that supports a **megasporangium** (the nucellus). One or two **integuments** develop from the surface cells of the nucellus, completely surrounding it except for a small opening, the **micropyle** (Figure 26.16*b*). The number of ovules produced in an ovary varies from genus to genus. Corn and sunflowers are examples of ovaries in which only one ovule is present; peas and beans produce a few ovules in each ovary, tulips and orchids produce progressively many more.

In most cases, a **megaspore mother cell** (megasporocyte) differentiates from a single cell within the nucellar tissue (Figures 26.16*a* and *b*). This cell increases greatly in size, develops a larger nucleus, denser cytoplasm, and is easily recognizable. The megasporocyte divides by meiosis producing a tetrad of haploid megaspores that are usually arranged in a linear fashion (Figures 26.16*c* and *d*). Development within the ovule from this point varies considerably. However, at least 70% of those angiosperms studied produce their female gametophytes (embryo sacs) from a single megaspore that survives after the other three abort. Because this type of embryo sac development was first described in *Polygonum*, it is known as the *Polygonum* type of development, and is the most common type of development among the angiosperms (Figure 26.3). In this type of development, the surviving megaspore is the one farthest from the micropyle. The nucleus of the megaspore divides once by mitosis, and the two daughter nuclei move toward the two poles of the cell.

Figure 26.16
Development of the female gametophyte (*Lilium* type). (*a*) Large placentas from which ovules are developing (139×). (*b*) Ovule containing mature megasporocyte (156×). (*c*) Binucleate embryo sac at end of first meiotic division (130×). (*d*) Embryo sac containing four haploid nuclei at end of meiosis (130×). (*e*) A mature female gametophyte containing seven nuclei. The polar nucleus from the micropylar end does not appear in this section (136×).

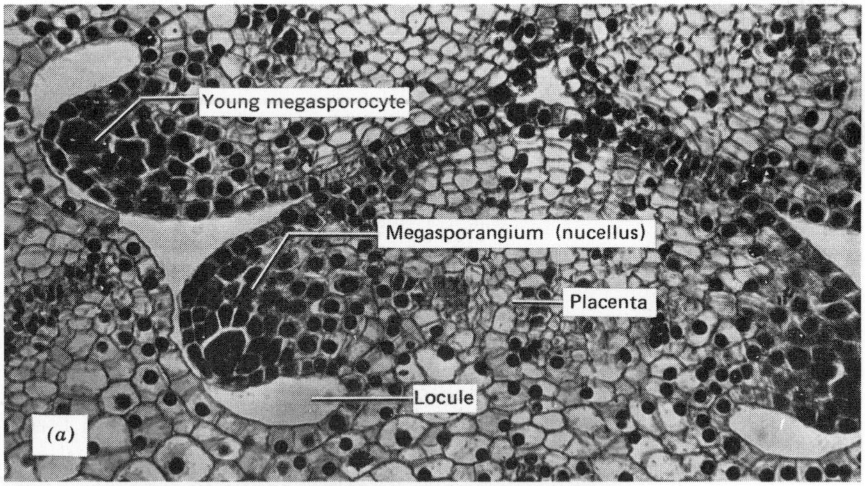

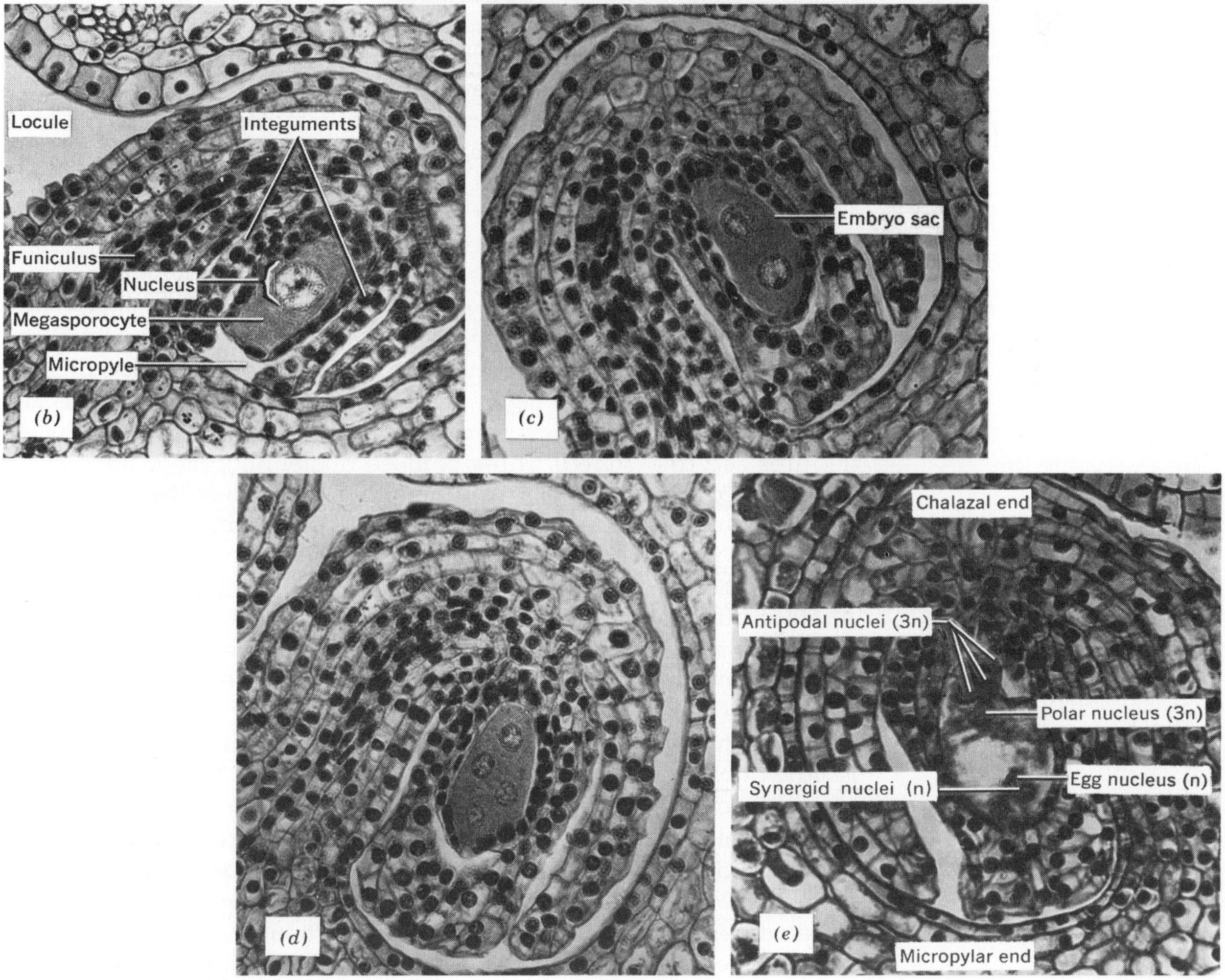

The cell enlarges during this time and becomes the embryo sac. Each of the two daughter nuclei then undergoes mitosis again producing two nuclei at each pole. A final duplication produces four nuclei at each pole of the now greatly enlarged embryo sac. Three of the nuclei at the micropylar end of the embryo sac become organized as the **egg cell** and two **synergid cells.** The fourth nucleus (**polar nucleus**) migrates toward the center of the embryo sac. At the opposite end of the embryo sac (the **chalazal end**), three of the nuclei become organized as **antipodal cells.** The fourth nucleus becomes a polar nucleus and also migrates toward the center of the embryo sac. The two polar nuclei may fuse when they meet and produce a diploid **fusion nucleus**, or they may lie together without fusion occurring. Thus the mature female gametophyte (embryo sac) consists of three antipodal cells, two synergid cells, an egg cell, and either two haploid polar nuclei or one diploid fusion nucleus.

The most easily and cheaply obtainable slides to illustrate the development of the embryo sac is lily (*Lilium* spp.). It is unfortunate, however, that lily does not possess the most common type of embryo sac development, although many authors use lily in their illustrations and the *Polygonum* embryo sac development in their discussions. In lily, four megaspore nuclei are produced by meiosis of the megasporocyte as in the usual type, but all four contribute to the further development of the embryo sac, and no cell walls are produced (Figures 26.2 and 26.16*d*). Three of these nuclei migrate to the chalazal end of the embryo sac and the other remains at the micropylar end. None of the nuclei abort. The nucleus at the micropylar end duplicates producing two nuclei; the three nuclei at the chalazal end fuse and then duplicate producing two triploid nuclei (Figure 26.2). Another duplication of the four nuclei produces four triploid nuclei at the chalazal end and four haploid nuclei at the micropylar end of the embryo sac. One nucleus migrates from each end of the sac toward the center where they usually immediately fuse producing a tetraploid fusion nucleus. Thus, in lily (Figure 26.16e), the mature female gametophyte also consists of eight nuclei but the ploidy numbers are quite different from those in the more common *Polygonum* type (Figure 26.3). The egg and the two synergids are each haploid, the three antipodal nuclei are each triploid, and the fusion nucleus is tetraploid. At least ten different types of embryo sac development have been described, but the two kinds described here are the ones the beginning student is most likely to encounter. It

Figure 26.5
(*a*) Indian pipe (*Monotropa uniflora*). (*b*) Tulip (*Tulipa* sp.). A complete flower with one sepal and one petal removed to show pistil and stamens. (c) Corn (*Zea mays*). (*d*) Castor bean (*Ricinus communis*). (e) White oak (*Quercus alba*).

(a)

(b)

(c)

(d)

(e)

(f)

Figure 26.12
(a) Golden barrel cactus (*Echinocactus crusonii*) with a massive photosynthetic stem and leaves reduced to spines. (b) Banana (*Musa ensete*) with huge leaves and two stalks of fruits. (c) Black walnut (*Juglans nigra*) with large pinnately compound leaves and three fruits. (d) American beech (*Fagus grandifolia*) with simple, toothed leaves. (e) Mangrove swamp in the Florida Everglades. Note the long prop roots extending into the water. (f) White potato (*Solanum tuberosum*). Note the rhizomes with enlarged tubers at their tips.

(a)

(b)

Figure 26.13
(*a*) Onions (*Allium cepa*). Note the above and below ground portions of the leaves. (*b*) Strawberry (*Fragaria* sp.) Note the stolons with the young plant at the tip. (*c*) Female inflorescence of corn (*Zea mays*). Note the brown stigmas; the long pale green, silky styles; and the yellow ovaries that are developing into fruits. (*d*) Male inflorescence in corn (*Zea mays*). Note the yellow stigmas hanging out of the flowers. (e) Trillium (*Trillium erectum*). Note outermost whorl of green sepals and inner whorl of purple petals.

(c)

(d)

(e)

Figure 26.19
Fruit types. (*a*) Longitudinal and cross section of bean. (*b*) Longitudinal and cross sections of cherries; whole cherries. (*c*) Cross section of tomato. (*d*) Longitudinal and cross section of apple; surface view of section shows remains of sepals. (*e*) Longitudinal and surface view of strawberry. (*f*) Surface view of pineapple. (Photos *a*-e by Claire and John W. Cieslewicz).

Figure 26.20
(*a*) Beet (*Beta vulgaris*). Note the large taproot. (*b*) Sassafras (*Sassafras albidum*). A mature sporophyte with some leaves turning red. (*c*) Sassafras twig showing the three distinctive leaf types to be found on each plant. (*d*) White oak (*Quercus alba*). A mature sporophyte. (*e*) Shagbark hickory (*Carya ovata*). Note the compound leaves and young fruits. (*f*) Black walnut (*Juglans nigra*). Mature young sporophyte.

(*a*)

(*b*)

(*d*)

(e)

(*f*)

Figure 26.21
(*a*) Tulip tree (*Liriodendron tulipifera*). Note unusual shape of leaf and the "primitive" flower. (*b*) Sweet gum (*Liquidamber styraciflua*). Note palmately veined and deeply-lobed leaf. (*c*) Red ash (*Fraxinus pennsylvanica*). Note pinnately compound leaves and fruits. (*d*) Hemp (*Cannabis sativa*). A young sporophyte. (*e*) Agave (*Agave americana*). (*f*) Kale (*Brassica oleracea* var. *acephala*).

(*a*)

(*b*)

(*c*)
(*d*)

(*e*)

(a)

(b)

(c)

(d)

(e)

Figure 26.22
(*a*) Polkweed (***Phytolacca americana***). Young, edible plants with flowers. (*b*) Tobacco (***Nicotiana tabacum***). Mature sporophyte with flowers. (*c*) Broccoli (***Brassica oleracea***). Mature sporophyte with edible flowering buds. (*d*) Rice (***Oryza sativa***). A female infloresence that has produced fruit. (*e*) Peach (***Prunus persica***). Longitudinal and surface view of mature fruit. (Photo by Claire and John W. Cieslewicz).

(a)

(b)

(c)

(d)

Figure 26.23
(*a*) Orange (*Citrus* sp.) Mature sporophyte with fruits of different ages. (*b*) Orange (*Citrus* sp.). Longitudinal, cross section and surface views of a mature fruit. (Photo by Claire and John W. Cieslewicz). (*c*) Cotton (*Gossypium* sp.) Mature sporophyte with young (white), old (red), flowers and young fruits. (*d*) Cotton (*Gossypium* sp.). Longitudinal, cross section and surface views of young fruits Note the young seed hairs. (e) Coffee (*Coffea arabica*). Mature sporophyte with red fruits. (*f*) Chocolate tree (*Theobroma cacao*). Portion of main stem of mature sporophyte. Note many white flowers and young fruit.

(f)

should be noted that no archegonia are produced in any of the embryo sacs.

The embryo sac is now ready for fertilization. The pollen grain has, however, landed on the stigma which is usually some distance away from the ovule (Figure 26.2). The exosporic portion of the male gametophyte, the pollen tube, begins to grow from the pollen grain through the style toward the ovule. The tube nucleus and the generative nucleus (if the two sperm nuclei have not already been produced) move into the pollen tube. It was once thought that the tube nucleus was necessary for the continuation of the pollen tube and must precede the generative nucleus. It has now been demonstrated that the tube nucleus is often behind the generative nucleus and, indeed, often aborts before it leaves the pollen grain. The generative nucleus soon duplicates producing two aflagellate sperm nuclei. The method by which the sperm nuclei move through the pollen tube has never been satisfactorily explained, but they do proceed toward the tip of the pollen tube as it moves through the style. The tip of the pollen tube usually enters the ovule through the micropyle, but it may enter through some part of the ovule wall. In either case, the pollen tube penetrates the embryo sac wall, ruptures, and discharges the sperm nuclei. The tube nucleus soon aborts (if it has not already done so). One of the sperm nuclei unites with the egg producing a diploid zygote; the other sperm nucleus unites with the fusion nucleus, if the two polar nuclei have already fused, producing an endosperm nucleus that is polyploid (Figures 26.2 and 26.17). The exact ploidy number of the endosperm nucleus depends on how the fusion nucleus was produced, but in the examples cited the endosperm nucleus would be 3n in the *Polygonum* type and 5n in the lily type. The participation of two sperm nuclei in the fertilization of the embryo sac is known as double fertilization and is unique to angiosperms.

The diploid zygote is the first cell of the new sporophyte generation; the fusions of the polar nuclei and the other sperm nucleus is the first cell of a tissue unique to angiosperms, the **endosperm** (Figures 26.2 and 26.18). The endosperm nucleus undergoes a number of rapid mitoses that may or may not be followed immediately by cytokineses. In some plants the endosperm tissue thus produced is used up by the developing embryo. In other plants, some of the endosperm remains as a food supply that is used by the young seedling. In this manner, it functions in much the same way as the female

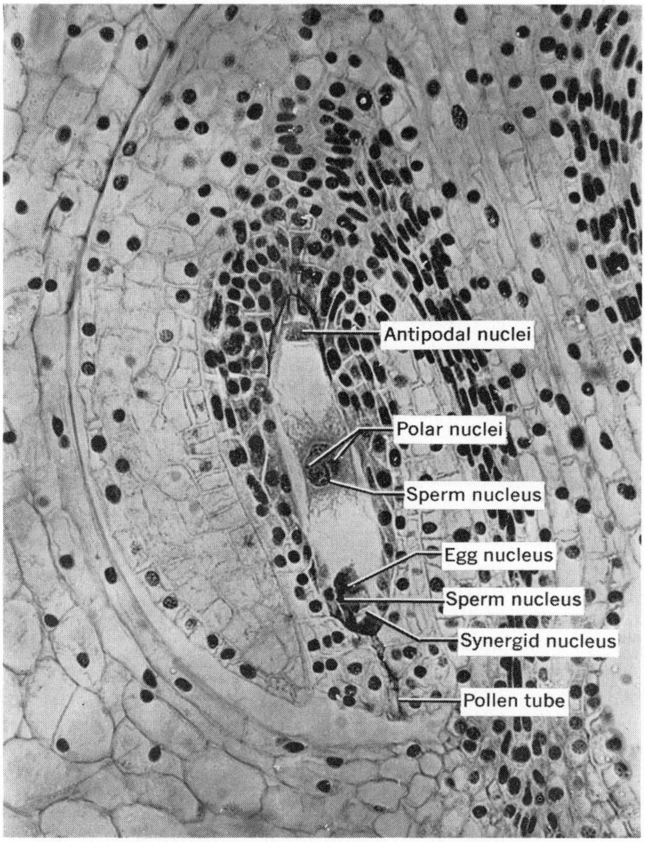

Figure 26.17
Lilium sp. Double fertilization. Not all of the nuclei appear in this section (117×).

gametophytic tissue of the gymnosperms.

The zygote nucleus begins to undergo cellular duplications soon after the endosperm begins to form. There are no free-nuclear stages in the angiosperm embryos as there are in the gymnosperms. The developing embryo soon produces a recognizable axis that produces a **radicle** at one end and a **plumule** at the other. The radicle becomes the first root when the seed germinates and the plumule produces the first true leaves of the plant as well as contains the apical meristem of the stem (Figure 26.4). These two embryonic structures are connected by the rest of the **embryonic axis.** The cotyledons or seed leaves are attached to the axis at **cotyledonary nodes.** The portion of the axis from these nodes to the radicle is the hypocotyl; the portion from the nodes to the plumule is the epicotyl.

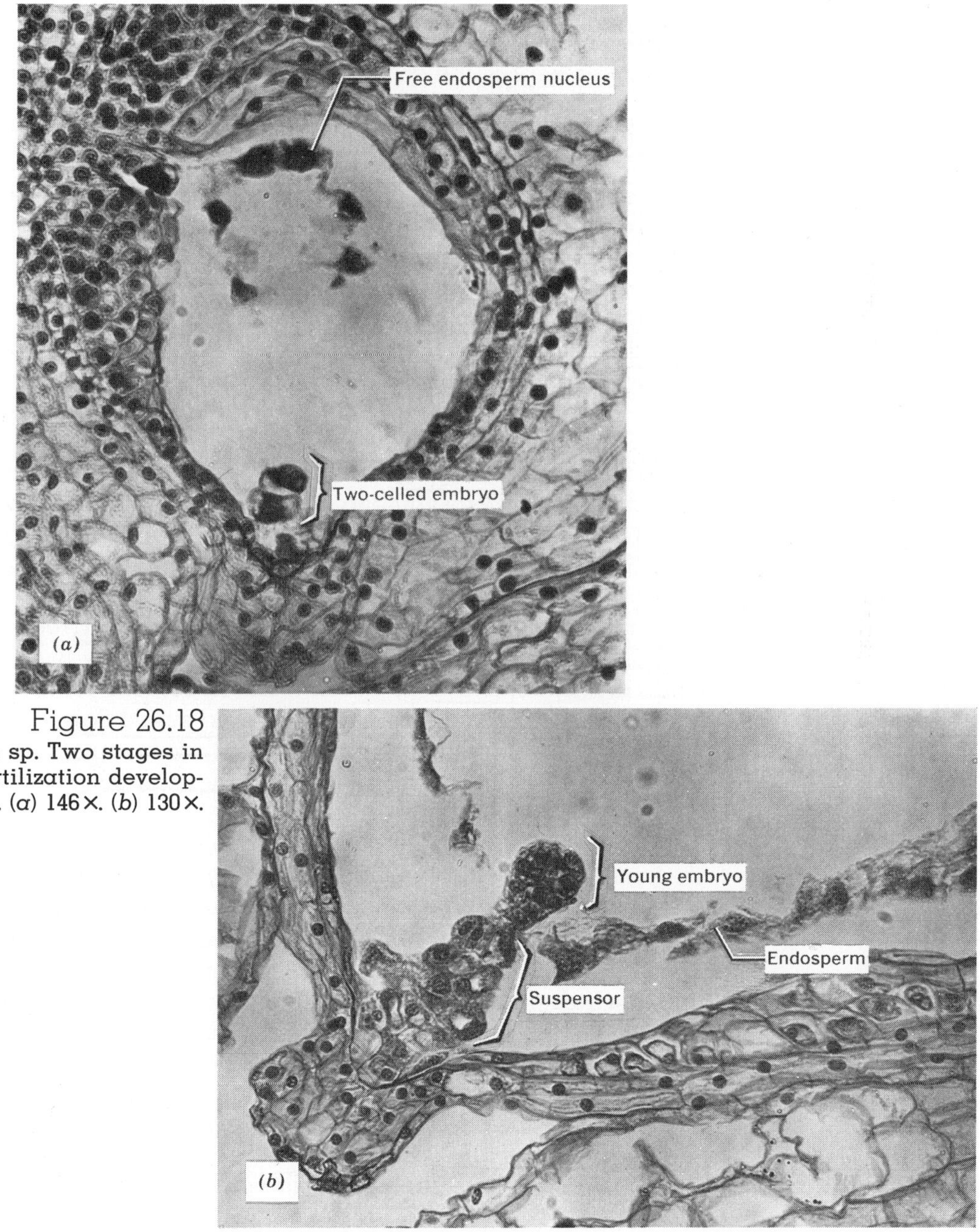

Figure 26.18
Lilium sp. Two stages in postfertilization development. (*a*) 146×. (*b*) 130×.

The relative lengths of these structures varies from species to species. In the monocotyledons (Figures 26.2 and 26.4), only one cotyledon is present at the cotyledonary node, but in the dicotyledons, two are present (Figure 26.4).

During all of this development, the integuments mature into the **testa** or **seed coat.** Thus the angiospermous seed is a mature ovule and consists of the plant axis, the food supply, and the seed coat (Figures 26.2 and 26.4). The food supply may vary in the mature seed. In those plants in which the mature seed possesses endosperm (e.g., corn), the cotyledonary leaves do not contain much stored food material but may function for some time after germination as photosynthetic structures (Figure 26.4). In many plants (e.g., peas, beans) the mature seed contains little or no endosperm because it was used by the developing embryo. In these plants the cotyledons (Figure 26.4) usually contain a large quantity of food that is eventually absorbed by the developing seedling. This type of cotyledon is not very important as a photosynthetic structure. Still other seeds (e.g., Castor bean) contain endosperm and cotyledons at maturity (Figure 26.4).

Pollination, fertilization, and the subsequent development of the embryo stimulates changes in the ovary. The ovary wall, often the receptacle, and sometimes other floral structures increase in size and produce the **fruit.** Fruit structure varies greatly in size, shape, color, and structures from which it develops depending on the plant. Those fruits that arise from the ovary of a single flower are called simple fruits. The ovary may contain only one carpel (**simple**) or many (**compound**). Examples of the former are peas and beans (Figure 26.19*a*, p. 446); the latter, lily (Figure 26.2).

These fruits are dry at maturity, but other simple fruits may be fleshy. Examples of these are grapes, cherries (Figure 26.19*b*, p. 446), tomatoes (Figure 26.19*c*), and apples (Figure 26.19*d*).

Aggregate fruits are those that are produced from several simple ovaries of a single flower and its receptacle. Examples of these are strawberries (Figure 26.19e) and blackberries. Finally, some fruits are produced from a number of ovaries from separate flowers that are closely associated on the same axis. These are called **multiple fruits** and are exemplified by pineapple (Figure 26.19*f*).

Eventually, the seed is released by either dehiscence or decay of the fruit. The length of time after release and before germination is quite variable. Some seeds may germinate soon after release while others require long ripening periods. Some seeds lose their viability shortly after maturation; others may germinate many years after they mature. Most seeds retain their viability for at least one year.

HUMAN RELEVANCE

The angiosperms is the group of plants that is the most important to man. They provide him with most of his oxygen through photosynthesis and either directly or indirectly with most of his food. Many are important sources of medicinals. They provide him with vast quantities of building materials, and provide the various niches whereby the numerous varieties of land animals develop. Finally, the esthetic pleasures that angiosperms give to man must not be overlooked. Many angiosperms are suspected of having certain mystical powers that guide the actions of superstitious peoples.

The angiosperms have provided most of the basis for the sciences of agriculture, conservation, and economic botany. They have been variously discussed from many viewpoints, both economically and culturally. It is not important in a treatment such as this to consider all of the ways in which angiosperms are relevant to man. However, much is presently being published about the importance of "natural" foods, "natural" vitamins, and "organic" gardening. Many people regard these things as highly desirable and newly discovered. Health-conscious individuals are being charged inflated prices for foods that are supposed to be "natural" or "organically" grown, and thus are purported to be much more beneficial than other available foods. It is important that the student recognize that man has long used the angiospermous plant parts in a vast number of ways, and that there is nothing new about his present usage. With his present technology, he has simply increased production. Therefore, the rest of this chapter will be devoted to a discussion of examples of ways in which man has used and is using angiospermous plant structures. For purposes of discussion, the plant will be divided into root, stem, leaf, flower, seed, and fruit.

THE ROOT

The most obvious way in which man has learned to use the root systems of plants is as food. The list of edible roots is a long one that varies from one part of the world to another, but the root has long been a source of starch and other carbohydrates. Many root crops also possess quantities of proteins, vitamins, iron, calcium, and other minerals.

The common carrot (*Daucus carota*) has been under cultivation in Europe for at least 2000 years. It is a biennial that stores large amounts of carbohydrates in its large taproot during its first year of growth. This stored food supply is used by the plant during the second year when it produces its large, umbrella-shaped flowering head. Man plants the seeds and then harvests the root at the end of the first year's growth for the carbohydrate that is stored in the large cortex and phloem areas. These cells also contain large amounts of **carotene** which is a precursor for vitamin A. The stele does not store very much food and may become quite woody. The carrot has often been used as a food for livestock as well as for man himself. There are many wild and cultivated forms of carrots. One of the most common wild forms is Queen Ann's Lace which is a wildflower of old fields and roadbeds.

Another biennial that produces a valuable taproot is the beet (*Beta vulgaris*, Figure 26.20*a*, p. 446), which may have originated in southern Europe. The beet is harvested as an annual and its food is stored mostly in phloem and xylem parenchyma. The taproots are eaten directly by man as a fresh vegetable, or they may be canned or pickled. A variety of the same species is the source of beet sugar. This variety probably originated in northern Europe, but selective breeding produced plants that stored more and more sugar until the sugar content has gradually increased from the original 2 to 4% to the present 15 to 20%. The plant has been an important crop in the United States since about 1875 and is especially important in Colorado, Utah, and Montana where the cool, late summers stimulate maximum sugar formation in the taproot. Cultivation of the plant is world-wide and after the sugar is extracted the pulp can be used as livestock feed, fertilizer, and a pectin source.

Other biennials that produce edible taproots are parsnip, horseradish, radish, rutabaga, oyster plant (salsify), and turnip. These roots are eaten by man in a variety of ways either raw or cooked. They are also often used as livestock feed. Parsnips have been known since well before the 16th century and were introduced into North America about 1610. The phloem cells are filled with starch and become sweet only after exposure to low temperatures. The radish is often eaten raw although many Japanese and Chinese varieties are cooked or pickled. The turnip stores its food in phloem and xylem parenchyma and the plant has been in cultivation long before Christianity. It has been eaten raw, boiled, baked, or made into kraut. It was also used extensively for stock food. It was introduced into North America along with the parsnip and both plants were soon adopted by the American Indian.

The sweet potato (*Ipomoea batatas*) is a plant that was first found in the hot, moist lowlands of Central and South America by the Spanish explorers who took it to Spain in the 16th and 17th centuries. From there it spread throughout southern Europe to the Far East. It became an important food in the Pacific Islands, the East Indies, China, and India but gained great popularity in Japan where it is now the second most important crop (after rice). From the Far East, the sweet potato was reintroduced to the United States where it finds great popularity in the south. The sweet potato is a trailing vine that is a perennial in its native range but is harvested as an annual. It does not produce a taproot but some of its roots become swollen with accumulated starch that is deposited in the xylem and phloem parenchyma. The sweet potato contains many important food materials. It is about 70% water, 26% starch and other carbohydrates, 2% protein and 1% minerals. It is also rich in iron and calcium. Many varieties are white or pale yellow, but the deep yellow to orange types contain carotene. The potato is eaten directly by man as a boiled, baked, or mashed dish. In Japan the potato is also dried and ground into a meal. The Japanese also use them in the production of alcohol. They are used as a cattle food wherever they are grown.

An important source of starch in the low, tropical lands of Central and South America is the manioc plant (*Manihot* spp.), also called mandioca, cassava, yucca, and tapioca. The roots of this plant are about 30% starch and contain almost no protein. The roots are boiled or pulverized and dried into a meal. The manioc is quite tasteless and may be mixed with sauces, beans, or fats. The boiled roots may have their juices squeezed out of them. The juices are used to produce an alcoholic beverage and in meat sauces. The starch is also the source of tapioca which is exported to Europe and North America. During the preparation of tapioca, the starch is partially hydrolyzed to sugar. The crude starch is also used in sizings and mucilages. Sassafras (*Sassafras albidum* Figure 26.20*b* and *c*, p. 446) is a shrubby plant whose dried roots (and to some extent the stems) are boiled to produce a drink called sassafras tea. With sugar added, this liquid tastes like root beer and is thought by some people of the southern and central United States where it grows to have the healthy quality of "thinning the blood" after the winter.

In addition to being an important food source for man and his animals, many fibrous root systems are extremely important in preventing erosion. Many of the grasses produce fibrous root systems that are dense interlocking mats and that allow water to soak into the soil instead of washing the soil away. Such root systems are extremely important in some areas in maintaining a watershed.

THE STEM

The stem is extremely important to man for a variety of reasons. He has made use of it as a source of building material, medicinals, esthetic appreciation, and, of course, as a food.

The value to man as a building material is mostly dependent on the woody stem as was the case with the gymnosperms. The angiospermous woody stem is usually known as a "hardwood," although the wood of some is softer than some gymnosperms. The angiospermous forests are divided into major groups, the tropical and the temperate forests. These categories are often further subdivided.

The tropical hardwood forests contain many diverse species. Because they do not have a dormant period as occurs in the temperate zones, they are usually "evergreen." The leaves are usually shed annually but are either replaced just before they are shed or very soon after. In some areas there is a seasonal variation that is a result of changes in moisture availability rather than temperature changes. The lack of moisture may cause the trees to remain barren of leaves for a few months.

Most of the tropical forests are in central Africa and the Amazon Valley of South America. Others are in Malaysia and the Far East. These represent regions of low elevation, high rainfall, warm climate, and continuous growing season. Although the soils are usually quite poor for agriculture, the other environmental factors are sufficient to produce a dense, lush growth that is a jungle. Most of the forested land in Africa is tropical hardwoods that contain a number of trees known as the "mahoganies." Many species of several genera fall into this group. Also found here are ebony (*Diospyros* spp.), rosewood (*Pterocarpus* spp.), oil palm (*Elaeis*), and many others. Several of these species are used for cabinet woods, oil, and other valuable products. The Amazon Valley contains many tropical hardwoods. Because the area has not been completely explored, much of the timber remains unexploited. There has been selective cutting of some of the more valuable trees used for cabinet wood and lumber that is extremely resistant to rotting. Also, some areas have had a great deal of exploitation because of the rubber industry. However, the South American jungle still retains much of its primitive condition and most of the trees in the area have remained safe from man. Although other valuable tropical hardwood forests exist, the most relevant hardwood forests for man are those found in the temperate zones of the world.

Because of space limitations the discussion of temperate hardwood forests will be limited to those found in the United States and Canada. Although many angiospermous stands occur among the generally pure forest belts of gymnosperms as discussed in the previous chapter, there are only two hardwood forest belts on the North American continent. By far the largest and most important of these is the deciduous hardwood belt that extends from New England south and west to northern Georgia, northeastern Texas and southern Minnesota in the United States and to southern Ontario and the St. Lawrence Valley in Canada. This area contains about 48 million acres of hardwood forest and all of the significant hardwood forests of the North American continent. The only other hardwood forest is the minor subtropical hardwood belt that is found in southern Florida and along a portion of the Gulf Coast to Texas.

The number of species found in the deciduous forest belt is large and varied. There are relatively few pure stands of any species. Some of the more important members of the forest are oaks (*Quercus* spp.), hickories (*Carya* spp.), poplars (*Populus* spp.), walnuts (*Juglans* spp.), birches (*Betula* spp.), beeches (*Fagus* spp.), chestnut (*Castanea dentata*), elms (*Ulmus* spp.), tulip tree (*Liriodendron tulipifera*), sweet or red gum (*Liquidambar styraciflua*), sycamore (*Platanus occidentalis*), wild cherry (*Prunus serotina*), locust (*Robinia* spp.) maples (*Acer* spp.), basswood (*Tilia* spp.), and ash (*Fraxinus* spp.). Many other interesting and locally important species exist, but the ones listed most often produce stems that are used by man. Other species will be discussed when certain products are discussed.

The oaks (Figures 26.5e, p. 446 and 26.20*d*, p. 446) are the most economically important hardwoods. Various species can be found throughout the hardwood forests, but all of them fall into two general categories, the white oak or the black oak group. Of these two, the white oak group contains the most important species. The wood is heavy, durable, and attractive and is extensively used in cabinet work, flooring, veneers, poles, and general construction. The wood is used in charcoal manufacture, railroad ties, mine timbers, railroad-car construction, ships and boats, caskets and fence posts. The black oaks do not produce timber that is as heavy or as durable as the white oaks, but they are used for many of the same purposes. The cork oak (*Quercus suber*) is native to North Africa, Spain, Portugal, and other Mediterranean areas but it has been successfully introduced to southern California and other subtropical areas in the United States. It is from the bark of this tree that is produced the natural commercial cork. The tree produces an unusually continuous cork cambium that produces a consistent layer of cork that may be stripped from the tree every 5 to 10 years. The cork is extremely valuable because of the **suberin** that is in the cell walls. This material makes the wall impervious to water. Each cell of the cork becomes a water-tight compartment of air that has very good insulative qualities. The cork also has good frictional qualities which make it useful for nonslip handles and rollers. It is also lightweight, stable, tasteless, odorless, and of pleasing appearance. Each of these qualities makes the product useful in some way such as in cushioning machinery, decorative effects, linoleum, floats, and, of course, for "corking" wine.

Many of the hickories (Figure 26.20*e*, p. 446) are found with oaks in the dry hardwood forest stands known as the oak-hickory stands. Hickory wood is tough, hard, and shock-resistant and is used mostly for handles and sporting goods. The poplars have soft wood that has never been very valuable as timber, but it is widely used as a paper pulp wood. Some species are planted as windbreaks because they grow rapidly. Although the wood warps easily when exposed to moisture, it is odorless and tasteless and therefore is used for crates for vegetable shipping.

The walnuts (Figure 26.12*c*, p. 446 and 26.20*f*, p. 446) are important as cabinet wood and black walnut (*Juglans nigra*) is the most valuable cabinet and furniture wood found in North America. Solid walnut furniture is very popular, but very expensive; most walnut furniture is now veneer. The beautifully grained and colored wood is used for interior trim and is the leading wood used in gunstocks. The birches are also used for furniture wood, but they are used extensively for boxes, baskets, crates, wooden ware (bowls, etc.), handles, spools, flooring, interior finish, clothespins, toothpicks, and as pulp. The bark of the paper birch (*Betula papyrifera*) was used by the Indians for their birchbark canoes because of its buoyancy.

The American beech (*Fagus grandifolia*) is the only native beech (Figure 26.12*d*, p. 446). It is a handsome tree with very smooth bark. Because it is very shade tolerant, beech forests are dense and dark. It is used for food containers, furniture, handles, flooring, novelties and woodenware, railroad ties, veneer, and pulp. The American chestnut was exterminated by the ascomycete *Endothia parasitica*, the causative organism of chestnut blight. Large stands of this species were found throughout Appalachian Mountains, Ohio Valley, and southern Ontario regions. The dead stems are extremely resistant to decay, and are still being harvested as a source of tannin. Their decay resistance also makes them valuable for general construction and caskets.

Another tree that is threatened with extinction because of a fungus is the American elm (*Ulmus americana*, Figure 6.15, p. 446). These trees are being killed by the millions by the ascomycete *Ceratocystis ulmi*, the causative organism of Dutch Elm disease. There are several other species of elms, however, and they are used in furniture, caskets, containers, hoops and staves, posts, and shipbuilding. The tulip tree (Figure 26.21*a*, p. 446) is a close relative of the magnolias and is a rapidly growing tree that is valuable for construction, interior finish, boxes, veneers, excelsior, pulp, furniture, siding, and musical instruments.

Sweet gum (Figure 26.21*b*, p. 446) is a hardwood of the wet bottomlands of the southern United States. The wood warps easily but has an attractive grain and is easily worked. It is the leading furniture wood and second in veneer production. It is used in boxes, cabinets, interior trim, and millwork. The American sycamore or planetree is a very large tree that also does well in wet soils. Because its wood is difficult to split, it is often used for butcher's blocks, boxes, and musical instruments. It is tasteless, which makes it useful for food containers and may be used for interior trim, veneer, and furniture.

Cherry is a tree that produces a wood demanded for furniture almost to the same degree as walnut. Because cherry wood is so expensive, most furniture made from it is veneered. Locust produces a heavy wood that does not expand or contract when exposed to moisture. The wood is used for posts, mine timbers, and railroad ties, but is the principal wood for dowels and pins to support glass insulators.

Many species of maple occur throughout the hardwood forest. The most important species is the hard or sugar maple (*Acer saccharum*). The wood is lighter in weight than oak but stronger. It, with other species of maple, ranks third in production of hardwood timber. It is used for furniture, flooring, boxes, handles, woodenware, spools, motor vehicle parts, veneers, and pulpwood. It is from the sugar maple that the sap is collected for processing into maple sugar. This was first practiced by the Indians; the white man did not start doing it until about 1673. In the early spring the sap from the stem is about 95% water and 2 to 6% sucrose. The sugar results from the conversion of starches stored during the previous growing season. The sap flows about 35 days and is collected during this time. The sap is poured into shallow evaporator pans and boiled down to a particular thickness. After filtering to remove precipitates, the product is sold as maple syrup. Further boiling produces a paste that crystallizes and forms maple sugar. The tobacco industry uses a great deal of maple syrup to flavor its product.

Basswood produces a soft, light, and easily worked wood that is tasteless and thus is often used for food containers. It is also used for furniture, veneer, and excelsior. Several species of ash (Figure 26.21c, p. 446) are found throughout the hardwood forest, but the most valuable of them is the white ash (*Fraxinus americana*). The wood of this tree is used mostly for handles, furniture, motor vehicle parts, boxes, and athletic goods as well as railroad ties and veneer.

Although the preceding discussion appears to be a long one, it barely touches on the ways in which tree stems are relevant to human beings. Many of these species are highly valued as shade trees and ornamentals. Which tree is used is determined by the location. Many of them are also used to control erosion and preserve watersheds. In some areas of the United States, the woods are still used as a necessary source of fuel; in other areas they are used as a luxury in fireplaces.

One stem should be mentioned because its bark contributes a very important medicine. The tree from which the bark is obtained is the *Cinchona* spp. and the medicine is quinine. This genus was first discovered in the high Andean Mountains of South America and was probably used by the Indians there as an antimalarial. The Jesuits introduced it to Europe. Four alkaloids in the bark are useful as antimalarials and by selective breeding strains have been developed that yield up to 17% of the alkaloid.

The Para rubber tree (*Hevea brasiliensis*) is the plant that produces most of the world's supply of natural rubber. The plant and its product were known by the Mexican Aztecs long before the white man came to South America but was used by the Spanish conquerors by at least 1600 for waterproofing shoes and other clothes. With the discovery of the vulcanization process (the addition of sulfur or sulfur compounds to crude rubber under proper temperature to make it nonplastic and increase its strength and elasticity) in 1869, the way was opened to the exploitation of the rubber tree. The tree is native to the Amazon Valley in South America. It has now been introduced into various parts of the world including Ceylon, Sumatra, and Java where it is grown under plantation conditions. The latex contains about 30% rubber and is found in vessels in the bark. Cuts are made in the bark as close to the cambium as possible. However, the cambium is not cut. When the vessels are cut, the latex runs into a collecting cup. This material is then treated in various ways to produce the vast variety of rubber products used by modern man.

Although he finds many uses for woody stems, man uses relatively few angiospermous stems as food materials. The white potato (*Solanum tuberosum*) is undoubtedly the most important stem food (Figure 26.12*f*, p. 446). The potato is the enlarged tip of the underground stem (rhizome). This particular portion of the rhizome is called a tuber and is composed mostly of parenchymatous pith cells that contain starch. The tuber has a thin layer of conducting cells just under its epidermis. The epidermis becomes suberized with age and this allows the potato to be successfully stored for later use. The potato contains about 19% carbohydrate, 2% protein, and very little fat.

The tuber has the same type of external morphology as does any other stem. The "eyes" are at nodes and are composed of a leaf scale and two or three buds from which new stems will grow. Indeed, tubers are often cut in such a way as to separate the nodes; the pieces are planted, and each piece produces a new potato plant. If a whole potato is planted, the end bud (terminal bud) will assume terminal dominance and will be the only bud to "sprout."

The white potato is another plant that had its origin in the Andes Mountains of South America. However, it went from there to Spain by way of the Spanish conquerors, then slowly through Europe and to Ireland. The story of the Irish potato famine of 1845–1847 is told in Chapter 5 on phycomycetes. The potato was not easily accepted as a food fit for Christian tables because it was not mentioned in the Bible and was often the inspiration for sermons denouncing it. However, the Prussians began to accept the plant at the instigation of

Frederick the Great. Its acceptance in France was aided when Marie Antoinette wore a wreath of potato flowers in her hair to a state ball. It was introduced to North America by the early settlers who came from Europe and it quickly became a staple food here.

The potato plant is an extremely efficient producer of starch, even on relatively poor soils, and is often used as a substitute for bread. In many parts of its range it is used as animal food. In addition to its uses as a food, it has been used as sources of flour, syrup, and some industrial alcohols. Several alcoholic beverages are also derived from the fermentation of potatoes. Among the best known of these is vodka, although it may also be made from wheat or rye.

The short, upright, underground stem (corm) of *Colocasia esculenta* (taro; dasheen) also stores large quantities of starch. The entire stem is the corm and the only above-ground portion of the plant is the huge heart-shaped leaves that have large petioles. Several hundred varieties of this plant exist in the Pacific Islands and southeastern Asia, and the corm is a staple in the diet of the native people. It is eaten in much the same way as the white potato but contains a greater starch concentration than does the potato. The cooking process degrades the calcium oxalate crystals that make it unfit for raw consumption. In addition to eating the plant boiled, baked, or mashed, it is sometimes steamed, crushed, and fermented to produce the famous Hawaiian poi.

One of the few above-ground stems that is widely used for human food is asparagus. This plant is probably native to the Mediterranean area, but has spread throughout the world to become a popular "vegetable." The plant is perennial and, in the spring, produces aerial stems that have small, scalelike leaves. These early stems may be removed and the plant produces another "crop." The stems may be removed several times before they are allowed to mature and carry on the work of the plant. Asparagus contains about 94% water, more protein than most plants, and several vitamins and minerals.

In addition to the root of the sugar beet, man also derives sugar from the stem of a grass, the sugar cane (*Saccharum* spp.). Although the plant is native to the Far East, it has been introduced all over the tropical areas. The plants are usually propagated vegetatively by planting sections of stem. Roots are produced adventitiously at the nodes, and a new stem grows from the buds. The stems ordinarily grow to a height of 20 feet and may be harvested in 1 to 1½ years.

Production of new aerial stems from the rhizomes may continue for several years without replanting. Extraction of the juice is usually accomplished by squeezing the stems between rollers. The juice then undergoes evaporation and the crude sugar that results contains sucrose, glucose, gum, protein, organic acids, and other impurities. This sugar is part of the staple diet of some people, and is more nourishing than much of the highly refined white sugar. Further processing of the crude sugar removes all of the "impurities," and allows the sucrose to crystallize. The liquid portion of the crude sugar is molasses which may also be used for human consumption, or for cattle feeds, or for fermentation into rum, ethyl alcohol, or vinegar. The stem residue may be used in the manufacture of paper or cardboard.

Man has made use of the elongate, thick-walled cells that are found in the cortex and bark of many angiospermous stems. The walls are composed of cellulose, but also contain lignins and other materials. These cells usually occur in bundles and are tightly attached to each other by means of the middle lamella. The cells of the most useful fibers interlock at their ends.

The use of the stem fibers of flax (*Linum usitatissimum*) for the production of linen predates recorded history. Linen has been found among the artifacts of Swiss Lake dwellers and it was used to wrap Egyptian mummies several thousand years ago. It reached a high degree of popularity during the Middle Ages and, until cotton began to be widely used, it was the most important fiber plant. Flax is still used in the highest quality and most expensive fabrics.

The stems are pulled and undergo **retting** which is a process that allows the softer tissues of the stem to decay. In fine flax, the wall of the fiber cell is composed of about 90% cellulose and 2 to 5% lignin and pectic materials. The fibers are spun into yarn and the yarn is then woven into cloth or twisted into threads. Linen threads are used for sewing clothes and shoes and in the production of fish lines and nets because of their strength. The yarns are used in making cloth, firehoses, and carpets.

Another plant fiber that is similar to flax and that is known to have been cultivated by the Chinese at least 4500 years ago is hemp (*Cannabis sativa,* Figure 26.21*d*, p. 446). The fiber was also used in Greece before the time of Christ. The cell walls contain more pectin than do the walls of flax, but the two fibers are often interchangeable. The alkaloids produced by the leaves and flowers produce an hallucinogenic effect in many people. The plant parts are the source of marijuana; because of the problems of security and licensing, hemp is being grown much less often as a fiber plant.

THE LEAF

A plant that produces leaves containing an important fiber is the sisal or agave (*Agave* spp., Figure 26.21e, p. 446) plant that is native to tropical America. The plant produces a rosette of large leaves from an underground rhizome. A large flower stalk is produced after several years' growth and the plant then dies. The older, outer leaves are harvested as they increase in size and produce the long fibrovascular bundles that are embedded in their parenchyma. Harvesting of the outer leaves may continue from 6 to 10 years. The fibers are removed from the leaves by machines that crush the parenchymatous tissues. The fibers are then washed, dried, and usually made into cordage. The cordage is used in the production of twine and rope. The sisal fibers are also used for brushes, stuffings, paper, rugs, and hats.

Other leaf fibers that are harvested in some areas are the yucca (*Yucca* spp.) and mother-in-law's tongue (*Sansevieria* spp.). The fibers may be used for twine and brushes. The natives of the area use the fibers for sandals, robes, and baskets.

The fermented juice from crushed agave leaves is also used in the production of pulque, an alcoholic drink common in Mexico. When this fermented product undergoes distillation, the widely popular alcoholic beverage, tequila, is produced.

In the past, leaves of several angiosperms have been used to produce vegetable dyes. These are of lesser importance now because of the coal tar dyes, but the brilliant blue indigo was derived from the leaves of *Indigofera* spp., a legume.

Although vegetable dyes are no longer widely used, there is still a large demand for perfume oils from plant sources. These belong to a class of oils called **essential oils,** that have smaller molecules, are usually liquid, and have an aromatic fragrance when they evaporate. Some of these are geranium oil which is derived from the foliage of several of the ornamental geraniums, verbena oil from the leaves of *Verbena* spp., and lemon grass oil from the foliage of lemon grass (*Cymbopogon*). Essential oils are also important for those plants that are used as spices, condiments, and flavoring. Among those leaves that are used in this fashion are parsley, sage, tarragon, wintergreen, thyme, and bay leaf.

Wintergreen is also used in medicinals as is the peppermint oil that is derived from the foliage of various species of *Mentha*. Menthol is derived from the peppermint oil and is used in many antiseptics, lotions, nasal drops and sprays, and soothing throat lozenges. Some of the more important medicinals derived from leaves do not depend upon the essential oils, but upon various alkaloids.

Cocaine is a local anesthetic that is derived from the leaves of a plant that is native to the Andes Mountains. The leaves are chewed by the natives to reduce emotion and perception and to relieve pain. It is used in modern medicine as a local anesthetic and is important in dentistry and minor surgery. It is also used to reduce the pain of soldiers suffering wounds produced in battle. Its close chemical relatives, novocain and procaine have been artificially synthesized and are now widely used instead of cocaine.

Atropine is one of the important alkaloids derived from the deadly nightshade (*Atropa belladonna*), a relative of the white potato and tomato. The toxic properties of the plant have been recognized for many hundreds of years, and **belladonna** was the first alkaloid drug produced from the plant by the early herbalists. Atropine, and several other alkaloids, were later separated from belladonna. Atropine is used as a stimulant to the sympathetic nervous system, to stimulate circulation, dilate the pupils of the eye, and as a local anesthetic. Another of the alkaloids derived from belladonna (scopolamine) has helped to remove much of the pain and fear from childbirth. This drug is known to many a mother as "twilight sleep."
Digitalis is derived from the leaves of a plant (*Digitalis purpurea*) that was used medicinally at least 1250 years ago. This drug is very important as a heart stimulant and is used to aid in the regulating of heart contractions, blood pressure, and pulse.

Man uses many of the leaves and stems of legumes and grasses as an indirect food source in the form of forage crops for his farm animals. Not only do these animals eat the plants while they are still growing, but the plants may be cut, dried, and stored away as hay for winter feeding. Among the most important legume forage crops are the lespedezas, alfalfas, and clovers.

The lespedezas or bush clovers (*Lespedeza* spp.) can grow on very poor soil, and increase the soil fertility by adding nitrogen. They are used as erosion control plants and are especially important as hay. Because they sometimes grow as bushes or shrubs, because some species are perennials, and because others reseed themselves, this group is especially important as a source of food and cover for wildlife. The alfalfas (*Medicago* spp.) were originally western Asian plants that were introduced into the United States by Spanish priests from Mexico to southern California. This perennial, low-growing plant is the most valuable hay plant in the United States and other parts of the world where dairy and beef cattle are important. The plant grows rapidly during a good growing season and may be cut several times during one season. The plant greatly improves the soil fertility and is often known among farmers as "green manure." The clovers (*Melilotus* spp. and *Trifolium* spp.) are also important forage plants and soil builders. They are well known over the world, and some have high protein and mineral content.

Many kinds of grasses are used directly for forage or for hay. The species of grass used varies greatly with geography, but many of the grasses that produce fruits useful to man and his animals are also used for forage, hay, or as a stored green fodder (**silage**). Examples of these are sorghum, corn (Figure 26.5*a*, p. 446), oats, wheat, barley, rye, and many others. Many other grasses are cultivated because they make good hay crops, or because they produce good grazing. Some of these are timothy, wheat, bent, bluestem, brome, Bermuda, or rye grasses. Many of these grasses when ploughed under are called **green manure** and greatly enrich the soil.

Not only are angiospermous leaves important to man as an indirect food source, but they are, of course, extremely important as a direct food source. Some of the most obvious of these are the ones in which the expanded part of the leaf (the blade) is the main food source. The list is almost endless, but it includes such things as many cultivated varieties of lettuce, cabbage, mustard, spinach, kale (Figure 26.21*f*, p. 446), and the "soul food" collards. Preferences vary greatly according to the location and culture of the inhabitants. Many wild varieties of "greens" are eaten in various parts of the world. These include dandelions, pokeberry (Figure 26.22*a*, p. 446), and watercress. Most of these plants are important as sources of a variety of vitamins and minerals but do not contain many carbohydrates, fats, or proteins. They are often prepared by cooking, but are also eaten raw as salad plants.

The petiole of the leaf is the most important food source from some leaves. Two of the most common of these are celery and rhubarb. Celery was first cultivated in Europe as a medicinal but later found much favor as a raw salad plant. The petioles are green if left exposed to the sun, but they are often bleached by piling paper or other materials around them to remove the sun's influence. White celery was often prized, but the trend now seems to be the acceptance of green celery. Rhubarb petioles were probably used as a medicinal in China by 2500 B.C. The blade of the plant contains a mild toxin, but the petioles, which are thick and long, are boiled with sugar to produce a delicious sauce. They are also used in pies, either alone or mixed with other foods such as strawberries. Their popularity for use in pies is demonstrated by their common name, "pie plant."

One of the most common leaf plants that is eaten by man is the onion (*Allium cepa.*, Figure 26.13*a*, p. 446). The photosynthetic parts of the leaves extend above the soil, but the thick, storage part of the leaf is below the soil. The leaves are attached to a much reduced stem, and the combination of the stem and the storage part of the leaf is called a **bulb.** The bulb is the part of the onion that is most often eaten, but the photosynthetic parts are sometimes used in salads. The leaves store sugar in fairly large quantities, but they also contain **allyl sulfide,** the onion "flavor." The onion is native to middle Asia but has been widely cultivated and hybridized throughout the world. A close relative of the onion is garlic, a bulb that was fed to the Roman soldiers to make them "strong." Garlic is used as a flavoring in much the same manner as onions, but in much smaller quantities.

Two products of angiosperm leaves have created great debate in the medical profession about their harmful effects upon human health. One of these products has long been socially acceptable and creates relatively unemotional reactions; the other is presently gaining widespread popularity but creates explosive emotional reactions, both pro and con, among some sectors of the population. The first of these products is tobacco (Figure 26.22*b*, p. 446), the second is marijuana (Figure 26.21*d*, p. 446).

Most of the tobacco currently used comes from the plant *Nicotiana tabacum* although *N. rustica* is the native North American plant that was introduced to Europe. Tobacco was widely used throughout the Americas long before any of the conquerors reached the area. The plant was smoked during sacred rituals and sometimes chewed or used as snuff with other plants. The Spanish introduced the first tobacco, in the form of cigars, to Europe from the West Indies. Sir

John Hawkins later (1565) took tobacco to England from Florida, and in 1579 Jean Nicot (after whom the plant was named) introduced it to the French court. The use of tobacco spread rapidly throughout Europe and then into the oriental countries. Cigars and pipes were the way in which tobacco was first used; then the French began to powder the leaf to produce an inhalent called snuff. The soldiers of the Crimean War introduced the cigarette to Europe from Turkey.

The tobacco plant is a difficult one to cultivate. It requires a well-drained soil and will deplete it of potash and phosphorous. The leaf, after harvesting, requires an elaborate process of aging, conditioning, fermenting, curing, and blending with other strains. Various flavorings such as sugars, licorice, and rum, and substances to retain moisture such as glycerine and concentrated apple juice are added. The mixtures are then processed to be used in cigars, pipe tobacco, chewing tobacco, snuff, or the ever-popular cigarette.

The attitudes toward the use of tobacco have progressed from the idea that it had great medicinal value to the realization that it has no medicinal value whatever, and is even extremely detrimental to the health of the user. At one time tobacco was thought to cure a long list of ills that included the common cold, toothache, rheumatism, indigestion and, surprisingly enough, in view of modern findings, blood poisoning, cancer, and halitosis. We now know that cigarette smoking, and to a lesser extent, the use of tobacco in other forms, is strongly implicated as the contributory, if not the actual causal agent, in a number of disease relationships. This list rivals the previous one. The evidence is very strong that smoking increases the blood pressure, causes the heart to work harder, slows blood circulation, irritates the bronchial tubes, and contributes to lung cancer, shortness of breath, coughing, and chest pains. Evidence continues to accumulate that smoking probably contributes to even a longer list of illnesses. No physician recommends that a patient smoke to improve his health, and almost all physicians will recommend that a patient quit or reduce his smoking regardless of the disease for which he is being treated. Withdrawal is very difficult for some patients, and the symptoms experienced during withdrawal suggest that tobacco produces a psychological habituation and probably a mild addiction.

The marijuana plant (*Cannabis sativa*) is the Indian hemp plant discussed in the section about fibers. Its impact upon human behavior, is, however, of much greater importance than its use as a fiber.

The plant has been known since at least 2737 B.C. when the Chinese Emperor Shen-Nung wrote about it. It was then used as a cure for beriberi, absentmindedness, and as a mind-liberator. Since that time, it has been suggested as treatment for a number of illnesses and also used as a surgical anesthetic. Because of its psychic properties, it spread from China to India and then throughout eastern Europe and Africa, and was probably introduced to western Europe by Napoleon's Army as it was returning from Egypt in the early 1800s. The widespread use of marijuana in the United States began when Mexican laborers brought it into the southwestern states during the early 1900s. Since that time, its use has spread throughout the country.

Marijuana (which is probably a corruption of a Portuguese word meaning intoxicant), is a beautiful plant. It grows to be 18 feet tall, and has dark green, deeply dissected, palmately compound leaves. It is dioecious, and some of the early definitions limited the term marijuana to the female plant because it was the only one thought to produce the active ingredient, **tetrahydrocannabinol** (THC). It is now known that the male plant also produces THC. THC is found in residues produced by the leaves and flowers, but the stems and seeds seem to lack the substance. The residues are probably produced as a response to stimulation by the sun and the potency of the THC is positively correlated with increased temperature. The plant is an annual and must depend on yearly reseeding to survive.

Much of the emotion surrounding the use of marijuana centers around the controversial issues concerning the effect of THC on the human body and whether or not the use of marijuana leads to the use of more dangerous drugs. Many of these questions cannot be satisfactorily answered because of the lack of proper research. Because preparations vary in concentration according to content and location, dosages are difficult to ascertain. There also appears to be individual reactions to the use of the drug, and the form in which it is used—either smoked or ingested in some fashion. The product usually referred to as marijuana is often made from the whole plant and is not usually very potent. Refinements of processing can produce stronger products. The dried resin is known as **hashish** and is about five times stronger than marijuana.

Proponents of the use of the drug insist that sensual experiences are greatly increased, that perception is improved and that imagery is enhanced. Increases in memory, understanding, emotion, and identity are cited as important consequences of its use. Detractors of marijuana use cite studies that demonstrate that marijuana adversely affects memory of recent events, results in slight decrease in highest cortical functions, and produces arthritis in the extremities. Some evidence suggests that the drug may be a liver toxin and, of course, many people contend that its use leads directly to drugs that have been proved to be more dangerous. The major problem appears to be with those people who have personality disorders or who are under great stress. It has been shown that the use of only a small quantity of the drug may cause acute or chronic psychotic reactions.

It should be noted that the complete story about marijuana use is not yet known. To use or not to use seems to be a personal decision (disregarding the stringent legal sanctions) at this stage of our knowledge. However, one wonders if there would be so many tobacco users today if the effects of tobacco on the body had been thoroughly tested and the populace properly educated. Will we discover many years hence that the use of marijuana has an extremely detrimental effect that we do not yet recognize? Until the many questions concerning its effect have been satisfactorily explored, the use of marijuana should certainly not be encouraged.

THE FLOWER

Plants that produce flowers that are unusual in shape, brightly colored, or fragrant, have long been used as ornamentals. Entire industries have been built on these sexual reproductive structures of angiospermous plants. Millions of dollars are spent each year on flowers to express love, desire, sympathy, or good wishes. It is a time-honored courtship ritual that the human male will present flowers to the person whom he finds attractive and desirable. It is hoped that the flowers will act as an inducement to continued favorable consideration from the object of his attentions. Without flowers the world would, indeed, lose much of its beauty.

Flowers are very important to man for several reasons in addition to the aesthetic ones. Some are used, especially while they are still buds, directly as food. Squash buds and young flowers are prepared by frying lightly in butter with various spices. Two examples of more widespread usage are cauliflower and broccoli (Figure 26.22*c*, p. 446). Both of these plants are different varieties of the same species (*Brassica oleracea*), and they and other edible members of the species have been known for centuries. Cauliflower was known in the Near East before the time of Christ and was introduced into northern Europe during the Middle Ages. The part that is eaten is the large modified, stemless mass of flower buds. The cauliflower head is white because the large outer leaves are tied over the head preventing the formation of chlorophyll. Broccoli is another food plant that we derived from wild cabbages (as was cauliflower). It, too, is cultivated for its flower heads, but in this case they are not bleached as is done with cauliflower. Many other members of the same species contribute to man's larder, although parts other than the flower are used. Other examples of varieties of the same species are cabbage and its many varieties, kale, brussel sprouts, kohlrabi, and rutabaga.

Man has also learned to use some flowers as spices. One of the most famous of these is cloves which come from the clove tree (*Eugenia caryophyllata*). The **clove** is really an unopened, dried, flower bud with the cylindrical part being the ovary surrounded by the calyx. The corolla is between the two structures, but is not easily seen. The tree is native to the Molucca or Spice Islands of the Malay Archipelago and was at one time confined to one of these islands by the Dutch. The spice was used to perfume the breath of court officials in China long before the time of Christ. It was later highly esteemed by the Romans and sold for several hundreds of dollars per ounce. Clove plantations have now been extended to the West Indies and Mauritius Islands as well as other suitable geographic locations. The flower buds may be used as the dried spice or distilled and used as clove oil in disinfectants, toothpastes, mouth washes, and chewing gums.

Hops, the essence of the brewing industry, is really the female flower clusters of an herbaceous climbing vine (*Humulus lupulus*). Their value is provided by resins, tannins, and volatile oils that are produced in yellow, glandular hairs on the leaves surrounding the flower. These materials flavor and clarify the beer. Hops are used in making malt extract, as a vegetable, and as a substitute for yeast cake in bread making in addition to brewing.

Some members of the genus *Chrysanthemum* have long been known to be remarkably free of insect pests. Some of these are known as the pyrethrum daisies, and their flower heads have been used as the source of commercial **pyrethrum,** an important "natural" insecticide. Some of the volatile oils act as contact poisons and others paralyze the insect. The insecticide has been extensively used in mosquito control.

Because this section about the flower was opened by a discussion of the way in which the human male uses flowers to expand his influence over the human female, it is probably appropriate to close the section with a short discussion of the reverse attempt at influence. The perfume industry is presently aimed mostly at the human female who wants to smell attractive. Perfume usage has not, of course, always been limited to the female. Sweet-smelling essences have long been used in religious ceremonies and the gods of Babylonia, Persia, India and other countries were often propitiated by the burning of perfumes. Later, during more extravagant periods of history, perfumes were lavishly applied to baths, bodies, and clothing. Through many parts of man's history, even in the most royal courts, personal cleanliness of the body was not very highly esteemed. Indeed, bathing too often was considered to be unhealthy, if not downright sinful. Therefore, it became quite fashionable to add more perfume from day to day to cover up the body odor. To one who is accustomed to a modern concept of cleanliness, the mind reels to consider the unique odors that must have been accomplished by the continual application of perfumes to stale perfumes and even staler body odors.

Some of the most common perfume oils are derived from the flowers of roses, gardenias, jasmine, and violets. These volatile oils quickly evaporate, however, and are therefore mixed with other substances that keep them from rapid evaporation. Also, many oils are blended to provide a variety of perfumes, and substances that keep them evaporating at the proper speeds and in the correct proportions are also added. The perfume industry is an old one, and the masters of the craft of blending perfumes are much in demand. Modern woman, and the even more modern man, find that the task of every-day living is made just a little bit better by the proper use of a bit of perfume in the form of bath lotions, sweet-smelling soaps, deodorants, colognes, and after-shave lotions.

THE FRUIT

As has been previously discussed, the fruit is the ovary and various accessory structures of a flower. Maturation of the fruit occurs after fertilization of the ovules contained within it. In many cases, the entire fruit is used by man but in many others only parts of the fruit are used. Some of man's foods that are called seeds are in reality fruits (e.g., corn, Figure 26.13c, p. 446), and some that are called fruits are in reality seeds (e.g., coconut). Many other fruits are referred to as "vegetables," a meaningless term botanically because it is used to describe any plant part that is eaten. These include roots, stems, leaves, flowers, fruits, and seeds.

As has been previously discussed, fruits may be classified according to the way in which they are produced. That is, they may be simple, aggregate, or multiple depending on whether they are produced from the single ovary of a flower, from several ovaries of the same flower, or from the ovaries of several flowers that are closely associated on the same receptacle. Several examples of these various types will be discussed.

The simple fruit may contain only a single cavity (**locule**) in which the seeds are produced or it may contain many. The mature fruit may be dry or fleshy. Most of the simple fruits that man uses fall into the categories of single locules that are either dry or fleshy, or multiple locules that are fleshy.

The simple dry fruit with the single locule is probably one of man's most important fruits. This group contains the legumes, grasses, nuts, sunflower, celery, and parsley. As has been previously mentioned, many of the legumes (e.g., beans, peas, alfalfa, and clovers) are often used either alone or in mixtures with grasses as graze or hay for livestock. Most of this food material is composed of the stems and leaves, but the mature fruits are also often used. Man uses only a few of these fruits, and the most common of these is the green bean (Figure 26.19*a*, p. 446). The bean pod containing the seed is a fruit. When the pod is very young, before the seeds have had enough time to mature, the pods may be eaten as the "French style" green beans. Later, the bean fruits may be eaten although the seeds are more mature. The fruits may also be pickled or dilled in various ways. When the fruit completely matures, the wall becomes tough and dry. It eventually splits along two sides and releases the seeds. Peas of various types may also be eaten in the same manner.

Grass fruits have long been an important food source for man and his animals and it is these cereal grains that are probably the most important group of food plants. The grain is often called a seed although it is really a fruit in which the ovary wall fuses with the seed coat. At maturity, the two walls are indistinguishable. The grains store large amounts of carbohydrates and some proteins, fats, vitamins, and minerals. The list of cereal grains used throughout the world is a long one that includes rice, corn, wheat, barley, oats, rye, millet, sorghum, and many others that are of local importance. Of these, the first three constitute the basis of most of the world's food crop.

Rice (*Oryza sativa*, Figure 26.22*d*, p. 446) has been cultivated in the tropics of eastern and southern Asia for at least 5000 years and is still used by about ½ of the world's population as the mainstay of their diet. Thousands of strains of the plant are known; most are grown and consumed in the Orient. The plant was introduced into the United States in 1647, but is not as widely used here as are wheat and corn. Rice is a swamp plant and has high water requirements for germination of its seed and growth of the plant. Rice is the grain that is most often used directly by man without first feeding it to an animal or milling it into flour. It is commonly boiled until tender and eaten. The outer covering of the rice fruit is the brown ovary wall that contains proteins and minerals. Rice is more nutritious in this form than in the highly polished state that is so popular in the western world. The polishing process produces the white grains (which are the seeds), but leaves mostly the carbohydrate portion of the grain by removing the ovary wall. However, "brown rice" seems to be gaining popularity over polished rice, even in the United States. In addition to being an extremely important food plant, rice also provides the fermentation base for the Oriental liquor "sake."

Corn (*Zea mays*, Figures 26.4, 26.5*c*, p. 446, 26.13*c* and *d*, p. 446), is an ancient plant whose ancestry is not satisfactorily known. Although several theories as to its origin have been advanced, the most widely accepted one is that the plant originated in the Andes of northwestern South America and it was developed in the central valley of Mexico by 4000 B.C. Modern corn varieties have been so selected and hybridized that none of them could survive unless planted and cultivated by man. Because of the care needed in its cultivation, corn is planted in rows that are kept free of all other plants. This practice of clean cultivation encourages erosion and increases the demands on the soil.

Corn possesses its male flowers in the **tassel** (Figure 26.13*d*, p. 446) which is produced at the top of the stem. Its female flowers are grouped together in an inflorescence called the **"ear"** (Figure 26.13*c*, p. 446). The "ear" is composed of a central axis, the "cob," on which are produced the female flowers. The "ear" is surrounded by a group of specialized leaves, the "husk." The styles of the female flower extend beyond the "husk" and the pollen grains stick to them. Individual pollen tubes grow through each elongate style to the ovary where fertilization of the single ovule occurs. Thus, each corn grain is a fruit.

Corn is used in a multiple of ways, but by far the greatest usage is as a food for livestock. Most corn is grown on farms where it is fed directly to animals that produce meat, milk, butter, cheese, and eggs. The "sweet" corn varieties are eaten directly by man and some corn is made into hominy. Corn does not make a good flour for bread because it lacks the **gluten** necessary to bind the bread into a loaf. However, ground corn, cornmeal, is used in a number of ways by many peoples of the world to produce "cornbread," tortillas, tamales, mush, breakfast foods, and other such dishes. Cornstarch is made by separating the embryo from the endosperm (Figure 26.4). The endosperm is processed to produce the cornstarch as well as dextrins and gums that are used in the glues on stamps and envelopes and in the textile industry. The starch is also fermented to produce alcohol and, by hydroysis, to produce various corn syrups. The embryo contains a large quantity of oil that is extracted to be used as a cooking oil. Various strains of corn are grown for the various products desired, but "popcorn" is grown because it has a heavy, tough fruit wall. This wall retains the moisture when the grain is heated until it explodes and turns the endosperm wrong side out.

Wheat (*Triticum* spp.) is one of the oldest cultivated plants known. It has been grown in the Nile Valley and around the Mediterranean Sea since at least 2000 B.C., but was not introduced to the Americas until about 1520. This is the grain that is used to produce the flour needed for bread. The gluten in the flour holds the bread together in a loaf. Many varieties of wheat are grown for rather specific purposes.

The grain is the fruit of the wheat plant. The grains are produced in a terminal spike of the plant. The ovary wall, known as the **bran,** is rich in proteins. This layer is often removed during milling to produce the white flour produced from the endosperm of the seed. However, many flours are produced that contain all of the ovary wall, or at least part of it. These are the more nutritious of the two flours. Some wheat is used for livestock feed, but most is consumed by man in the form of bread. Other products from wheat flour include pastries, pastas (macaroni, spaghetti, noodles), crackers, and other baked goods. Much wheat is fermented as the base for beer and other alcoholic beverages and many kinds of breakfast foods are produced from wheat. Wheat starches are sometimes used instead of cornstarch and wheat gluten is important in some adhesives, emulsifiers, vitamins, and polishes. The **wheat "germ"** is the embryo of the seed and is often removed during the milling process. It is high in oils, vitamins and minerals, and is used in the production of artificial vitamins. The bread would be much more wholesome if the "germ" were left in the flour.

The fruits of the "big three" cereal grains (rice, corn, and wheat) provide most of man's food either directly or indirectly. Because they are dry, they can be easily stored for future consumption or shipped to different locations. When we realize that these are grasses, it is not difficult to understand the importance of grasses to man, and it is easy to explain the horror and worldwide starvation that would occur without them. This problem has been very nicely explored in the novel *No Blade of Grass* by John Christopher.

The coconut (*Cocos nucifera*) that is commonly sold in the modern supermarket is the seed. The fruit consists of the seed enclosed in a very thick, fibrous husk. It is this husk, which provides a great deal of buoyancy to the fruit, that allowed the coconut palm to become distributed by ocean waves to almost all of the coastal areas of the tropical islands. It is thought that the palm originated in South America and spread from there. The outer husk of the fruit is retted in salt water and the fibers refined to produce **coir,** a very good cordage fiber. Coir is also used in making doormats, stuffing materials, and other materials requiring a light, stiff fiber.

Other types of nuts are one-seeded dry fruits with a very hard inner portion of the ovary wall that enclosed the seed or "meat." Examples of this type of nut are chestnut, walnut (Figure 26.12*c*, p. 446), hickory nut, hazelnut, and beechnut. Most of these fruits have an outer layer that is usually composed of bracts, sepals, and outer ovary wall layers. This layer is usually removed before the fruit reaches the market. None of these nuts form a major source of man's diet, but are used as flavorings, and in specialty items such as candies, cakes, cookies, ice creams, and other such confections. Nuts contain a great deal of oil rather than carbohydrates. Other dry fruits that are often used for flavoring are celery and parsley "seeds."

Among those fruits that are simple and fleshy with but a single locule are olives, plums, cherries (Figure 26.19*b*, p. 446), peaches (Figure 26.22*e*, p. 446), and apricots. Unlike the rest, olives are rarely eaten fresh because they have a very bitter flavor. To produce acceptable flavor, the olive is treated with lye or soda and then pickled in a salt solution. If picked before it is ripe, the olive will remain green; if picked after ripening, the olive will be black. Many olives are used solely for their oil of which they contain 14 to 40%. High-grade olive oil is used in cooking and medicinals; other grades are used in soaps and lubricants.

Because the other fruits mentioned (e.g., cherry, etc.) usually contain large quantities of sugar and a great deal of water, they are extremely difficult to store. Therefore, they are eaten raw only when "in season." However, they are often used in jams, jellies, and preserves. They may be canned or dried for later use, and a number of them (if not all) are fermented to produce various wines, brandies, and cordials.

Among those few, simply dry fruits with multiple locules that man uses are okra and opium poppy. Okra (*Hibiscus esculentus*) is a member of the same family (Malvaceae) to which cotton belongs. It probably originated in the Nile region of northern Africa and arrived in the Americas about 1600. The fruit is not extremely popular today, but is used throughout the southern United States as a"vegetable." It may be fried or added to soups, stews, and "gumbos." If boiled, it becomes very gelatinous and is unpalatable to most people. If properly prepared, it is delicious and provides a very unique flavor.

Opium is derived from the opium poppy (*Papaver somniferum*) and is both a blessing and a bane to man's existence. Opium is obtained by performing a shallow cut on the immature fruit. The milky juice that flows from the cuts is collected and allowed to dry. The dried material is then scraped from the fruit and rolled into a ball of crude opium. Various alkaloids are found in the opium, but the most important of these are **morphine** and **codeine.** These drugs are, of course, extremely important medicinally where they are used to render specific areas of the body insensitive to pain. They are also used to cause the individual to relax and fall asleep. These narcotic effects are not an unmixed blessing. Many people chew or smoke opium for these same narcotic effects and become addicted to it. In this way, the plant contributes to the growing problem of drug misuse.

Opium has been known since the beginning of recorded time and is thought to have originated around the Mediterranean. It is now extensively cultivated in Egypt, India, Southeast Asia, Asia Minor, and Persia, although most of the legal, medicinal opium comes from

Turkey. The drug was introduced to China by the English in the 1600s where it found great popular usage.

The list of multiple loculate, simple, fleshy fruits that man uses is an extremely long one and includes such diverse examples as apples, citrus fruits, grapes, bananas, tomatoes, watermelon, squash, pumpkin, cantaloupe, cucumbers, green peppers, eggplant, pears, and many others. It would, of course, be impossible to discuss each of these fruits. They are used for a great diversity of purposes, and are prepared in many different ways. Many are eaten raw and others are eaten only after cooking.

The apple (*Malus* spp., Figure 26.19*d*, p. 446) probably originated in western Asia where it has been cultivated for thousands of years. Many horticultural varieties have been produced since that time and they have spread throughout the world. The fruit is composed of the fused bases of the sepals, petals, and stamens in addition to the ovary wall and its contents. Apples are one of the few fleshy fruits that can be successfully stored. They are, therefore, used as a "fresh fruit" throughout the year.

In addition to their value in this manner, they are often squeezed for their juice. The juice contains 11 to 15% sugar and may be used immediately as "sweet cider." If allowed to stand, natural fermentation by yeasts will produce an alcoholic content that makes "hard cider." This may be followed by a bacterial fermentation that will yield vinegar. The residue from the apples is a commercial source of **pectin,** an important constituent of middle lamellae. Pectin is extremely important to insure the "jelling" of jellies, jams, and marmalades. It is used in medicines, cosmetics, textiles, adhesives, fibers, and films. It is important in the production of confections such as gum drops, ice cream, sherbets, and syrups. It is used during the production of frozen juices, catsups, and salad dressings as an emulsifier and stabilizer. It is used in the same manner in cheese, milk, bakery goods, salves, lotions, pastes, creams, and pills.

The citrus fruits (*Citrus* spp., Figure 26.23*a*, p. 446) include such important items as oranges, lemons, and grapefruits. They are native to southeastern Asia and have spread throughout the world where they have been cultivated and hybridized.

The citrus fruit is really a type of berry that has a thick, leathery peel. The peel contains numerous oil glands that are the source of essential oils. Each of the "sections" of a citrus fruit represents one locule that has been filled with pulp sacs that grow from the inner wall layer of the ovary. The juice is contained in many living cells that are in each of the pulp sacs.

Citrus fruits are usually consumed fresh although many are now being extracted or "juiced" for the frozen juice industry. Citric acid, used for flavoring and pharmaceuticals, is derived from lemon juice. The juice of all of the citrus fruits contains high concentrations of vitamin C and sugar. The peels and pulp "waste" from the juicing process can be used as a source of pectin.

Entire books have been written about the importance of the grape (*Vitis* spp.). The plant probably originated in southwestern Asia and has been in cultivation for thousands of years. Cultivation and hybridization have produced many species and varieties of this plant. Grapes are used fresh as a table fruit or squeezed to produce grape juice. They are made into jellies and jams and used as flavorings for candies, and other confections. Many varieties are dried to produce raisins.

It is, however, in the production of wine that grapes are truly appreciated. Although many fruits may be fermented to produce wine, grapes are certainly the choice fruit for most people. Indeed, by far, most of the grapes produced are used for this delectable drink of the ancient god Bacchus. Wines vary greatly according to the variety of grape from which they are produced, their sugar and alcohol content, and their color. Naturally fermented wines contain only 12 to 14% alcohol because the yeast plants are killed at that concentration. However, some wines are "fortified" by the addition of more alcohol. Brandy is produced by distilling wine.

The banana (*Musa* spp., Figure 26.12*b*, p. 446) has many varieties that are used as a staple food source among many tropical peoples. Some varieties are boiled, baked, or fried, but most are eaten raw. The banana "tree" is not really a tree because the so-called "trunk" or "stem" is really the overlapping bases of the leaf petioles. The flower stalk is produced from the rhizome and grows through the center of the petioles to come out the top of the plant. After fruiting once, the plant dies. The ovaries of the common banana develop into fruits without fertilization of their ovules. The aborted ovules are present in the mature fruit only as small brown spots. All reproduction of this plant is entirely vegetative.

The aggregate fruits contain such members as strawberries (Figure 26.19e, p. 446), raspberries, blackberries, and dewberries. The flowers have many separate ovaries on one receptacle. Each ovary matures into a fruit with a fleshy wall and their walls partially fuse producing the aggregate fruit. The strawberry (*Fragaria* spp.) is probably native to eastern North America. The early settlers of North America were so happy to find these plants with their tasty fruits on the banks when they landed their ships that they named their settlement Strawberry Bank. This name was later changed to Portsmouth, New Hampshire. Strawberries and the other aggregate fruits mentioned above are all easily bruised and difficult to store. They are usually eaten fresh although they may be made into jellies and jams. They are now often frozen for later consumption, and sometimes they are fermented into a fruity wine. They are also used in ice creams and in the flavoring of other confections.

The multiple fruits include the pineapple (Figures 26.19*f*, p. 446 and 26.24), fig, and mulberry. In these, several flowers grow close together. When fertilization occurs, the ovaries mature and the fleshy ovary walls grow together and partially fuse. The pineapple (*Ananas* spp.) originated in the tropics of the Western Hemisphere. When ripened on the plant, the pineapple contains about 15% sugar and many vitamins and minerals. The first fruit produced by the plant is the largest and considered to be prime. Two secondary fruits are then produced. These three fruits are the ones used in fresh and canned pineapple. Successive fruits are smaller and are used for juice. The plant finally produces fruits that are of no commercial value and it must then be replaced. In addition to being eaten fresh and canned, pineapple is often used as a flavoring for confections.

The commercial fig (*Ficus* spp.) is an enlarged fleshy receptacle that is curved in such a manner that it contains small female and male flowers on its inner wall. Both sexes may be present on the same receptacle or the receptacle may contain either male or female flowers. Several of the figs (e.g., Kadota) develop without fertilization, but others (e.g., Smyrna) require fertilization. In the latter case, the receptacle contains only female flowers and depend upon a particular wasp for fertilization. The wasp lives within the receptacle of a fig (the caprifig) that is not commercially valuable. However, because the Smyrna fig requires the pollen of the caprifig and the wasp to transfer that pollen, the Smyrna fig and the caprifig must be grown in close association before the commercially valuable fruit is produced. The fig is eaten fresh as a table fruit, or it is canned or dried. In addition, it is used as a flavoring agent in many pastries, as livestock feed, and sometimes as a food for silver foxes.

Figure 26.24
Pineapple (*Ananas* sp.). A young sporophyte in flower. (Photo courtesy of the Chicago Field Museum of Natural History).

THE SEED

As has already been discussed, the seed is a fertilized ovule that contains an embryonic plant that is provided with a food supply and bounded by a seed coat or testa. Seeds in the angiosperms are produced inside of fruits and are therefore inadvertently eaten by many animals. The cereal grains were treated in the section on fruits because they are often used as fruits. It should be noted here, however, that the milling process of the grains often removes the ovary wall and leaves only the seed to be used. This is not a very economical practice because the ovary wall often contains high concentrations of proteins, vitamins, and minerals. Man sometimes further processes the seeds by removing the endosperm from the embryo ("germ"). This reduces the part of the seed used to starch, although he may sometimes use the embryo separately.

Man's most important fiber crop is derived from the seed hairs of the cotton plant (*Gossypium* spp., Figure 26.23*c* and *d*, p. 446). This plant has influenced civilization both culturally and economically more than any other single plant. It is probably the only plant of any economic importance to be native to and cultivated independently in both the old world and the new world. Cotton fibers 5000 years old have been found in East Indian ruins. Columbus and the early Spanish explorers found the plant already under cultivation when they reached the Americas.

The seed hair or lint is a single cell that is 1000 to 6000 times as long as it is wide. As the seed matures, the fibers produce a twisted filament of almost pure cellulose. It is this fiber that makes cotton so valuable to man. It was not until the invention of the cotton gin in 1792 that the fibers could be cheaply removed from the seed. Until then, the hand labor required to remove the tenaciously clinging fibers drove the price beyond the means of many people. The seeds are produced in a fruit ("boll") that opens when mature exposing the fiber-covered seeds. The mass of seeds and fibers were, for many years, picked by hand. This hand labor has now been greatly reduced by the invention of "picking machines." These cheap long fibers can be spun into thread that is then woven into cloth.

For many years, the long fibers were considered to be the only important part of the seed. However, it was noticed that many short cotton filaments still remained on the seed after "ginning." These are called linters and are now removed by special machines and are used for making rayons, plastics, photographic film, and paper. Man also found ways of removing the oil from the seed. Cottonseed oil now finds many uses such as in making margarine, cooking oils, and soaps. The residue after the oil is expressed is pressed into a "cake," and used as livestock feed. In spite of these uses, however, the long fibers still remain of primary importance. In addition to fabrics, cotton threads are used in automobile tires, carpeting, cordage, and plastics.

As long as the production of cotton required great amounts of hand labor, slavery was considered necessary in those areas of the country that had become dependent economically on that one crop. The slavery and economic issue touched off the American Civil War which so greatly changed the United States. After mechanization (particularly the invention of the cotton gin) "King" cotton cultivation greatly aided in bringing about the Industrial Revolution.

The coconut seed also contributes in an important way to man's welfare. The endosperm of the coconut seed occurs in two forms, as a solid (the "coconut") and as a liquid (the "milk"). The endosperm contains a high concentration of oil that may be extracted as "coconut oil." The "nut" is cracked and the solid endosperm is allowed to dry. This dried coconut endosperm (**copra**) is then expressed to retrieve the oils. This oil is used in the manufacture of soaps, glycerin, candles, margarines, lubricants, cosmetics, and dentifrices. The residue after pressing may be used as a food for livestock or as a fertilizer. Some of the solid endosperm is, of course, dried for use in confections. The liquid endosperm is quite good as a refreshing drink, and has been used in various media for the culturing of fungi.

The peanut (*Arachis* spp.) also contributes an important oil. The nuts contain 30 to 50% oil, which can be used in margarines, salads, cooking, soaps and various industries. The peanut contains large amounts of protein, and the press cake is valuable as a livestock food. Much of the peanut production is eaten directly. The nut may be eaten raw, roasted, or in candies and other confections. Large quantities are used in the preparation of the ever-popular peanut butter.

Other important oils are produced from sources other than the palm: linseed, soybean, castor beans (Figures 26.4 and 26.5*d*, p. 446), and many others. Any person who has ever been subjected to taking doses of castor oil because it is "good for you" can appreciate the flavor of the 30 to 60% concentration of the oil in the seed. It has long been used as a medicinal as well as a lubricant. It is more importantly used in soaps, imitation leathers, linoleum, oil cloth, plastics, typewriter ink, nylons, paints, and dyeing. The castor bean is toxic, and the oil cannot be used until properly treated to get rid of the toxic alkaloid **ricin.**

Another seed that produces toxic alkaloids is the seed of *Strychnos* spp., which contains the toxin **strychnine.** The toxin is used, in proper dosage, to relieve paralysis and stimulate the central nervous system. The seed is also used with other plants by the native people to produce **curare** which is capable of paralyzing either man or lower animal.

The most important nonalcoholic drink to most Americans is the internationally famous coffee. It is firmly believed by many people that the day does not begin until after their first cup, and that it does not progress unless cups of the brew are consumed intermittently throughout the day. The "coffee break" has become so important to the American working man or woman that it is now often written into labor contracts. The members of the genus *Coffea* (Figure 26.23e, p. 446) that give us this potion are thought to have originated in northeastern Africa. They were then carried into the Arab countries from which they spread into Europe. Coffee houses sprang up in the Near East but often attracted seditious members of the societies and were usually quickly closed. The first coffee houses in Europe first appeared in Paris in the early 1600s, but they attracted the literary peoples and became somewhat snob clubs of the intelligentsia. Various religious oppressions failed to slow the popularity of the drink, but rumors that coffee slowed or stopped the sex drive ran many coffee houses into bankruptcy. Lloyds of London opened as a coffee house in the late 1600s, and the first such establishment opened in North America in Boston in 1669. During this same period, the plant was spreading throughout the Orient and West Indies. After severe fungal diseases wiped out most of the coffee in Ceylon, tea became the more important drink among many of these countries.

The coffee seeds are produced inside of small, red, fleshy fruit on a small, bright green shrub. The seeds are removed from the fruit, washed, dried, and roasted. The roasting varies from place to place, but is necessary to bring out the flavor. The "beans" are then ground. The stimulating effect of coffee is because of the **caffeine** that it contains. The caffeine content ranges to 3% and an average content is about 1.5%. The original use of coffee was as a medicinal because of its effect upon blood pressure, pulse, and sensory perceptions. After roasting and grinding, coffee gradually loses its flavor and the oils become rancid. Much of this problem has been overcome by the use of vacuum packing and the now-popular instant coffees.

The source of chocolate and cocoa is the seed of the cacao plant (*Theobroma cacao*, Figures 26.23*f*, p. 446 and 26.25), which is native to central and northern South America. The seed was used before the Spaniards got to the area, and was considered quite valuable by the rulers of the Aztec Empire who called it chocolatl. Chocolatl was a cold frothy drink that was seasoned with peppers and spices. The Spaniards added sugar, and the English added milk. Somewhere the spices and peppers were left out, and chocolate houses then spread throughout London and Amsterdam. When the Swedish botanist, Linnaeus tasted the chocolate, he is said to have exclaimed "Ah, tea for the Gods" and gave it the Greek generic name *Theobroma*, which represents this expression. Much cacao is still cultivated in Brazil, Venezuela, Trinidad, Ecuador, and Panama but the plant is now grown commercially in the Gold Coast of Africa, Nigeria, and the Far East.

The seeds and pulp (Figure 26.25) are removed from the fruit and allowed to ferment. This fermentation process, with its corresponding increase in temperature, cures the seeds and improves their flavor. After fermentation, the seeds are then washed and dried. The seeds are then cracked and the seed coat is separated from the rest of the seed. The embryo and endosperm are pressed to extract their oils and fats. These products are known as cacao butter, which is edible. The cacao butter is used in the finest chocolates, cosmetics, ointments, and soaps. The press cake contains **theobromine,** which is used in cola drinks. The embryo and endosperm may be treated with milk, sugar, and press cake to produce chocolate and cocoa.

One of the most important seeds that man uses directly as a food material is the bean. Beans are members of the legume family which also includes other important seeds such as the pea. Beans and peas store large quantities of food in their cotyledons, and it is this stored food that makes them valuable as a food source.

There are many varieties of beans and peas and the group crosses a number of generic lines. The bean pod containing immature beans was discussed under the section dealing with fruit, but the seed of many of those same beans also constitutes an important

Figure 26.25
Chocolate fruit (*Theobroma cacao*). Longitudinal and surface view of mature fruit. (Photo courtesy of the Chicago Field Museum of Natural History).

food source. Some beans are products of Europe and Asia, but most originated in the Americas. Most of the commonly eaten beans are members of the genus *Phaseolus*. These include the widely consumed lima, string, white, navy, black, and kidney beans. Soybeans (*Glycine max*) are now finding wider usage as human food. The bean contains large amounts of carbohydrate, fats, proteins, vitamins, and minerals, and because they may be dried are easily stored for future use.

The soybean from eastern Asia is becoming more and more important to modern man. It has long been a livestock food and used as a builder of soil, but its potential as an oil and food crop is only recently being exploited. The soybean provides an extremely valuable source of protein. Thirty to 50% of the bean is protein and about 20% is oil. The uses of soybean oil constitute a long list. It is used as a salad or cooking oil, soy sauce, in margarine, paint, linoleum, lubricants, soaps, plastics, and rubber substitutes. The press cake is valuable as an animal feed and fertilizer. The soybean has long been directly consumed by Asians, but was not palatable to most western peoples, but more "edible" varieties are now being developed. They may be eaten green when boiled or baked, or they may be roasted. They are used as additives to flour, used in breakfast foods, bread, and pastries.

The most often eaten pea is *Pisum sativum* which probably originated in central Asia. This seed may be eaten green while it contains a relatively high sugar content, or it may be allowed to ripen then dried to be eaten at a later time. Many other genera of peas exist and they are often used as livestock food, sometimes mixed with beans.

Several "nuts" are not really fruits but the seeds of various plants. These include the almond and Brazil nut. The almond is the seed of a fleshy, unpalatable fruit. The nut may be consumed raw, roasted, or used in confections. Some varieties are toxic because they contain **prussic acid,** but the toxin is removed during proper processing. The Brazil nut is a seed produced in clusters of 10 to 20 contained in a large fruit in tall trees of the Amazon basin.

Suggested References

The following books and papers should be consulted by students who desire more information about the plant kingdom.

Ahmadjian, Vernon. 1967. *The lichen symbiosis*. Blaisdell Publishing Co. 152 pp.

Alexopoulos, Constantine J. 1962. *Introductory mycology*. 2nd. ed. John Wiley and Sons, New York. 613 pp.

Allegro, John M. 1970. *The sacred mushroom and the cross*. Bantam Books. 355 pp.

Arthur, Joseph C. 1934. *Manual of the rusts in United States and Canada*. Purdue Research Foundation. Reprinted 1962. Hafner Publishing Company, New York. 438 pp.

Bierhorst, D. W. 1971. *Morphology of vascular plants*. Macmillan. 560 pp.

Bold, Harold C. 1973. *Morphology of plants*. 3rd. ed. Harper and Row. 668 pp.

Bolton, Eileen M. 1960. *Lichens for vegetable dyeing*. Studio Books, Longacre Press, Ltd., London. 63 pp.

Bonner, John T. 1967. *The cellular slime molds*. 2nd ed. Princeton University Press, Princeton, N.J. 205 pp.

Burnett, J. H. 1968. *Fundamentals of mycology*. St. Martin's Press. 546 pp.

Byrd, Oliver E., ed. 1970. *Medical readings on drug abuse*. Addison-Wesley Publishing Co., Reading, Mass.

Carefoot, G. L., and E. R. Sprott. 1967. *Famine on the wind: Man's battle against plant disease*. Rand McNally & Co. 229 pp.

Chamberlain, Charles J. 1966. *Gymnosperms: structure and evolution*. Dover Publications. 484 pp.

Chapman, V. J. 1968. *The algae*. St. Martin's Press, New York. 472 pp.

Christensen, C. M. 1951. *The molds and man*. University of Minnesota Press, Minneapolis. 244 pp.

Christopher, John. 1956. *No blade of grass*. Simon & Schuster, Inc. Avon Books, New York.

Cobb, Boughton. 1956. *A field guide to the ferns*. Houghton Mifflin Co. 281 pp.

Conant, Norman F., D. T. Smith, R. D. Baker, J. L. Callaway, and D. S. Martin. 1954. *Manual of clinical mycology*. W. B. Saunders Company. 456 pp.

Conrad, Henry S. 1956. *How to know the mosses and liverworts*. Wm. C. Brown, Dubuque, Iowa. 226 pp.

Dale, Paul W. 1969. *Seventy north to fifty south*. Prentice-Hall, Inc., Englewood Cliffs, N.J.

DeBary, A. 1887. *Comparative morphology and biology of the fungi, mycetozoa, and bacteria*. (Trans. by H. E. F. Garnsey. Revised by I. B. Balfour.) Clarendon Press, Oxford. 525 pp.

Dennis, R. W. G. 1968. *British ascomycetes*. J. Cramer Publ. 455 pp. of text followed by about 70 pp. illus. unnumbered.

Dodd, John D. 1962. *Form and function in plants*. Iowa State University Press. 233 pp.

Doyle, W. T. 1965. *Nonvascular plants: form and function*. Wadsworth Pub. Co., Inc. Fundamentals of Botany Series, Belmont, Calif. 147 pp.

Emmons, C. W., C. H. Binford, and J. P. Utz. 1963. *Medical mycology*. Lea and Febiger, Philadelphia. 375 pp.

Esau, Katherine. 1960. *Anatomy of seed plants*. John Wiley & Sons. 376 pp.

Esau, Katherine. 1965. *Vascular differentiation in plants*. Holt, Rinehart and Winston. 160 pp.

Fergus, Charles L. 1960. *Illustrated genera of wood decay fungi*. Burgess Pub. Co., Minneapolis, Minn. 132 pp.

Fink, Bruce. 1935. *The lichen flora of the United States*. University of Michigan Press, Ann Arbor. 426 pp.

Foster, A. S., and E. M. Gifford, Jr. 1959. *Comparative morphology of vascular plants*. W. H. Freeman. 555 pp.

Fuller, John G. 1968. *The Day of St. Anthony's Fire*. Macmillan, New York.

Garrett, S. D. 1963. *Soil fungi & soil fertility*. Pergamon Press. The Macmillan Co. 165 pp.

Gäumann, E. A. 1928. *Comparative morphology of the fungi*. McGraw-Hill Book Co., New York. 701 pp.

Gilman, Joseph C. 1957. *A manual of soil fungi*. Iowa State University Press. 450 pp.

Gray, William D., and C. J. Alexopoulos. 1968. *Biology of the myx-*

omycetes. The Ronald Press, New York. 288 pp.

Gray, William D. 1959. *The relation of fungi to human affairs*. Henry Holt and Co., Inc. 510 pp.

Groves, J. Walton. 1962. *Edible and poisonous mushrooms of Canada*. Research Branch Canada Department of Agriculture, Ottawa, Ontario. 298 pp.

Hale, E. M. 1961. *Lichen handbook*. Pub. 4434 Smithsonian Institution, Washington, D.C. 178 pp.

Hard, M. E. 1908. *The mushroom, edible and otherwise*. Reprinted 1961. Hafner Pub. Co. 609 pp.

Haupt, Arthur W. 1953. *Plant morphology*. McGraw-Hill. 464 pp.

Heald, Frederick D. 1933. *Manual of plant diseases*. McGraw-Hill. 953 pp.

Hylander, C. J. 1944. *The world of plant life*. Macmillan Co. 722 pp.

Jackson, Daniel F., ed. 1964. *Algae and man*. Plenum Press, New York. 434 pp.

Jackson, Daniel F., ed. 1968. *Algae, man, and the environment*. Syracuse University Press, Syracuse, New York. 554 pp.

Kavaler, Lucy. 1965. *Mushrooms, molds, and miracles. The strange realm of fungi*. John Day Co. 318 pp.

Konar, R. N., and Sundri Ramchandani. 1958. "The morphology and embryology of *Pinus wallichiana* Jack." *Phytomorphology*. *8*:328–346.

Luttrell, E. S. 1951. *Taxonomy of the pyrenomycetes*. University of Missouri Study 24, No. 3. University of Missouri, Columbia. 120 pp.

Macbride, T. H. 1899. *North American slime molds*. 2nd ed., Macmillan Company New York. 1922.

MacDonald, J. A. 1951. *Introduction to mycology*. Butterworth's Scientific Publications, London. 177 pp.

Maio, Joseph J. 1958. "Predatory fungi." *Scientific American*. *199*(1): 67–72.

Martin, G. W. 1932. "Systematic position of the slime molds and its bearing on the classification of the fungi." *Bot. Gaz.* *93*:421–435.

Martin, G. W. 1960 (1961). "The systematic position of the Myxomycetes." *Mycologia*. *52*:119–129.

Martin, G. W., and C. J. Alexopoulos. 1969. *The myxomycetes*. University of Iowa Press, Iowa City, Iowa. 560 pp.

Palmer, C. Mervin. 1962. *Algae in water supplies*. Public Health Service Publication No. 657. U.S. Govt. Printing Office, Washington, D.C.

Parsons, Frances T. 1961. *How to know the ferns*. Dover Publications, New York. 215 pp.

Plant diseases: Yearbook of Agriculture. 1953. United State Dept. of Agriculture, Washington, D.C. 940 pp.

Powell, James B. 1971. "A review of marijuana." *Carolina Tips*. *34*:40. Carolina Biological Supply Company, Burlington, N.C.

Prescott, G. W. 1962. *Algae of the western great lakes area. Rev. ed.* Wm. C. Brown Company, Dubuque, Iowa. 977 pp.

Prescott, G. W. 1968. *The algae: A review*. Houghton Mifflin Co., Boston. 436 pp.

Prescott, G. W. 1970. *How to know the fresh-water algae*. Wm. C. Brown, Dubuque, Iowa. 348 pp.

Prescott, Samuel C., and Cecil G. Dunn. 1959. *Industrial microbiology*. 3rd ed. McGraw-Hill. 945 pp.

Ramsbottom, John. 1953. *Mushrooms and toadstools*. Collins Clear-Type Press. 306 pp.

Raper, Kenneth B., and Dorothy I. Fennell. 1965. *The genus aspergillus*. The Williams and Wilkins Company. Baltimore. 686 pp.

Raper, Kenneth B., and Charles Thom. 1949. *A manual of the penicillia*. The Williams and Wilkins Company. Baltimore. 875 pp.

Renault, Mary. 1972. *The persian boy*. Pantheon Books. 436 pp.

Ricketts, E. F., and J. Calvin. 1952. *Between pacific tides*. Stanford University Press. 501 pp.

Round, F. E. 1965. *The biology of the algae*. Edward Arnold Ltd. London. 269 pp.

Schery, Robert W. 1952. *Plants for man*. Prentice-Hall, Inc. Englewood Cliffs, N.J. 564 pp.

Smith, Gilbert M. 1944. *Marine algae of the Monterey Peninsula, California*. Stanford Univ. Press, Stanford, Calif. 622 pp.

Smith, Gilbert M. 1955. *Cryptogamic botany*. Vol. I. Algae and fungi. McGraw-Hill. New York. 546 pp.

Smith, Gilbert M. 1955. *Cryptogamic botany*. Vol. II. Bryophytes and pteridophytes. McGraw-Hill, New York. 399 pp.

Taylor, C. J. 1967. *The lichens of Ohio*. Pt. I. foliose lichens. Biological Notes #3. The Ohio Biological Survey. Ohio State Univ. Press. 151 pp.

Taylor, William T., and Richard J. Weber. 1956. *General botany*. D. Van Nostrand Co., Inc. 376 pp.

Taft, Clarence E. 1965. *Water and algae—world problems*. Educational Publishers, Inc. Chicago. 236 pp.

Tiffany, Lewis H. 1968. *Algae. The grass of many waters*. 2nd. ed. Chas. C. Thomas, Springfield, Ill. 199 pp.

Wasson, Valentina Pavlovna, and R. Gordon Wasson. 1957. *Mushrooms, Russia, and history*. 2 vol. Pantheon Books. 433 pp.

Wasson, R. Gordon. ca. 1969. *Soma: devine mushroom of immortality*. Harcourt Brace Javanovich, Inc. 381 pp.

United States Dept. of Agriculture. 1949. *Trees: The yearbook of agriculture*. U.S. Government Printing Office. 944 pp.

United States Dept. of Agriculture. 1953. *Plant diseases: The yearbook of agriculture*. U.S. Government Printing Office. 940 pp.

Glossary

A

abaxial facing away from the axis of the plant, for example, the under-surface of a leaf.

acervulus a small cluster of hyphae that is under the cuticle, epidermis, or other host tissues, but is never entirely superficial; has no specialized wall structures but bears conidiophores on the upper surface.

acrasin a chemotactically active substance produced by amoebae of cellular slime molds and causes their aggregation.

adaxial facing toward the axis of the plant, for example, the upper surface of a leaf.

aecial initial a coil of hyphae produced by the haploid mycelium resulting from the germination of a basidiospore of some rust fungi.

aeciospore a binucleate spore produced in an aecium.

aecium a cup-shaped structure consisting of dikaryotic hyphal cells produced by some rust fungi on host plants.

aerobic capable of growing in the presence of free oxygen.

aethalium a sessile generally cushion-shaped fructification formed by a massing of all or most of the plasmodium in some of the myxomycetes.

akinete a vegetative cell transformed by wall thickening into a nonmotile spore.

algae photosynthetic organisms that reproduce by unicellular sex organs when gametangia are present or if the gametangium is multicellular, each cell is fertile.

algin a commercially valuable gelatinous substance obtained from certain algae.

alkaloid an organic substance containing nitrogen and with alkaline properties; the organic base in many plants that are medicinal or poisonous.

amoeboid like an amoeba, without a cell wall and changing in form, showing a creeping movement; applied to the plasmodium and swarm cells of myxomycetes and the zoospores of some phycomycetes.

amphithecium an external cell layer or layers differentiated in young sporophytes from an internal endothecium.

anaerobic capable of growing in the absence of free oxygen.

anastomosis the union of parts as in the fibrovascular bundles of a leaf.

androsporangium the sporangium in which androspores are produced.

androspore a zoospore that produces a dwarf-male plant in the Oedogoniales.

anisogamy the union of two similar motile gametes, one being larger than the other; see heterogamy.

annual living only one year; with a life history that requires only one year.

annual ring a layer of wood produced each year by the vascular cambium in a woody dicotyledonous stem; applies primarily to plants growing in the temperate zones.

annulus in mushrooms, a remnant of the inner veil left as a ring on the stipe; in mosses, a specialized layer of cells in the capsule; in ferns, a specialized ring of cells around the sporangium.

antheridiophore a fertile branch that produces antheridia.

antheridium a sperm-producing organ or reproduction; a single cell in fungi and algae; many cells in bryophytes and vascular plants.

antherozoid a male sex cell: a sperm.

anthracnose a common type of plant disease caused by several species of Fungi Imperfecti; characterized by ulcerlike areas on the host.

antibiotic a substance that is capable of damaging microorganisms and is derived from a living microorganism.

aphanoplasmodium a plasmodium consisting of a network of very fine transparent, undifferentiated strands in which the protoplasm is not granular; characteristic of *Stemonitis* spp.

aplanogamete a nonmotile gamete that may or may not be released.

aplanospore a nonmotile spore.

apogamy production of an organism without fusion of gametes.

apospory production of a gametophyte from a sporophyte without meiosis or the production of spores.

apothecium an open ascocarp, often shaped like a dish, cup, or bowl; the asci produced on the inner surface.

archegonial chamber a cavity or groove at the apex of the female gametophyte in certain gymnosperms, especially the cycads.

archegoniophore a fertile branch that produces antheridia.

archegonium the flask-shaped female reproductive organ in bryophytes, pterdophytes, and some gymnosperms; composed of a narrow neck pierced by a canal that leads to a basal venter containing one egg.

arthrospore spore formed by fragmentation of a hypha; an oidium.

ascocarp in ascomycetes, the sporocarp producing asci and ascospores. There are three kinds: the apothecium, the perithecium, and the cleistothecium.

ascogenous hyphae the hyphae that produce the asci in the ascomycetes.

ascogonium the female gametangium in the ascomycetes.

ascospore a haploid spore formed by free-cell formation following karyogamy and meiosis in an ascus.

ascostroma a stromatic ascocarp producing asci in locules within the stroma.

ascus a saclike structure, characteristic of the ascomycetes, in which a particular number of haploid ascospores (usually 8) develop following karyogamy, meiosis, and free-cell formation.

autocolony a daughter colony formed by coenobic algae.

autoecious completing the fungus life cycle on one host, used especially for rusts.

autospore a nonmotile, asexual daughter cell that usually assumes the character of the parent cell before being liberated.

autotroph a plant that uses only inorganic materials as a source of nutrients, carbon dioxide as a source of carbon, and sunlight as a source of energy.

autotrophic capable of producing food independently without using some preformed energy source.

B **bacillus** a rod-shaped bacterium.

bark an inclusive term for all tissues outside the vascular cambium of a stem.

basidiocarp in basidiomycetes, the sporocarp producing basidia and basidiospores. There are many kinds.

basidiospore a haploid spore resulting from karyogamy and meiosis and produced on the outside of, but attached to, the basidium.

basidium characteristic of the basidiomycetes, on which develop a particular number of haploid basidiospores (usually four) following karyogamy and meiosis.

biennial living two years; with a life history that requires two years; only vegetative growth occurs the first year; flowering and fruiting occur in the second year.

bipolar germination the germination of a spore by the production of a germ tube at each end.

bisexual a gametophyte that produces both antheridia and archegonia; hermaphroditic.

bitunicate ascus an ascus with two walls, the inner being elastic and expanding greatly beyond the rigid, outer wall at the time of spore liberation.

blade the flat expanded portion of the leaf; usually green in color.

blastospore an asexual spore produced by budding.

blepharoplast the granule from which arise the longitudinal fibers of a flagellum.

blight a common name for a number of different plant diseases, used especially when there is sudden wilting and/or death of leaves, stems, flowers, or entire plants.

botulism food poisoning caused by the toxin of the anaerobic bacterium *Clostridium botulinum*.

bract a modified and reduced leaflike structure subtending a flower or inflorescence.

bulb a short, usually subterranean, stem covered by fleshy leaf bases or scales as in onion; roots are produced at its base.

bunt a disease of wheat, caused by the smut fungus *Tilletia* spp., in which the grain is replaced by an ill-smelling, dark mass of teliospores.

C

calyptra the enlarged and modified archegonium that encloses the embryonic sporophyte of liverworts, mosses, and some lower vascular plants, often observed as the cap that partially or entirely covers the capsule after the calyptra has been ruptured by expansion of the sporophyte.

cambium, cork a lateral zone of meristematic cells, external to the phloem, that produces cork cells external to itself; derived from several cell types.

cambium, vascular a lateral zone or cylinder of meristematic cells, located between primary xylem and primary phloem, derived from provascular tissue; produces secondary xylem and phloem.

canker various fungal diseases of trees, usually having localized patches of dead bark surrounded by swollen margins.

capillitium threadlike sterile structures mixed among the spores in the fructification of many myxomycetes and gasteromycetes.

capsule a dry fruit that develops from a compound pistil; the sporangium of liverworts and mosses; the collidal sheath in algae and bacteria.

carpel in angiosperms, a modified leaf that encloses developing ovules; usually consists of a swollen basal portion (the ovary), an elongation of the ovary (the style), and a receptive tip (the stigma); a megasporophyll.

carpogonium the female gametangium of red algae.

carposporangium in red algae, a sporangium in which are produced carpospores; produced in the carposporophyte.

carpospore in red algae, a uninucleate nonmotile spore produced by division of the zygote; depending on origin may be haploid or diploid.

carposporophyte in red algae, a group of carposporangia and carpospores plus sterile covering cells; produced directly or indirectly from the zygote and attached to the female gametophyte.

carrageenin commercially valuable polysaccharides extracted by hot water from certain red algae; often used as an emulsifier in nonsettling chocolate drinks.

casparian strip secondary thickenings that form on the radial and transverse walls of the endodermis of some plants.

caulidium the central axis (sometimes incorrectly called a "stem") of the bryophyte gametophyte; supports the vegetative and reproductive organs.

cellulose a commercially valuable carbohydrate of unknown molecular structure that is the major substance composing the cell walls of most plant cells.

central nodule a thickened region in the center of the valves of motile, pennate diatoms.

centrosome a minute body in the cytoplasm believed to influence nuclear division.

centrum in ascomycetes, the structures enclosed by the ascocarp wall.

chalaza the base of an ovule where the megasporangium and the integuments are united.

chemotaxis the movement of cells or organisms toward or away from a chemical stimulus.

chemotroph an organism that derives its energy from inorganic compounds (e.g., chemosynthetic bacteria).

chitin a substance characteristically found in insect exoskeletons and other animal structures, but also in the cell wall of many fungi.

chlamydospore an asexual, thick-walled spore produced from a fungal hyphal cell; may become separated from the hyphae and act as a resting spore.

chlorenchyma parenchyma cells that contain chloroplasts.

chytrid a member of the *Chytridiales*, the simplest parasitic or saprophytic phycomycetes.

cirrhus a column of spores that have been extruded from an ostiole: a "spore-horn."

clamp connection in basidiomycetes, a lateral hyphal outgrowth that connects two adjacent cells in a secondary mycelium; probably allows for the maintenance of the dikaryon in some new cells.

cleistothecium in ascomycetes, an ascocarp that usually remains closed and produces its asci in an inordinate arrangement; if the ascocarp possesses an ostiole, the asci are still inordinately arranged.

coccus a spherical bacterium.

coenobium a colonial organism with the cell number fixed at time of origin; no further cell duplications occur.

coenocyte a multinucleate mass of protoplasm.

coleoptile a sheathlike structure covering the plumule of monocotyledons; probably the first leaf of the plant above the cotyledon.

coleorrhiza a sheathlike structure that covers the radicle of many monocotyledons.

collenchyma a tissue composed of cells that fit closely together and with walls that are thickened at their angles.

columella a sterile column or bulb extending into a sporangium, as in mosses and myxomycetes.

commensal an organism living with another; both benefiting by the association (e.g., algae and fungi in lichens).

cone a structure composed of modified leaves or branches that bear sporangia (see strobilus).

conidiophore in fungi, a specialized hypha that produces conidia.

conidium an asexual thin-walled fungus spore produced at the tip or side of a specialized hypha.

conifer a general term used to refer to cone-bearing trees.
conjugation union of gametes and/or gametangia in certain algae and fungi, especially those with gametes that are alike or nearly so.
conjugation tube the tube conecting two conjugating cells and through which one gamete moves toward another gamete or both gametes move toward each other.
coprophilous growing on dung.
corm a short, thickened, underground stem in which food is accumulated.
corona the crown of cells on the oögonium of the members of the stoneworts.
cortex the primary tissues of the stem or root, extending from the primary phloem to the epidermis; composed mostly of parenchyma cells, but many other cell types may be present; the endodermis, if present, is usually considered to be the inner layer.
corticating cell a cell that covers or surrounds the central filament in some algae (e.g., some species of *Chara*).
costae in diatoms, wall markings produced by ridges.
cotyledon an embryonic leaf within which food may or may not be stored and which may or may not become photosynthetic when the seed containing the embryo germinates.
cotyledonary node the node on the embryonic axis where the cotyledon or cotyledons are attached.
crozier body the hooks of an ascogenous hypha prior to the formation of asci in some ascomycetes; composed of an ultimate, penultimate, and basal cell.
cruciate basidium in heterobasidiomycetes, basidium whose base contains longitudinal septations that are at approximate right angles to each other.
cuticle a noncellular layer of waxy material (cutin) covering the surface of the epidermis, except for the stomates.
cyanophycean starch a unique starch produced by the blue-green algae.
cycad plants that resemble palms and are often mistaken for them; really members of ancient gymnosperm groups that produce male and female cones on separate plants and have motile sperm.
cyclosis the movement of protoplasm within the cell.
cystidium a sterile, enlarged cell in the hymenium of basidiomycetes.

cytokinesis division of the cytoplasm that often occurs during karyokinesis; if simultaneous with karyokinesis, cell division results.

D **diatom** a unicellular, filamentous, or colonial microscopic alga with cell wall composed of silica and divided into two parts.

diatom, centric diatoms in which the body symmetry is radial.

diatom, pennate diatoms in which the body symmetry is bilateral.

damping off a disease, caused by fungi, of seeds or young seedlings; often seen as a rotting of the seedling at soil level.

dermatomycosis a fungus that can invade the skin of man or other animals.

desoxyribonucleic acid a complex organic acid in the chromosomes; contains genetic information that is coded in specific sequences.

dichotomy the division of an axis into equal branches.

digitalis an alkaloid used as a powerful cardiac stimulant and diuretic; produced in the leaves of the common foxglove (*Digitalis* spp.).

dikaryon two independent, usually haploid nuclei closely associated in one cell; each nucleus usually comes from a different parent cell.

dimorphism the condition of having two forms.

dinoflagellate the motile members of the Pyrrophyta.

dioecious the condition whereby the spore-bearing organs are on different sporophytes; the microspores and megaspores are on separate plants.

diplanetism the condition in which there are two periods of motility, interrupted by a resting stage, in one life history.

diplobiontic having thalli of two kinds, one haploid, the other diploid; both free living.

diplococcus cocci regularly occurring in pairs; the separating wall being somewhat flattened.

diploid a nucleus having the $2n$ chromosome number that is twice the number found in the gametes; in some fungi may be used to indicate a cell containing the dikaryon.

disease the relationship established between a parasite and a host with detriment to the host.

E **early wood** the part of an annual ring produced early in the growing season; spring wood.

ecosystem a natural grouping of organisms that, with the physical environment, is self-sustaining and independent of other groups providing it receives radient energy.

egg female gamete.

elater sterile, hygroscopic filaments in the capsule of certain liverworts; the hygroscopic bands formed from the outer spore wall in *Equisetum*.

embryo the young plant developed from a fertilized egg.

embryo sac the female gametophyte of angiosperms.

enation a scalelike appendage on a stem or other plant surface.

endobiotic living inside another organism.

endoconidium conidium formed inside a hypha.

endodermis the innermost, differentiated layer of the cortex, present in roots, rhizomes, and some stems; cell walls may produce Casparian strips.

endophytic living in other plants.

endosperm a nutrient tissue produced during double fertilization of the angiosperm embryo sac and used by the developing embryo as a source of energy.

endospore a thick-walled spore produced inside the cells of bacteria and certain blue-green algae.

endothecium the internal tissue of an embryonic sporophyte; the inner portion of an another wall.

endozoic living in an animal.

epibasidium the upper portion of the basidium of the heterobasidiomycetes.

epibiotic living on the surface of another plant or animal.

epicotyl the portion of the axis of an embryo or seedling above the cotyledonary node; develops into the shoot.

epidermis the outer layer of cells of the leaf, young stems, and young roots.

epigean growing on or above the soil.

epiphyte a green plant that grows on another plant but is not parasitic.

epiphytotic the sudden and destructive development of a plant disease, usually involving an extensive area.

epitheca the larger half of the frustule of diatoms.

epithecium a thin layer over the asci in an apothecium; formed from the tips of the paraphyses.

ergot a fungus disease of ovaries of many grasses; caused by

Claviceps purpurea; produces several alkaloids that have poisonous and/or therapeutic uses.

estuary an inlet of the sea, especially the wise mouth of a river where the tide of the sea meets the current of the river.

eucarpic having only part of the thallus of a fungus used in sporocarp production.

eukaryotic nucleus nuclear material that is separated from the cytoplasm by a nuclear membrane.

eusporangium a sporangium whose wall develops from several superficial cells and whose spores develop from internal cells of the sporophyll.

excipulum the fleshy part of an apothecium that supports the hypothecium and hymenium.

- **ectal excipulum** the outer layer of an apothecium.
- **medullary excipulum** the inner portion of an apothecium.

exoconidium an asexual spore formed on the surface or at the tip of a hypha.

exogenous developing on the outside of another body; produced externally.

exospore a spore that is not produced within a sporangium, for example, conidia.

F **facultative** have the ability to live under different nutritional conditions, depending on the environment.

fauna animal life; usually the animals of a region, period, or geological stratum.

fertilization the union of sexual reproductive nuclei.

fertilization tube a tube or branch produced in the phycomycetes by which the male gametangium unites with the female gametangium.

filament the stalk of the stamen; a separate thread of mycelium; the threadlike bodies of certain algae.

fission the duplication of a cell or a one-called organism by the division of the cell or organism into two, usually equal, parts.

flagellum an elongate structure, the beating of which propels motile cells;

- **tinsel flagellum** with many fine hairlike processes (mastigonemes) produced along its length.
- **whiplash flagellum** with a smooth continuous surface.

floridean starch a carbohydrate resembling starch formed by red algae.

flower the reproductive structure of the angiosperms.

foot the organ of the embryonic sporophyte in liverworts, mosses, and some vascular plants that attaches the sporophyte to the gametophyte and that allows the sporophyte to absorb food from the gametophyte.

fragmentation a method of asexual reproduction whereby a thallus is broken into a number of parts, each of which is capable of producing a new thallus.

frond a specialized name for the leaf of a fern.

fructification a general, and misleading, term for spore-bearing organs of the fungi; properly used only when referring to the angiosperms.

fruit a mature ovary or ovaries and their associated parts, containing the seeds in angiosperms.

"fruiting" structure see *fructification*.

fucoxanthin the brown xanthophyll pigment that is characteristic of the brown algae and chrysophytes.

fungicide a substance that kills or inhibits the growth of fungi.

fungus a thallophyte that lacks chlorophyll and thus depends on some preformed carbon compound as a source of energy, it may be either saprobic or parasitic.

fusion to unite, to blend together as in the fusion of gametes.

G

gall an outgrowth produced by a plant as the result of attack by a fungus or other agent.

gametangium a structure in which gametes are produced; in algae and fungi, a single cell; in other plants, a multicellular structure with a sterile, outer layer.

gamete a reproduction cell that unites with another producing a zygote.

gametophore a branch-bearing gametes; a branch upon which are produced gametangia.

gametophyte the sexual phase of the life history of plants in which are produced gametes.

gametothallus a thallus that produces gametes, as opposed to one that produces spores (sporothallus).

gangrene local death of soft tissues due to loss of blood supply.

gelatinous sheath the jellylike covering around certain algae.

gemma a specialized, multicellular fragment of a plant that becomes detached and may asexually reproduce the plant.

gemmae cup a cuplike structure in which gemmae are produced.

generative cell the cell in male gametophytes of many gymnosperms that produces the stalk and body cell, and in angiosperms produces two sperm nuclei.

germ tube the first tubular outgrowth produced from a germinating spore.

germination the resumption of growth of an embryo or spore.

endosporic germination germination within a spore with the production of a gametophyte contained within the old spore wall.

exosporic germination germination of a spore with the production of a gametophyte produced outside of the old spore wall.

girdle band in diatoms, the structures to which hypotheca and epitheca are attached.

gleba the inner, spore-producing tissue of the fruiting body of the gasteromycetes.

globule the male organ of the stoneworts (Charophyta).

gonidium an asexual reproductive cell, especially in some algae (e.g., *Volvox* spp.).

gonimoblast filament a filament produced in the red algae from the carpogonium after fertilization; produces the carposporangia.

griseofulvin a chlorine-containing antibiotic produced by *Penicillium griseofulvum;* effective against ringworm fungi.

ground meristem the primary meristematic tissue other than protoderm and procambium.

guard cell specialized kidney-shaped epidermal cells surrounding a stomate.

gullet a longitidinal groove present in some algae, especially the euglenas.

gymnosperm a seed plant with seeds not enclosed by a megasporophyll or pistil.

gynandrosporous the production of both androsporangia and oögonia on the same filament in *Oedogonium* spp.

H **hallucination** perception of objects with no reality, usually arising from disorder of the nervous system.

haploid with only one member of each chromosome pair present; having the n number of chromosomes in the nucleus of the gamete; characteristic, but not exclusive to, the gametophyte.

haplobiontic having a free-living thallus of only one kind; may be either haploid or diploid.

haustorium an absorptive hypha of a fungus mycelium within a host cell; a modified root or shoot of a higher plant that serves as an attachment to the host through penetration of the host cells and through which food is obtained.

heartwood nonliving, nonfunctional, and commonly darker colored wood surrounded by sapwood in woody stems, especially trees.

hematochrome red pigments, probably related to xanthophyll, that occur in certain green algae and euglenas.

heterocyst a cell that differs in size and content from the other cells in the filament of certain blue-green algae.

heterobasidium a basidium often indefinite in form and usually sepatate that arises as a probasidium and is typically composed of two parts; characteristic of the heterobasidiomycetes.

heteroecious requiring two host species for the completion of the life history of certain parsitic fungi.

heterogamete one of a pair of gametes, both of which are flagellated but one of which is larger than the other although they are otherwise similar.

heterogamy the union of two similar motile gametes, one being larger than the other; see anisogamy.

heterokaryotic a condition in which genetically different nuclei are associated in the same protoplast.

heteromorphic having structure that varies from the usual or expected structure; having different structures (as in heteromorphic alternation of generations).

heterothallism self-incompatible; requiring two compatible strains of an organism before sexual reproduction can occur.

heterospory having spores of two kinds, usually designated as microspores and megaspores that upon germination produce male and female gametophytes, respectively.

heterotrophic requiring preformed organic compounds for nourishment, as a parasite or saprophyte that can not carry on photosynthesis.

heterozygote a zygote that has inherited unlike genes for a particular characteristic.

hermaphroditic see *bisexual.*

histoplasmin an antigen prepared from *Histoplasma capsulatum,* especially for skin testing.

histoplasmosis an intracellular mycosis caused by *Histoplasma capsulatum* and involving several organs of the body including the lymphatic tissues, lung, spleen, liver, adrenals, kidneys, skin, and central nervous system.

holdfast a unicellular or multicellular basal organ that attaches many algae to their substrates.

holocarpic having the entire thallus of a fungus used in sporocarp production.

holozoic devouring other organisms.

homobasidium a basidium usually definite in form and nonseptate; not divisible into hypobasidium and epibasidium, characteristic of the homobasidiomycetes.

homokaryotic a condition in which nuclei that are genetically alike are associated in the same protoplast.

homologous structures that have a common origin.

homosporous having spores of only one kind.

homothallism self-compatible; requiring only one strain of an organism before sexual reproduction can occur.

hormogonium a multicellular segment of some members of the filamentous blue-green algae.

hülle cells thick-walled cells that occur in large numbers in association with the perithecia of *Emericella nidulans*.

hyaline colorless or transparent.

hygroscopic easily absorbing moisture from the atmosphere.

hymenium the spore-bearing layer of asci or basidia in ascomycetes or basidiomycetes; may contain various sterile structures.

hymenomycetes a basidiomycete with an exposed hymenium.

hyperplasia an excessive, usually pathological, increase in the number of cells composing a tissue or organ.

hypertrophy an excessive, usually pathological enlargement of cell size in cells comprising a tissue or organ.

hypha the threadlike filament composing the fungus body or the mycelium.

hypobasidium the lower portion of the basidium of the heterobasidiomycetes.

hypocotyl the portion of the embryonic axis between the cotyledonary node and radicle; sometimes used to include the radicle.

hypogean growing below the soil.

hypothallus the thin film of transparent material left at the base of the sporophores on the substrate by some slime molds.

hypotheca the smaller half of the frustule of diatoms.

hypothecium the layer of hyphal tissue immediately beneath the hymenium of apothecia.

I **idioandrosporous** the production of androsporangia and oögonia on separate filaments in *Oedogonium* spp.

indigenous produced naturally in a region or country; native to an area.

indusium the thin covering layer, produced from the epidermis, of a group of sporangia in ferns.

false indusium the inrolled leaf margin when it covers marginal sporangia.

inflorescence a flower cluster; the arrangement of flowers on the floral axis.

inoculation the introduction of bacteria or some other material into living organisms to secure immunity to future infections; the putting of bacteria or some other material into soil, a culture medium, etc.

inoculum the material containing microorganisms and used for inoculation.

integument the tissues covering the megasporangium in ovules; the outer layer or layers of the ovule that develop into the seed coat.

intercarlary lying between other bodies in a row or placed somewhere along the length of a stem, filament, or hypha.

internode the portion of the stem between two nodes.

isogamete one of a pair of gametes of similar size and appearance; both usually are flagellated.

isogamy the sexual union of two like gametes both of which are usually flagellated and the fusion of which usually occurs outside of the parent plants.

isomorphic having structure that is expected; having similar structures (as in isomorphic alternation of generations).

isoplanogametes motile sex cells of equal size; see isogamete.

isthmus the constricted portion connecting the two semicells in many desmids.

K **karyogamy** the fusion of two sex nuclei following the fusion of protoplasts (plasmogamy).

karyokinesis nuclear duplications; mitosis.

kelp a general name for large marine algae, usually browns.

L **lamella** any thin or platelike structure; for example, the hymenium-covered plates on the underside of the pileus of some basidiomycetes.

lateral conjugation the conjugation of adjacent cells in the filament, as in *Spirogyra* spp.

late wood the part of the annual ring that is produced later in the growing season; the summer wood.

leaf gap a break in the continuity of the primary vascular cylinder above the departure of a leaf trace or traces; this break is filled with parenchyma tissue.

leaf trace vascular tissue that connects the vascular tissues of the stem with those of the leaf.

leptosporangium a sporangium produced from the periclinal duplication of a single superficial cell.

lesion a wound; a well-marked but limited diseased area.

leucosin a polysaccharide composed of glucose units; found in some chrysophytes.

lichen a symbiotic (mutualistic) relationship between an alga and a fungus; the two components often so closely associated as to form a distinctive plantlike individual or thallus.

- **crustose lichen** a lichen thallus that develops as a thin crust on the surface of rocks.
- **foliose lichen** a lichen thallus that is flattened, often ribbed or leaflike.
- **fruticose lichen** a lichen thallus that is attached at one point and growing away from the point of attachment; may be shrublike or beardlike in appearance.

lip cells thin-walled cells that interrupt the annulus at the point of dehiscence in certain fern sporangia.

littoral zone the area along the shore of a sea or lake that is the shallow water region within a depth to which light and wave action reach.

locule a compartment or cavity; in sac fungi, the stromatic chamber containing the asci; in flowering plants, the chamber that contains the ovules in the ovary.

lorica a hard case or shell that surrounds the protoplast of some algae.

lysergic acid diethylamide an alkaloid derived from the ascomycete *Claviceps purpurea*. Induces various kinds of hallucination and other toxic conditions when ingested by man.

M **macrandrous** antheridia produced in filaments the size of vegetative filaments, as in *Oedogonium* spp.

macroconidium produced by a fungus that also produces microconidia; usually the diagnostic spore of such fungi.

macrocyclic long cycles; applied to those species of rusts that produce all five of the possible spore types, including a dikaryotic spore in addition to the teliospore.

macrocyst in myxomycetes, the inactive form of the plasmodium in which multinucleate portions of the sclerotium has been separated by wall material.

macrophyll leaves whose vascular traces are associated with leaf gaps in the stem.

megagametophyte the gametophyte produced from the megaspore; the female gametophyte that produces the female gamete.

megaphyll a macrophyll.

megasporangium the structure in which megaspores are produced; the nucellus.

megaspore a spore arising by meiosis of the megasporocyte and that produces the female gametophyte.

megasporocyte the megaspore mother cell that produces megaspores as the result of meiosis.

megasporophyll a modified leaf bearing one or more megasporangia or ovules; a carpel.

megastrobilus the female cone of many lower vascular plants and gymnosperms in which are found megasporophylls or ovuliferous scales on which are produced ovules.

meiosis reduction division of the nuclear material; the reduction of the chromosome number from diploid ($2n$) to haploid (n); genetic segregation occurs.

melanin a dark brown or black pigment.

meristele a strand of vascular tissue consisting of a central strip of xylem enclosed by a sheath of phloem that is enclosed by a sheath of endodermis; is a component of a dictyostele.

meristem a tissue composed of undifferentiated, unspecialized cells that are capable of further cellular duplication.

mesome an internode close to a telome.

mesophyll the parenchyma tissue, usually photosynthetic, between the upper and lower epidermis of a leaf.

microaerophilic microorganisms that grow best in the presence of small amounts of atmospheric oxygen.

microcylic short-cycled; applied to those species of rusts that produce fewer than the five possible spore types and only one dikaryotic spore, the teliospore.

microcyst a small, encysted protoplast, often an encysted myxamoeba of the slime molds.

microgametophyte the gametophyte produced from the microspore; the male gametophyte that produces the male gamete.

microconidium the smaller of two conidia produced by a fungus that also produces macroconidia; sometimes functions as a spermatium.

micron 1/1000 of a millimeter; usually designated by the Greek letter mu (μ).

micropyle an opening between the integument or integuments of an ovule; usually remains present after the ovule matures into a seed.

microphyll leaves whose vascular traces are not associated with leaf gaps in the stem.

microspore a spore arising by meiosis of the microsporocyte and that produces a male gametophyte; a pollen grain.

microsporangium the structure in which microspores are produced.

microsporocyte the microspore mother cell that produces microspores as the result of meiosis.

microsporophyll a modified leaf bearing one or more microsporangia.

microstrobilus the male cone of many lower vascular plants and gymnosperms in which are found microsporophylls on which are produced microsporangia.

mitosis see *karyokinesis;* nuclear duplication in which the daughter nuclei receive genetically identifical material to that contained by the parent nucleus.

monocentric having one center of growth and differentiation at which is produced a reproductive organ.

monoecious the condition whereby the spore-bearing organs are on the same sporophyte; the microspores and megaspores are on the same plant.

monopolar germination the germination of a spore by the production of a germ tube at only one end.

monosporangium a sporangium that produces simple spores or monospores; usually produced in some red algae as the result of differentiation of a single vegetative cell.

monospore a spore, characteristic of some red algae, that may be amoeboid and will germinate and grow into a new plant.

morphogenesis the origin and structural changes during the development of organs or organisms.

morphology the study of form and structure, both from the study of the development of an individual (ontogeny) and from the study of the relationships of organisms through evolutionary history (phylogeny). Sometimes expanded to include some study of method of function of structure.

multilocular having many cells, chambers, or locules.

mutualism a form of symbiosis in which both organisms derive benefit from the association.

mycelium the mass of threadlike structures that constitute the thallus of a fungus.

mycetoma a fungal disease, caused by several fungi, that causes a localized swollen lesion, usually on the hand or foot and involving skin and other tissues; the lesion contains mycotic granules that vary in size, color, shape, and texture.

mycobiont the fungal component of a lichen association.

mycologist one who studies fungi.

mycophagy the eating of mushrooms.

mycorrhiza the symbiotic association of a fungus with the roots or rhizomes of a higher plant; the association may be mutualistic, simply nonparasitic, or slightly parasitic.

mycosis a fungus disease of man or other animals.

myxamoeba a uninucleate ameoboid cell that may precede or follow the zoospore (swarm cell) of a slime mold.

myxomyosin a rod-shaped contractile protein 4000–5000 Å long found in the plasmodium of slime molds; behaves like actomyosin in animal muscle.

N

nannandrous antheridia produced in specialized dwarf male filaments usually attached near the oögonium, as in *Oedogonium* spp.

nematode a group of worms with elongate, cylindrical, unsegmented bodies; contains both plant and animal parasites.

nucellus megasporangium; the tissue of the young ovule within which develops the embryo sac (female gametophyte).

nucule the female reproductive structure in the stoneworts; includes both the oögonium and the outer sterile cells.

O

oidium spermatia produced on hyphal branches; asexual spores produced by the separation of a hypha into individual cells; an arthrospore.

oögamy the sexual fusion of a large nonmotile egg and a small motile sperm; fusion usually occurs inside of the oögonium.

oögonium a unicellular female sex organ (gametangium).

oösphere a female gamete; an egg.

oöspore the thick-walled spore produced after the fertilization of an oösphere (egg).

operculum a lid or cover.

ostiole an opening or pore that permits spores or pollen to escape from a plant structure.

ovary the enlarged base of the pistil or carpel in which the ovules develop; may mature into the fruit.

ovule a megasporangium (nucellus) covered by an integument; after fertilization of the egg it may mature into the seed.

ovuliferous scale in some conifers, the structure in the female cone to which the ovule is attached.

P

palisade parenchyma elongated cells, usually containing many chloroplasts, found just beneath and at right angles to the upper epidermis of leaves.

palmella stage a nonmotile stage, surrounded by a gelatinous sheath, or some unicellular motile algae.

paradesmose a slender filament that connects the centers of two halves of a blepharoplast.

paramylum a starchy material found in the cytoplasm of some brown and red algae.

paraphysis sterile filaments occurring in the sporocarp of many lower plants.

parasexual cycle a process in which plasmogamy, karyogamy, and haploidization take place in sequence, but not at specified points in the life history of an individual; recombination of hereditary materials is based on the mitotic cycle rather than on sexual reproduction; of significance in some of the fungi.

parasite an organism that derives its nourishment by living in or on another organism to the detriment of that organism (the host).

parenchyma tissue composed of large, undifferentiated, cells with thin cellulose walls; the cells are usually loosely compacted leaving large intercellular spaces.

parthenogenesis the development of a single gamete into an embryo.
pathogen the causative organism of a disease relationship.
peat any semicarbonized plant tissue formed by partial decomposition; especially *Sphagnum* spp. mosses.
penicillus like a small brush; the branches bearing conidia in *Penicillium* spp.
perennial a plant that lives from year to year; the aerial portions may die to the ground during adverse conditions, but the roots continue to live.
pericarp the mature ovary wall in a fruit.
pericycle a layer of cells just outside the primary phloem and inside the endodermis; if the endodermis is lacking, the pericycle cannot be distinguished from the cortex.
peridiole a specialized division of the gleba which contains the basidiospores in some of the gasteromycetes (esp. the bird's nest fungi); often an organ of spore dispersal.
peridium the wall or limiting membrane of a sporangium or other sporocarp in a fungus.
periphysis a short, sterile filament extending into the opening of an ostiole or neck of a perithecium or pycnidium.
periplasm in the Peronosporales, the outer nonfunctional protoplasm of the oögonium or antheridium.
periplast a complex and often ornamented membrane surrounding the protoplasm of some algae.
peristome multicellular structures at the mouth of the capsule in many mosses.
perithecium a globose or flask-shaped ascocarp characteristic of the pyrenomycetes and some lichens; may or may not have an ostiole, but the asci are in an ordinate arrangement.
petal one of the sterile units of the corolla of the flower; usually colored.
petiole the stalk that attaches the leaf blade to the stem; the stalk of the leaf.
phagocyte a cell capable of ingesting other cells, or foreign particles.
phagotrophic ingesting solid food materials.
phaneroplasmodium a plasmodium consisting of a well-differentiated advancing fan and conspicuous thick strands in which ecto- and endo-plasmic regions are well differentiated and which the protoplasm is coarsely granular; characteristic of the Physarales.

phialide a short, flask-shaped structure from the apex of which thin-walled conidia are produced.

phialopore a small hole in the colony of members of the Volvocaceae.

phloem a complex part of the vascular tissues; composed of living thin-walled cells consisting of sieve tubes, companion cells, and their adjacent parenchyma; responsible for most of the translocation of dissolved food materials.

photosynthesis the process whereby green plants, using the radiant energy of the sun, combine carbon dioxide and water and produce carbohydrates, free oxygen, and water.

phototaxis the response or reaction of an organism to the stimulus of light.

phototroph an organism that derives its nutrition by photosynthesis.

phototrophism the bending of a plant part in response to the stimulus of light.

phycobiont the algal component of a lichen association.

phycocyanin a blue pigment produced by blue-green and red algae.

phycoerythrin a red pigment produced by blue-green and red algae.

phyllidium the flat lamellar structure (sometimes incorrectly called a "leaf") of the bryophyte gametophyte; carries on most of the photosynthesis of the plant.

piedra a fungus disease of the hair shaft.

pinna a subdivision (leaflet) of a pinnate leaf.

pinnule the ultimate division of a bipinnately compound leaf.

pistil the female organ of the flower, composed of one or more carpels and enclosing the ovules; may be differentiated into a stigma, style, and ovary; a megasporophyll.

pith the tissue in the center of the vascular cylinder; usually composed of parenchyma, but other types of cells may also occur; mostly in stems, but found also in some roots.

placenta the area of the ovary to which the ovules (or seeds) are attached.

plakea a flat, single-layered colony of flagellated algae: may develop into a spherical colony as in *Volvox* spp.

plankton the passively floating or weakly swimming organisms in bodies of water; includes both plants (phytoplankton) and animals (zooplankton).

plasmodiocarp a type of myxomycete sporocarp that develops from the thick strands of the plasmodium and resembles an elongate, branched, sessile sporangium.

plasmodium a multinucleate, motile mass of protoplasm having no firm wall; the vegetative phase of the acellular slime molds (myxomycetes).

plasmogamy the fusion of two protoplasts, not necessarily followed by nuclear fusion.

planogamete a motile sex cell (gamete).

planospore a motile asexual spore; a zoospore.

plumule the terminal bud of an embryo; that portion of the young embryonic axis above the cotyledonary nodes.

plurilocular gametangium a multicellular sex organ, each cell of which is fertile and produces a sex cell.

plurilocular zoösporangium a multicellular spore-producing organ, each cell of which produces an asexual motile spore.

polar nodule the wall thickening at each end of the groove or raphe of certain diatoms and the heterocysts of blue-green algae.

pollen the microspores or young male gametophytes of the seed plants.

pollen chamber a depression at the tip of the megasporangium of gymnosperms; the pollen settles and eventually germinates here.

pollination the transfer of pollen from the anther to the stigma in flowering plants, and from microsporangium to pollen chamber in gymnosperms.

pollination droplet a droplet of fluid at the micropyle of gymnosperms at the time of pollination; its evaporation draws the pollen grain into the pollen chamber.

polycentric having many centers of growth and development.

polymorphic having more than one form.

predacious living by preying on animals.

primary hypha the uninucleate, haploid hypha produced from an ascospore or a basidiospore.

primary vascular tissue the vascular tissues produced from the procambium.

procambium that primary meristem that produces the primary vascular tissues and the vascular cambium, where one is produced.

proembryo a mass of cells produced by duplication of the zygote before the organs of the embryo are differentiated.

prokaryotic nucleus nuclear material that is not separated from the cytoplasm by a nuclear membrane.

promycelium the basidium of the rusts and smuts; produced from the teliospore.

prosorus in the chytrids, a cell that produces a group of sporangia (the sorus).

protandry the production of male before female gametes in a bisexual plant.

prothallial cell vegetative, vestigial cells found in the male gametophyte of certain vascular plants.

prothallium the gametophyte of certain vascular plants.

protoderm the primary meristem from which the epidermis is produced.

protonema in mosses, the green filamentous gametophyte that produces "leafy" gametophores.

protoplasmodium a microscopic plasmodium with no differentiated fan-shaped region and that produces only a single, minute sporocarp.

protoplast the protoplasmic contents of a cell.

protostele a solid core of xylem surrounded by phloem.

pseudocapillitium an irregular plate, tube, or threadlike body in the fructification of some myxomyctes; resembles capillitium.

pseudomycelium a series of loosely connected cells adhering end to end.

pseudoplasmodium a group of closely associated myxamoeba that form the beginning of sorocarp formation in the cellular slime molds.

pseudopodium the elongating portion of the gametophore that elevates the sporophyte of some mosses (e.g., *Sphagnum* spp.).

pseudoseptum a septum that is perforated by pores.

puncta the markings that are produced as the result of fine pores on the valves of diatoms.

pycnidium the sporocarp of some of the Fungi Imperfecti and many lichens; usually globose or flask-shaped and lined on the inner surface with conidiophores.

pyrenoid a protenaceous center of starch formation or storage in the chloroplast of certain algae and bryophytes.

R **rachis** the axis from which leaflets are produced in a compound leaf.

radicle the basal end of the embryonic axis; produces the primary root.

raphe the longitudinal fissure in the valve of motile pennate diatoms, has a nodule at each end and one in the middle; also the ridge on some seeds.

receptive hypha a hypha that can be "spermatized" by a microconidium, sometime extends from a female gametangium.

reniform kidney-shaped.

resupinate a sporocarp that lies flat on the substratum with the hymenium on the outer side.

retting the destruction of pectin-binding material in the stems of plants such as flax and hemp so that the fibers can be separated and used for the manufacture of thread, rope, and cloth.

rhizoid a cellular, rootlike outgrowth from the undersurface of many mosses and thallophytes; also hyphae extending into the substratum from the mycelium (e.g., *Rhizopus* spp.).

rhizome a fleshy, elongate, nonerect stem usually subterranean, distinguished from a root by the presence of nodes and internodes.

rhizomorph a thread- or cord-like structure composed of many hyphal strands and sometimes quite tough.

rhizophore a leafless root-bearing organ or region (e.g., in some club mosses).

rhizoplast chromatic threads that join the blepharoplast to the nucleus.

ribosome submicroscopic granules either free in the cytoplasm or attached to the endoplasmic reticulum, composed of about 60 percent RNA and 40 percent protein in prokaryotic cells and about 50 percent RNA and 50 percent protein in eukaryotic cells, the site of protein synthesis.

ringworm a contagious, superficial infection of the skin in man and domestic animals, caused by various imperfect fungi, and which produce ring-shaped discolored patches on various parts of the body (e.g., feet, nails, groin).

root hair a tubular extension of an epidermal cell of the root in the zone of maturation.

S **saprobe** a saprophyte; an heterotrophic plant that uses dead organisms as an energy source and commonly causes their decay.

sapwood outer part of the wood of a stem, usually distinguished from the heartwood by its lighter color; functions in the transportation of water and dissolved minerals.

sarcinae spherical bacteria that are grouped in a cubical pocket.

scalariform conjugation conjugation between two neighboring filaments that are connected by conjugation tubes arranged in a ladderlike sequence (e.g., some *Spirogyra* spp.).

schizophrenic a psychotic disorder characterized by loss of contact with environment and by disintegration of personality.

sclerenchyma tissue composed of cells with uniform thick, lignified walls; usually dead at maturity; may be produced as fibers or as short isodiametric and often ramified cells.

sclerotium a firm mass of hyphae that may or may not contain host tissue or soil; usually contains no spores; in noncellular slime molds, the firm resting condition of a plasmodium.

secondary tissue tissue produced by lateral meristems (e.g., vascular cambium), and that results in increase in diameter.

seed a mature ovule containing an embryonic sporophyte, usually surrounded by a food supply, and covered complete by a seed coat. In gymnosperms, the food supply is female gametophyte tissue; in angiosperms the food supply may be incorporated into the cotyledons of the embryo or in a specialized tissue, the endosperm. Some seeds (e.g., orchids) contain no food supply.

semicell one half of a desmid.

sepal one of the parts of the calyx; one of lowermost sterile appendages, usually green, on a floral receptacle.

septum a wall or partition.

serology a science of disease diagnosis and treatment using serum.

sessile without a stalk; sitting directly on the base.

seta a bristlelike hair, generally thick-walled and dark-colored; in bryophytes, the stalk that supports the capsule.

siphonostele a cylinder composed of xylem and phloem; may or may not surround pith tissue.

- **amphiphloic siphonostele** a siphonostele with phloem on both sides of the xylem.
- **ectophloic siphonostele** a siphonostele with phloem only external to the xylem.

sirenin a hormone produced by the female gametes of some algae and fungi and one that attracts the male gametes.

slug in the cellular slime molds, the aggregation of vegetative cells from which is produced the sorocarp.

smut a disease relationship caused by one of the members of the Ustilaginales; a smut fungus.

smuts, covered a smut in which the mature spore mass remains for a time within the sorus, frequently till after the sorus becomes free from the host.

smuts, loose a smut in which the spores are as an uncovered mass of powder becoming free from the host by wind and rain.

somatic body refers to the vegetative phase in plants; excludes the reproductive structures.

somatogamy the fusion of somatic cells during plasmogamy.

soredia clusters of fungal and algal cells that are capable of vegetatively reproducing a lichen association.

sorocarp the undifferentiated fruiting structure of the cellular slime molds; without membrane or containing wall.

sorophore the stalk supporting the sorus in the cellular slime molds.

sorus a cluster of spores and/or sporangia.

sperm the male gamete.

spermatangium the cell that produces a spermatium in red algae.

spermatium a tiny nonmotile male gamete found in some sac fungi, rusts, and red algae.

spermatization plasmogamy by the union of a spermatium with a receptive structure.

spermogonium a structure in which are produced spermatia (e.g., in rusts).

spirillum a spiral or corkscrew-shaped bacterium.

spongy parenchyma the globose parenchyma, usually found in leaves, that contains large, conspicuous intercellular spaces.

spontaneous generation the idea that living organisms can originate from nonliving materials independent of other living matter.

sporangiospore a spore produced in a sporangium.

sporangium a structure in which spores are produced; is unicellular in algae, fungi, and bacteria, but multicellular in other plant groups.

sporangiole a small sporangium with few spores.

sporangiophore the stalk supporting the sporangium.

spore a general term for a type of unicellular or multicellular asexual reproductive structure; may be produced as a result of mitosis (mitospore) or meiosis (meiospore); many specialized types of spores with many specialized names are found throughout the plant kingdom.

sporocarp a hard structure containing sori in the water ferns; a spore-producing organ in some fungi (e.g., ascocarp, basidiocarp).

sporocyte a spore mother cell; a cell that produces a tetrad of meiospores as a result of meiosis.

sporodochium a cushionlike, compact mass of sporophores found in some of the imperfect fungi.

sporogenous tissue tissue composed of cells that produce spores of various kinds.

sporophore a spore-bearing branch or organ; the reproductive body in fleshy fungi (e.g., ascocarp, basidiocarp).

sporophyll a specialized leaf or leaflike organ that bears sporangia (e.g., fertile fronds of ferns; stamens and carpels of angiosperms).

sporophyte the asexual phase of the life history of plants in which are produced haploid spores (meiospores).

sporothallus a thallus that produces spores, as opposed to one that produces gametes (gametothallus).

spring wood see *early wood*.

stamen in angiosperms, a modified leaf from which are produced pollen grains; usually consists of apical sporangia (anthers) and an elongated stalk (filament); a microsporophyll.

staphylococcus a spherical bacterium occurring in irregular, grapelike clusters.

stele the vascular cylinder in stems and roots; composed of xylem and phloem; may sometimes include pith and other cell types such as pericycle (in roots).

sterigma a structure produced by a basidium and that supports a basidiospore.

stipe a stalk.

stipitate having a stalk or stipe.

stipule a leafy appendage at the base of some petioles.

stolon a stem or hypha that grows horizontally along the surface of the soil or other substrate; may be propagative when it produces roots and aerial stems (e.g., strawberry) or rhizoids and sporangiophores (e.g., *Rhizopus stolonifera*).

stomate an opening or pore in the epidermis, surrounded by two specialized epidermal cells, the guard cells.

stomium the thin-walled cells in the lip cell region where the rupture occurs in a fern sporangium.

streptococcus a spherical bacterium that divides in such a way that chains of cells are produced.

striae narrow lines, streaks, grooves, or channels; in some diatoms, rows of punctae.

strobilus a stem with short internodes and spore-bearing appendages.

stroma a mass of vegetative hyphae that may or may not include tissue from the host or substrate; the texture may be hard or soft; spores may be produced in or on the stroma.

suberin waxy material found in the cell walls of cork tissue, usually makes it waterproof.

subiculum a loose hyphal mat on which some sporocarps are produced.

substrate the material on or in which a plant is growing.

summer wood see *late wood*.

susceptibility the lack of ability to resist some extraneous agent such as a pathogen or a drug; capable of being infected.

suspensor in zygomycetes, a cell that supports a gametangium or zygospore; in seed plants, the structure of the embryo that pushes the terminal part of the embryo into the food supply.

swarm cell a motile cell that acts as an isogamete; sometimes produced by germination of the spores of acellular slime molds.

symbiosis the living together of unlike organisms.

synnema a compacted group of closely united and sometimes fused conidiophores; may bear conidia only at tips or on tips and at sides.

systemic relating to the entire organism; affecting the body generally.

T

tannin an astringent, bitter substance especially abundant in some barks (e.g., some oaks).

tapetum a nutritive cell layer within sporangia.

taxon a general term that can be applied to any category of taxonomic classification.

teliospore the spore of the rust or smut fungi from which the basidium is produced.

telium a structure in some rusts and smuts in which the teliospores are produced.

tetracoccus spherical bacterium occurring in groups of four cells.

tetrasporangium a sporangium in red algae that produces four spores after meiosis.

tetraspore a haploid spore produced by meiosis in a tetrasporangium.

tetrasporophyte in the red red algae, the diploid plant that produces tetraspores.

thallus a plant body without differentiation into root, stem, and leaf; the vegetative body of a thallophyte (e.g., fungi and algae).

tinea a skin disease caused by a fungus in man or other animals; a ringworm.

trabecula a row of cells bridging an intercellular space.

tracheid a single-celled, lignified, nonliving water-conducting element of xylem in which the walls are pitted but the elongate tapering ends have no perforations.

trama the hyphae in the central part of the gills or pileus of an agaric.

transduction transfer by viruses of genetic material from one bacterial cell to another.

transformation genetic change in bacteria caused by direct incorporation of extracellular DNA from another strain.

transversely septate basidium in heterobasidiomycetes, an elongate basidium divided into four cells by cross walls.

trichogyne the receptive, hairlike extension of the female sex structure in some sac fungi and red algae.

trichome in blue-green algae, a single row of cells in a filamentous colony; a hairlike outgrowth of the epidermis, as a hair or bristle.

tube cell one of the cells resulting from the germination of the microspore of gymnosperms and flowering plants; a vegetative cell, probably vestigial.

tumor the production of a mass of new tissue that grows independently of surrounding tissue, but that may invade it.

tuning fork basidium in heterobasidiomycetes, a deeply bifurcate basidium.

U **unilocular** having only one compartment or cavity (locule).

unilocular zoösporangium a zoöspore-producing sporangium containing only one cavity.

unisexual a gametophyte that produces only antheridia or archegonia.

unitunicate an ascus in which both the inner and outer wall are more or less rigid and do not separate or expand during spore ejection.

uredium in some rusts, a structure containing a group of binucleate cells that produce uredospores.

uredospore a binucleate spore in some rusts that may reinfect the same host species on which is is produced.

V

valve in diatoms, one half of the body wall; in liverworts and hornworts, a segment of the capsule wall at dehiscence.

vascular cambium the lateral meristem, one cell thick, that produces secondary phloem and secondary xylem.

vascular tissue tissue composed of primary and/or secondary xylem and phloem; the conducting tissues of the plant.

vasoconstrictor an agent that induces the narrowing of the lumen of blood vessels.

vector an animal, usually an insect, that carries a pathogen from one host to another.

venation the arrangement of the vascular bundles (veins) in a leaf blade.

venter the enlarged basal portion of an oögonium; contains an egg.

vesicle a bladderlike sac (e.g., the apex of conidiophore of *Aspergillus* spp.).

vessel a tubelike, water-conducting structure of the xylem, composed of cells (vessel elements) placed end to end and connected by perforations in their common end walls.

virulence the capacity or ability of a microorganism to produce disease.

W

watershed a region draining ultimately to a particular body of water.

X

xerophyte a plant that is very resistant to extremely dry environmental conditions and that lives in very dry habitats.

xylem a complex part of the vascular tissues; composed of nonliving, lignified tracheids, vessels, and fibers and their adjacent parenchyma; responsible for most of the movement of water and dissolved minerals; composes wood.

Z

zoösporangium a sporangium in which are produced flagellated or ciliated spores.

zoöspore a flagellated or ciliated asexual spore.

primary zoöspore a pip-shaped spore with two flagella located at the tip.

secondary zoöspore a kidney-shaped spore with two flagella located laterally.

zygospore the thick-walled resistant spore that results from the fusion of isogametes or the conjugation of gametangia.

zygote the cell resulting from the fusion of two gametes.

index

Those numbers in regular type indicate a mention or discussion of the subject; those in boldface italic indicate a black and white illustration, a line drawing, or a life history; those in boldface indicate a color illustration; those names in italics indicate a taxonomic category (a taxon).